ENGINEERING TOOLS, TECHNIQUES, AND TABLES

MECHANICS OF FUNCTIONALLY GRADED MATERIALS AND STRUCTURES

ENGINEERING TOOLS, TECHNIQUES AND TABLES

Additional books in this series can be found on Nova's website under the Series tab.

Additional e-books in this series can be found on Nova's website under the e-book tab.

ENGINEERING TOOLS, TECHNIQUES, AND TABLES

MECHANICS OF FUNCTIONALLY GRADED MATERIALS AND STRUCTURES

ZHENG ZHONG,
LINZHI WU
AND
WEIQIU CHEN
EDITORS

Nova Science Publishers, Inc.
New York

LIBRARY OF CONGRESS CATALOGING-IN-PUBLICATION DATA

ISBN: 978-1-61668-920-9

Published by Nova Science Publishers, Inc. † New York

CONTENTS

PREFACE

This new book reviews research results in the field of mechanics research. Also discussed herein are the most important areas in the mechanics of functionally graded materials and structures, including the analytical and the semi-analytical solutions of functionally graded beams, plates and shells as well as their simplified theories, fracture analysis of functionally graded materials, a micro-element method for the macro-micro scale analysis and the optimal design of functionally graded structures.

Chapter 1 - This chapter presents semi-analytical solutions for three-dimensional problems of multi-directional functionally graded magneto-electroelastic circular plates under various boundary conditions. The semi-analytical approach combines the state space method, the differential quadrature method and the Fourier series expansion to give the solution. In the thickness direction of the plate the state space method which takes unknown displacement components, electric potential, magnetic potential and their first order derivatives as the state variables is employed to obtain the analytical solution. For the in-plane domain differential quadrature method is adopted in the radial direction to give the approximate solution while the Fourier series expansion is used in the circumferential direction of plates to separate variables. The proposed semi-analytical method is validated and the superiority of the method is also illustrated. Three-dimensional static and dynamic behavior of multi-directional functionally graded magnetoelectroelastic plates is investigated under various boundary conditions by using the semi-analytical method. And the influence of the graded variation of material constants is illuminated.

Chapter 2 - Some traditional applications of structures and devices with homogeneous materials are being gradually replaced by functionally graded materials (FGM) with spatial variation of properties. The analysis of surface acoustic waves (SAW) propagating in FGM structures will be different primarily due to variations of material properties and resulted differential equations with variable coefficients. With the exception of a few patterns of variation, analytical solutions of wave propagation in FGM cannot be obtained. As a result, there will be challenges in the validation of numerical and approximate solutions. To meet the need for exact solutions, the authors studied the surface acoustic waves propagating in an FGM plate with the layered method, the Frobenius method, and the homotopy analysis method (HAM) for analytical and exact solutions. These three methods represent available techniques for analytical solutions of a large class of problems with important applications in the FGM structures. Numerical results based on the three methods are given for FGM plates with material property variations in a linear pattern. These analytical examples are presented along with detailed explanations of the methods and formulation to demonstrate the effectiveness of the methods and the effect of FGM on changes of plate deformation and frequency effects in SAW propagation with possible applications in devices and structures.

Chapter 3 - A microelement method for the analyses of functionally graded structures was suggested. In order to reflect micro-constitution of material components, the microelement method sets up the dense microelements in the normal finite element of the structure, and translates the degrees of freedom of every microelement node to the degrees of freedom of the same normal finite element by the compatibility conditions, which will be calculated by the computer. This microelement method can fully realize the transition from material micro-constitution to macro- response of structure and its computational elements and degrees of freedom are same as the normal finite element method. Therefore, this method is an effective means for the scale-span analysis of the functionally graded structures. In this paper, the macro-response can be immediately calculated by the material component distribution determined in the process of material production; the influence of gap on the macro-responses can be analyzed, the scale-span analysis between micro-constitution and macro-response of the functionally graded materials and structures can be achieved; the three-dimensional distributions of mechanical quantities of the functionally graded plates with different complex shapes are given. Instead of material parameter input of traditional mechanical analysis, the mechanical analyses are based on the input of metallographic diagram information by microelement method, the mechanics responses of functionally graded plates are calculated and 3-D distributions of macro-mechanical variables and identical stress lines tendency on microstructure as well as 3-D distributions of macro natural mode shapes and identical stress mode shape lines tendency on microstructure are given.

Chapter 4 - Methods are described in this chapter to acquire analytical solutions of anisotropic functionally graded magneto-electro-elastic (FGMEE) beams under polynomial and arbitrary loads. The material properties are functions of the thickness coordinate. The partial differential equations governing stress function, electric displacement function and magnetic induction function are derived for plane problems of anisotropic FGMEE materials. These functions are assumed in terms of unknown functions of thickness coordinate only, which are determined subsequently. The analytical expressions of axial force, bending moment, shear force, average electric displacement, average magnetic induction, displacements, electric potential and magnetic potential are then deduced. The integral constants are determinable completely from boundary conditions. A series of analytical solutions are thus obtained. The methods can also be used to find analytical solutions for FG piezoelastic beams and FG elastic beams under polynomial and arbitrary loads.

Chapter 5 - The state-space description has been widely adopted in the modern control theory, while it is only occasionally employed for seeking the solutions of elasticity problems until recently. This chapter presents a brief overview of the state-space method for elasticity, with particular emphasis on its applications in analyses of functionally graded materials plates and structures. Because of the material inhomogeneity, it is generally very difficult to obtain exact static/dynamic solutions for these structures directly based on three-dimensional (or even two-dimensional) elasticity equations. Except for some particular cases, a layerwise scheme thus has been incorporated into the analysis, facilitating us to derive analytical solutions. These though approximate solutions can be made as accurately as possible by involving more layers in the approximation, and hence can be benchmarks for clarifying other approximate and numerical solutions. A newly developed semi-analytical method, which combines the state-space method and the differential quadrature method, is also presented.

Chapter 6 - This work surveys the application of the finite element method in functionally graded materials and structures firstly. Some graded isoparametric elements are then formulated based on the variational principle of piezoelectricity. This work also presents a graded harmonic element for axisymmetric piezoelectricity with non-symmetric loads. Finally some numerical results are shown to demonstrate the proposed elements.

Chapter 7 - Transient response is an important physical quantity in modern engineering analysis. In this chapter, the transient responses of multi-field coupled functionally graded cylindrical and spherical shells under radial deformation are investigated. The cylindrical and spherical shells are assumed to be graded along the radial direction as a power function. For functionally graded magneto-electro-elastic (MEE) cylindrical and spherical shells, the governing equations are finally transformed into two Volterra integral equations of the second kind with respect to two functions of time. The recursive formulae are constructed and the integral equations are solved effectively. The present methodology can also be employed to solve the transient responses of functionally graded cylindrical and spherical shells made of piezoelectric and purely elastic materials. For functionally graded piezoelectric cylindrical and spherical shells, the governing equations are transformed into a Volterra integral equation of the second kind with respect to a function of time. For functionally graded purely elastic cylindrical and spherical shells, no Volterra integral equation is involved and the obtained solution is a closed-form solution.

Chapter 8 - In this chapter, the dynamic behavior of an interface crack and the basic solution of a Mode-I limited-permeable crack in functionally graded piezoelectric/piezomagnetic materials were investigated. At the same time, the electric permittivity and the magnetic permeability of air in crack are considered in the present paper. By using the Fourier transform technique, these problems can be solved with the help of a pair of dual integral equations in which the unknown variable is the displacement jump across the crack surfaces. These equations were solved by using the Schmidt method. The relations among the electric filed, the magnetic flux field and the stress field near the crack tips can be obtained. From the solutions of the present paper, it is found that the singular stresses, the singular electric displacements and the singular magnetic flux at crack tips in functionally graded piezoelectric/piezomagnetic materials carry the same forms as those in homogeneous piezoelectric/piezomagnetic materials but the magnitudes of the intensity factors are dependent on the gradient of functionally graded piezoelectric/piezomagnetic materials. It can be also obtained that the effects of the electric and magnetic boundary conditions on the electric displacement and magnetic flux fields near crack tips are large from the solution of the present paper.

Chapter 9 - Fracture mechanics evaluation provides a deeper insight into the physics of crack initiation and propagation in FGMs, which can aid in the design, optimization and application of these novel materials. Stationary crack problem in FGMs have received considerable attention from a number of researchers around the world. However, it should be mentioned that most FGMs will be used in critical situations, such as dynamic loading. Because one additional dimension, time, enters when going from a static to a corresponding dynamic case, analysis of the latter is more difficult. Therefore, the dynamic fracture responses of FGMs have received little attention from the scientific community. This chapter summarizes the recent developments of elastodynamic crack problem of functionally graded materials and structure. It covers crack moving at constant velocity in FGMs; Scattering waves about stationary cracks in FGMs and Numerical analysis of dynamic crack problems in FGMs.

Chapter 10 - In this chapter, the modeling methods for the crack problems of functionally graded materials (FGMs) with general (or arbitrary) properties are reviewed and discussed. The focus of this chapter is on the analytical and numerical modeling methods for the crack problem of FGMs with general properties. Firstly, the basic problem (a "particular" crack problem of a functionally graded layered structure with exponential properties and a crack crossing the interface) for establishing the piecewise-exponential model (PE model) is presented. Then the PE model for FGMs with general mechanical properties is introduced. In the PE model, the FGM is divided into some nonhomogeneous layers along the gradient direction of the properties, with each layer's properties varying exponentially. By this method, the real material properties can be approached by a series of exponential functions. Moreover, considering the significance of the investigation of the thermal fracture behavior for FGMs, the PE model is extended to the crack problems of FGMs with general thermomechanical properties. It should be mentioned that: in past decades, there are very limited literatures on the analytical study of crack problems in the FGMs with non-exponential properties. However, for the FGMs with exponential properties, there have been lots of investigations regarding many types of crack problems. These achievements will provide a wide support for the PE model to be applied in solving many types of crack problems in FGMs with general properties. On the other hand, the analytical modeling methods for the crack problems of FGMs with general properties are very difficult to consider the influences of the complex internal structures (such as the component particles and interfaces) on the fracture behavior of FGMs. Therefore, numerical modeling methods for the crack problems of FGMs with general properties should be developed together with the analytical methods. The key problem in the numerical modeling methods lies in how to obtain the crack tip quantities with high efficiency. An interaction integral and its associated domain form without any derivatives of the material properties are introduced. It can be used to analyze the crack problems of FGMs with general properties, considering the effects of the component particles and internal interfaces on the fracture behavior.

In: Mechanics of Functionally Graded Materials and Structures ISBN 978-1-61668-920-9
Editors: Z. Zhong, L. Wu and W. Chen

Chapter 1

SEMI-ANALYTICAL SOLUTIONS FOR THREE-DIMENSIONAL PROBLEMS OF MULTI-DIRECTIONAL FUNCTIONALLY GRADED MAGNETO-ELECTRO-ELASTIC CIRCULAR PLATES

***Guojun Nie*[1] *and Zheng Zhong*[2]**
School of Aerospace Engineering and Applied Mechanics
Tongji University, Shanghai, China

ABSTRACT

This chapter presents semi-analytical solutions for three-dimensional problems of multi-directional functionally graded magnetoelectroelastic circular plates under various boundary conditions. The semi-analytical approach combines the state space method, the differential quadrature method and the Fourier series expansion to give the solution. In the thickness direction of the plate the state space method which takes unknown displacement components, electric potential, magnetic potential and their first order derivatives as the state variables is employed to obtain the analytical solution. For the in-plane domain differential quadrature method is adopted in the radial direction to give the approximate solution while the Fourier series expansion is used in the circumferential direction of plates to separate variables. The proposed semi-analytical method is validated and the superiority of the method is also illustrated. Three-dimensional static and dynamic behavior of multi-directional functionally graded magnetoelectroelastic plates is investigated under various boundary conditions by using the semi-analytical method. And the influence of the graded variation of material constants is illuminated.

[1] e-mail: ngj@tongji.edu.cn.
[2] e-mail: zhongk@tongji.edu.cn.

1. INTRODUCTION

Functionally graded materials (FGMs) are nonhomogeneous composites which are composed of two or more phases. By gradually changing the composition of the constituent materials along one or more directions during manufacturing, one can obtain the smooth spatial variations of macroscopic material properties of FGMs. The advantages of this kind of materials over laminated composites include the elimination of interfaces between different layers thereby avoiding failures from interfacial stress concentrations. Furthermore, one can get optimum performance of the components to externally applied loading by tailoring the spatial variations of material properties. In recent years, FGMs have gained considerable attention as advanced structural materials in many engineering applications [1-6]. However the analysis of FGMs structures is more complicated than that of homogeneous material structures, owing to the spatial variations of material properties. In order to investigate the stress and displacement fields of FGMs body, one needs to solve the partial differential equations with variable coefficients. This class of problem is challenging and thus there is a lot of literature on the mechanical behavior of FGMs structures.

Many researchers have studied the stress and displacement fields of a circular plate made of FGMs. Based on the classical plate theory, Ebrahimi and Rastgo [7] studied analytically the free vibration behavior of FGMs thin circular plates integrated with two uniformly distributed actuator layers made of piezoelectric material; Hosseini-Hashemi et al. [8] investigated buckling and free vibration behavior of radially functionally graded circular and annular sector thin plates subjected to uniform in-plane compressive loads and resting on the Pasternak elastic foundation. Based on the first order shear deformation plate theory, Reddy et al. [9], Sahraee [10] and Jomehzadeh et al. [11] investigated axisymmetric bending and stretching of FGMs circular and annular plates; Nosier and Fallah [12] investigated the non-linear axisymmetric and asymmetric behaviour of FGMs circular plates under transverse mechanical loading; Efraim and Eisenberger [13] presented the vibration analysis of thick annular plates with variable thickness made of isotropic FGMs by exact element method; Prakash and Ganapathi [14] analyzed asymmetric free vibration characteristics and thermoelastic stability of FGMs circular plates using finite element procedure; for the cases where electrodes on the piezoelectric layers are shortly connected, Liu et al. [15] studied free vibration of piezoelectric coupled moderately thick circular plate for two boundary conditions of the plate, that is, clamped edge and simply supported. Based on the third-order shear deformation plate theory, Saidi et al. [16], Najafizadeh and Heydari [17], Ma and Wang [18] solved the axisymmetric bending and buckling problems of FGMs circular plates. Based on fourth-order shear deformation plate theory, Sahraee and Saidi [19] studied axisymmetric bending and stretching of FGMs circular plates subjected to uniform transverse loading. The three-dimensional solutions of FGMs circular plates are relatively few at present. Li et al. [20-21] considered the axisymmetric bending of transversely isotropic, functionally graded, elastic and magneto-electro-elastic circular plates with simply supported or clamped edge based on the three-dimensional theory of elasticity. Dong [22] investigated three-dimensional free vibration of FGMs annular plates with different boundary conditions using the Chebyshev-Ritz method, in which a set of duplicate Chebyshev polynomial series multiplied by the boundary function satisfying the boundary conditions are chosen as the trial functions of the displacement components. Nie and Zhong [23-25] presented the semi-analytical solutions for three-dimensional problem of elastic or magnetoelectroelastic circular and annular FGM plates with various boundary conditions.

The aforementioned works have presumed that the material constants vary only in one direction, usually in the thickness direction. However there are practical occasions which require the materials to be varied continuously in two or three directions. Problems of multi-directional FGMs structures have been studied in [26-30]. Lü et al. [26-27] gave the semi-analytical elasticity solutions for multi-directional functionally graded beams and plates by using the state-space-based differential quadrature method. Nie and Zhong [28-30] put forward an improved state-space-based differential quadrature method to study the axisymmetric and asymmetric bending and vibration of bi-directional FGMs circular and annular plates.

In this chapter, we will introduce the improved semi-analytical method and its application in three-dimensional static and dynamic analysis of multi-directional functionally graded magnetoelectroelastic circular and annular plates.

2. Semi-Analytical Method

Consider a circular plate composed of functionally graded materials (FGMs) with radius a and thickness h, as shown in Figure 1. A cylindrical coordinate system r,θ,z with the origin at the center of the bottom plate is employed to describe the displacement field of the plate.

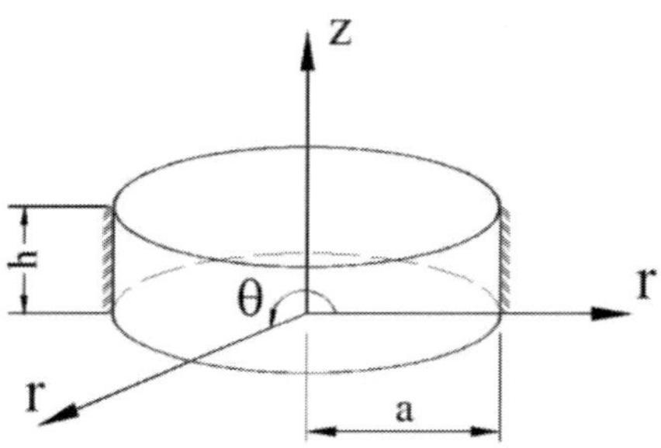

Figure 1. Multi-directional functionally graded magneto-electro-elastic circular plates

For a linear, orthotropic, and magneto-electro-elastic solid, the coupled constitutive relations can be written in the following form

$$
\begin{aligned}
\sigma_r &= c_{11}\frac{\partial u_r}{\partial r} + c_{12}\left(\frac{1}{r}\frac{\partial u_\theta}{\partial \theta} + \frac{u_r}{r}\right) + c_{13}\frac{\partial u_z}{\partial z} + e_{31}\frac{\partial \phi}{\partial z} + q_{31}\frac{\partial \psi}{\partial z} \\
\sigma_\theta &= c_{12}\frac{\partial u_r}{\partial r} + c_{22}\left(\frac{1}{r}\frac{\partial u_\theta}{\partial \theta} + \frac{u_r}{r}\right) + c_{23}\frac{\partial u_z}{\partial z} + e_{32}\frac{\partial \phi}{\partial z} + q_{32}\frac{\partial \psi}{\partial z} \\
\sigma_z &= c_{13}\frac{\partial u_r}{\partial r} + c_{23}\left(\frac{1}{r}\frac{\partial u_\theta}{\partial \theta} + \frac{u_r}{r}\right) + c_{33}\frac{\partial u_z}{\partial z} + e_{33}\frac{\partial \phi}{\partial z} + q_{33}\frac{\partial \psi}{\partial z} \\
\tau_{\theta z} &= c_{44}\left(\frac{1}{r}\frac{\partial u_z}{\partial \theta} + \frac{\partial u_\theta}{\partial z}\right) + e_{24}\left(\frac{1}{r}\frac{\partial \phi}{\partial \theta}\right) + q_{24}\left(\frac{1}{r}\frac{\partial \psi}{\partial \theta}\right) \\
\tau_{rz} &= c_{55}\left(\frac{\partial u_r}{\partial z} + \frac{\partial u_z}{\partial r}\right) + e_{15}\frac{\partial \phi}{\partial r} + q_{15}\frac{\partial \psi}{\partial r} \\
\tau_{r\theta} &= c_{66}\left(\frac{\partial u_\theta}{\partial r} + \frac{1}{r}\frac{\partial u_r}{\partial \theta} - \frac{u_\theta}{r}\right)
\end{aligned} \tag{1a}
$$

$$D_r = e_{15}\left(\frac{\partial u_r}{\partial z}+\frac{\partial u_z}{\partial r}\right)-\alpha_{11}\frac{\partial \phi}{\partial r}-d_{11}\frac{\partial \psi}{\partial r}$$

$$D_\theta = e_{24}\left(\frac{1}{r}\frac{\partial u_z}{\partial \theta}+\frac{\partial u_\theta}{\partial z}\right)-\alpha_{22}\left(\frac{1}{r}\frac{\partial \phi}{\partial \theta}\right)-d_{22}\left(\frac{1}{r}\frac{\partial \psi}{\partial \theta}\right)$$

$$D_z = e_{31}\frac{\partial u_r}{\partial r}+e_{32}\left(\frac{1}{r}\frac{\partial u_\theta}{\partial \theta}+\frac{u_r}{r}\right)+e_{33}\frac{\partial u_z}{\partial z}-\alpha_{33}\frac{\partial \phi}{\partial z}-d_{33}\frac{\partial \psi}{\partial z} \tag{1b}$$

$$B_r = q_{15}\left(\frac{\partial u_r}{\partial z}+\frac{\partial u_z}{\partial r}\right)-d_{11}\frac{\partial \phi}{\partial r}-\mu_{11}\frac{\partial \psi}{\partial r}$$

$$B_\theta = q_{24}\left(\frac{1}{r}\frac{\partial u_z}{\partial \theta}+\frac{\partial u_\theta}{\partial z}\right)-d_{22}\left(\frac{1}{r}\frac{\partial \phi}{\partial \theta}\right)-\mu_{22}\left(\frac{1}{r}\frac{\partial \psi}{\partial \theta}\right)$$

$$B_z = q_{31}\frac{\partial u_r}{\partial r}+q_{32}\left(\frac{1}{r}\frac{\partial u_\theta}{\partial \theta}+\frac{u_r}{r}\right)+q_{33}\frac{\partial u_z}{\partial z}-d_{33}\frac{\partial \phi}{\partial z}-\mu_{33}\frac{\partial \psi}{\partial z} \tag{1c}$$

where $\sigma_r,\sigma_\theta,\sigma_z,\tau_{\theta z},\tau_{rz},\tau_{r\theta}$ are stress components, D_r,D_θ,D_z are electric displacement components, B_r,B_θ,B_z are magnetic induction components, u_r,u_θ,u_z are displacement components, ϕ and ψ are the electric and magnetic potentials, respectively. $c_{ij}(i,j=1,2,\cdots 6)$, $e_{kl}(k=1,2,3,l=1,2,\cdots 6)$, $q_{kl}(k=1,2,3,l=1,2,\cdots 6)$, $\alpha_{kk}(k=1,2,3)$, $d_{kk}(k=1,2,3)$ and $\mu_{kk}(k=1,2,3)$ are the elastic, piezoelectric, piezomagnetic, dielectric, magnetoelectric, magnetic permeability constants, respectively.

For the problem considered in this chapter, the body forces, electric charge and current density are assumed to be zero. Thus, the equations of motion and the governing equations of electric and magnetic fields under the cylindrical coordinate system are

$$\frac{\partial \sigma_r}{\partial r}+\frac{1}{r}\frac{\partial \tau_{r\theta}}{\partial \theta}+\frac{\partial \tau_{rz}}{\partial z}+\frac{\sigma_r-\sigma_\theta}{r}=\rho\frac{\partial^2 u_r}{\partial t^2}$$

$$\frac{\partial \tau_{r\theta}}{\partial r}+\frac{1}{r}\frac{\partial \sigma_\theta}{\partial \theta}+\frac{\partial \tau_{\theta z}}{\partial z}+\frac{2\tau_{r\theta}}{r}=\rho\frac{\partial^2 u_\theta}{\partial t^2}$$

$$\frac{\partial \tau_{rz}}{\partial r}+\frac{1}{r}\frac{\partial \tau_{\theta z}}{\partial \theta}+\frac{\partial \sigma_z}{\partial z}+\frac{\tau_{rz}}{r}=\rho\frac{\partial^2 u_z}{\partial t^2} \tag{2a}$$

$$\frac{\partial D_r}{\partial r}+\frac{1}{r}\frac{\partial D_\theta}{\partial \theta}+\frac{\partial D_z}{\partial z}+\frac{D_r}{r}=0 \tag{2b}$$

$$\frac{\partial B_r}{\partial r}+\frac{1}{r}\frac{\partial B_\theta}{\partial \theta}+\frac{\partial B_z}{\partial z}+\frac{B_r}{r}=0 \tag{2c}$$

where ρ denotes material density and t is time.

For a multi-directional functional graded magnetoelectroelastic material, the material constants are continuous functions of the coordinates and we assume that they have the following exponent-law distributions along the thickness and radial direction of the plate

$$k_{ij}(r,z)=k_{ij}^{0}\mathrm{e}^{\lambda_1\left(\frac{z}{h}\right)+\lambda_2\left(\frac{r}{a}\right)} \tag{3}$$

where k_{ij} denotes $c_{ij}, e_{kl}, q_{kl}, \alpha_{kk}, d_{kk}, \mu_{kk}$ and ρ. $k_{ij}^{0}=k_{ij}(0,0)$ is the corresponding value of material constants at the central point of the bottom plane, λ_1 and λ_2 are the graded indexes in the thickness and radial direction, respectively.

Take $u_r, u_\theta, u_z, \phi, \psi, \frac{\partial u_r}{\partial z}, \frac{\partial u_\theta}{\partial z}, \frac{\partial u_z}{\partial z}, \frac{\partial \phi}{\partial z}$ and $\frac{\partial \psi}{\partial z}$ as state variables. In view of Eq.(3), the governing equations in Eqs. (1) and (2) along with the following equation (4) can be written as the state space formulation in Eq. (5).

$$\frac{\partial}{\partial z}(u_r)=\frac{\partial u_r}{\partial z},\ \frac{\partial}{\partial z}(u_\theta)=\frac{\partial u_\theta}{\partial z},\ \frac{\partial}{\partial z}(u_z)=\frac{\partial u_z}{\partial z},\ \frac{\partial}{\partial z}(\phi)=\frac{\partial \phi}{\partial z},\ \frac{\partial}{\partial z}(\psi)=\frac{\partial \psi}{\partial z} \tag{4}$$

$$\frac{\partial}{\partial z}\begin{bmatrix}\mathbf{\Pi}\\ \mathbf{\Gamma}\end{bmatrix}=\mathbf{A}\begin{bmatrix}\mathbf{\Pi}\\ \mathbf{\Gamma}\end{bmatrix} \tag{5}$$

where

$$\mathbf{\Pi}=\begin{bmatrix}u_r & u_\theta & u_z & \phi & \psi\end{bmatrix}^T,\ \mathbf{\Gamma}=\begin{bmatrix}\frac{\partial u_r}{\partial z} & \frac{\partial u_\theta}{\partial z} & \frac{\partial u_z}{\partial z} & \frac{\partial \phi}{\partial z} & \frac{\partial \psi}{\partial z}\end{bmatrix}^T,$$

and $\mathbf{A}=\begin{bmatrix}\mathbf{0} & \mathbf{I}\\ \mathbf{D}_2 & \mathbf{D}_3\end{bmatrix}$ is the state matrix. $\mathbf{I}$ is the unity matrix of 5×5. The expressions of $\mathbf{D}_2=d_{ij}\,(i,j=1,2,\cdots 5)$ and $\mathbf{D}_3=d'_{ij}\ (i,j=1,2,\cdots 5)$ are

$$d_{11}=-\frac{c_{11}^0}{c_{55}^0}\left(\frac{1}{r}\frac{\partial}{\partial r}+\frac{\partial^2}{\partial r^2}+\frac{\lambda_2}{a}\frac{\partial}{\partial r}\right)-\frac{c_{12}^0}{c_{55}^0}\frac{\lambda_2}{a}\frac{1}{r}+\frac{c_{22}^0}{c_{55}^0}\frac{1}{r^2}-\frac{c_{66}^0}{c_{55}^0}\frac{1}{r^2}\frac{\partial^2}{\partial\theta^2}+\frac{\rho^0}{c_{55}^0}\frac{\partial^2}{\partial t^2},$$

$$d_{12}=-\frac{c_{12}^0}{c_{55}^0}\left(\frac{1}{r}\frac{\partial^2}{\partial r\partial\theta}+\frac{\lambda_2}{a}\frac{1}{r}\frac{\partial}{\partial\theta}\right)+\frac{c_{22}^0}{c_{55}^0}\frac{1}{r^2}\frac{\partial}{\partial\theta}-\frac{c_{66}^0}{c_{55}^0}\left(\frac{1}{r}\frac{\partial^2}{\partial r\partial\theta}-\frac{1}{r^2}\frac{\partial}{\partial\theta}\right),\ d_{13}=-\frac{\lambda_1}{h}\frac{\partial}{\partial r},$$

$$d_{14}=-\frac{\lambda_1 e_{15}^0}{hc_{55}^0}\frac{\partial}{\partial r},\ d_{15}=-\frac{\lambda_1 q_{15}^0}{hc_{55}^0}\frac{\partial}{\partial r},$$

$$d_{21}=-\frac{c_{12}^0}{c_{44}^0}\left(\frac{1}{r}\frac{\partial^2}{\partial r\partial\theta}\right)-\frac{c_{22}^0}{c_{44}^0}\frac{1}{r^2}\frac{\partial}{\partial\theta}-\frac{c_{66}^0}{c_{44}^0}\left(\frac{1}{r^2}\frac{\partial}{\partial\theta}+\frac{1}{r}\frac{\partial^2}{\partial r\partial\theta}+\frac{\lambda_2}{a}\frac{1}{r}\frac{\partial}{\partial\theta}\right),$$

$$d_{22}=-\frac{c_{22}^0}{c_{44}^0}\frac{1}{r^2}\frac{\partial^2}{\partial\theta^2}-\frac{c_{66}^0}{c_{44}^0}\left(\frac{1}{r}\frac{\partial}{\partial r}+\frac{\partial^2}{\partial r^2}-\frac{\lambda_2}{a}\frac{1}{r}+\frac{\lambda_2}{a}\frac{\partial}{\partial r}-\frac{1}{r^2}\right)+\frac{\rho^0}{c_{44}^0}\frac{\partial^2}{\partial t^2},\ d_{23}=-\frac{\lambda_1}{h}\frac{1}{r}\frac{\partial}{\partial\theta},$$

$$d_{24} = -\frac{\lambda_1 e_{24}^0}{hc_{44}^0 r}\frac{\partial}{\partial\theta}, \quad d_{25} = -\frac{\lambda_1 q_{24}^0}{hc_{44}^0 r}\frac{\partial}{\partial\theta}, \quad d'_{11} = -\frac{\lambda_1}{h}, \quad d'_{12} = 0,$$

$$d'_{13} = -\frac{c_{13}^0}{c_{55}^0}\left(\frac{\partial}{\partial r} + \frac{1}{r} + \frac{\lambda_2}{a}\right) + \frac{c_{23}^0}{c_{55}^0}\frac{1}{r} - \frac{\partial}{\partial r}, \quad d'_{14} = -\frac{e_{15}^0}{c_{55}^0}\frac{\partial}{\partial r} - \frac{e_{31}^0}{c_{55}^0}\left(\frac{1}{r} + \frac{\lambda_2}{a} + \frac{\partial}{\partial r}\right) + \frac{e_{32}^0}{c_{55}^0}\frac{1}{r},$$

$$d'_{15} = -\frac{q_{15}^0}{c_{55}^0}\frac{\partial}{\partial r} - \frac{q_{31}^0}{c_{55}^0}\left(\frac{1}{r} + \frac{\lambda_2}{a} + \frac{\partial}{\partial r}\right) + \frac{q_{32}^0}{c_{55}^0}\frac{1}{r}, \quad d'_{21} = 0, \quad d'_{22} = -\frac{\lambda_1}{h},$$

$$d'_{23} = -\frac{c_{23}^0}{c_{44}^0}\frac{1}{r}\frac{\partial}{\partial\theta} - \frac{1}{r}\frac{\partial}{\partial\theta}, \quad d'_{24} = -\frac{e_{24}^0 + e_{32}^0}{rc_{44}^0}\frac{\partial}{\partial\theta}, \quad d'_{25} = -\frac{q_{24}^0 + q_{32}^0}{rc_{44}^0}\frac{\partial}{\partial\theta},$$

and other elements of $\mathbf{D}_2$ and $\mathbf{D}_3$ involve lengthy expressions and are thus omitted.

It is seen that the differential operators with respect to r, θ, t are contained in the right-hand side of Eq. (5). To eliminate the differential operators with respect to θ, t, we expand the displacement components, electric and magnetic potentials in Fourier series in the angular coordinate θ and consider a common time factor $e^{i\omega t}$ as follows

$$u_r(r,\theta,z,t) = \sum_{m=0}^{\infty}\left[u_{rm}^c(r,z)\cos(m\theta) + u_{rm}^s(r,z)\sin(m\theta)\right]e^{i\omega t}, \tag{6a}$$

$$u_\theta(r,\theta,z,t) = \sum_{m=0}^{\infty}\left[u_{\theta m}^c(r,z)\cos(m\theta) + u_{\theta m}^s(r,z)\sin(m\theta)\right]e^{i\omega t}, \tag{6b}$$

$$u_z(r,\theta,z,t) = \sum_{m=0}^{\infty}\left[u_{zm}^c(r,z)\cos(m\theta) + u_{zm}^s(r,z)\sin(m\theta)\right]e^{i\omega t}, \tag{6c}$$

$$\phi(r,\theta,z,t) = \sum_{m=0}^{\infty}\left[\phi_m^c(r,z)\cos(m\theta) + \phi_m^s(r,z)\sin(m\theta)\right]e^{i\omega t}, \tag{6d}$$

$$\psi(r,\theta,z,t) = \sum_{m=0}^{\infty}\left[\psi_m^c(r,z)\cos(m\theta) + \psi_m^s(r,z)\sin(m\theta)\right]e^{i\omega t}, \tag{6e}$$

where the integer m equals the circumferential wave number, and superscripts c and s signify quantities associated with the cosine and the sine terms. For an axisymmetric problem for a circular plate, $u_{r0}^c \neq 0$, $u_{z0}^c \neq 0$, $\phi_0^c \neq 0$, $\psi_0^c \neq 0$, and coefficients of the remaining cosine and the sine terms in Eq. (6) can be taken as zeros. For non-axisymmetric problems, we take $u_{r0}^s = u_{\theta 0}^s = u_{z0}^s = \phi_0^s = \psi_0^s = 0$ since they do not contribute to u_r, u_θ, u_z, ϕ and ψ, respectively. The displacement components, electric and magnetic potentials

$u_{rm}^{c}, u_{\theta m}^{s}, u_{zm}^{c}, \phi_{m}^{c}, \psi_{m}^{c}$ and $u_{rm}^{s}, u_{\theta m}^{c}, u_{zm}^{s}, \phi_{m}^{s}, \psi_{m}^{s}$ for a certain circumferential wave number m are two different sets of deformation or vibration modes.

Furthermore we use the differential quadrature method (DQM) to eliminate the partial derivatives with respect to r in the right-hand side of Eq. (5). Let $f(r)$ denote the unknown displacement components, electric and magnetic potentials, $u_r(r), u_\theta(r), u_z(r), \phi(r)$, and $\psi(r)$. The n-th order partial derivative of $f(r)$ at a given point r_i is approximated as a linear sum of weighted function values at all of the discrete points in the domain of r:

$$\left.\frac{\partial^n f(r)}{\partial r^n}\right|_{r=r_i} = \sum_{j=1}^{M} g_{ij}^{(n)} f(r_j), \quad (n=1, 2, \cdots, N-1;\ i=1, 2, \cdots, N) \tag{7}$$

where N is the number of discrete points, $g_{ij}^{(n)}$ are weighted coefficients depending on r_i $(i=1, 2, \cdots, N)$ only, and the corresponding expressions are [31]

$$g_{ij}^{(1)} = \frac{\prod_{v=1, v\neq i}^{N} (r_i - r_v)}{(r_i - r_j) \prod_{v=1, v\neq j}^{N} (r_j - r_v)} \tag{8a}$$

$$g_{ij}^{(n)} = n\left[g_{ii}^{(n-1)} g_{ij}^{(1)} - \frac{g_{ij}^{(n-1)}}{r_i - r_j} \right], \quad (n=2, 3, \cdots, N-1) \tag{8b}$$

where $i, j = 1, 2, \cdots, N$, but $i \neq j$, $g_{ii}^{(n)}$ are determined by

$$g_{ii}^{(n)} = -\sum_{v=1, v\neq i}^{N} g_{iv}^{(n)}, \quad (i=1, 2, \cdots, N;\ n=1, 2, \cdots, N-1) \tag{8c}$$

Before applying the rule in Eq. (7), we non-dimensionalize displacement components u_{rm}^{c}, $u_{\theta m}^{s}$, u_{zm}^{c}, u_{rm}^{s}, $u_{\theta m}^{c}$, u_{zm}^{s} and the thickness coordinate z by h, the radial coordinate r by a, electric potentials ϕ_m^c, ϕ_m^s by $h\sqrt{c_{11}^0/\alpha_{33}^0}$, magnetic potentials ψ_m^c, ψ_m^s by $h\sqrt{c_{11}^0/\mu_{33}^0}$, elastic constants c_{ij} and stress components $\sigma_r, \sigma_\theta, \sigma_z, \tau_{\theta z}, \tau_{rz}, \tau_{r\theta}$ by c_{11}^0, piezoelectric constants e_{kl} and electric displacement components D_r, D_θ, D_z by $\sqrt{c_{11}^0 \alpha_{33}^0}$, piezomagnetic constants q_{kl} and magnetic induction components B_r, B_θ, B_z by $\sqrt{c_{11}^0 \mu_{33}^0}$, dielectric constants α_{kk} by α_{33}^0, magnetic permeability constants μ_{kk} by μ_{33}^0, magnetoelectric constants d_{kk} by $\sqrt{\alpha_{33}^0 \mu_{33}^0}$, and the circular frequency ω by $\sqrt{c_{11}^0/(\rho^0 h^2)}$. Henceforth we use non-dimensional variables and denote them by the same symbols as before.

For a certain circumferential wave number m we can derive the state equation with respect to the state variables at discrete points as follows

$$\frac{\partial}{\partial z}\begin{bmatrix}\mathbf{\Pi}^i \\ \mathbf{\Gamma}^i\end{bmatrix} = \mathbf{A}_i \begin{bmatrix}\mathbf{\Pi}^i \\ \mathbf{\Gamma}^i\end{bmatrix}, \quad i = 1, 2, \cdots, N \tag{9}$$

where $\mathbf{A}_i = \begin{bmatrix}\mathbf{0} & \mathbf{I}_N \\ \mathbf{D}_2^i & \mathbf{D}_3^i\end{bmatrix}$ is the state matrix of $10N \times 10N$

and

$$\mathbf{\Pi}^i = \left[u_{rm1}^c, \ldots, u_{rmN}^c \quad u_{\theta m1}^s, \ldots, u_{\theta mN}^s \quad u_{zm1}^c, \ldots, u_{zmN}^c \quad \phi_{m1}^c, \ldots, \phi_{mN}^c \quad \psi_{m1}^c, \ldots, \psi_{mN}^c\right]^T \tag{10a}$$

$$\mathbf{\Gamma}^i = \left[\frac{\partial u_{rm1}^c}{\partial z}, \ldots, \frac{\partial u_{rmN}^c}{\partial z} \quad \frac{\partial u_{\theta m1}^s}{\partial z}, \ldots, \frac{\partial u_{\theta mN}^s}{\partial z} \quad \frac{\partial u_{zm1}^c}{\partial z}, \ldots, \frac{\partial u_{zmN}^c}{\partial z} \quad \frac{\partial \phi_{m1}^c}{\partial z}, \ldots, \frac{\partial \phi_{mN}^c}{\partial z} \quad \frac{\partial \psi_{m1}^c}{\partial z}, \ldots, \frac{\partial \psi_{mN}^c}{\partial z}\right]^T \tag{10b}$$

N is the number of discrete points. While interchanging the superscripts c and s in Eq. (10), we get another set of deformation or vibration modes. $\mathbf{I}_N$ is the unity matrix of $5N \times 5N$. The expressions of $\mathbf{D}_2^i$ and $\mathbf{D}_3^i$ are obtained from $\mathbf{D}_2$ and $\mathbf{D}_3$ and they have been transformed to algebraic matrices. Eq. (9) is the state equation for multi-directional FGMs magneto-electro-elastic circular plates. We can get the solution of the state equation as follows.

$$\begin{bmatrix}\mathbf{\Pi}^i(z) \\ \mathbf{\Gamma}^i(z)\end{bmatrix} = \mathrm{e}^{\mathbf{A}_i \cdot z} \begin{bmatrix}\mathbf{\Pi}^i(0) \\ \mathbf{\Gamma}^i(0)\end{bmatrix} \tag{11}$$

where $\mathrm{e}^{\mathbf{A}_i \cdot z}$ is the matrix exponential function, which can be calculated by various methods [32]. $\mathbf{\Pi}^i(z), \mathbf{\Gamma}^i(z)$ and $\mathbf{\Pi}^i(0), \mathbf{\Gamma}^i(0)$ are the values of the state variables at arbitrary plane z and the bottom plane $z = 0$. Setting $z = 1$ in Eq. (11) gives a relation between the state variables on the top and bottom surfaces of the plate as follows

$$\begin{bmatrix}\mathbf{\Pi}^i(1) \\ \mathbf{\Gamma}^i(1)\end{bmatrix} = \mathrm{e}^{\mathbf{A}_i} \begin{bmatrix}\mathbf{\Pi}^i(0) \\ \mathbf{\Gamma}^i(0)\end{bmatrix} \tag{12}$$

After taking account of certain boundary conditions, we can determine the values of unknown variables on the bottom and the top surfaces of the plate. Furthermore, we can obtain the values of state variables at any point from Eq. (11). Thus we achieve the semi-analytical analysis for three-dimensional problems of multi-directional FGMs magneto-electro-elastic circular plates. While the material constants do not vary with the coordinates z, r as the exponential form in Eq. (3) but in other functions of z and r, the layerwise approximation proposed by Chen et al. [33-34] can be used in the thickness direction to derive the solution of the problem. In the following section, we will illustrate the application of this semi-analytical method to three-dimensional static and dynamic problems of multi-directional FGMs magneto-electro-elastic circular, annular and annular sectorial plates.

3. APPLICATION TO MULTI-DIRECTIONAL FGMS ELASTIC PLATES

For a FGMs elastic plate only displacement components and their first order derivatives are the state variables, that is, $\mathbf{\Pi}=\begin{bmatrix}u_r & u_\theta & u_z\end{bmatrix}^T$, and $\mathbf{\Gamma}=\begin{bmatrix}\frac{\partial u_r}{\partial z} & \frac{\partial u_\theta}{\partial z} & \frac{\partial u_z}{\partial z}\end{bmatrix}^T$. In the following, we will investigate the static and dynamic behavior of multi-directional FGMs elastic circular, annular and annular sectorial plates.

3.1. Multi-Directional FGMs Circular and Annular Plates

For a solid circular plate, the boundary conditions can be expressed as
Clamped:

$$\text{at } r=a,\ u_{rm}^{c}=0, u_{\theta m}^{s}=0, u_{zm}^{c}=0;\ \text{at } r=0,\ u_{rm}^{c}=0, u_{\theta m}^{s}=0, \frac{\partial u_{zm}^{c}}{\partial r}=0. \qquad (13a,b)$$

Simply supported:

$$\text{at } r=a,\ \sigma_r=0, u_{\theta m}^{s}=0, u_{zm}^{c}=0;\ \text{at } r=0,\ u_{rm}^{c}=0, u_{\theta m}^{s}=0, \frac{\partial u_{zm}^{c}}{\partial r}=0. \qquad (14a,b)$$

For an annular plate with the inner edge b and outer edge a, the boundary conditions are
Clamped-Clamped:

$$\text{at } r=a \text{ and } r=b,\ u_{rm}^{c}=0, u_{\theta m}^{s}=0, u_{zm}^{c}=0. \qquad (15a,b)$$

Clamped-Simply supported:

$$\text{at } r=a,\ u_{rm}^{c}=0, u_{\theta m}^{s}=0, u_{zm}^{c}=0;\ \text{at } r=b,\ \sigma_r=0, u_{\theta m}^{s}=0, u_{zm}^{c}=0. \qquad (16a,b)$$

Clamped-Free:

$$\text{at } r=a,\ u_{rm}^{c}=0, u_{\theta m}^{s}=0, u_{zm}^{c}=0;\ \text{at } r=b,\ \sigma_r=0, \tau_{r\theta}=0, \tau_{rz}=0. \qquad (17a,b)$$

For static problem, boundary conditions on the top and bottom surfaces are

$$\text{at } z=0,\ \tau_{rz}=K_{rz}^{-}(r,\theta),\quad \tau_{\theta z}=K_{\theta z}^{-}(r,\theta),\quad \sigma_z=K_{zz}^{-}(r,\theta) \qquad (18a)$$

$$\text{at } z=h,\ \tau_{rz}=K_{rz}^{+}(r,\theta),\quad \tau_{\theta z}=K_{\theta z}^{+}(r,\theta),\quad \sigma_z=K_{zz}^{+}(r,\theta) \qquad (18b)$$

The loads $K_{rz}^{-}(r,\theta), K_{\theta z}^{-}(r,\theta), K_{zz}^{-}(r,\theta)$ and $K_{rz}^{+}(r,\theta), K_{\theta z}^{+}(r,\theta), K_{zz}^{+}(r,\theta)$ can be further expanded in Fourier series in the angular coordinate θ. Then boundary conditions are satisfied in the sense of Fourier series.

For free vibration problem, boundary conditions on the top and bottom surfaces are

at $z=0$ and $z=h$, $\tau_{rz}=0, \tau_{\theta z}=0, \sigma_z=0$. (19a,b)

For forced vibration problem, normal and tangential tractions on the top and bottom surfaces can be assumed as follows:

at $z=0$, $\tau_{rz}=K_{rm}^{-}\cos(m\theta)\mathrm{e}^{\mathrm{i}\omega_m t}, \tau_{\theta z}=K_{\theta m}^{-}\sin(m\theta)\mathrm{e}^{\mathrm{i}\omega_m t}, \sigma_z=K_{zm}^{-}\cos(m\theta)\mathrm{e}^{\mathrm{i}\omega_m t}$ (20a)

at $z=h$, $\tau_{rz}=K_{rm}^{+}\cos(m\theta)\mathrm{e}^{\mathrm{i}\omega_m t}, \tau_{\theta z}=K_{\theta m}^{+}\sin(m\theta)\mathrm{e}^{\mathrm{i}\omega_m t}, \sigma_z=K_{zm}^{+}\cos(m\theta)\mathrm{e}^{\mathrm{i}\omega_m t}$ (20b)

where ω_m denotes the circular frequency of the tractions.

According to the rule of DQM, the boundary conditions in Eqs. (13)-(20) can be written as the discretized forms expressed by the state variables at discrete points. For instance, the discretized forms of boundary conditions in Eq. (13) for a clamped solid circular plate are

at $r=1$, $u_{rmN}^{c}=0, u_{\theta mN}^{s}=0, u_{zmN}^{c}=0$; (21a)

at $r=0$, $u_{rm1}^{c}=0, u_{\theta m1}^{s}=0, u_{zm1}^{c}=-\sum_{s=2}^{N}\frac{g_{1s}^{(1)}}{g_{11}^{(1)}}u_{zms}^{c}$. (21b)

where $g_{ij}^{(1)}$ is the weighted coefficient matrix of first-order derivative.

Substituting the corresponding boundary conditions into Eq. (12) and eliminating the first derivatives of the displacement components Γ_i, we get the following equation for free vibration problem

$$\mathbf{B}\cdot\begin{bmatrix}\bar{\mathbf{\Pi}}^{i}(0)\\ \bar{\mathbf{\Pi}}^{i}(1)\end{bmatrix}=0,\quad i=2,\cdots,N-1 \tag{22}$$

where $\mathbf{B}$ is the matrix of $6(N-2)\times 6(N-2)$ obtained from the state matrix $\mathbf{A}_i$ along with the corresponding transfer matrix and boundary conditions. $\bar{\mathbf{\Pi}}^{i}(1)$ and $\bar{\mathbf{\Pi}}^{i}(0)$ are the values of the displacement components at the top and bottom surfaces of the plate. And

$$\bar{\mathbf{\Pi}}^{i}(0)=\left[u_{rm2}^{c}(0),\ldots,u_{rmN-1}^{c}(0)\quad u_{\theta m2}^{s}(0),\ldots,u_{\theta mN-1}^{s}(0)\quad u_{zm2}^{c}(0),\ldots,u_{zmN-1}^{c}(0)\right]^{T},$$

$$\bar{\mathbf{\Pi}}^{i}(1)=\left[u_{rm2}^{c}(1),\ldots,u_{rmN-1}^{c}(1)\quad u_{\theta m2}^{s}(1),\ldots,u_{\theta mN-1}^{s}(1)\quad u_{zm2}^{c}(1),\ldots,u_{zmN-1}^{c}(1)\right]^{T}.$$

While setting the determinant of $\mathbf{B}$ matrix to zero for nontrivial solutions, the characteristic frequency equation of FGMs circular plate can be gotten, that is,

$$|\mathbf{B}|=0 \tag{23}$$

Solving the above frequency equation, we can get an infinite number of frequencies and corresponding mode shapes. Similar to free vibration analysis, we get the following algebraic equations for static problem and the forced vibration problem

$$\mathbf{B}\cdot\begin{bmatrix}\bar{\mathbf{\Pi}}^i(0)\\ \bar{\mathbf{\Pi}}^i(1)\end{bmatrix}=\mathbf{Q},\quad i=2,\cdots,N-1 \qquad (24)$$

where $\mathbf{Q}$ is a traction force vector obtained from boundary conditions on the top and bottom surfaces of the plates. By solving Eq. (24), we get the displacement components at $z=0$ and $z=1$ planes. Then the displacement and stress of FGMs elastic circular plate are obtained by Eq. (6) and Eq. (1).

3.1.1. Verification of Results

Example 1: Let us first validate the semi-analytical solution. Consider a clamped circular plate with a = 1.0 m and h = 0.1a. Young's modulus, the density and Poisson's ratio at the center of the bottom plane of the plate are E = 380 GPa, ρ = 3800 kg/m^3, ν= 0.3, respectively. The material constants are assumed to be the exponential distributions both in the thickness and in radial direction of the plate, that is, $\lambda_1=1, \lambda_2=1$. The first two frequencies of the plate are calculated and compared with those obtained from the commercial finite element software, ANSYS, shown in Table 1. While analyzing the problem with ANSYS, we use three-dimensional eight-node solid element (Solid5). The plate is divided into ten layers both in the thickness and in the radial direction, and the element size in the circumferential direction is less than 0.05a. The total numbers of the elements and nodes are 14390 and 29070, respectively. The value of the elastic constants in each layer is a constant and equals that obtained from Eq. (3) at the mid-point of the layer. From Table 1, it is observed that the first two frequencies approach the results of ANSYS with the increase of the number of discrete points along the radial direction of the plate. The semi-analytical solution presented here is validated. Furthermore, the deflection of FGMs plate is calculated under the following boundary conditions,

$$\text{at } z=0,\ \sigma_z=\tau_{rz}=\tau_{\theta z}=0,\ \text{at } z=h,\ \sigma_z=-1.0\text{MPa}, \tau_{rz}=0, \tau_{\theta z}=0.$$

And the results are compared with those obtained from ANSYS, shown in Table 2. While using DQM, equally spaced discrete points are adopted and the number of discrete points in the radial direction is eleven. It is clear that results from the two approaches agree well each other. It suggests that the proposed semi-analytical method provides a reasonably accurate solution of the problem. On the other hand, it is noticed that the computational efficiency of the present model is better than the numerical solution depending on the mesh of the model. The proposed semi-analytical method gives analytical solution both in the thickness and in circumferential direction and numerical solution in the radial direction.

Table 1. Comparison of the first two frequencies of FGMs plate

Graded indexes	Frequency	Number of the discrete points along the radial direction while using DQM				
		6	7	8	9	ANSYS
$\lambda_1=1$	ω_1	0.0298	0.0296	0.0298	0.0296	0.0296
$\lambda_2=1$	ω_2	0.0573	0.0560	0.0569	0.0561	0.0558

Table 2. Comparison of the deflection of FGMs plate (z = 0) (Unit: mm)

Graded indexes		r/a				
		0.0	0.2	0.4	0.6	0.8
$\lambda_1 = 1$	Present	-0.1521	-0.1373	-0.1003	-0.0550	-0.0168
$\lambda_2 = 1$	ANSYS	-0.1515	-0.1370	-0.1002	-0.0551	-0.0169

3.1.2. Free Vibration Problem

Example 2: We take the same material parameters at the center of the bottom plane of the plate as Example 1 to investigate the dynamic characteristic of a FGMs annular plate with b = 0.3 m, a = 1.0 m, h = 0.1 m, and clamped on the inner and outer circumferential edges. The variations of the first three frequencies with the graded indexes λ_1 and λ_2 are shown in Figs. 2a and 2b, respectively. Note that the effect of different gradient direction of FGMs plate on the dynamic behavior is different. And the radial graded variation of material constants is helpful for the improvement of the stiffness of the plate.

3.1.3 Static Problem

Example 3: Take all of the geometry and material parameters of circular plate the same as Example 1 except the graded indexes λ_1, λ_2 and external pressure. The non-axisymmetric loads applied on the top surface of the plate are

$$\text{at } z = h, \ \sigma_z = -1.0 \times \cos(2\theta)\text{MPa}, \tau_{rz} = 0, \tau_{\theta z} = 0.$$

The results of the displacement and stress are shown in Figure 3. It is found from Figure 3(a) that the transverse displacement at the location (θ = 0, z = 0) for different graded variations has great difference. And the transverse displacement of bi-directional FGMs plate is the least. Similarly, it is observed from Figure 3(b) that the maximum of the stress in bi-directional FGMs plate can be reduced and the distribution of the stress can approach to uniformity by changing the graded indexes in the thickness and radial directions. It shows that bi-directional FGMs plate may have better performance than homogeneous material plate.

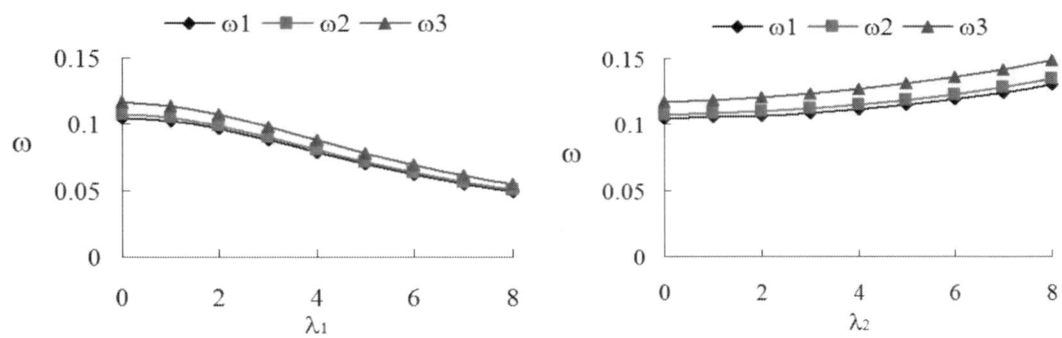

(a) (b)

Figure 2. Variation of the first three frequencies of FGMs plate with the graded index. (a) material constants varying only in the thickness, (b) material constants varying only in the radial direction.

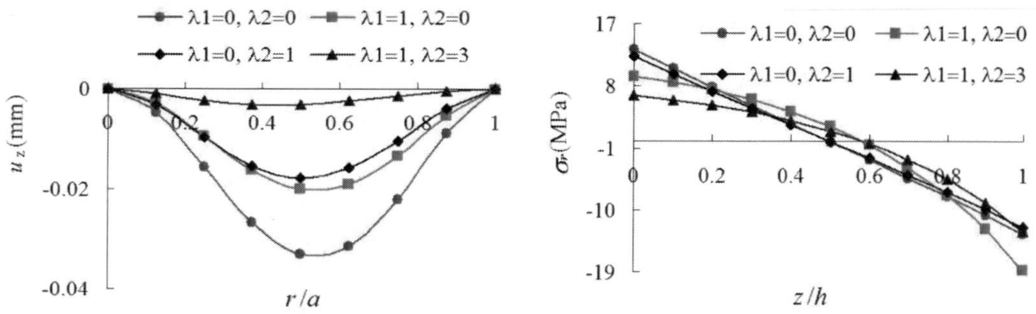

(a) u_z at $\theta = 0$, $z = 0$ (b) σ_r at $r = 0.5a$, $\theta = 0$

Figure 3. Variations of (a) displacement u_z (mm) at the location ($\theta = 0$, $z = 0$), (b) stress σ_r (MPa) at the location ($r = 0.5a$, $\theta = 0$).

3.2. FGM Annular Sectorial Plates

Consider an annular sectorial plate having inner radius b, outer radius a, and thickness h which is made of a transversely isotropic FGM, as shown in Figure 4. For certain types of vibration we assume that the displacement components have the following forms for the plate with a sector angle α and simply supported radial edges.

$$
\begin{aligned}
u_r(r,\theta,z,t) &= u_{rm}^s(r,z)\sin(m\pi\theta/\alpha)\mathrm{e}^{\mathrm{i}\omega t} \\
u_\theta(r,\theta,z,t) &= u_{\theta m}^c(r,z)\cos(m\pi\theta/\alpha)\mathrm{e}^{\mathrm{i}\omega t} \\
u_z(r,\theta,z,t) &= u_{zm}^s(r,z)\sin(m\pi\theta/\alpha)\mathrm{e}^{\mathrm{i}\omega t}
\end{aligned} \tag{25}
$$

This results in the satisfaction of the simply supported boundary conditions along $\theta = 0$ and $\theta = \alpha$ exactly. That is,

$$
\begin{aligned}
u_r(r,0,z,t) &= u_r(r,\alpha,z,t) = 0 \\
u_z(r,0,z,t) &= u_z(r,\alpha,z,t) = 0 \\
\sigma_\theta(r,0,z,t) &= \sigma_\theta(r,\alpha,z,t) = 0
\end{aligned} \tag{26}
$$

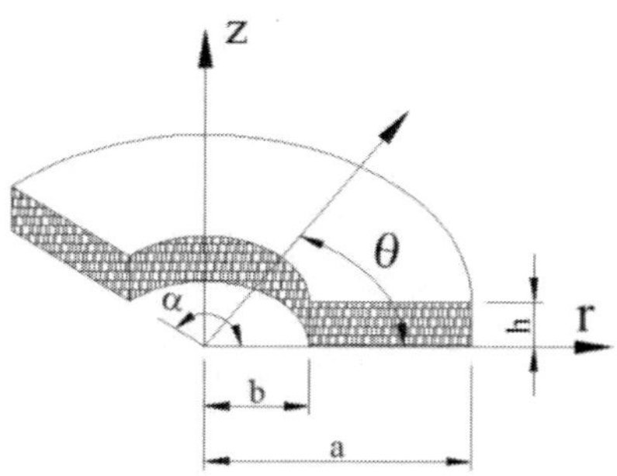

Figure 4. A FGMs annular sectorial plate with simply supported radial edges.

The boundary conditions along the inner ($r = b$) and outer ($r = a$) circumferential edges can be expressed as those in Eqs. (15)-(17). For the free vibration problems boundary conditions at the top and bottom surfaces are the same as those in Eqs. (19a,19b). For forced vibrations normal and tangential tractions at the top and bottom surfaces can be assumed as

$$\tau_{rz}^{0} = K_{rm}^{-} \sin\left(\frac{m\pi\theta}{\alpha}\right)e^{i\omega_m t}, \tau_{\theta z}^{0} = K_{\theta m}^{-} \cos\left(\frac{m\pi\theta}{\alpha}\right)e^{i\omega_m t}, \sigma_{z}^{0} = K_{zm}^{-} \sin\left(\frac{m\pi\theta}{\alpha}\right)e^{i\omega_m t}, \tag{27a}$$

$$\tau_{rz}^{h} = K_{rm}^{+} \sin\left(\frac{m\pi\theta}{\alpha}\right)e^{i\omega_m t}, \tau_{\theta z}^{h} = K_{\theta m}^{+} \cos\left(\frac{m\pi\theta}{\alpha}\right)e^{i\omega_m t}, \sigma_{z}^{h} = K_{zm}^{+} \sin\left(\frac{m\pi\theta}{\alpha}\right)e^{i\omega_m t}, \tag{27b}$$

for $z = 0$ and $z = h$, respectively. And $m = 0,1,2,\cdots,\infty$ is the circumferential wave number, ω_m denotes the circular frequency.

Considering the displacements in Eq. (25) and the corresponding boundary conditions along the circumferential edges and on the top and bottom surfaces, we can derive the state equation with respect to the state variables at discrete points similar to Eq. (9). Then we can carry out dynamical analysis of FGMs annular sectorial plates with simply supported radial edges. We give below a numerical example to illustrate the effect of the material graded index and geometrical parameters on the frequency of FGMs annular sectorial plates.

The material constants of FGMs annular sectorial plates are assumed as the exponential distribution along the thickness but uniform distribution in the radial direction, shown in Eq. (3), and Young's modulus, the density and Poisson's ratio at the center of the bottom plane of the plate are the same as those in Example 1.

The first frequency of the plate is calculated and listed in Table 3. It is seen from Table 3 that the first frequency decreases with the increase of the graded index λ_1 for different thickness ratios, radii ratios and sector angles and it increases with the increase of thickness ratios and radii ratios.

Table 3 The first frequency of FGMs annular sectorial plate with clamped circumferential edges

α	h/a	b/a	λ_1			
			0	1	2	3
135°	0.1	0.1	0.0717	0.0701	0.0658	0.0598
		0.2	0.0866	0.0847	0.0797	0.0726
	0.2	0.1	0.2389	0.2343	0.2220	0.2047
		0.2	0.2776	0.2727	0.2592	0.2402
225°	0.1	0.1	0.0671	0.0656	0.0616	0.0561
		0.2	0.0834	0.0815	0.0767	0.0698
	0.2	0.1	0.2197	0.2157	0.2047	0.1892
		0.2	0.2657	0.2610	0.2483	0.2303

4. Application to Magneto-Electro-Elastic FGMs Plates

The boundary conditions of magnetoelectroelastic circular plate on the circumferential edge are expressed as mechanically clamped, electrically shorted and magnetically open,

at $r=a$, $u_r=0, u_\theta=0, u_z=0, \phi=0, \psi=0$ (28a)

mechanically clamped, electrically open and magnetically shorted,

at $r=a$, $u_r=0, u_\theta=0, u_z=0, D_r=0, B_r=0$ (28b)

mechanically simply-supported, electrically shorted and magnetically open,

at $r=a$, $\sigma_r=0, u_\theta=0, u_z=0, \phi=0, \psi=0$ (28c)

mechanically simply-supported, electrically open and magnetically shorted,

at $r=a$, $\sigma_r=0, u_\theta=0, u_z=0, D_r=0, B_r=0$ (28d)

For free vibration problem, the boundary conditions on the top and bottom surfaces of the plate can be expressed as

at $z=0$ and $z=h$, $\tau_{rz}=0, \tau_{\theta z}=0, \sigma_z=0, D_z=0, B_z=0$. (29)

Table 4. Material constants of the piezoelectric PZT-4

c_{11}	c_{12}	c_{13}	c_{33}	c_{44}	ρ
139	77.8	74.3	115	25.6	7500
e_{15}	e_{31}	e_{33}	α_{11}	α_{33}	
12.7	-5.2	15.1	6.46×10^{-9}	5.62×10^{-9}	

* c_{ij} (GPa); α_{ij} (C^2/Nm^2); e_{ij} (C/m^2); ρ (kg/m^3).

Table 5. Material constants of the magnetostrictive $CoFe_2O_4$

c_{11}	c_{12}	c_{13}	c_{33}	c_{44}	ρ
286	173	170.5	269.5	45.3	5300
q_{15}	q_{31}	q_{33}	μ_{11}	μ_{33}	
550	580.3	699.7	-590×10^{-6}	157×10^{-6}	

* c_{ij} (GPa); μ_{ij} (Ns^2/C^2); q_{ij} (N/Am); ρ (kg/m^3).

In view of Eqs. (28) and (29), we can get the semi-analytical solution from Eq. (12) for the dynamic behavior of bi-directional FGMs magneto-electro-elastic circular plate. As an example, we calculate the first frequency of FGMs magneto-electro-elastic circular plate under the boundary conditions in Eq. (28a). The material constants at the center of the bottom of the plate are shown in Tables 4 and 5. And the graded index λ_2 in the radial direction is set zero. The variations of the first frequency of the plate with the graded index λ_1 and the ratio of h and a are shown in Figure 5.

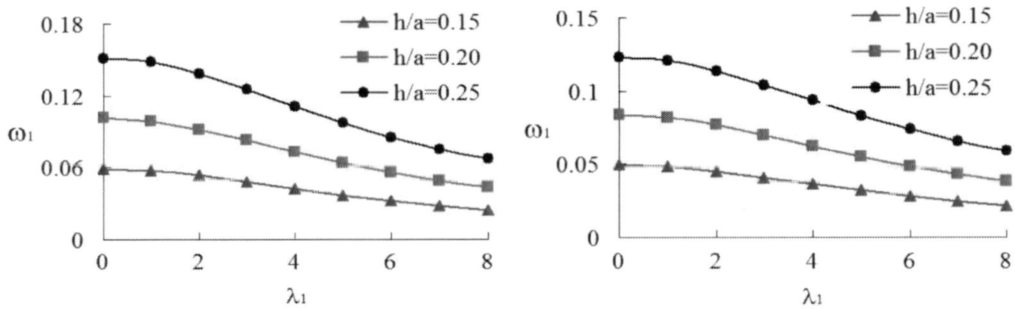

(a) piezoelectric material plate (b) magnetostrictive material plate

Figure 5. The first frequency of FGMs circular plates. (a) piezoelectric material plate, (b) magnetostrictive material plate.

It is noted from Figure 5 that the first frequency of FGMs piezoelectric and magnetostrictive circular plates decreases with the increase of the graded index λ_1 and it increases with the increase of h/a.

CONCLUSION

Owing to the material nonhomogeneity of FGMs, it is very difficult to obtain exact static/dynamic solutions for three-dimensional problems of multi-directional FGMs circular plates. Accordingly, we present an improved semi-analytical method namely state-space-based differential quadrature method for multi-directional FGMs elastic and magneto-electro-elastic circular plates in this chapter. A detailed procedure to achieve the method is given and the corresponding application to the three-dimensional static and dynamic problem is also exhibited. Numerical results show that the proposed method is accurate and effective and it can be used to study the three-dimensional problem in cylindrical coordinates.

ACKNOWLEDGMENTS

This work was supported by the National Natural Science Foundation of China (No. 10432030) and the Research Fund for the Doctoral Program of Higher Education (20070247029).

REFERENCES

Cui, L., Forero Rueda, MA., Gilchrist, MD., Optimisation of energy absorbing liner for equestrian helmets. Part II: Functionally graded foam liner *Materials and Design* 2009 *Vol* 30, 3414-19.

Jin, X., Wu, LZ., Sun, YG., et al., Microstructure and mechanical properties of ZrO2/NiCr functionally graded materials *Materials Science and Engineering A* 2009 *Vol* 509, 63-68.

Ghorbanpour Arani, A., Salari, M., Khademizadeh, H. and Arefmanesh, A., Magnetothermoelastic transient response of a functionally graded thick hollow sphere subjected to magnetic and thermoelastic fields *Arch Appl Mech* 2009 *Vol* 79, 481-97.

Lee, KH., Lee, YJ., Cho, SB., Characteristics of a transiently propagating crack in functionally graded materials *Journal of Mechanical Science and Technology* 2009 *Vol* 23, 1306-22.

Genin, GM., Birman, V., Micromechanics and structural response of functionally graded, particulate-matrix, fiber-reinforced composites *International Journal of Solids and Structures* 2009 *Vol* 46, 2136-50.

Wu, CP., Lu, YC., A modified Pagano method for the 3D dynamic responses of functionally graded magneto-electro-elastic plates *Composite Structures* 2009 *Vol* 90, 363-72.

Ebrahimi, F., Rastgo, A., An analytical study on the free vibration of smart circular thin FGM plate based on classical plate theory *Thin-Walled Structures* 2008 *Vol* 46, 1402-08.

Hosseini-Hashemi, Sh., Akhavan, H., Taher, H., Rokni, Damavandi, Daemi, N., Alibeigloo, A., Differential quadrature analysis of functionally graded circular and annular sector plates on elastic foundation *Materials and Design* 2010 *Vol* 31, 1871-80.

Reddy, JN., Wang, CM., Kitipornchai, S., Axisymmetric bending of functionally graded circular and annular plates *European Journal of Mechanics A/Solids* 1999 *Vol* 18, 185-99.

Sahraee, S., Bending analysis of functionally graded sectorial plates using Levinson plate theory *Composite Structures* 2009 *Vol* 88, 548-57.

Jomehzadeh, E., Saidi, AR., Atashipour, SR., An analytical approach for stress analysis of functionally graded annular sector plates *Materials and Design* 2009 *Vol* 30, 3679-85.

Nosier, A., Fallah, F., Non-linear analysis of functionally graded circular plates under asymmetric transverse loading *International Journal of Non-Linear Mechanics* 2009 *Vol* 44, 928-42.

Efraim, E., Eisenberger, M., Exact vibration analysis of variable thickness thick annular isotropic and FGM plates *Journal of Sound and Vibration* 2007 *Vol* 299, 720-38.

Prakash, T., Ganapathi, M., Asymmetric flexural vibration and thermoelastic stability of FGM circular plates using finite element method *Composites: Part B* 2006 *Vol* 37, 642-49.

Liu, X., Wang, Q., Quek, ST., Analytical solution for free vibration of piezoelectric coupled moderately thick circular plates *International Journal of Solids and Structures* 2002 *Vol* 39, 2129-51.

Saidi, AR., Rasouli, A., Sahraee, S., Axisymmetric bending and buckling analysis of thick functionally graded circular plates using unconstrained third-order shear deformation plate theory *Composite Structures* 2009 *Vol* 89, 110-19.

Najafizadeh, MM., Heydari, HR., An exact solution for buckling of functionally graded circular plates based on higher order shear deformation plate theory under uniform radial compression *International Journal of Mechanical Sciences* 2008 *Vol* 50, 603-12.

Ma, LS., Wang, TJ., Relationships between axisymmetric bending and buckling solutions of FGM circular plates based on third-order plate theory and classical plate theory *International Journal of Solids and Structures* 2004 *Vol* 41, 85-101.

Sahraee, S., Saidi, AR., Axisymmetric bending analysis of thick functionally graded circular plates using fourth-order shear deformation theory *European Journal of Mechanics A/Solids* 2009 Vol 28, 974-84.

Li, XY., Ding, HJ., Chen, WQ., Elasticity solutions for a transversely isotropic functionally graded circular plate subject to an axisymmetric transverse load q_r^k I*nternational Journal of Solids and Structures* 2008 *Vol* 45, 191-210.

Li, XY., Ding, HJ., Chen, WQ., Three-dimensional analytical solution for functionally graded magneto-electro-elastic circular plates subjected to uniform load *Composite Structures* 2008 *Vol* 83, 381-90.

Dong, CY., Three-dimensional free vibration analysis of functionally graded annular plates using the Chebyshev-Ritz method *Materials and Design* 2008 *Vol* 29, 1518-25.

Nie, GJ., Zhong, Z., Semi-analytical solution for three-dimensional vibration of functionally graded circular plates *Comput. Methods Appl. Mech. Engrg.* 2007 *Vol* 196, 4901-10.

Nie, GJ., Zhong, Z., Vibration analysis of functionally graded annular sectorial plates with simply supported radial edges *Composite Structures* 2008 *Vol* 84, 167-76.

Nie, GJ., Zhong, Z., Three-dimensional dynamical analysis of functionally graded magnetoelectroelastic circular plates *Journal of Tongji University(Natural Science)* 2009 *Vol* 37(6), 749-54. (in Chinese)

Lü, CF., Chen, WQ., Xu, RQ., Lim, CW., Semi-analytical elasticity solutions for bi-directional functionally graded beams *International Journal of Solids and Structures* 2008 Vol 45, 258-75.

Lü, CF., Lim, CW., Chen, WQ., Semi-analytical analysis for multi-directional functionally graded plates: 3-D elasticity solutions *Int. J. Num. Meth. Engng.* 2009 *Vol* 79, 25-44.

Nie, GJ., Zhong, Z., Axisymmetric bending of two-directional functionally graded circular and annular plates *Acta Mechanica Solida Sinica* 2007 *Vol* 20(4), 289-95.

Nie, GJ., Zhong, Z., Asymmetric Bending of Bi-directional Functionally Graded Circular Plates *Advances in heterogeneous material mechanics* 2008, 1013-1016.

Nie, GJ., Zhong, Z., Dynamic analysis of multi-directional functionally graded annular plates *Applied Mathematical Modelling* 2010 Vol 34, 608-16.

Bert, CW., Malik, M., Differential quadrature method in computational mechanics: a review *Appl. Mech. Rev.* 1996 *Vol* 49, 1-28.

Moler, C., Loan, CV., Nineteen dubious ways to compute the exponential of a matrix, twenty-five years later *SIAM Rev.* 2003 *Vol* 45, 3-49.

Chen, WQ., Ding, HJ., Bending of functionally graded piezoelectric rectangular plates *Acta Mech. Solida Sinica* 2000 *Vol* 13, 312-19.

Chen, WQ., Ding, HJ., On free vibration of a functionally graded piezoelectric rectangular plate *Acta Mech.* 2002 Vol 153, 207-16.

ABOUT THE AUTHORS

Zheng Zhong got Bachelor of Engineering at Tsinghua University in 1985, Master of Engineering at Tsinghua in 1987 and Ph. D at Tsinghua University in 1991. He joined Tongji University as an assistant professor in 1991 and then promoted to an associate professor in 1994, a full professor in 1998. He worked as a visiting scholar at University of Toronto between March 1995 and September 1996. He is now the dean of School of Aerospace Engineering and Applied Mechanics at Tongji University. His main research interests include mechanical behavior of advanced materials, smart materials and structures, structural vibration and noise, and finite element method, etc. He has published over 100 technical papers. He obtained Chinese National Outstanding Young Teacher Award in 2000 and Chinese National Excellent Young Investigator Fund in 2001.

Guojun Nie received the Ph.D. degree from Tongji University, China, 2002. She is an Associate Professor of the School of Aerospace Engineering and Applied Mechanics at Tongji University, Shanghai, China. Her research interests include the mechanics of inhomogeneous materials, structural mechanics and computational mechanics.

In: Mechanics of Functionally Graded Materials and Structures ISBN 978-1-61668-920-9
Editors: Z. Zhong, L. Wu and W. Chen

Chapter 2

ANALYTICAL STUDIES OF SURFACE ACOUSTIC WAVES IN SOLIDS OF FUNCTIONALLY GRADED MATERIALS

Ji Wang[1]1, Zheng Zhong[2], Rongxing Wu[1] and Liming Gao[2]

[1]Piezoelectric Device Laboratory, School of Mechanical Engineering and Mechanics, Ningbo University, Ningbo, Zhejiang, China

[2]School of Aerospace Engineering and Mechanics, Tongji University, Shanghai, China

ABSTRACT

Some traditional applications of structures and devices with homogeneous materials are being gradually replaced by functionally graded materials (FGM) with spatial variation of properties. The analysis of surface acoustic waves (SAW) propagating in FGM structures will be different primarily due to variations of material properties and resulted differential equations with variable coefficients. With the exception of a few patterns of variation, analytical solutions of wave propagation in FGM cannot be obtained. As a result, there will be challenges in the validation of numerical and approximate solutions. To meet the need for exact solutions, we studied the surface acoustic waves propagating in an FGM plate with the layered method, the Frobenius method, and the homotopy analysis method (HAM) for analytical and exact solutions. These three methods represent available techniques for analytical solutions of a large class of problems with important applications in the FGM structures. Numerical results based on the three methods are given for FGM plates with material property variations in a linear pattern. These analytical examples are presented along with detailed explanations of the methods and formulation to demonstrate the effectiveness of the methods and the effect of FGM on changes of plate deformation and frequency effects in SAW propagation with possible applications in devices and structures.

[1] e-mail: wangji@nbu.edu.cn.

1. INTRODUCTION

Surface acoustic waves (SAW), also known as Rayleigh waves, exist in a semi-infinite solid of homogeneous materials with a free plane surface. Since its discovery by Lord Rayleigh in 1885 [1], surface acoustic waves have found wide engineering applications including piezoelectric surface acoustic wave resonators [2]. More general applications in structural and building foundations for vibration control and isolation caused by moving loads are also important because surface waves have been the destructive wave component due to its lower velocity (frequency) and energy concentration in the vicinity of surface [3-6]. Although these applications are drastically different by nature, the common feature we can summarize is that both surface wave velocity and displacements are the major concerns for given structures. For SAW devices, which are widely used in communications to provide a highly stable high frequency sources, it is always desirable to have a relatively large surface deformation to generate relatively high electrical voltage. On the other hand, for vibration isolation foundations under the excitation of Rayleigh waves, it is expected that a reduction of surface waves is preferred for protective purpose. In summary, we need to realize the wave velocity and deformation modification in a solid or a structure through design or selection of materials and variation schemes, which can be treated as a mechanical system from a conventional point of view. Since extensive studies have been done with homogeneous materials and structures with homogeneous elements, an alternative based on the change of material properties in one or more spatial coordinates, or the utilization of functionally graded materials (FGMs), is worth to explore.

Although the FGMs were first conceived for thermal protection of space structures and similar mechanical applications, they have been quickly adopted for devices and intelligent structures with piezoelectric elements [7-16]. The basic idea of these applications is that by allowing material properties varying along one spatial coordinate, which is the thickness direction in most applications, the wave velocity and deformation will be modified to meet performance requirements. In analysis, except a few cases such as the exponential variation of material properties which guarantee analytical solutions, the structure has to be treated with a layered model for more general variation patterns. It is generally conceivable that by using thin layers we should be able to analyze plates of any property variation schemes provide that the computational challenges can be overcome for large number of layers and accompanying variables. This has been a general practice in the analysis of structures of FGMs [17-26]. In our study, this method is also employed to obtain simple and direct solutions for a few property variation schemes we are interested in [27].

As for the general effect on the propagation of surface acoustic waves in a special FGM with exponential variation of material properties, we have obtained analytical solutions of phase velocity and displacements [21]. It is found that by allowing material properties varying exponentially, we can adjust the phase velocity in a range with corresponding changes in displacements. It is clear that there are benefits such as the perturbation of frequency and deformation will be important in applications. For instance, separated frequencies can be used for frequency selectivity in devices and the increased deformation can be used for the enhancement of electrical charge generation. In applications related to structural engineering, the reduced deformation is preferable in the vibration isolation and control, which is commonly required in civil and mechanical engineering. Consequently, a general analytical procedure for FGM solids is of practical importance, although only the approximation in layered structures [13] and the finite element method [15-18] can be employed to obtain useful results.

We concentrate on the frequency behavior, or the phase velocity change and the deformation, or displacements, of an FGM solid with materials of linear property variation in the depth direction. We start with a layered structure with homogeneous material properties in each layer and the overall material properties change along the thickness coordinate in a prescribed pattern [27]. For comparison and validation purpose, we studied a layered structure with material properties approximated as an exponential function in the thickness direction in general and each layer has constant material properties. In another analytical approach aimed at exact solutions, we employed the Frobenius method to solve the resulted differential equations with variable coefficients [28]. For certain cases of material property variation in the thickness direction, such as the linear variation, exact solutions of displacements in the form of power series are obtained. The frequency solution is then derived from the displacement solutions. The solutions of the layered and Frobenius methods are exactly the same as expected, providing a validation needed for the general cases which can be only solved with one method of them. To add another solution method after the accurate solutions have been validated, we utilized the homotopy analysis method (HAM) for the solutions of the differential equations with variable coefficients [29]. The straightforward implementation of the HAM for solutions in power series has been shown that it is equivalent to the Frobenius method [28]. As a result, we have three distinct methods for the exact solutions of surface acoustic waves propagating in an FGM solid of linear variation of material properties. These solutions, undoubtedly, proven that the methods we used are effective in solving this seemingly challenging problem with the given differential equations. Furthermore, it points out that the FGM solids of arbitrary scheme can be solved with any one of the three methods we have presented for accurate characteristics of wave propagation. Our work based on current results meets multiple objectives of research on FGM solids, namely 1) provide reliable methods for analysis, 2) provide accurate results as references, and 3) evaluate material property variation patterns for optimal selection and design.

2. Surface Acoustic Waves in an FGM Solid: A Layered Plate Model

2.1. A Layered Plate Model Formulation

First we consider an infinite plate with isotropic material properties $\lambda, \mu, \text{and}\, \rho$ as the Lamé constants and density, respectively. For an infinite plate with uniform thickness and waves traveling in the x_1 direction, we assume the displacements are [19-21]

$$u_1 = A\exp(k\beta x_2)\exp ik(x_1 - ct),$$
$$u_2 = B\exp(k\beta x_2)\exp ik(x_1 - ct), \tag{1}$$

where $u_i\,(i=1,2), A(B), x_i\,(i=1,2), k, \beta, c, \text{and}\, t$ are displacements, amplitudes, coordinates, wavenumber, index, velocity, and time, respectively.

From (1), corresponding strains in abbreviated notations are

$$S_1 = \frac{\partial u_1}{\partial x_1} = ikA\exp(k\beta x_2)\exp ik(x_1 - ct),$$
$$S_2 = \frac{\partial u_2}{\partial x_2} = k\beta B\exp(k\beta x_2)\exp ik(x_1 - ct),$$
$$S_6 = \frac{\partial u_1}{\partial x_2} + \frac{\partial u_2}{\partial x_1} = k(\beta A + iB)\exp(k\beta x_2)\exp ik(x_1 - ct). \tag{2}$$

Similarly, stresses in abbreviated notations are

$$T_1 = k\left[(\lambda + 2\mu)iA + \lambda\beta B\right]\exp(k\beta x_2)\exp ik(x_1 - ct),$$
$$T_2 = k\left[(\lambda + 2\mu)\beta B + \lambda iA\right]\exp(k\beta x_2)\exp ik(x_1 - ct),$$
$$T_6 = k\mu(\beta A + iB)\exp(k\beta x_2)\exp ik(x_1 - ct). \tag{3}$$

The differential equations of motion are

$$T_{1,1} + T_{6,2} = \rho\ddot{u}_1,$$
$$T_{6,1} + T_{2,2} = \rho\ddot{u}_2. \tag{4}$$

With stresses from (3), we have

$$\left(\beta^2 c_T^2 - c_L^2 + c^2\right)A + i\beta\left(c_L^2 - c_T^2\right)B = 0,$$
$$i\beta\left(c_L^2 - c_T^2\right)A + \left(\beta^2 c_L^2 - c_T^2 + c^2\right)B = 0, \tag{5}$$

where

$$c_L = \sqrt{\frac{\lambda + 2\mu}{\rho}}, c_T = \sqrt{\frac{\mu}{\rho}}, \tag{6}$$

are velocities of the longitudinal and transverse waves, respectively.

For nontrivial solutions of the amplitudes in (5), we must have

$$\begin{vmatrix} \beta^2 c_T^2 + c^2 - c_L^2 & i\beta\left(c_L^2 - c_T^2\right) \\ i\beta\left(c_L^2 - c_T^2\right) & \beta^2 c_L^2 - c_T^2 + c^2 \end{vmatrix} = 0, \tag{7}$$

which gives the indices

$$\beta_1 = \sqrt{1 - \frac{c^2}{c_L^2}}, \beta_2 = -\sqrt{1 - \frac{c^2}{c_L^2}}, \beta_3 = \sqrt{1 - \frac{c^2}{c_T^2}}, \beta_4 = -\sqrt{1 - \frac{c^2}{c_T^2}}. \tag{8}$$

By substituting (8) into (7), we have the amplitude ratios as

$$\frac{B_1}{A_1}=-i\beta_1,\frac{B_2}{A_2}=-i\beta_2,\frac{B_3}{A_3}=-\frac{i}{\beta_3},\frac{B_4}{A_4}=-\frac{i}{\beta_4}. \tag{9}$$

As a result, the displacements are

$$u_1=\sum_{j=1}^{4}A_j\exp(k\beta x_2)\exp ik(x_1-ct),$$
$$u_2=-i\left[\sum_{j=1}^{2}\beta_j A_j\exp(k\beta x_2)+\sum_{j=3}^{4}\frac{A_j}{\beta_j}\exp(k\beta x_2)\right]\exp ik(x_1-ct). \tag{10}$$

It is important to emphasize that there are four undetermined amplitudes $A_i\left(i=1,2,3,4\right)$ in (10).

With displacements in (10), stresses in (3) can be rewritten as

$$T_1=ik\sum_{j=1}^{2}\left[(\lambda+2\mu)-\lambda\beta_j^2\right]A_j\exp(k\beta_j x_2)\exp ik(x_1-ct)+ik\sum_{j=3}^{4}2\mu A_j\exp(k\beta_j x_2)\exp ik(x_1-ct),$$
$$T_2=ik\sum_{j=1}^{2}\left[\lambda-(\lambda+2\mu)\beta_j^2\right]A_j\exp(k\beta_j x_2)\exp ik(x_1-ct)\ -ik\sum_{j=3}^{4}2\mu A_j\exp(k\beta_j x_2)\exp ik(x_1-ct),$$
$$T_6=k\mu\left\{\sum_{j=1}^{2}2\beta_j A_j\left(\exp k\beta_j x_2\right)+\sum_{j=3}^{4}\left(\beta_j+\frac{1}{\beta_j}\right)A_j\exp(k\beta_j x_2)\right\}\exp ik(x_1-ct). \tag{11}$$

For our analysis, a layered structure is shown in Figure 1. It has N distinctive layers with isotropic material properties of the ith layer, $\lambda^{(i)},\mu^{(i)}$, and $\rho^{(i)}$, as the Lamé constants and density. We assume the material in each layer is homogeneous and the surface acoustic waves travel in the x_1 direction. We now start our formulation of the problem from each layer.

From (12) and (13), we have displacements and stresses on the upper face (denoted with U) and lower face (denoted with L) of the nth layer with $e^{ik(x_1-ct)}$ omitted as

$$u_{1U}^{(n)}=\sum_{j=1}^{4}A_j^{(n)},$$
$$u_{1L}^{(n)}=\sum_{j=1}^{4}A_j^{(n)}\exp\left(k\beta_j^{(n)}h^{(n)}\right),$$
$$u_{2U}^{(n)}=-\sum_{j=1}^{2}i\beta_j^{(n)}A_j^{(n)}-\sum_{j=3}^{4}\frac{i}{\beta_j^{(n)}}A_j^{(n)},$$
$$u_{2L}^{(n)}=-\sum_{j=1}^{2}i\beta_j^{(n)}A_{nj}\exp\left(k\beta_j^{(n)}h^{(n)}\right)-\sum_{j=3}^{4}\frac{i}{\beta_j^{(n)}}A_j^{(n)}\exp\left(k\beta_j^{(n)}h^{(n)}\right); \tag{12}$$

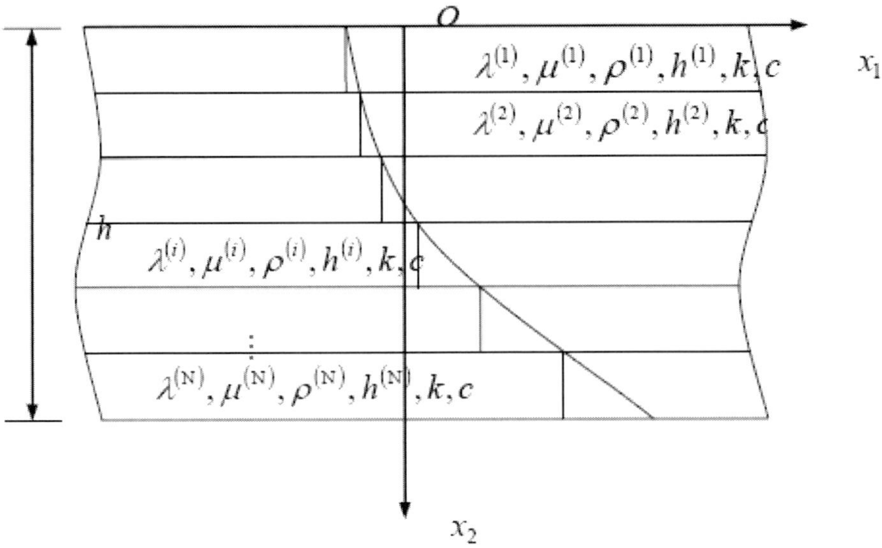

Figure 1. An N-layer plate with distinct isotropic material properties in each layer.

$$
\begin{aligned}
T_{2U}^{(n)} &= ik\left\{\sum_{j=1}^{2}\left[\lambda^{(n)}-\left(\lambda^{(n)}+2\mu^{(n)}\right)\beta_j^{(n)2}\right]A_j^{(n)}-\sum_{j=3}^{4}2\mu^{(n)}\beta_j^{(n)2}A_j^{(n)}\right\},\\
T_{2L}^{(n)} &= ik\sum_{j=1}^{2}\left[\lambda^{(n)}-(\lambda^{(n)}+2\mu^{(n)})\beta_j^{(n)2}\right]\exp\left(k\beta_j^{(n)}h^{(n)}\right)A_j^{(n)}-ik\sum_{j=3}^{4}2\mu^{(n)}\beta_j^{(n)2}\exp\left(k\beta_j^{(n)}h^{(n)}\right)A_j^{(n)},\\
T_{6U}^{(n)} &= k\mu^{(n)}\left\{\sum_{j=1}^{2}2\beta_j^{(n)}A_j^{(n)}+\sum_{j=3}^{4}\left(\beta_j^{(n)}+\frac{1}{\beta_j^{(n)}}\right)A_j^{(n)}\right\},\\
T_{6L}^{(n)} &= k\mu^{(n)}\sum_{j=1}^{2}2\beta_j^{(n)}\exp\left(k\beta_j^{(n)}h^{(n)}\right)A_j^{(n)}+k\mu^{(n)}\sum_{j=3}^{4}\left(\beta_j^{(n)}+\frac{1}{\beta_j^{(n)}}\right)\exp\left(k\beta_j^{(n)}h^{(n)}\right)A_j^{(n)},
\end{aligned}
\tag{13}
$$

where the superscript n denotes variables in the nth layer and $h^{(n)}$ is the layer thickness.

On the upper and lower surfaces of the layered structure shown in Figure 1, we have the traction-free boundary conditions

$$
T_{2U}^{(1)} = T_{6U}^{(1)} = T_{2L}^{(N)} = T_{6L}^{(N)} = 0, \tag{14}
$$

where N is the total number of layers.

At the interfaces of two layers, we must have the continuity of displacements and stresses, or

$$
T_{2L}^{(n)} = T_{2U}^{(n+1)},\ T_{6L}^{(n)} = T_{6U}^{(n+1)},\ u_{1L}^{(n)} = u_{1U}^{(n+1)},\ u_{2L}^{(n)} = u_{2U}^{(n+1)}. \tag{15}
$$

These are the boundary conditions we need for the complete formulation of the problem.

2.2. Free Vibrations of a Layered Plate in the Surface Acoustic Wave Mode

We now apply the boundary conditions and continuity conditions from (14) and (15) to displacements and stress expressions in (12) and (13) for each layer to establish the equation for free vibrations of the layered structure. The resulted equation can be written as

$$\mathbf{KA} = \mathbf{0}, \tag{16}$$

where the amplitude vector is

$$\mathbf{A} = \left\{ A_1^{(1)}, A_2^{(1)}, A_3^{(1)}, A_4^{(1)}, \cdots A_1^{(j)}, A_2^{(j)}, A_3^{(j)}, A_4^{(j)}, \cdots A_1^{(N)}, A_2^{(N)}, A_3^{(N)}, A_4^{(N)} \right\},$$

and the coefficient matrix is

$$\begin{bmatrix}
-T_{2\mathrm{U}1}^{(1)} & -T_{2\mathrm{U}2}^{(1)} & -T_{2\mathrm{U}3}^{(1)} & -T_{2\mathrm{U}4}^{(1)} & 0 & 0 & 0 & 0 & \cdots & & & & & 0 \\
-T_{6\mathrm{U}1}^{(1)} & -T_{6\mathrm{U}2}^{(1)} & -T_{6\mathrm{U}3}^{(1)} & -T_{6\mathrm{U}4}^{(1)} & 0 & 0 & 0 & 0 & \cdots & & & & & 0 \\
T_{2\mathrm{L}1}^{(1)} & T_{2\mathrm{L}}^{(1)} & T_{2\mathrm{L}3}^{(1)} & T_{2\mathrm{L}4}^{(1)} & -T_{2\mathrm{U}1}^{(2)} & -T_{2\mathrm{U}2}^{(2)} & -T_{2\mathrm{U}3}^{(2)} & -T_{2\mathrm{U}4}^{(2)} & & & & & & 0 \\
T_{6\mathrm{L}1}^{(1)} & T_{6\mathrm{L}2}^{(1)} & T_{6\mathrm{L}3}^{(1)} & T_{6\mathrm{L}4}^{(1)} & -T_{6\mathrm{U}1}^{(2)} & -T_{6\mathrm{U}2}^{(2)} & -T_{6\mathrm{U}3}^{(2)} & -T_{6\mathrm{U}4}^{(2)} & & & & & & 0 \\
u_{1\mathrm{L}1}^{(1)} & u_{1\mathrm{L}2}^{(1)} & u_{1\mathrm{L}3}^{(1)} & u_{1\mathrm{L}4}^{(1)} & -u_{1\mathrm{U}1}^{(2)} & -u_{1\mathrm{U}2}^{(2)} & -u_{1\mathrm{U}3}^{(2)} & -u_{1\mathrm{U}4}^{(2)} & & & & & & 0 \\
u_{2\mathrm{L}1}^{(1)} & u_{2L2}^{(1)} & u_{2L3}^{(1)} & u_{2\mathrm{L}4}^{(1)} & -u_{2\mathrm{U}1}^{(2)} & -u_{2\mathrm{U}2}^{(2)} & -u_{2\mathrm{U}3}^{(2)} & -u_{2\mathrm{U}4}^{(2)} & & & & & & 0 \\
0 & 0 & 0 & 0 & 0 & 0 & 0 & 0 & \ddots & & & & & 0 \\
 & & & \vdots & & & & & & \ddots & & & & 0 \\
 & & & & & & & & & 0 & -T_{2\mathrm{U}1}^{(N)} & -T_{2\mathrm{U}2}^{(N)} & -T_{2\mathrm{U}3}^{(N)} & -T_{2\mathrm{U}4}^{(N)} \\
 & & & & & & & & & 0 & -T_{6\mathrm{U}1}^{(N)} & -T_{6\mathrm{U}2}^{(N)} & -T_{6\mathrm{U}3}^{(N)} & -T_{6\mathrm{U}4}^{(N)} \\
 & & & & & & & & & 0 & -u_{1\mathrm{U}1}^{(N)} & -u_{1\mathrm{U}2}^{(N)} & -u_{1\mathrm{U}3}^{(N)} & -u_{1\mathrm{U}4}^{(N)} \\
 & & & & & & & & & 0 & -u_{2\mathrm{U}1}^{(N)} & -u_{2\mathrm{U}2}^{(N)} & -u_{2\mathrm{U}3}^{(N)} & -u_{2\mathrm{U}4}^{(N)} \\
 & & & & & & & & & 0 & T_{2\mathrm{L}1}^{(N)} & T_{2\mathrm{L}2}^{(N)} & T_{2\mathrm{L}3}^{(N)} & T_{2\mathrm{L}4}^{(N)} \\
0 & & & & & & & & & 0 & T_{6\mathrm{L}1}^{(N)} & T_{6\mathrm{L}2}^{(N)} & T_{6\mathrm{L}3}^{(N)} & T_{6\mathrm{L}4}^{(N)}
\end{bmatrix}, \tag{17}$$

with elements

$$T_{2\mathrm{Lj}}^{(n)} = \begin{cases} ik\left[\lambda^{(n)} - \left(\lambda^{(n)} + 2\mu^{(n)}\right)\beta_j^{(n)2}\right]\exp\left(k\beta_j^{(n)}h^{(n)}\right), & j = 1,2, \\ -2ik\mu^{(n)}\exp\left(k\beta_j^{(n)}h^{(n)}\right), & j = 3,4; \end{cases}$$

$$T_{2\mathrm{Uj}}^{(n)} = \begin{cases} ik\left[\lambda^{(n)} - (\lambda^{(n)} + 2\mu^{(n)})\beta_j^{(n)2}\right], & j = 1,2, \\ -2ik\mu^{(n)}, & j = 3,4; \end{cases}$$

$$T_{6\mathrm{Lj}}^{(n)} = \begin{cases} 2k\mu^{(n)}\beta_j^{(n)}\exp\left(k\beta_j^{(n)}h^{(n)}\right), & j = 1,2, \\ k\mu^{(n)}\left(\beta_j^{(n)} + \dfrac{1}{\beta_j^{(n)}}\right)\exp\left(k\beta_j^{(n)}h^{(n)}\right), & j = 3,4; \end{cases}$$

$$T_{6\mathrm{Uj}}^{(n)} = \begin{cases} 2k\mu^{(n)}\beta_j^{(n)}, & j = 1,2, \\ k\mu^{(n)}\left(\beta_j^{(n)} + \dfrac{1}{\beta_j^{(n)}}\right), & j = 3,4; \end{cases} \tag{18}$$

and

$$u_{1Lj}^{(n)} = \exp\left(k\beta_j^{(n)} h^{(n)}\right), j = 1,2,3,4;$$
$$u_{1Uj}^{(n)} = 1, j = 1,2,3,4;$$
$$u_{2Lj}^{(n)} = \begin{cases} -i\beta_j^{(n)} \exp\left(k\beta_j^{(n)} h^{(n)}\right), & j = 1,2, \\ -\dfrac{i}{\beta_j^{(n)}} \exp\left(k\beta_j^{(n)} h^{(n)}\right), & j = 3,4; \end{cases}$$
$$u_{2Uj}^{(n)} = \begin{cases} -i\beta_j^{(n)}, & j = 1,2, \\ -\dfrac{i}{\beta_j^{(n)}}, & j = 3,4. \end{cases} \quad (19)$$

Now this equation is capable of studying free vibrations of an N-layer structure shown in Figure 1 in the surface acoustic wave mode. As usual, the equation for the velocity can be obtained from (17) by setting the determinant to vanish. By calculating the amplitude ratios, we can also obtain mode shapes. These results, with material property variation pattern given, can be used evaluate the effect of FGMs on the wave velocity and mode shapes of a particular scheme. An optimal variation scheme can be selected from a series of calculations in conjunction with the required velocity and deformation.

2.3. Numerical Examples

First, we divide a homogeneous isotropic plate with Poisson's ratio of 0.2 and thickness of two wavelengths into ten layers. We found that the two surface acoustic wave modes have velocities normalized by the transverse wave velocity c_T as 0.90717921 and 0.91541065, which are exact results from the earlier study [19]. The displacements of the two modes are shown in Figs. 2 and 3.

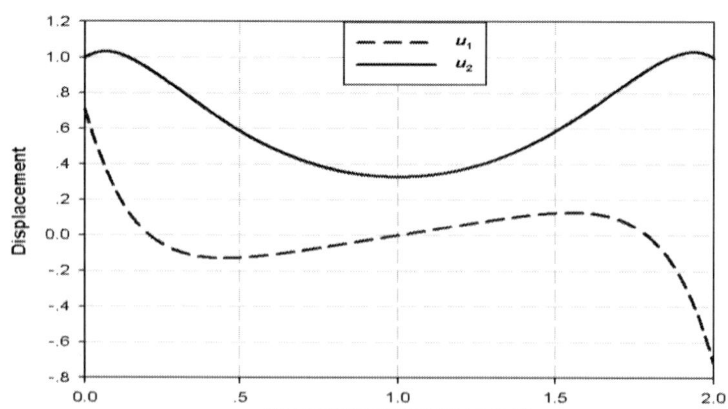

Figure 2. Displacements of surface acoustic waves in a homogeneous plate with thickness of two wavelengths at $c = 0.90717921\ c_T$.

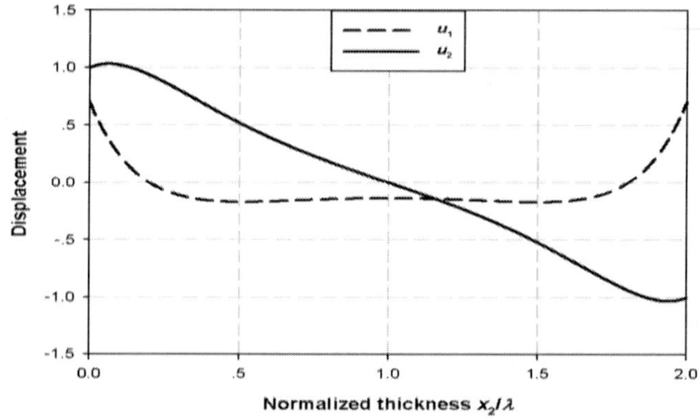

Figure 3. Displacements of surface acoustic waves in a homogeneous plate with thickness of two wavelengths at $c = 0.91541065\, c_T$.

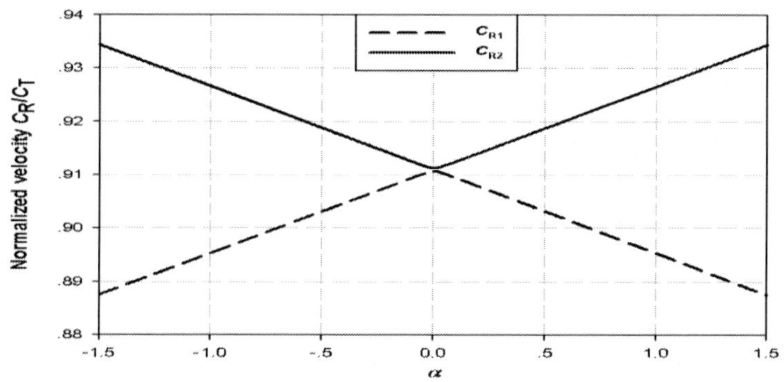

Figure 4. Surface acoustic wave velocity versus the material grading index α in an exponential FGM plate with thickness of three wavelengths.

Next, we consider a plate with material properties varying along the thickness direction in an exponential function like $(\lambda, \mu, \rho) = (\lambda_0, \mu_0, \rho_0)\exp(\alpha x_2)$ with α as a constant, and the material properties are taken from the top of each layer for approximation. By dividing a plate with thickness of three wavelengths into ten layers, we obtained the wave velocity and displacements for different material property variations parameters. From the velocity in Figure 4, we see two branches of the velocity curve showing surface waves in a plate with two free faces. The velocity curves are actually two straight lines because the exponential decaying nature of surface waves. This has also been obtained in our earlier study for the semi-infinite case [21]. Displacements shown in Figs. 5 and 6 have the same pattern. In addition, it is clear that the hardening of the material along the thickness direction (positive α) from the top will enlarge the displacements of the lower face in the first mode but reduce them in the second mode. Similarly, the softening of the material from the top of the plate (negative α) will reduce the displacements of the lower face in the first mode but enlarge them in the second mode. These essential results can be the fundamental guidelines in the selection of material variation scheme for practical applications. It demonstrates that the effect of material property variation depends on the particular vibration mode, implying it is important to identify the working mode and analyze the effect before the FGM is utilized.

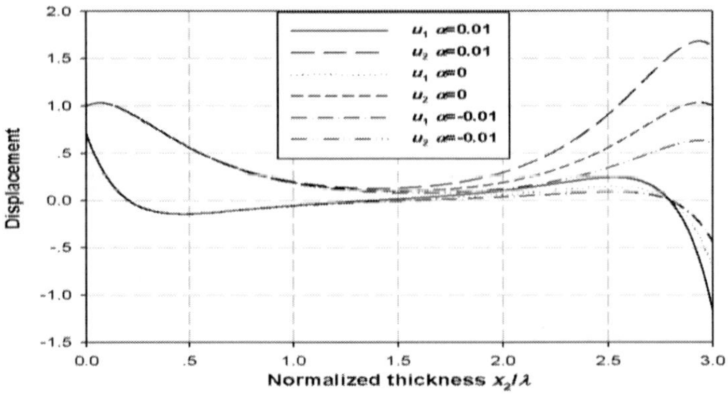

Figure 5. Displacements of the first mode of surface acoustic waves in an exponential FGM plate with thickness of three wavelengths for property grading indices α .

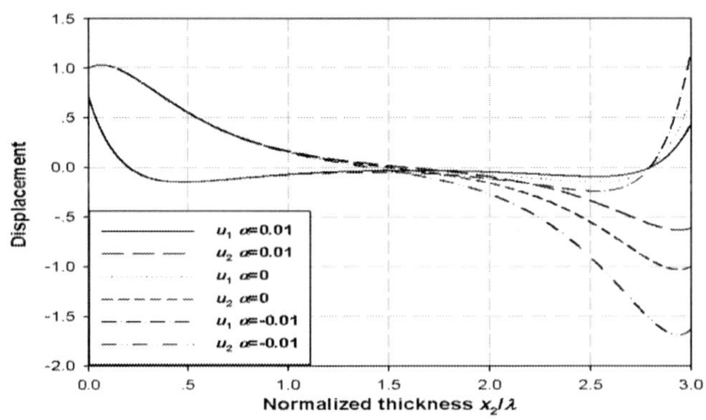

Figure 6. Displacements of the second mode of surface acoustic waves in an exponential FGM plate with thickness of three wavelengths for property grading indices α.

We now examine the effect of the variation of material properties in the form of

$$(\lambda,\mu,\rho)=(\lambda_0,\mu_0,\rho_0)(1+\alpha x_2^n), \tag{20}$$

with α as a constant in a plate. For linear variation case with $n=1$, velocity results are plotted in Figure 7 for different variation parameters. There are also two distinctive branches in the velocity with one is linear. The displacements for the first mode in Figs. 8 and 9 basically show the same conclusion we draw from the previous exponential case except both hardening and softening of material properties have much strong effect on displacements of the two faces, which again imply the reduction and amplification of the deformation in such a layered structure. The displacements in the second mode are shown in Figs. 10 and 11 and they show that the hardening of material from the top face to bottom face will reduce the deformation on the bottom. Now for different n, we divided a plate with thickness of three wavelengths into ten layers and the wave velocity and displacements are obtained. From Figure 12, the displacements in the lower surface of the first mode are enlarged along with the increase of n. From Figure13, displacements in the lower surface of the second mode are reduced along with the increase of n. For the velocity in Figure 14, the second velocity of the plate with material varying along the thickness direction in polynomial gradually approaches the velocity of semi-infinite solid along with the increase of n.

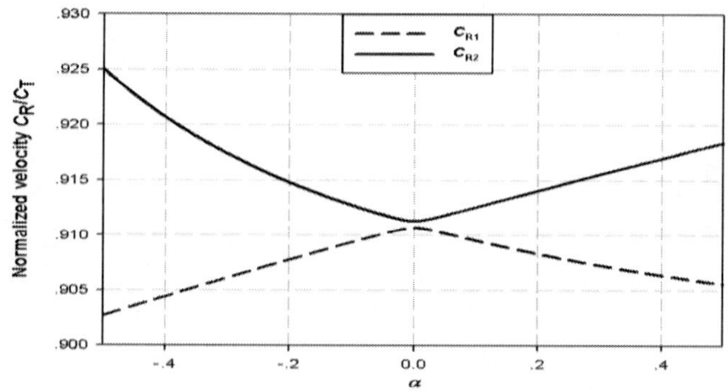

Figure 7. Surface acoustic wave velocity of a layered plate with thickness of three wavelengths and material property of $1+\alpha x_2$.

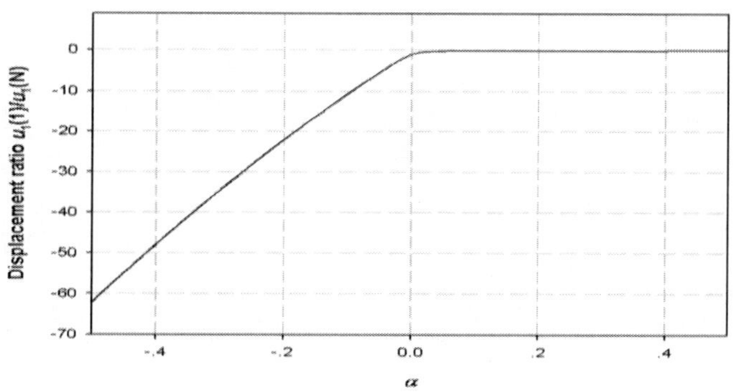

Figure 8. Ratios of displacement u_1 of the upper surface to lower surface of the first mode in an FGM plate with thickness of three wavelengths material properties of $1+\alpha x_2$.

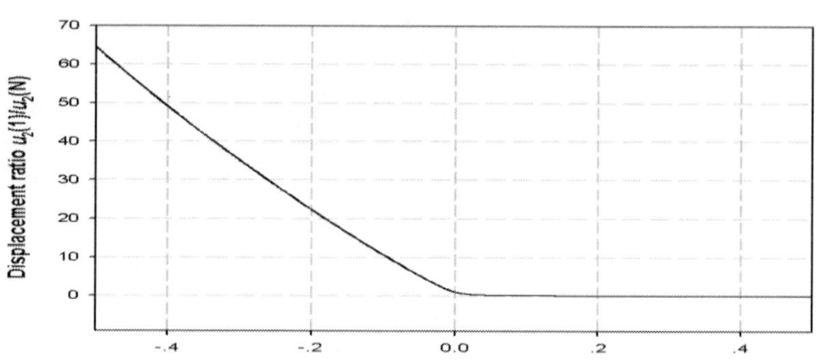

Figure 9. Ratios of displacement u_2 of the upper surface to lower surface of the first mode in an FGM plate with thickness of three wavelengths with material properties of $1+\alpha x_2$.

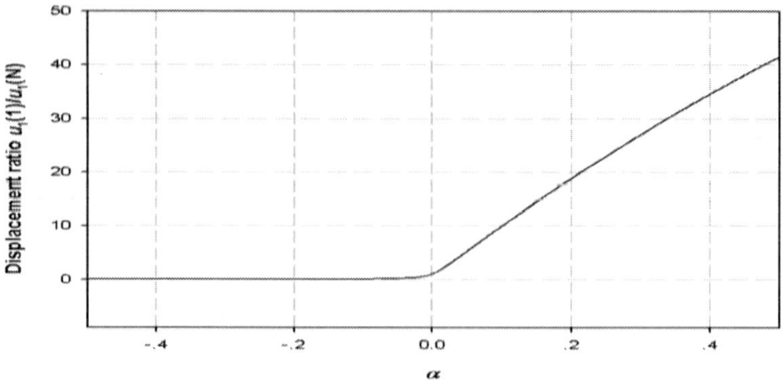

Figure 10. Ratios of displacement u_1 of the upper surface to lower surface of the second mode in an FGM plate with thickness of three wavelengths material properties of $1+\alpha x_2$.

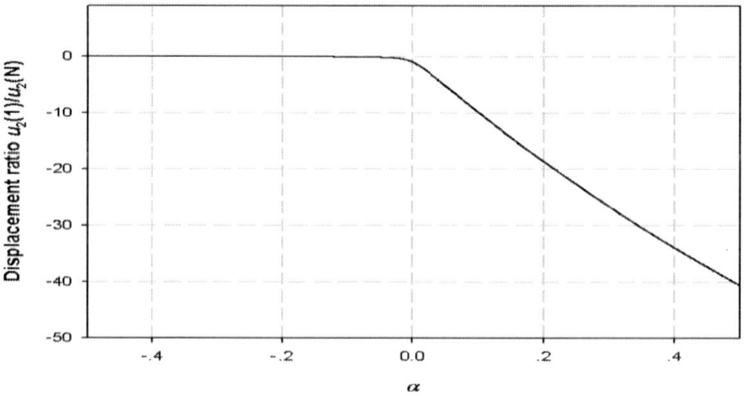

Figure 11. Ratios of displacement u_2 of the upper surface to lower surface of the second mode in an FGM plate with thickness of three wavelengths and material properties of $1+\alpha x_2$.

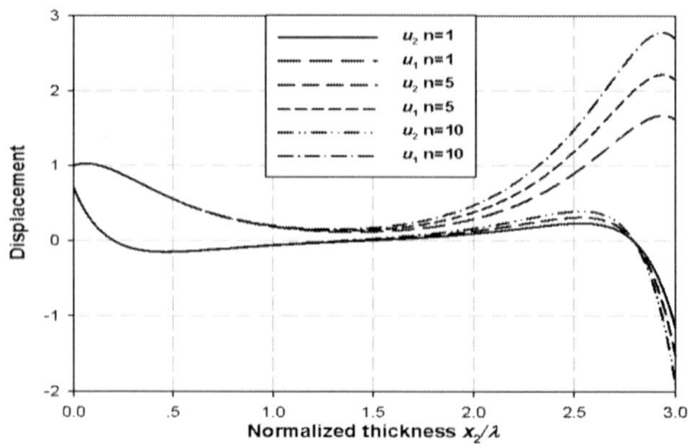

Figure 12. Displacements of the first mode of surface acoustic waves in an FGM plate with thickness of three wavelengths and material properties of $1+\alpha x_2^n$ $(\alpha = 0.01)$.

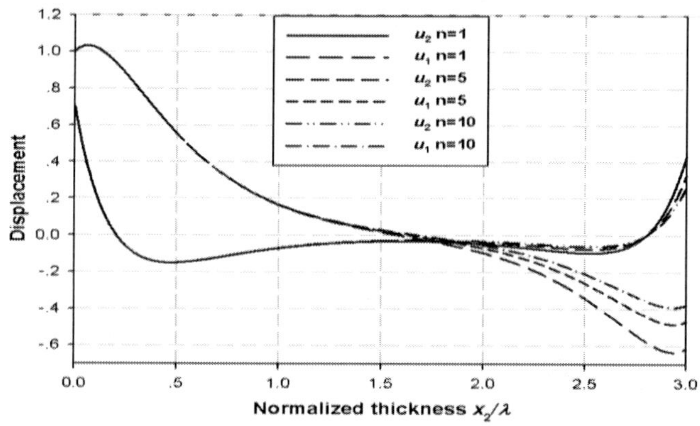

Figure 13. Displacements of the second mode of surface acoustic waves in an FGM plate with thickness of three wavelengths and material properties of $1+\alpha x_2^n\ (\alpha = 0.01)$.

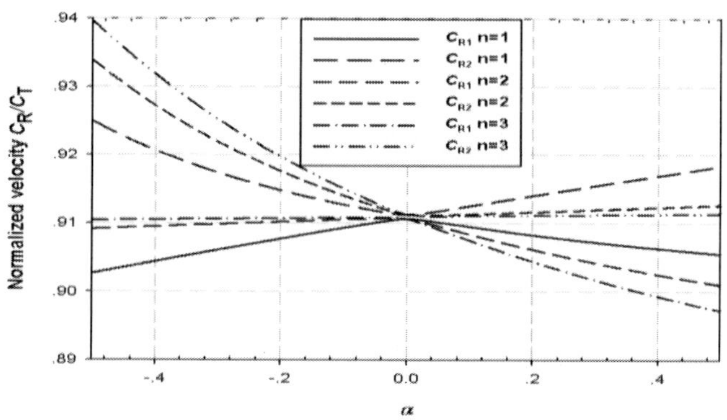

Figure 14. Velocity of surface acoustic waves in an FGM plate with thickness of three wavelengths and material properties of $1+\alpha x_2^n\ (\alpha = 0.01)$.

The above results show that the layered model for the analysis of surface acoustic waves in an infinite plate of functionally graded materials (FGMs) is capable for the calculation of displacements and wave propagation velocity in cases where the analytical solutions can not be obtained. The process for the model development is straightforward with the known structure and only minimum efforts in extracting essential parameters of the surface acoustic wave are required. After equations based on the layered model are formed, a simple and direct procedure for the validation of results based on the layer refinement is needed. Our experience shows that the convergence of deformation and wave velocity can be easily achieved.

2.4. Conclusions

In order to analyze surface acoustic waves in a plate of functionally graded materials (FGMs), a layered model consisting of layers of thin homogeneous plates is proposed and tested. By using the fundamental solutions of surface acoustic waves in a thin plate and applying the boundary and continuity conditions, a formulation for free vibrations of the layered structure in surface acoustic wave mode is completed. The wave velocity and mode shapes are directly solved from this equation with given material property variation schemes. The equations and solution procedure are validated by comparing the solutions from the layered model with earlier analytical solutions with good agreement.

From both the exponential and general variation of material properties cases, we can see two velocity branches of layered plates. It can be clearly seen that there are different effects on the deformation, or the displacements on the two faces of the plate with material property variations schemes. While the hardening of the material properties from the top to the bottom will generally enlarge the displacements on the bottom face at the lower velocity mode, it will reduce the bottom displacements in the higher velocity mode. Vice versa, the effect on the displacements will be reversed for the softening of material properties from the top to the bottom. As a result, if we need to select the material variation to achieve displacement amplification or reduction on a particular face, we have to consider the velocity, or the working mode of the structure. Otherwise, the desired effect may not be obtained and adversary results may be induced.

In general, the utilization of functionally graded materials in surface acoustic wave modes of a plate has advantages in many known applications where the modifications of frequency and deformation are desired. We need to use this simple model and analytical procedure to evaluate the effect and select the proper configuration to take advantages of widely utilized structures of FGMs.

3. Frobenius Method for Surface Acoustic Waves in FGM Solids

3.1. Introduction

Before we carry out mathematical formulation and analysis for analytical solutions with the Frobenius method, it has to be emphasized that since we are dealing with wave propagation in a plate with finite thickness, the surface wave mode we studied is no longer the known Rayleigh waves existing in semi-infinite solids. What we shall have is one of the complex wave modes of Lamb waves in a plate, but the concentration of deformation and energy in the vicinity of the surfaces also demonstrating this is a surface wave mode, or generalized Rayleigh waves [30]. Such a characterization of the wave mode is necessary because the known SAW applications like resonators are generally based on wave propagations in finite solids with complications and they are also working at generalized Rayleigh wave modes. As a result, such generalized SAW modes are addressed as SAW modes without further clarification in this paper.

In our study, the Frobenius method is applied to analyze SAW in an FGM plate with properties variations of a polynomial type [28]. As a systematic method for solving ordinary differential equations without known solutions in elementary functions, the Frobenius method has been widely used for analytical solutions in power series for certain types of differential equations [31-33]. The method has been used in the analysis of wave propagation in inhomogeneous solids [34] and layered structures [35]. Lee and Wang investigated the vibrations of AT-cut quartz strips of narrow width and finite length with variable thickness [36]. They also studied thickness-shear and flexural vibrations of contoured crystal strip resonators [37]. For certain classes of beams and plates, solutions of vibration frequencies and mode shapes are obtained with Jacobi orthogonal polynomials by Caruntu [38-39]. In solving these vibration problems, displacements are usually expanded in power series and substituted into governing equations in order to obtain a recursive relation for the undetermined coefficients. This is equivalent to express the solutions in power series with unknown parameters. Finally, by applying the appropriate boundary conditions, all the unknowns are obtained and the solutions are given. For SAW in FGM solids, with known solutions of phase velocity, displacements are also evaluated for the mode shapes of wave propagation. Finally, the solution procedure and numerical solutions are presented for an FGM plate with linear variation of material properties in the thickness direction. Although the material property variation is limited to polynomials of the thickness coordinate and the examples are given in linear variation cases with this method, the rather restrictive solutions are actually a few of the available exact results for the evaluation of material property variation schemes for possible engineering applications. Of course, these results are also the fundamental references in additional to the widely known exponential scheme for the validation of numerical methods.

3.2. Formulation of Surface Acoustic Waves in a Functionally Graded Plate

First, we consider a functionally graded plate shown in Figure 15 with isotropic material properties $\lambda, \mu,$ and ρ as the Lamé constants and density varying along the thickness coordinate, respectively. In this study, they are assumed to be linear functions of x_2, i.e. the thickness coordinate

$$(\lambda, \mu, \rho) = (\lambda_0, \mu_0, \rho_0) f(x_2), \tag{21}$$

where $\lambda_0, \mu_0,$ and ρ_0 are constants, and

$$f(x_2) = 1 + \alpha x_2, \tag{22}$$

where α is a parameter denoting the material property variation. Although this assumption is rather restricted in appearance because all the material properties have the same variation pattern, the method we shall demonstrate in this study can be used to investigate variations of each material property as general polynomial functions. The only difference is the associated complexity involved in the mathematical formulation and manipulation.

In this case, displacements are assumed as

$$\begin{aligned} u_1 &= u(x_2) e^{ik(x_1 - ct)}, \\ u_2 &= v(x_2) e^{ik(x_1 - ct)}, \end{aligned} \tag{23}$$

where $u, v, x_i\ (i = 1, 2), k, c$ and t are unknown functions, coordinates, wavenumber , phase velocity, and time, respectively.

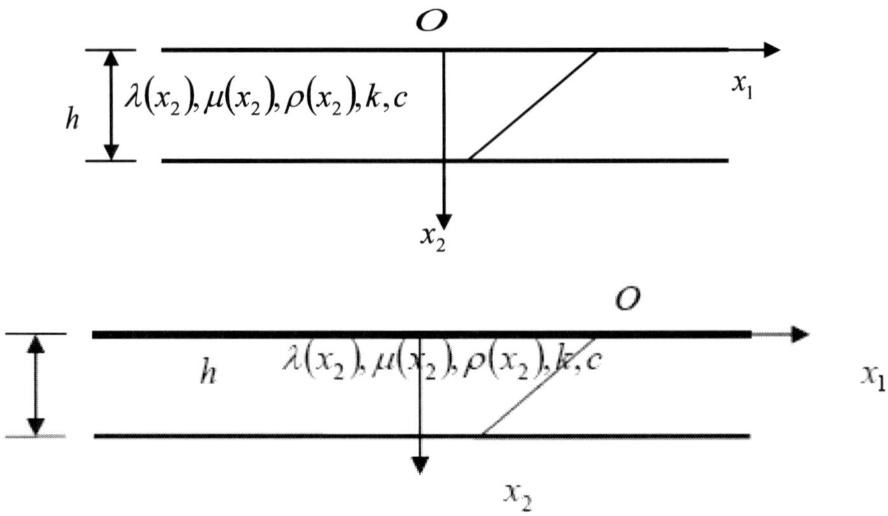

Figure 15. A functionally graded plate.

With (23), the corresponding strain components can be obtained in abbreviated notations as

$$
\begin{aligned}
S_1 &= \frac{\partial u_1}{\partial x_1} = ikue^{ik(x_1-ct)}, \\
S_2 &= \frac{\partial u_2}{\partial x_2} = v'e^{ik(x_1-ct)}, \\
S_6 &= \frac{\partial u_1}{\partial x_2} + \frac{\partial u_2}{\partial x_1} = (u' + ikv)e^{ik(x_1-ct)}.
\end{aligned}
\tag{24}
$$

Consequently, the stresses can be obtained in displacements from constitutive equations as

$$
\begin{aligned}
T_1 &= (\lambda + 2\mu)S_1 + \lambda S_2 = \left[ik(\lambda + 2\mu)u + \lambda v'\right]e^{ik(x_1-ct)}, \\
T_2 &= (\lambda + 2\mu)S_2 + \lambda S_1 = \left[(\lambda + 2\mu)v' + ik\lambda u\right]e^{ik(x_1-ct)}, \\
T_6 &= \mu S_6 = \mu(u' + ikv)e^{ik(x_1-ct)}.
\end{aligned}
\tag{25}
$$

By substituting stresses from (25) into equations of motion in (4), we have

$$
\begin{aligned}
& c_T^2 f(x_2)u'' + c_T^2 f'(x_2)u' + k^2\left(c^2 - c_L^2\right)f(x_2)u + ik\left(c_L^2 - c_T^2\right)f(x_2)v' + ikc_T^2 f'(x_2)v = 0, \\
& ik\left(c_L^2 - c_T^2\right)f(x_2)u' + ik\left(c_L^2 - 2c_T^2\right)f'(x_2)u + c_L^2 f(x_2)v'' + c_L^2 f'(x_2)v' + k^2\left(c^2 - c_T^2\right)f(x_2)u = 0,
\end{aligned}
\tag{26}
$$

where

$$
c_L = \sqrt{\frac{\lambda_0 + 2\mu_0}{\rho_0}}, \quad c_T = \sqrt{\frac{\mu_0}{\rho_0}},
\tag{27}
$$

are the phase velocities of the longitudinal and transverse waves of homogeneous materials with given material properties, respectively. By solving (26), we should be able to obtain displacements, consequently stresses for boundary conditions and eventually the phase velocity of wave propagation. However, even with a simple linear function of $f(x_2)$ as given in (22), analytical solutions of (26) cannot be obtained in elementary functions. To investigate SAW in such an FGM plate, in addition to known numerical methods, the Frobenius method has to be employed for power series solutions [36-39].

To obtain the Frobenius solutions, displacement functions u and v in (23) are written as

$$u = \sum_{n=0}^{\infty} a_n x_2^{n+s}, \qquad v = \sum_{n=0}^{\infty} b_n x_2^{n+t}, \tag{28}$$

where coefficients a_n and b_n are to be determined along with parameters s and t. By substituting (28) into (26), the two equations can be rewritten as

$$\sum_{n=0}^{\infty}\left[\sum_{i=-2}^{1} f_i(n,s)a_n x_2^{n+s+i} + \sum_{j=-1,0} g_j(n,t)b_n x_2^{n+t+j}\right] = 0,$$
$$\sum_{n=0}^{\infty}\left[\sum_{i=-2}^{1} \bar{f}_i(n,s)a_n x_2^{n+s+i} + \sum_{j=-1,0} \bar{g}_j(n,t)b_n x_2^{n+t+j}\right] = 0, \tag{29}$$

where

$$f_{-2}(n,s) = c_T^2(n+s)(n+s-1),$$
$$f_{-1}(n,s) = c_T^2(n+s)^2\alpha,$$
$$f_0(n,s) = k^2(c^2 - c_L^2),$$
$$f_1(n,s) = k^2(c^2 - c_L^2)\alpha,$$
$$g_{-1}(n,t) = ik(c_L^2 - c_T^2)(n+t),$$
$$g_0(n,t) = ik\alpha\left[(c_L^2 - c_T^2)(n+t) + c_T^2\right],$$
$$\bar{f}_{-1}(n,s) = ik(c_L^2 - c_T^2)(n+s),$$
$$\bar{f}_0(n,s) = ik\left[(c_L^2 - c_T^2)(n+s) + (c_L^2 - 2c_T^2)\alpha\right],$$

$$\bar{g}_{-2}(n,t) = c_L^2 (n+t)(n+t-1),$$
$$\bar{g}_{-1}(n,t) = c_L^2 (n+t)^2 \alpha,$$
$$\bar{g}_0(n,t) = k^2 (c^2 - c_T^2),$$
$$\bar{g}_1(n,t) = k^2 (c^2 - c_T^2)\alpha. \quad (30)$$

By setting the coefficients of the terms with the lowest power of x_2 in (29) to zero, i.e.

$$f_{-2}(n,s) = 0, g_{-1}(n,t) = 0, \bar{f}_{-1}(n,s) = 0, \bar{g}_{-2}(n,t) = 0, \quad (31)$$

we have following indicial equations [31-33]

$$c_T^2 s(s-1) = 0 \quad \text{or} \quad s = 0,1,$$
$$ik(c_L^2 - c_T^2)t = 0 \text{ or } t = 0,$$
$$ik(c_L^2 - c_T^2)s = 0 \text{ or } s = 0,$$
$$c_L^2 t(t-1) = 0 \quad \text{or} \quad t = 0, t = 1. \quad (32)$$

It can be seen from (29) that the choice of $s = 0$ and $t = 0$ makes the coefficients of terms x_2^{-2} and x_2^{-1} to vanish for arbitrary values of a_0, b_0, a_1 and b_1. It is important to emphasize that there are four undetermined constants, a_0, b_0, a_1 and b_1 in (28). Hence, the remaining constants in (28) can be expressed in terms of these four arbitrary constants by the recursive relations

$$a_{n+3} = \sum_{j=0}^{2} \left(\Lambda_{n,j} a_{n+j} + \Gamma_{n,j} b_{n+j} \right),$$
$$b_{n+3} = \sum_{j=0}^{2} \left(\bar{\Lambda}_{n,j} a_{n+j} + \bar{\Gamma}_{n,j} b_{n+j} \right), n = 0,1,2,3,\cdots,\infty, \quad (33)$$

where

$$\Lambda_{n,2} = -\alpha \frac{n+2}{n+3},$$
$$\Lambda_{n,1} = -k^2 \frac{c^2 - c_L^2}{c_T^2 (n+3)(n+2)},$$
$$\Lambda_{n,0} = -k^2 \alpha \frac{c^2 - c_L^2}{c_T^2 (n+3)(n+2)},$$

$$\Gamma_{n,2} = -ik\frac{\left(c_L^2 - c_T^2\right)}{c_T^2\left(n+3\right)},$$

$$\Gamma_{n,1} = -ik\alpha\frac{c_T^2 + \left(c_L^2 - c_T^2\right)\left(n+1\right)}{c_T^2\left(n+3\right)\left(n+2\right)}.$$

$$\Gamma_{n,0} = 0.$$

$$\bar{\Lambda}_{n,2} = -ik\frac{c_L^2 - c_T^2}{c_L^2\left(n+3\right)},$$

$$\bar{\Lambda}_{n,1} = -ik\alpha\frac{c_L^2 - 2c_T^2 + \left(c_L^2 - c_T^2\right)\left(n+1\right)}{c_L^2\left(n+3\right)\left(n+2\right)},$$

$$\bar{\Lambda}_{n,0} = 0,$$

$$\bar{\Gamma}_{n,2} = -\alpha\frac{n+2}{n+3},$$

$$\bar{\Gamma}_{n,1} = -k^2\frac{c^2 - c_T^2}{c_L^2\left(n+3\right)\left(n+2\right)},$$

$$\bar{\Gamma}_{n,0} = -k^2\alpha\frac{c^2 - c_T^2}{c_L^2\left(n+3\right)\left(n+2\right)}. \tag{34}$$

Then inserting (34) back to the right side of (33), we have

$$a_n = a_0 h_0\left(n\right) + a_1 h_1\left(n\right) + b_0 h_2\left(n\right) + b_1 h_3\left(n\right),$$
$$b_n = a_0 k_0\left(n\right) + a_1 k_1\left(n\right) + b_0 k_2\left(n\right) + b_1 k_3\left(n\right), \tag{35}$$

where h_i and $k_i (i = 0,1,2,3)$ are known functions of integer n and phase velocities c, c_T, and c_L .Substitution of (34) and (35) into (28) yields

$$u\left(x_2\right) = a_0 W_0\left(x_2\right) + a_1 W_1\left(x_2\right) + b_0 W_2\left(x_2\right) + b_1 W_3\left(x_2\right),$$
$$v\left(x_2\right) = a_0 G_0\left(x_2\right) + a_1 G_1\left(x_2\right) + b_0 G_2\left(x_2\right) + b_1 G_3\left(x_2\right), \tag{36}$$

where

$$W_i = \sum_{n=0}^{\infty} h_i x_2^n,$$
$$G_i = \sum_{n=0}^{\infty} k_i x_2^n, i = 0,1,2,3. \tag{37}$$

If the four arbitrary constants can be obtained, the formulation of free vibrations of the FGM plate in SAW mode is completed. Then further substitution of (37) into (25) gives required stresses in abbreviated notations (with $e^{ik(x_1 - ct)}$ omitted) as

$$
\begin{aligned}
T_1 &= a_0 P_0(x_2) + a_1 P_1(x_2) + b_0 P_2(x_2) + b_1 P_3(x_2),\\
T_2 &= a_0 Q_0(x_2) + a_1 Q_1(x_2) + b_0 Q_2(x_2) + b_1 Q_3(x_2),\\
T_6 &= a_0 R_0(x_2) + a_1 R_1(x_2) + b_0 R_2(x_2) + b_1 R_3(x_2),
\end{aligned} \tag{38}
$$

where

$$
\begin{aligned}
&P_i(x_2) = \hat{X}\left[ik(\lambda_0 + 2\mu_0)W_i(x_2) + \lambda_0 G_i'(x_2)\right],\\
&Q_i(x_2) = \hat{X}\left[ik\lambda_0 W_i(x_2) + (\lambda_0 + 2\mu_0)G_i'(x_2)\right],\\
&R_i(x_2) = \mu_0 \hat{X}\left[W_i'(x_2) + ikG_i(x_2)\right], i = 0,1,2,3,\\
&\hat{X} = 1 + \alpha x_2.
\end{aligned} \tag{39}
$$

On the upper and lower surfaces of the plate shown in Figure 15, we have the traction-free boundary conditions

$$
T_2(-h/2) = T_6(-h/2) = T_2(h/2) = T_6(h/2) = 0. \tag{40}
$$

By substituting (38) into (40), the resulting equation can be written as

$$
\mathbf{KA} = \mathbf{0}, \tag{41}
$$

where the amplitude vector is

$$
\mathbf{A} = \{a_0 \quad a_1 \quad b_0 \quad b_1\}, \tag{42}
$$

and the coefficient matrix is

$$
\mathbf{K} = \begin{bmatrix}
Q_0(-h/2) & Q_1(-h/2) & Q_2(-h/2) & Q_3(-h/2)\\
R_0(-h/2) & R_1(-h/2) & R_2(-h/2) & R_3(-h/2)\\
Q_0(h/2) & Q_1(h/2) & Q_2(h/2) & Q_3(h/2)\\
R_0(h/2) & R_1(h/2) & R_2(h/2) & R_3(h/2)
\end{bmatrix}. \tag{43}
$$

The phase velocity of SAW propagating in the FGM solid can be obtained by setting the determinant of (42) to vanish. With the amplitude ratios from (42), we can also obtain the mode shapes from (36).

A procedure for the Frobenius solution of displacements for SAW in an FGM plate with linear variation of properties in the thickness coordinate is now established. Again, we want to emphasize that the same procedure can be applied to material parameters varying in other polynomials of the thickness coordinate.

3.3. Numerical Examples

With power series solutions given in the previous section, it is straightforward to calculate the phase velocity and displacements in an infinite FGM plate with known linear variation of material properties and plate thickness. As an example, we calculated displacement solutions of an FGM plate with material properties varying as a linear function of the thickness coordinate as shown in (22).

For a functionally graded plate with thickness of three wavelengths, material properties varying along thickness direction as a linear function $f(x_2)=1+\alpha x_2$ and the Poisson's ratio $\nu=0.2$, displacements are written in power series as shown in (28). From calculations, we found the convergence of displacements and wave velocity can be achieved with number of terms $n=80$, which ensures that the neglected terms of the infinite series are smaller than 10^{-8}. Velocity and displacements, shown in Figs. 16, 17 and 18, are obtained from (41).

From the velocity in Figure 16, two branches of the velocity curve, with the solid line denoting as the first mode of smaller velocity c_{R1} and the dashed line as the second mode of larger velocity c_{R2}, are shown with the gap between each other with the increase of $|\alpha|$. As we know, separated frequencies can be used as an important feature of frequency selectivity in devices. It is important to note that the two velocities at $\alpha=0$ are not exactly the same because they are expected to be the same only for plates with larger thickness.

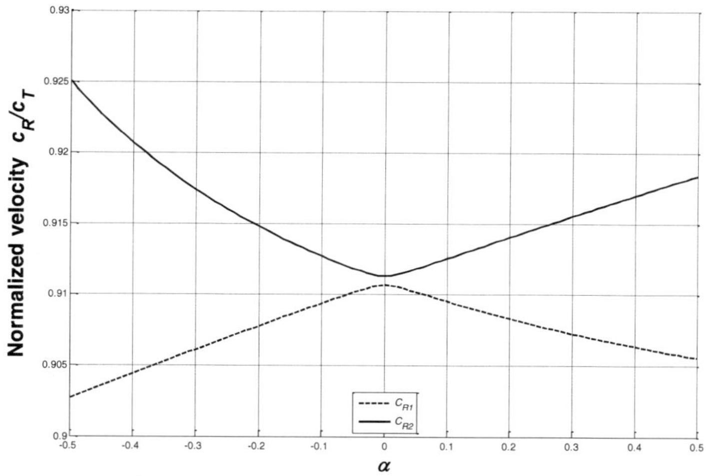

Figure 16. Velocity of surface acoustic waves in an FGM plate with thickness of three wavelengths and material properties variation of $f(x_2)=1+\alpha x_2$.

From Figs. 17 and 18, it is clear that the hardening of the material along the thickness direction (positive α) from the top will enlarge the displacements of lower face in the first mode but reduce them in the second mode. Similarly, the softening of the material from the top of the plate (negative α) will reduce the displacements of the lower face in the first mode but enlarge them in the second mode. The increased deformation in actual SAW resonators of piezoelectric materials, as can be expected, will be advantageous for enhancement of electrical charge generation under electrodes in SAW devices. On the other hand, the reduced deformation is preferable in the vibration isolation and control in foundations for precision equipments.

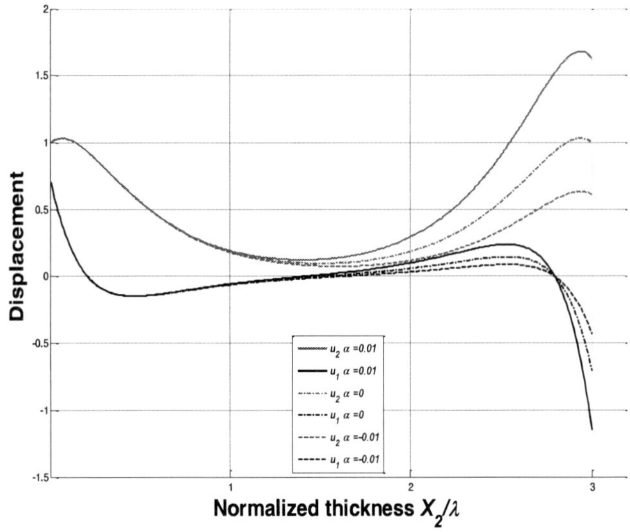

Figure 17. Displacement of the first mode of surface acoustic waves in an FGM plate with thickness of three wavelengths and material properties variation of $f(x_2)=1+\alpha x_2$.

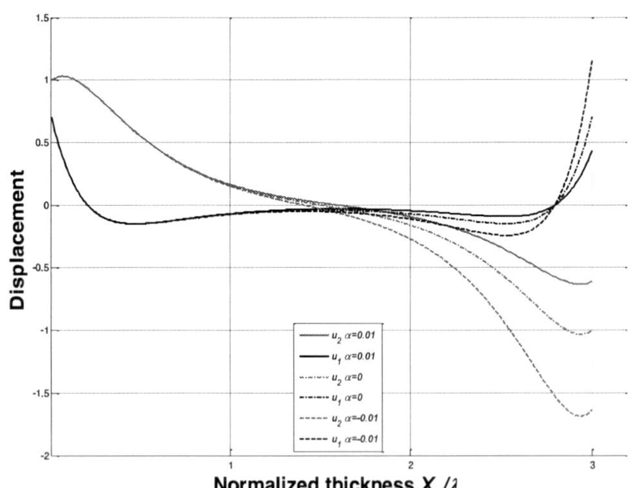

Figure 18. Displacement of the second mode of surface acoustic waves in an FGM plate with thickness of three wavelengths and material properties variation of $f(x_2)=1+\alpha x_2$.

From Figure 19, the two velocities in an FGM plate with property variation parameter $\alpha = 0.5$ show the typical trend that of homogeneous infinite plates with the increase of thickness.

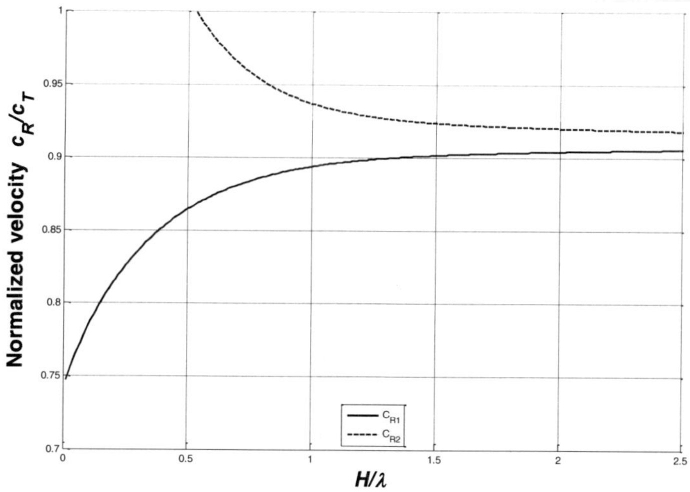

Figure 19. Normalized velocity of surface acoustic waves versus thickness of an FGM plate with linear property variation parameter $\alpha = 0.5$.

To check the convergence of power series solutions and computing procedure for larger thickness of plates, we calculated the velocities of an FGM plate with thickness of five wavelengths in Figure 20. The merge of two velocities with $\alpha = 0.5$ is clearly shown because of the larger thickness (five wavelengths) of plate in comparison to the case in Figure 16 and the velocity dependence on α is following the same trend.

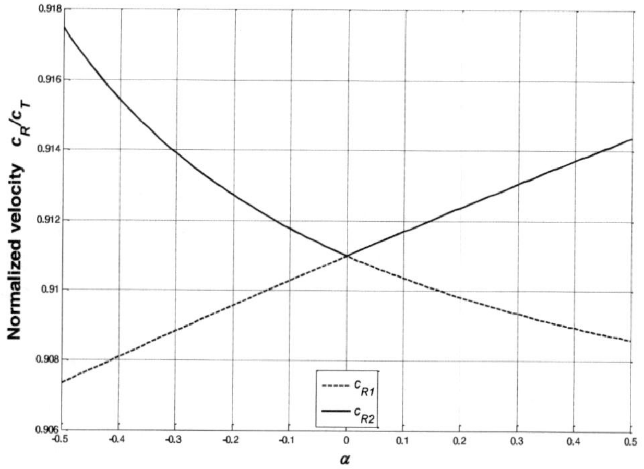

Figure 20. Velocity of surface acoustic waves in an FGM plate with thickness of five wavelengths and material properties variation of $f(x_2) = 1 + \alpha x_2$.

The propagating characteristics of SAW in the FGM plate with linear variation of material properties are the same with the layered model in our earlier study [40]. The convergence of numerical computation has been studied by examining the evaluation procedure of the power series solutions and compared with the other methods. The straightforward method can be applied to wave propagation problems in general FGM solids for analytical solutions which are important for benchmarking and practical applications. The method has been validated in this study through following the standard procedure and making comparison with earlier results.

3.4. Conclusions

The Frobenius method is used to solve differential equations with variable coefficients for SAW propagating in infinite plates of FGM with property variation in a linear function of the thickness coordinate. The numerical examples have been shown that the presence of material variation in the thickness will create a noticeable gap between two SAW modes. In addition, the displacements on the faces will also be different on magnitudes, posing possible applications for deformation amplification or reduction depending on the objectives and modes utilized. The equations and solution procedure are validated by comparing the solutions from the Frobenius method with earlier layered model solutions with good agreement.

From above results, we found that if plate properties vary along the thickness, the vibration modes and frequency behavior are perturbed from the homogeneous plates. It is clear that there are different effects on deformation of the two faces of an FGM plate. It is also clear that the FGMs can be used effectively to modify the characteristics of SAW in broad engineering applications.

4. Homotopy Analysis Method for the Surface Acoustic Waves in FGM Solids

4.1. Introduction

For structural components of FGMs in many practical engineering applications, we need to study the effect of material property variation patterns before a particular design is adopted. It is obvious that the cost of analysis is an important concern. In other words, we need to find a fast and robust analytical method for the evaluation of FGMs, especially if the material property variation may cause difficulties in using analytical methods we are familiar with. Recently, a general method for solving differential equations with strong nonlinearity and variable coefficients, the homotopy analysis method (HAM), has been proposed and improved by Liao [41] with successful applications in solving some typical nonlinear problems in fluid mechanics [42-47] and similar problems in solid mechanics such as the large deformation of a beam [48]. From these examples, we see a great potential in applying this method to the deformation and vibrations of structural members of FGMs. As an extension of the HAM to vibration problems, or eigenvalue problems, we studied surface acoustic waves in an elastic plate of FGM, which have coupled differential equations with variable coefficients. Such problems have been solved by a layered model [49] and Frebenius method [28] with satisfactory results, but it is now studied with the HAM [29] to demonstrate its effectiveness and simplicity in dealing with more general material property variation patterns which are difficult to solve with the either the layered model or Frobenius method.

Although the FGM components are created for the thermal protection of space structures [50-55], it has been proliferating in other engineering applications as functions and properties of structural components have been precisely specified to meet more stringent requirements. To improve the functions of small- and micro-scale structures like electronic devices which are widely encountered in sensors and actuators, one of the possibilities is to use the FGM components based on the modified mechanical behavior and associated properties. For instance, wave propagation in an FGM solid will be perturbed by the material property variation, and the modifications of mechanical deformation and electrical charge distribution will produce obvious function changes in acoustic wave devices [10,12,14,16,49,56-58,64]. Based on this observation, the study of surface acoustic waves in an FGM plate is designated as long term efforts toward the advantageous adoption of the FGM in the future generation of acoustic wave devices [49, 57, 58]. Our earlier study with layered model has shown that the modified wave propagation in FGM plates can be used for the control and isolation of surface acoustic waves, which are important components existing in earthquake and various dynamic disruptions, with improved efficiency [49]. These implications, along with other foreseeable applications, have promoted our in-depth studies on the wave propagation in FGM structures for both solution method and property variation pattern evaluation [9]. Our studies and comparisons on the effectiveness of analytical methods, which include the layered model, the Frobenius method, and the HAM in this paper, will be helpful in identifying the most effective one for accurate and efficient solutions in practical problems, while the numerical results from these studies with unusual material property variation patterns will also be useful references in selecting FGM materials for actual applications. The systematic studies will also contribute to the expansion of applications and studies of the FGM structures, especially in new and emerging fields such as acoustic wave devices.

4.2. Homotopy Analysis of Surface Acoustic Waves in Functionally Graded Plates

Plates of different materials and configurations are typical structural components with wide engineering applications. While plates of homogeneous materials have been serving engineering applications well, the emerging FGM plates have the potential to bring many unique features to traditional functions we are familiar with. For instance, the thermal protection capability of FGM components is widely known as an addition to many important applications and has motivated subsequent studies. Particularly, plates of piezoelectric materials serving as resonant elements are required to have relatively large deformation to fulfill the requirement of electrical charge generation in intelligent applications suitable for the expected functions of FGM structures. This, naturally, promotes the study of wave propagation in FGM components with plates as the most popular one in all applications. In addition to the usual approach in obtaining the vibration frequencies of primary modes, we are concentrating on the particular vibration modes with higher vibration frequencies for acoustic wave device property estimation. Naturally, the formulation and solution procedure can be extended to other vibration modes for structural, sensor, and actuator applications we frequently encounter in intelligent structures nowadays. While earlier studies are using exponential variation of material properties to get analytical solutions [14, 59], polynomial variation of material properties are assumed in many practical applications [24, 52, 60-63]. It is apparent that the HAM employed in this study is more effective in dealing with differential equations of variable coefficients in polynomials because the Frobenius method will be more complicated in such cases.

In this study, we are interested in surface acoustic waves in an FGM plate with material property variation in the thickness direction. The primary motivation of this study is to show the effect of material property variation of an FGM plate on the surface wave propagation in terms of vibration frequency and deformation change to enable proper utilization of the FGM components in acoustic wave devices. The optimization of such devices based on deformation and frequency can be performed with the available solutions. With this particular objective, we consider a functionally graded plate shown in Figure 15 with isotropic material properties λ, μ, and ρ as the Lamé constants and density varying along the thickness coordinate, respectively. These variables are assumed to be functions of thickness coordinate x_2 as (21), and

$$f(x_2)=1+\frac{\alpha x_2}{h}, \tag{44}$$

with α and h are a nondimensional constant and the thickness of plate, respectively. It should be emphasized that the material properties are the base constants when the thickness coordinate is at its origin. The analytical procedure we shall outline, the homotopy analysis method (HAM), can be applied to any variation scheme of the FGM with only changes in the algebraic operations involved. The linear variation model we present in this study is for the demonstration of the procedure, not implying the limitation of the method itself.

For wave propagation in an FGM plate, the governing differential equations of motion with abbreviated stress notations are

$$\begin{aligned} &T_{1,1}+T_{6,2}=\rho \ddot{u}_1, \\ &T_{6,1}+T_{2,2}=\rho \ddot{u}_2, \end{aligned} \tag{45}$$

where $T_p\,(p=1,2,6)$ and $u_j\,(j=1,2)$ are stresses and displacements, respectively. The displacements of surface acoustic waves can be assumed as

$$\begin{aligned} &u_1=u(x_2)e^{ik(x_1-ct)}, \\ &u_2=v(x_2)e^{ik(x_1-ct)}, \end{aligned} \tag{46}$$

where $u(v), k, c,$ and t are displacement functions, wavenumber, phase velocity, and time, respectively. From (46), the corresponding strains in abbreviated notations can be obtained as

$$\begin{aligned} &S_1=ikue^{ik(x_1-ct)}, \\ &S_2=v'e^{ik(x_1-ct)}, \\ &S_6=(u'+ikv)e^{ik(x_1-ct)}. \end{aligned} \tag{47}$$

Then, stresses can be obtained from constitutive equations in terms of displacements as

$$\begin{aligned} &T_1=\left[ik(\lambda+2\mu)u+\lambda v'\right]e^{ik(x_1-ct)}, \\ &T_2=\left[(\lambda+2\mu)v'+ik\lambda u\right]e^{ik(x_1-ct)}, \\ &T_6=\mu(u'+ikv)e^{ik(x_1-ct)}, \end{aligned} \tag{48}$$

where λ and μ are Lamé constants which are functions of the thickness variable in this study. Substituting stresses from (48) into (45), we have the equations of motion in displacements

$$f(x_2)u''+f'(x_2)u'+k^2(\bar{c}^2-\bar{c}_L^2)f(x_2)u+ik(\bar{c}_L^2-1)f(x_2)v'+ikf'(x_2)v=0,$$
$$ik(\bar{c}_L^2-1)f(x_2)u'+ik(\bar{c}_L^2-2)f'(x_2)u+\bar{c}_L^2 f(x_2)v''+\bar{c}_L^2 f'(x_2)v'+k^2(\bar{c}^2-1)f(x_2)v=0, \tag{49}$$

Boundary conditions of the upper and lower traction-free surfaces can be written as following

$$\begin{aligned}
&(\lambda_0+2\mu_0)v'(0)+ik\lambda_0 u(0)=0,\\
&u'(0)+ikv(0)=0,\\
&(\lambda_0+2\mu_0)v'(h)+ik\lambda_0 u(h)=0,\\
&u'(h)+ikv(h)=0,
\end{aligned} \tag{50}$$

where h is the thickness of the plate shown in Figure 15.

For the HAM procedure [41], the displacement solutions in (49) and (50) are expressed in power series as

$$\begin{aligned}
u(x_2)&=\sum_{n=0}^{+\infty}c_n x_2^n,\\
v(x_2)&=\sum_{n=0}^{+\infty}d_n x_2^n,
\end{aligned} \tag{51}$$

where coefficients c_n and $d_n (n=1,2,3,\cdots,\infty)$ are to be determined.

Because boundary conditions are so complicated that it is difficult to find the initial guesses needed in the HAM solutions, auxiliary boundary conditions for homotopy analysis should be applied as

$$u(0)=a_0, u'(0)=a_1, v(0)=b_0, v'(0)=b_1, \tag{52}$$

where a_0, a_1, b_0, and b_1 are arbitrary constants, again, to be determined.

Then, it is natural to choose the initial guesses for homotopy analysis as

$$\begin{aligned}
u_0(x_2)&=a_0+a_1x_2,\\
v_0(x_2)&=b_0+b_1x_2.
\end{aligned} \tag{53}$$

Consequently, we choose the auxiliary linear operator

$$L=\frac{\partial^2}{\partial x_2^2}, \tag{54}$$

with the property

$$L(\beta_0+\beta_1 x_2)=0, \tag{55}$$

where β_0 and β_1 are also arbitrary constants.

With the governing equations (49), we further define the nonlinear operators

$$\begin{aligned}
N_u\left[U(x_2;q),V(x_2;q)\right]&=f(x_2)\frac{\partial^2 U(x_2;q)}{\partial x_2^2}+f'(x_2)\frac{\partial U(x_2;q)}{\partial y}\\
&+k^2\left(\bar{c}^2-\bar{c}_L^2\right)f(x_2)U(x_2;q)+ik\left(\bar{c}_L^2-1\right)f(x_2)\frac{\partial V(x_2;q)}{\partial x_2}+ikf'(x_2)V(x_2;q),\\
N_v\left[U(x_2;q),V(x_2;q)\right]&=\frac{1}{\bar{c}_L^2}\Bigg\{ik\left(\bar{c}_L^2-1\right)f(x_2)\frac{\partial U(x_2;q)}{\partial x_2}+ik\left(\bar{c}_L^2-2\right)f'(x_2)U(x_2;q)\\
&+\bar{c}_L^2 f(x_2)\frac{\partial^2 V(x_2;q)}{\partial x_2^2}+\bar{c}_L^2 f'(x_2)\frac{\partial V(x_2;q)}{\partial x_2}+k^2\left(\bar{c}^2-1\right)f(x_2)V(x_2;q)\Bigg\},
\end{aligned} \tag{56}$$

where $U(V)$ are the approximate solutions of the exact displacement $u(v)$ and $q\in[0,1]$ is the embedding parameter to be specified later. Consequently, N_u and N_v are the residuals of the governing equations with approximate displacement solutions.

The zeroth-order deformation equations are

$$\begin{aligned}
(1-q)L\left[U(x_2;q)-u_0(x_2)\right]&=q\hbar N_u\left[U(x_2;q),V(x_2;q)\right],\\
(1-q)L\left[V(x_2;q)-v_0(x_2)\right]&=q\hbar N_v\left[U(x_2;q),V(x_2;q)\right],
\end{aligned} \tag{57}$$

where $\hbar$ are the nonzero auxiliary parameter. The auxiliary initial conditions from (52) now are

$$U(0;q)=a_0,\frac{\partial U(0;q)}{\partial x_2}\Big|_{x_2=0}=a_1,V(0;q)=b_0,\frac{\partial V(0;q)}{\partial x_2}\Big|_{x_2=0}=b_1. \tag{58}$$

The zeroth-order deformation equation gives

$$\begin{aligned}
U(x_2;0)&=u_0(x_2),\\
V(x_2;0)&=v_0(x_2).
\end{aligned} \tag{59}$$

when $q=0$.

Similarly, when $q=1$, solutions of (57) are exactly the same form with that of (49), so that we have

$$\begin{aligned}
U(x_2;1)&=u(x_2),\\
V(x_2;1)&=v(x_2).
\end{aligned} \tag{60}$$

Therefore, $U(x_2;q)$ and $V(x_2;q)$ vary continuously from the initial approximation $u_0(x_2)$ and $v_0(x_2)$ to the exact solutions $u(x_2)$ and $v(x_2)$ of equations of motion (49) with q from 0 to 1. Clearly, (57) and (58) give an indirect relation between the initial approximation $u_0(x_2)$ and $v_0(x_2)$ and the general solutions $u(x_2)$ and $v(x_2)$. The HAM depends on finding a direct relationship between the two solutions which can be described as follows.

We assume that (57) has solutions $U(x_2;q)$ and $V(x_2;q)$ that converge for $0 \le q \le 1$ with properly selected parameter $\hbar$. We further assume that $U(x_2;q)$ and $V(x_2;q)$ are infinitely differentiable at $q = 0$, that is

$$u_0^{[m]}(x_2) = \left.\frac{\partial^m U(x_2;q)}{\partial q^m}\right|_{q=0},$$
$$v_0^{[m]}(x_2) = \left.\frac{\partial^m V(x_2;q)}{\partial q^m}\right|_{q=0},$$
$$m = 1, 2, 3, \cdots. \tag{61}$$

This reminds us the fact that the Maclaurin series of $U(x_2;q)$ and $V(x_2;q)$ are

$$U(x_2;q) = U(x_2;0) + \sum_{m=1}^{+\infty} \frac{\partial^m U(x_2;q)}{\partial q^m} \frac{q^m}{m!},$$
$$V(x_2;q) = V(x_2;0) + \sum_{m=1}^{+\infty} \frac{\partial^m V(x_2;q)}{\partial q^m} \frac{q^m}{m!}. \tag{62}$$

If the auxiliary linear operator L, the initial guesses $u_0(v_0)$, and auxiliary parameter $\hbar$ are properly chosen, the series, as shown in (62), converges at $q = 1$. With the assumption that the series above also converges, from (59), (60) and (62), we have relationships

$$u(x_2) = U(x_2;1) = u_0 + \sum_{m=1}^{+\infty} u_m(x_2),$$
$$v(x_2) = V(x_2;1) = v_0 + \sum_{m=1}^{+\infty} v_m(x_2), \tag{63}$$

where

$$u_m(x_2) = \frac{u_0^{[m]}}{m!},$$
$$v_m(x_2) = \frac{v_0^{[m]}}{m!},$$
$$m = 1, 2, 3, \cdots. \tag{64}$$

(63) gives the direct relationship between general solutions $u(x_2)$ and $v(x_2)$ and initial approximation $u_0(x_2)$ and $v_0(x_2)$ through introducing $u_m(x_2)$ and $v_m(x_2)$. To derive the governing equations of $u_m(x_2)$ and $v_m(x_2)$, we differentiate (57) m times with respect to q, divided them by $m!$, then set $q=0$. As a result of this operation, we get the mth-order deformation equations

$$
\begin{aligned}
&L\left[u_m(x_2)-\chi_m u_{m-1}(x_2)\right]=\hbar R_m^u(u_{m-1},v_{m-1}),\\
&L\left[v_m(x_2)-\chi_m v_{m-1}(x_2)\right]=\hbar R_m^v(u_{m-1},v_{m-1}),\\
&\chi_m=\begin{cases}0, & m=1,\\ 1, & m>1,\end{cases}
\end{aligned}
\tag{65}
$$

with

$$
\begin{aligned}
R_m^u(u_{m-1},v_{m-1})&=\frac{1}{(m-1)!}\frac{\partial^{m-1}N_u}{\partial q^{m-1}}\Big|_{q=0}\\
&=f(x_2)u_{m-1}''+f'(x_2)u_{m-1}'+ik(\bar{c}_L^2-1)f(x_2)v_{m-1}'\\
&\quad+ikf'(x_2)v_{m-1}+k^2(\bar{c}^2-\bar{c}_L^2)f(x_2)u_{m-1},\\
R_m^v(u_{m-1},v_{m-1})&=\frac{1}{(m-1)!}\frac{\partial^{m-1}N_v}{\partial q^{m-1}}\Big|_{q=0}\\
&=\frac{1}{\bar{c}_L^2}\{ik(\bar{c}_L^2-1)f(x_2)u_{m-1}'+ik(\bar{c}_L^2-2)f'(x_2)u_{m-1}+\bar{c}_L^2f(x_2)v_{m-1}''\\
&\quad+\bar{c}_L^2f'(x_2)v_{m-1}'+k^2(\bar{c}^2-1)f(x_2)v_{m-1}\},
\end{aligned}
\tag{66}
$$

as the mth-order error associated with the approximation.

The auxiliary initial conditions for the mth-order approximation are

$$
u_m(0)=0,u_m'(0)=0,v_m(0)=0,v_m'(0)=0. \tag{67}
$$

The solutions of (65) are

$$
\begin{aligned}
u_m(x_2)&=\chi_m u_{m-1}(x_2)+\sum_{n=0}^{3m+2}\frac{r_{m-1,n}}{(n+1)(n+2)}x_2^{n+2}+\beta_1+\beta_2x_2,\\
v_m(x_2)&=\chi_m v_{m-1}(x_2)+\sum_{n=0}^{3m+2}\frac{s_{m-1,n}}{(n+1)(n+2)}x_2^{n+2}+\beta_3+\beta_4x_2,
\end{aligned}
\tag{68}
$$

where $r_{m-1,n}$ and $s_{m-1,n}$ are coefficients of x_2^n in the expressions of u_{m-1} and v_{m-1}.

Applying the auxiliary initial condition in (67), $\beta_1, \beta_2, \beta_3$ and β_4 can be obtained as

$$\beta_1 = \beta_2 = \beta_3 = \beta_4 = 0. \tag{69}$$

As a result, the solution of (65) can be rewritten as

$$u_m(x_2) = \chi_m u_{m-1}(x_2) + \sum_{n=0}^{3m+2} \frac{r_{m-1,n}}{(n+1)(n+2)} x_2^{n+2},$$

$$v_m(x_2) = \chi_m v_{m-1}(x_2) + \sum_{n=0}^{3m+2} \frac{s_{m-1,n}}{(n+1)(n+2)} x_2^{n+2}. \tag{70}$$

Then, the Nth -order approximate solutions of (49) can be written as

$$u(x_2) = u_0 + \sum_{m=1}^{N} u_m,$$

$$v(x_2) = v_0 + \sum_{m=1}^{N} v_m. \tag{71}$$

Finally, for arbitrary variables a_0, a_1, b_0 and b_1, applying the boundary conditions (50) and letting the determinant of matrix be zero, wave velocity $\bar{c}$ can be obtained.

Through the analysis of surface acoustic waves in an FGM plate with linear variation of material properties in the thickness direction, an HAM procedure is established for the calculation of wave propagation velocity and displacements. Based on the nature of the HAM, similar solutions for a general variation of material properties can be obtained. The solutions from the HAM, with the Frobenius solutions as a special case, are also considered exact and accurate in infinite power series with known coefficients. In applications, such analytical solutions are always needed for the validation of numerical results and methods. In addition, the analytical solutions are generally more cost-effective in computing, thus offering a fast way to obtain solutions to satisfy immediate needs in material selection and design evaluation. Furthermore, new solution methods and techniques such as the HAM need to be introduced into traditional problems such as wave propagation for more accurate and robust solutions which are considered as first available ones or an improvement with practical importance. The is a process without limit, and the solutions, methods, techniques, will always contribute to the in-depth understanding and continuing accumulation of knowledge for both the specific subject and method itself required for engineering and scientific problems.

4.3. Numerical Examples

In the previous section, the solutions of surface wave propagating in an FGM plate with linear variation of material properties are obtained with the HAM. The displacements are given in power series with the Frobenius method as a special case, thus guaranteeing that those analytical solutions for general variation of material properties in an FGM plate can be obtained without being restricted by the solution methods we are familiar with. Such solutions are needed in problems involving wave propagation in FGM solids in almost all aspects like characterization, evaluation, validation, and verification. This will certainly expand the solutions and methods in the study of wave propagation in the increasingly complex structures and utilization of newer components such as the FGM types. As the procedure is well exhibited in obtaining analytical solutions based on the HAM, we now present a few numerical examples to demonstrate the numerical evaluation of the HAM solutions.

First, for comparison and validation purpose, we consider a homogeneous isotropic plate with Poisson's ratio of 0.2 and thickness of two wavelengths. In this case, we set the nondimensional parameter $\alpha = 0$.

To study the effect of parameter $\hbar$ on solutions (69), we should also study the convergence of another series defined as follows. Applying the boundary conditions in (48) to (69), a set of linear algebraic equations can be obtained. The determinant of the coefficient matrix, denoting as $D(0)$ of the parameter $\hbar$, must be zero with exact solutions of displacements and velocity. Apparently, for approximate solutions, we can get a series of $D(0)$ which depends on $\hbar$. In Figure21, the $D(0)$ versus $\hbar$ curve is plotted for the approximation to the 50th -order. It is obvious that the range for the admissible value is $-1.2 \le h \le -0.6$. When $\hbar = -0.95$, we found that the two surface acoustic wave modes have velocities as 0.90717921 and 0.91541065, which are exact results from our earlier study [27].

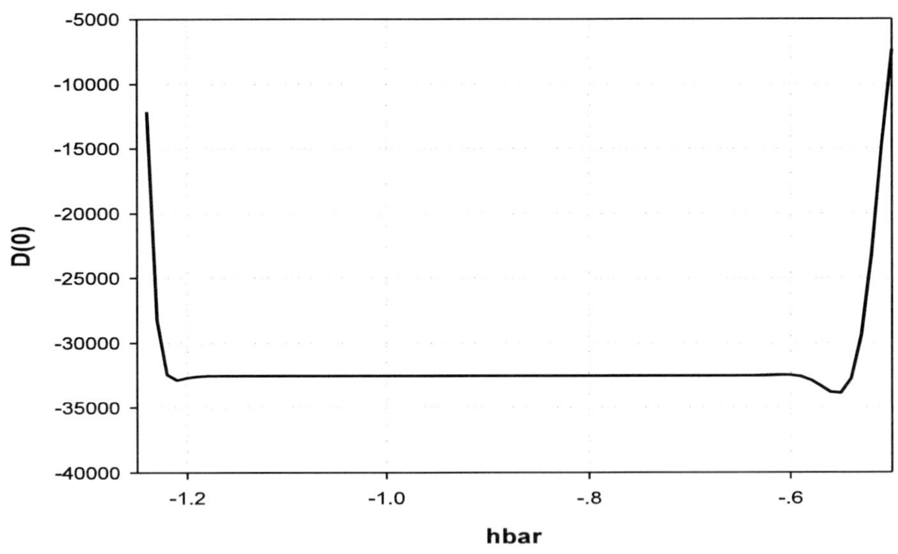

Figure 21. D versus $\hbar$ for a homogeneous plate with thickness of two wavelengths.

Next, we consider an FGM plate with $\alpha = -0.1$, Poisson's ratio of 0.2 , and thickness of three wavelengths. In Figure 22, $D(0)$ versus $\hbar$ curve is plotted with the 80th-order approximation. Clearly, the range for the admissible value is $-1.1 \le \hbar \le -0.6$. Then the relationship between normalized phase velocity $\bar{c}$ and α shown in Figure 23 can be obtained.

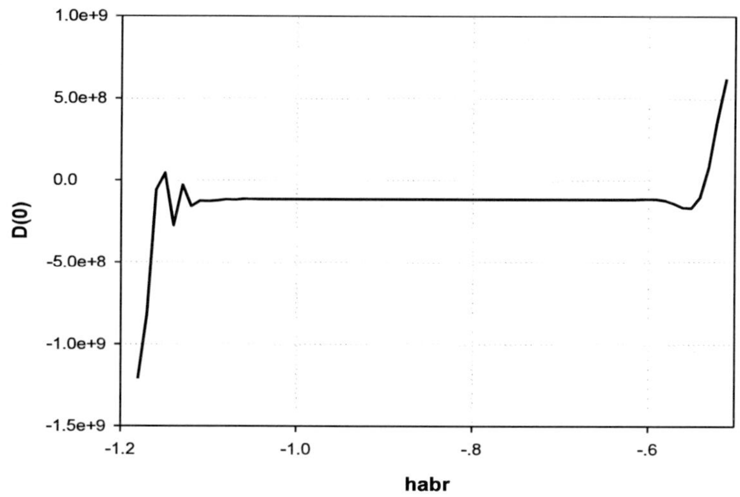

Figure 22. D versus $\hbar$ for a functionally graded plate with thickness of three wavelengths and $\alpha = -0.1$.

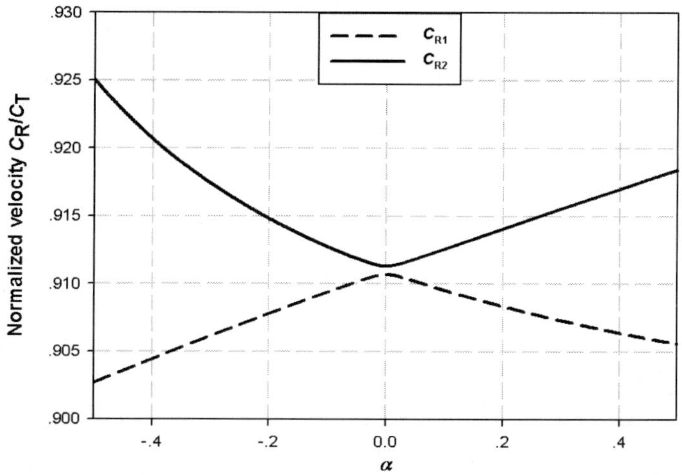

Figure 23. Normalized phase velocity versus graded index α for a functionally graded plate with Poisson's ration $\mu = 0.2$ and thickness of three wavelengths.

These results are exactly the same in comparison to the earlier results by the layered model method [27] and Frobenius method [28]. It proves that the HAM can be applied to solve many problems in engineering applications involving differential equations with variable coefficients and nonlinear terms such as surface acoustic waves in functionally graded plates of various variation schemes. The numerical examples presented in this section are more on the validation and verification of known results in order to justify the adoption of the HAM in wave propagation problems. The procedure, solutions, and numerical results show that the HAM is a straightforward method in solving differential equations with variable coefficients and nonlinear terms pertaining to the FGM structures and vibrations we are familiar with. If general variation of material properties and nonlinear deformation are considered, the HAM will provide the analytical solutions which are hard or even impossible to obtain with other known methods. This, of course, demonstrates the effectiveness of the HAM in many similar engineering problems involving nonlinear differential equations and complication of FGM structures.

4.4. Conclusions

In this study, the homotopy analysis method (HAM) is applied to obtain the solution of surface acoustic waves in a functionally graded plate. The HAM provides us with a convenient way to control the convergence of approximation series by adopting the parameter $\hbar$, which shows a fundamental difference between HAM and other methods. Also, we found out that the solution with $\hbar = -1$ is exactly the same with that from Frobenius method [28].

As we can see from the solution process presented above, the HAM can be applied to solve other nonlinear or inhomogeneous problems which are frequently encountered in various engineering problems. Particularly, when the material properties and configurations of structural members are not the simple functions which can guarantee analytical solutions forming basic functions or the Frobenius method, the HAM can provide solutions which are hard to obtain with computational efficiency and adequate approximation. If the large deformation is to be considered also with the FGM solids and nonuniform configurations, the HAM offers a chance to get the only approximate solutions for practical applications and validation of the easy-to-come numerical solutions. This implies, of course, the HAM can be used for a large class of nonlinear problems related to the wide applications of functionally graded materials as structural and functional members.

Conclusion

The never-ending innovation in the applications of acoustic waves concerning the surface acoustic waves is coming to a new era with the emergence of the functionally graded materials (FGMs) which posses the properties to change the known pattern of waves, thus providing possible improvement and enhancement of current acoustic wave devices such as resonators. Undoubtedly, the first challenge we have is the choice of right pattern of material property variation, which, as shown here, requires accurate analysis of surface waves propagating in FGMs with further complications such as presence of electrodes and mounting structures that are essential in devices. The analysis can be difficult due to the variation of material properties with methods we have, analytical or numerical. As a result, efforts have to be made to obtain accurate solutions whose values are multifold, revealing the effect of FGM and serving as references to validating other methods which are being developed for the new problems.

All the methods presented in this paper giving the same solutions for the problem with linear variation of material properties, demonstrating that these methods are effective in treating wave propagation in arbitrary FGMs with accurate solutions. The layered method, which is straightforward based on the continuing refinement of complicated materials with simple elements, is universally applicable at the cost of larger number of layers involving time- and memory-intensive computation. Even at this computer era, necessary efforts have to be made in securing adequate resources and verifying the convergence. The Frobenius method, as a traditional method for analytical solutions of differential equations in power series, is also effective in solving these coupled differential equations, and the evaluation of the solutions have also been proven to be straightfoarwarded. The validated results show that the method can be used for materials with variation is polynomials more effectively. It is always good to have exact solutions in closed form for comparison, and the Frobenius method does just that. Finally, with the newly developed Hpomotopy Analysis Method (HAM), we can obtain the solutions of wave propagation in an arbitrary FGM solid with solutions in power series but without a lengthy procedure to obtain the solutions. Actually, the solutions given by the HAM are easy to obtain and evaluate because the linear solutions as the basis function will make the computation much more efficient. The convergence of the solutions to the exact ones demonstrated that the HAM indeed is an effective technique which can be used for much more complicated variation of material properties in the FGMs. As a result, the application of HAM and the generalization of the procedure should be prompted for prompt adoption and improvement. The methods introduced in this paper demonstrates that the new challenges of wave propagation in newly conceived materials and structures can be well treated with both traditional and new techniques and the focus can be shifted to the development of applications as the solutions are quickly obtained with the methods in different forms of the same results. Apparently, the search of solution techniques and methods never has an end as the problems emerging with applications.

ACKNOWLEDGMENTS

This research is supported National Natural Science Foundation of China (Project Nos.10432030, 10125209, 10572065 and 10932004) and the Teaching and Research Award Program for Outstanding Young Teachers in High Education Institutions, Ministry of Education, China.

This research is also supported by Qianjiang River Fellow Fund established by Zhejiang Provincial Government and Ningbo University and administered by Ningbo University.

The authors are also grateful for the discussions with Professors Shijun Liao, Shanghai Jiaotong University and Jiankang Chen, Ningbo University.

REFERENCES

Rayleigh, L., On waves propagated along the plane surface of an elastic solid Proceedings of the London Mathematical Society 1885 Vol 17, 4-11.

Hashimoto, K.Y., Surface acoustic wave devices in telecommunications: modelling and simulation; Springer: Berlin, 2000.

Peplow, A.T., Jones, C.J.C., Petyt, M., Surface vibration propagation over an elastic half-space with an inclusion Applied Acoustics 1999 Vol 56, 283-96.

Ju, S.H., Lin, H.T., Analysis of train-induced vibrations and vibration reduction schemes above and below critical Rayleigh speeds by finite element method Soil Dynamics and Earthquake Engineering 2004 Vol 24, 993-1002.

Celebi, E., Schmid, G., Investigation of ground vibrations induced by moving loads Engineering Structures 2005 Vol 27, 1981-98.

Hildebrand, R., Asymptotic analysis of hard wave barriers in soil Soil Dynamics and Earthquake Engineering 2003 Vol 23, 143-58.

Zhong, Z., Shang, E.T., Exact analysis of simply supported functionally graded piezothermoelectric plates Journal of Intelligent Material Systems and Structures 2005 Vol 16, 643-51.

Wang, J., Yang, J.S., Li, J.Y., Energy trapping of thickness-shear vibration modes of elastic plates with functionally graded materials IEEE Trans. Ultrason., Ferroelect, and Freq. Contr. 2005, submitted.

Lefebvre, J. E., Zhang, V., Gazalet, J., Gryba, T., Sadaune, V., Acoustic wave propagation in continuous functionally graded plates: an extension of the legendre polynomial approach IEEE Trans. Ultrason., Ferroelect., Freq. Contr. 2001 Vol 48, 1332-40.

Chen, W. Q., Ding, H. J., On the free vibration of a functionally graded piezoelectric rectangular plates Acta Mechanica 2002 Vol 153, 207-16.

Berezovski, A., Engelbrecht, J., Maugin, G. A., Numerical simulation of two-dimensional wave propagation in functionally graded materials *European Journals of Mechanics, A/Solids* 2003 Vol 22, 257-65.

Liu, G. R., Dai, K. Y., Han, X., Ohyoshi, T., Dispersion of waves and characteristic wave surfaces in functionally graded piezoelectric plates Journal of Sound and Vibration 2003 Vol 268, 131-47.

Liu, G. R., Elastic waves in anisotropic laminates, CRC Press: Florida, 2002.

Zhong, Z., Shang, E. T., Three-dimensional exact analysis of a simply supported functionally gradient piezoelectric plate *International Journal of Solids and Structures* 2003 Vol 40, 5335-52.

Liu, G. R., Tani, J., Ohyoshi, T., Lamb waves in a functionally gradient material plates and its transient response,Part 1: Theory, Part 2: Calculation results Transactions of the Japan Society of Mechanical Engineers 1991 Vol 57, 603–11.

Liu, G. R., Tani, J., Surface waves in functionally gradient piezoelectric plates Trans. ASME 1994 Vol 116, 440-48.

Santare, M. H., Thamburaj, P., Gazonas, G. A., The use of graded finite elements in the study of elastic wave propagation in continuously nonhomogeneous materials *International Journal of Solids and Structures* 2003 Vol 40, 5621-34.

Chakraborty, A., Gopalakrishnan, S., A spectrally formulated finite element for wave propagation analysis in functionally graded beams International Journal of Solids and Structures 2003, Vol 40, 2421–48.

Wang, J., Du, J. K., Pan, Q. Q., A two-dimensional analysis of surface acoustic waves in finite plates with eigensolutions Science in China Series G 2007 Vol 37, 89-105.

Wang, J.; Du, J. K.; Lu, W. Q.; Mao, H. M. Exact and approximate analysis of surface waves in an infinite elastic plate with a thin metal layer. Ultrasonics 2006, 44s1, e941-e945.

Wang, J., Du, J. K., Zhou, L. H., Gao, L. M., Zhong, Z., Surface acoustic waves in solids of functionally graded materials IEEE TUFFC, submitted.

Wang, J., Lin, J. B., Li, Z., Two-dimensional analysis of the effect of an electrode layer on surface acoustic waves in a finite anisotropic plate Ultrasonics 2006 Vol 44(s1), e935-e939.

Ng, T. Y., He, X. Q., Liew, K. M., Finite element modeling of active control of functionally graded shells in frequency domain via piezoelectric sensors and actuators Computational Mechanics 2002 Vol 28, 1-9.

Han, X., Liu, G. R., Effects of SH waves in a functionally graded plate Mechanics Research Communications 2002 Vol 29, 327-38.

Chiu, T. C., Erdogan. F., One-dimensional wave propagation in functionally graded elastic medium Journal of Sound and Vibration 1999 Vol 222, 453-87.

Han, X., Xu, D., Liu, G. R., A computational inverse technique for material characterization of a functionally graded cylinder using a progressive neural network Neurocomputing 2003, Vol 51, 341-60.

Gao, L. M., Wang, J., Zhong, Z., Du, J. K., An analysis of surface acoustic wave propagation in a plate of functionally graded materials with a layered model Science in China (Ser. G) 2008, Vol 51, 165-75.

Gao, L. M., Wang, J., Zhong, Z., Du, J. K., An exact analysis of surface acoustic waves in a plate of functionally graded materials IEEE Trans. Ultrason., Ferroelect, and Freq. Contr. 2009, Vol 56, 2693-2700.

Gao, L. M., Wang, J., Zhong, Z., Du, J. K., An analysis of surface acoustic wave propagation in functionally graded plates with homotopy analysis method Acta Mechanica 2009, Vol 208, 249-58.

Auld, B. A., Acoustic Fields and Waves in Solids (second ed.), Krieger Publishing Company: Malabar, Florida, 1990.

Cakmak, A. S., Botha, J. F., Gray, W. G., Computational and Applied Mathematics for Engineering Analysis, Springer-Verlag: Berlin, 1987.

Weisstein, E.W., "Frobenius Method," MathWorld-A Wolfram Web Resource. http://mathworld.wolfram.com/FrobeniusMethod.html.

"Frobenius Method," Wikipedia-The Free Encyclopedia. http://en.wikipedia.org/wiki/Frobenius_method.

Martin, P. A., Berger, J. R., Waves in wood: free vibration of wooden pole. Journal of the Mechanics and Physics of Solids 2001, Vol 49, 1155-78.

Markus, S., Mead, D. J., Wave motion in three-layered orthotropic-isotropic-orthotropic, composite shell *Journal of Sound and Vibration* 1995, Vol 181, 149-67.

Lee, P. C. Y., Wang, J., Vibrations of AT-cut quartz strips of narrow width and finite length *Journal of Applied Physics* 1994, Vol 75, 7681-95.

Lee, P. C. Y., Wang, J., Thickness-shear and flexural vibrations of contoured crystal strip resonators *Journal of Applied Physics* 1996, Vol 79, 3403-10.

Caruntu, D. I., Self-adjoint differential equation for classical orthogonal polynomials Journal of Computational and Applied Mathematics 2005, Vol 180, 107–18.

Caruntu, D. I., Classical Jacobi polynomials, closed-form solutions for transverse vibrations *Journal of Sound and Vibration* 2007, Vol 306, 467-94.

Wang, J., Zhou, L. H., Du, J. K., Surface acoustic waves in an infinite plate of functionally graded materials Proceedings of the IEEE International Ultrasonics Symposium 2006, 2242-45.

Liao, S. J., Beyond Perturbation: Introduction to the Homotopy Analysis Method, Chapman and Hall/CRC: Boca Raton, 2003.

Liao, S. J., An explicit totally analytic approximation of Blasius's viscous flow problem *International Journal of Non-Linear Mechanics* 1999, Vol 34, 759-79.

Allan, F. M., Syam, M. I., On the analytic solution of non-homogeneous Blasius problem Journal of Computational and Applied Mathematics 2005, Vol 182, 362-71.

Hayat, T., Sajid, M., On analytic solution for thin film flow of a fourth grade fluid down a vertical cylinder Physics Letter A 2007, Vol 361, 316-22.

Sajid, M., Hayat, T., Non-similar series solution for boundary layer flow of a third-order fluid over a stretching sheet Applied Mathematics and Computation 2007, Vol 189, 1576-85.

Hayat, T., Shahead, F., Ayub, M., Analytic solution for steady flow of the third grade fluid in a porous half space Applied Mathematical Modelling 2007, Vol 31, 2424-32.

Abbasbandy, S., The application of Homotopy Analysis Method to nonlinear equations arising in heat transfer Physics Letter A 2006, Vol 360, 109-13.

Wang, J., Chen, J. K., Liao, S. J., An explicit solution of the large deformation of a Cantilever beam under point load at the free tip *Journal of Computational and Applied Mathematics* 2008, Vol 212, 320-30.

Gao, L. M., Wang, J., Zhong, Z., Du, J. K., An analysis of surface acoustic wave propagation in a plate of functionally graded materials with a layered model Science in China (Ser. G) 2007, Vol 37, 649-58.

Wu, X. H., Shen, Y. P., Chen, C. Q., An exact solution for functionally graded piezothermoelastic cylindrical shell as sensors and actuators Materials Letters 2003, Vol 57, 3532–42.

Prakash, T., Ganapathi, M., Asymmetric flexural vibration and thermoelastic stability of FGM circular plates using finite element method Composites: Part B 2006, Vol 37, 642–49.

Dai, K. Y., Liu, G. R., Han, X., Lim, K. M., Thermomechanical analysis of functionally graded material (FGM) plates using element-free Galerkin method Computers and Structures 2005, Vol 83, 1487–1502.

Zhang, N. H., Wang, M. L., A mathematical model of thermoviscoelastic FGM thin plates and Ritz approximate solutions Acta Mechanica 2006, Vol 181, 153–67.

Zhang, N. H., Wang, M. L., Thermoviscoelastic deformations of functionally graded thin plates *European Journal of Mechanics A/Solids* 2007 Vol 26, 872–86.

Nemat-Alla, M., Reduction of thermal stresses by developing two-dimensional functionally graded materials, International Journal of Solids and Structures 2003, Vol 40, 7339–56.

Chen, W. Q., Ding, H. J., Bending of functionally graded piezoelectric rectangular plates, Acta Mechanica Solida Sinica 2000, Vol 13, 312-19.

Mahapatra, D. R., Singhal, A., Gopalakrishnan, S., Lamb wave characteristics of thickness-graded piezoelectric IDT Ultrasonics 2005, Vol 43, 736–46.

Gao, L. M., Wang, J., Zhong, Z., Du, J. K., Propagation of surface waves in plate of functionally graded materials. In: Pizoelectricity, Acoustic Waves and Device Applications (Ji Wang and Weiqiu Chen, editors) 2007, World Scientific, 184-88.

Chakraborty, A., Gopalakrishnan, S. Wave propagation in inhomogeneous layered media: solution of forward and inverse problems. Acta Mechanica 2004, Vol 169, 153–85.

Chen, W. Q., Bian, Z. G., Lv, C. F., Ding, H. J., 3D free vibration analysis of a functionally graded piezoelectric hollow cylinder filled with compressible fluid *International Journal of Solids and Structures* 2004, Vol 41, 947–64.

Zenkour, A. M., On vibration of functionally graded plates according to a refined trigonometric plate theory *International Journal of Structural Stability and Dynamics* 2005, Vol 5, 279–97.

Zenkour, A. M., A comprehensive analysis of functionally graded sandwich plates: Part 1—Deflection and stresses *International Journal of Solids and Structures* 2005, Vol 42, 5224–42.

Zenkour, A. M., A comprehensive analysis of functionally graded sandwich plates: Part 1—Buckling and free vibration *International Journal of Solids and Structures* 2005, Vol 42, 5243–58.

Cao, X., Jin, F., Wang, Z., On dispersion relations of Rayleigh waves in a functionally graded piezoelectric material (FGPM) half-space Acta Mechanica 2008, Vol 200, 247-61.

Han, X., Liu, G. R., Lam, K. Y., A quadratic layer element for analyzing stress waves in FGMs and its application in material characterization *Journal of Sound and Vibration* 2000, Vol 236, 307-21.

Abramowitz, M., Stegun, I. A., Handbook of Mathematical Functions with Formulas, Graphs, and Mathematical Tables, Dover: New York, 1965.

Liu, G. R., Han, X., Lam, K. Y., An integration technique for evaluating confluent hypergeometric functions and its application to functionally graded materials Computers and Structures 2001, Vol 79, 1039-47.

Wu, X. H., Chen, C. Q., Shen, Y. P., Tian, X. G., A high order theory for functionally graded piezoelectric shells *International Journal of Solids and Structures* 2002, Vol 39, 5325–5344.

Yu, J. G., Wu, B., He, C. F., Characteristics of guided waves in graded spherical curved plates International Journal of Solids and Structures 2007, Vol 44, 3627–37.

Yu, J. G., Ma, Q. J., Circumferential wave in functionally graded piezoelectric cylindrical curved plates Acta Mech 2008, Vol 198, 171–90.

Wu, B., Yu, J. G., He, C. F., Wave propagation in non-homogeneous magneto-electro-elasitc plates *Journal of Sound and Vibration* 2008, Vol 317, 250-64.

Yu, J. G., Ma, Q. J., Su, S., Wave propagation in non-homogeneous magneto-electro-elasitc hollow cylinders Ultrasonics, in press.

Yu, J. G., Wu, B., Circumferential wave in magneto-electro-elasitc functionally graded cylindrical curved plates *European Journal of Mechanics A/Solids*, in press. DOI: 10.1016/j.euromechsol.2008.07.011

Zhang, V. Y., Lefebvre, J. E., Bruneel, C., Gryba, T., A unified formalism using effective surface permittivity to study acoustic waves in various anisotropic and piezoelectric multilayers *IEEE Transactions on Ultrasonics, Ferroelectrics, and Frequency Control* 2001, *Vol* 48, 1449-61.

Shaker, A., Abdelrahman, W., Tawfik,M., Sadek, E., Stochastic finite element analysis of the free vibration of functionally graded material plates *Comput. Mech.* 2008, *Vol* 41, 707–14.

Mindlin, R. D., High frequency vibrations of piezoelectric crystal plates *International Journal of Solids and Structures* 1972, *Vol* 8, 891-906.

Altay, G., Dökmeci, M. C., Variational principles and vibrations of a functionally graded plat *Computers and Structures* 2005, *Vol* 83, 1340–54.

ABOUT THE AUTHORS

Ji Wang received his B.S. in structural engineering in 1983 from Gansu University of Technology (currently Lanzhou University of Technology). From 1983 to 1988, he was a structural engineer at the 11th Institute of Project Planning and Research in Xi'an, Shaanxi, China. From 1988 to 1990, he was a visiting scientist at Argonne national Laboratory working on structural analysis. From 1990 to 1995, he was a graduate student at Princeton University studying high frequency vibrations of piezoelectric plates. He received his M.S. and PhD in civil engineering from Princeton in 1993 and 1996, respectively. He has been employed by Epson Palo Alto Laboratory from 1995 to 1999, NetFront Communications from 1999 to 2001, and SaRonix from 2001 to 2002. From 2002, he has been a Qian River Fellow Professor at Ningbo University, Ningbo, China and the founding director of the Piezoelectric Device Laboratory. He has been associated with the IEEE Ultrasonics, Ferroelectrics, and Frequency Control Society through Symposium Technical Program Committees and is the funding Chair of the Committee on Electromagnetic Devices of the Chinese Society of Theoretical and Applied Mechanics. His research has been on physical acoustic waves in piezoelectric resonators and computational methods.

In: Mechanics of Functionally Graded Materials and Structures ISBN 978-1-61668-920-9
Editors: Z. Zhong, L. Wu and W. Chen

Chapter 3

3-D AND SCALE-SPAN ANALYSES OF FUNCTIONALLY GRADED MATERIALS AND STRUCTURES BY MICROELEMENT METHOD

Cao Zhi Yuan[1], ***Cheng Hong Mei***[2], ***Tang Shou Gao***[3] ***and Wang Hua Ning***[4]

School of Aerospace Engineering and Applied Mechanics, Tongji University, Shanghai, China

ABSTRACT

A microelement method for the analyses of functionally graded structures was suggested. In order to reflect micro-constitution of material components, the microelement method sets up the dense microelements in the normal finite element of the structure, and translates the degrees of freedom of every microelement node to the degrees of freedom of the same normal finite element by the compatibility conditions, which will be calculated by the computer. This microelement method can fully realize the transition from material micro-constitution to macro- response of structure and its computational elements and degrees of freedom are same as the normal finite element method. Therefore, this method is an effective means for the scale-span analysis of the functionally graded structures. In this paper, the macro-response can be immediately calculated by the material component distribution determined in the process of material production; the influence of gap on the macro-responses can be analyzed, the scale-span analysis between micro-constitution and macro-response of the functionally graded materials and structures can be achieved; the three-dimensional distributions of mechanical quantities of the functionally graded plates with different complex shapes are given. Instead of material parameter input of traditional mechanical analysis, the mechanical analyses are based on the input of metallographic diagram information by microelement method, the mechanics responses of functionally graded plates are calculated and 3-D distributions of macro-mechanical variables and identical stress lines

[1] e-mail: 1zycao@tongji.edu.cn.
[2] hm2963_cn@163.com.
[3] tangsg@tongji.edu.cn
[4] wanghn@tongji.edu.cn

tendency on microstructure as well as 3-D distributions of macro natural mode shapes and identical stress mode shape lines tendency on microstructure are given.

1. INTRODUCTION

The functionally graded material (FGM) is a new composite material developed recently in which the material property graded distribution is formed by the composition of different material media in different ratios, so as to meet the requirements of material properties for different parts of a member. In the meanwhile, as each part of the material of structure varies in continuous way without boundary surface and sudden change in material properties, there is more advantage of the material properties than ordinary laminated and composed ones. Such kind of composite material composed non-uniformly and continuously in structure by different materials into desired way makes the development of material stride forward towards a higher level [1]. In recent years, more attention from international academic circles is paid to FGM and its structures [2], and the development of FGM in various regions has been made in succession by scientists of Japan, America, Russia, Germany, England, Switzerland, Finland, and et al.

Bing different from ordinary homogeneous members, the mechanical analysis of FGM members possesses a series of specific properties and characteristic which can be described as the following three aspects. (1) The parameters of FGM are related to spatial coordinates, and the correspondent dominant equations are of variable coefficient, and therefore it is very difficult to find analytic solution directly; (2) The macro properties of FGM members are related with not only the component material properties, but mainly microstructure as well. So, scale-span analyses are needed. It is presently a key research subject to calculate a member's macro response directly via its micro structure; (3) As the material properties of FGM member change in gradient style along spatial domain (especially across thickness), the mechanical variables are 3-D distribution in commonly used 1-D and 2-D members in which 3-D analysis is necessary.

The research work on FGM started as a specific functional material of aerospace engineering not long ago. Therefore, more attention of study was paid to thermal stresses, crack, optimizing design, etc [3-8]. There is still much work needed to do and many vulnerable spots existing in various fields of mechanics, even in static and dynamic analyses of ordinary structural members[9]. Presently, several methods, such as laminated model method [10], asymptotic approach [11], 3-D analysis method [12], simplified model technique [13], etc, are mainly applied to the analysis of FGM plate and shell structures. As the difficulty for mathematical analysis, exact analytic solutions can only be obtained for some specific problems with regular geometry under very limited kind of boundary conditions. For numerical solutions, such as the finite element method, the computation is extremely tedious and low efficient as the graded distribution of material properties inside the structure members requires each member to be divided by highly concentrated element net[14] [15].

Two different material components of FGM vary continuously in some direction along a functionally graded member. If one component is in dominant position, the other one is suspended in graininess in it. Net structure will be formed in the region where component of the materials is the same. Therefore, the microstructure of FGM is extremely complicated as a result of the difference of the components from point to point within the member[16]. The present analytic method in micromechanics is not valid for the scale-span analysis with such kind of microstructure [17-19] which has received wide attention [20].

In this paper, a microelement method is proposed for the science problems mentioned above. The 3-D and scale-span analysis of functionally graded member with complicated inner microstructure can be exactly realized. As the elements and the degrees of freedom are corresponding to those of the ordinary FEM, the computation is less and the analysis more efficient. Moreover, the metallographic diagrams of FGM can be directly taken as input data for solid mechanics analysis.

The gist of the method is a rational combination of two types of elements, that is, the member is firstly divided into 'macroelements' in the ordinary way of FEM, and then, each macroelement is further divided into a large amount of 'microelements' to describe precise FGM microstructure in it. The material properties can be different from microelement to microelement which itself is homogeneous and contains correspondent node degrees of freedom. The key point of the method is to utilize the deformation compatibility conditions between macroelement and microelement to transfer the large amount of the node degrees of freedom of microelements into the macroelement's node degrees of freedom without solving the equations. Such kind of combination, in which complicated microstructure is presented by microelements while computation is done on the macroelement, attains our goal mentioned above. The basic principle of microelement method will be discussed in the following.

2. Basic Principle of Microelement Method

2.1. Displacement Model of Macroelement

For various members and structures made of FGM with different kinds of microstructures, the ordinary FEM can be employed to divide them into finite number of elements. There is some kind of 1-D, 2-D or 3-D mechanical properties for each element which is named "macroelement". According to the FEM theory, the displacements within a macroelement can be expressed as

$$\{u\} = [N]\{\delta\} \tag{1a}$$

$$\{\delta\} = [\cdots\cdots u_i v_i w_i \cdots\cdots]^T \tag{1b}$$

where u_i, v_i, w_i and their derivations are node degree of freedom of the macroelement, and $[N]$ is the matrix of shape function which can be determined by ordinary FEM according to the mechanical properties of macroelement[21].

2.2. Microstructure and Mechanical Analysis of Microelement

The microstructure of each macroelement is composed of different materials, and its shape is complicated. Each composing part of the microstructure can usually be divided into a series of precise 3-D elements, named "microelement", which contains only one kind of material. The microstructures with various shapes can be obtained by assembling microelements of the same material in different position. FEM of micro-level can be applied to compatibility, equilibrium and micromechanics analyses for each microelement.

The displacement vector of each microelement can be expressed as

$$\{u^e\} = [N^e]\{\delta^e\} \tag{2a}$$

where u^e, N^e, δ^e are microelement's 3-D displacements, shape function and node degree of freedom respectively. The node degrees of freedom are composed of the microelement's node displacements in every direction and their corresponding derivatives, i.e.

$$\{\delta^e\} = [\cdots\cdots u_{kj}^e \cdots\cdots]^T \tag{2b}$$

where u_{kj}^e is the kth generalized displacement of the jth node of the microelement.

Being further analyzed by 3-D FEM, the corresponding strain, stress vectors, stiffness matrix, mass matrix, and load vector of the microelement are

$$\{\varepsilon^e\} = [B^e]\{\delta^e\} \tag{3a}$$

$$\{\sigma^e\} = [D^e]\{\varepsilon^e\} \tag{3b}$$

$$[K]^* = \int [B^e]^T [D^e][B^e] d\Omega \tag{3c}$$

$$[M]^* = \int [N^e]^T \rho^e [N^e] d\Omega \tag{3d}$$

$$\{F\}^* = \int [N^e]^T \{q\} d\Omega \tag{3e}$$

where, the meaning of $[B^e]$, $[D^e]$, ρ^e and $\{q\}$ is shown in reference [21].

2.3. The Operational Matrices Transfer for Microelements

Considering the compatibility and common operation between macroelement and microelements, the spatial coordinate value (inside macroelement) of the jth node of the microelement can be substituted into the macroelement's displacement expression (1). According to the definition of u_k , each u_{kj}^e could be obtained to form vector (2b). The node degree of freedom vector of the microelement in terms of the node degree of freedom $\{\delta\}$ of the macroelement is

$$\{\delta^e\} = [\cdots\cdots u_{kj}^e \cdots\cdots]^T = [A]\{\delta\} \tag{4}$$

Substituting formulation (4) into (2a) and (3a), the displacement vector and strain vector of microelement are expressed in terms of the node degree of freedom $\{\delta\}$ of the macroelement as

$$\{u^e\} = [N^e]\{\delta^e\} = [N^e][A]\{\delta\} = [\overline{N}]\{\delta\} \tag{5a}$$

$$\{\varepsilon^e\} = [B^e]\{\delta^e\} = [B^e][A]\{\delta\} = [\overline{B}]\{\delta\} \tag{5b}$$

The corresponding stiffness matrix of microelement is

$$[K]^e = \int[\overline{B}]^T[D^e][\overline{B}]d\Omega = \int[A]^T[B^e]^T[D^e][B^e][A]d\Omega = [A]^T[K]^*[A] \tag{6a}$$

Similarly, the mass matrix and load vector can be given as following

$$[M]^e = [A]^T[M]^*[A] \tag{6b}$$

$$\{F\}^e = [A]^T\{F\}^* \tag{6c}$$

where $[K]^*$, $[M]^*$ and $\{F\}^*$ are given by Eq. (3); and $[A]$ is expressed by Eq. (4). Eq. (6) shows that the operational matrix of a microelement has already been expressed by the node degree of freedom of the macroelement. The operation does not involve solving linear algebraic equations; it transforms a great amount of degrees of freedom of microelement to a small number of degrees of freedom of macroelement with only the product of matrices.

2.4. The Operational Matrices for Macroelements and Basic Formulae of Members

A macroelement has a very complicated microstructure, each part of which contains numerous microelements. No matter what kind of material or microstructure the microelement contains, $[K]^e$, $[M]^e$ and $\{F\}^e$ can always be obtained by formulae (6). As all these operational matrices correlate with the node degrees of freedom $\{\delta\}$ of the same macroelement, the operational matrices of the whole macroelement are simple superposition of the operational matrices of the microelements which all microstructures of the macroelement include, i.e.,

$$[K] = \sum_e [K]^e, \quad [M] = \sum_e [M]^e, \quad \{F\} = \sum_e \{F\}^e \tag{7}$$

Assembling these operational matrices according to the position of macroelement in the member (similar to ordinary FEM[21]), the basic computational formula of members and structures which contains microstructure materials can be established as

$$[M]\{\ddot{\delta}\} + [K]\{\delta\} = \{F\} \tag{8}$$

After having obtained the degrees of freedom $\{\delta\}$ for the members and the macroelements from Eq.(8), the detailed mechanical variables (i.e., micromechanics solutions) of each component part of microstructure can be calculated by the following formulae

$$\{u^e\} = [N^e][A]\{\delta\} \tag{9a}$$

$$\{\varepsilon^e\} = [B^e][A]\{\delta\} \tag{9b}$$

$$\{\sigma^e\} = [D^e][B^e][A]\{\delta\} \tag{9c}$$

Summing up Eq.(8), (7), (6), (4) and (3), it can be found that Eq. (8) reflects the microelement's geometric and material parameters (Eq.(3)) and its relative position (Eq.(4)). So, the solution of expression (8) , i.e., the macro response of a member, can be related directly to material microstructure to realize macro-micro scale-span analysis.

Besides, expression (8) shows that calculation on computer is based only on the degrees of freedom of the macroelements (no matter how many microelements are divided in a macroelement). While Eq. (9) shows that the solutions are detailed mechanical quantity $\{u^e\}$, $\{\varepsilon^e\}$ and $\{\sigma^e\}$ in microstructure. So, the microelement method is a practical and efficient approach to the analysis of new materials (with microstructures) and corresponding structures.

3. Numerical Check of Microelement Method

Considering a homogeneous square plate, simply-supported in four sides, the span of plate is $a = b = 1m$, Young's modulus $E = 5GPa$, and Poisson ratio $\mu = 0.3$, the uniform load on the above surface of plate is $q = 1000N/m^2$.The whole plate was divided to 20×20 macroelements, there are 5000 microelements in every macroelement. The curve of dimensionless maximum deflection $w_m^* = \frac{D}{qa^4} w\big|_{\substack{x=a/2\\ y=b/2}}$ $(D = \frac{Eh^3}{12(1-\mu^2)})$ versus thickness-span ratio $\frac{h}{a}$ calculated by the microelement method are given in Figure1, In which the thin plate theory[22], the thick plate theory[23] and the three-dimensional FEM (ANSYS) solutions are also be listed. Obviously, the accuracy of the microelement method is higher, and the three-dimensional characteristic of its solution can also be reflected.

4. Three-Dimensional Analyses of Functionally Graded Plates with Different Component Distributions

4.1. Implementing the Spatial Distribution of Material Components by the Microelement Method

Only the spatial distribution of the different material components can be controlled in the course of material production. In this microelement method, the macroresponse of FGM structures can be immediately calculated by the known material components, and different microelements in the course of calculation possess different material characters, which are determined by the local value of component. Therefore, the three-dimensional analysis of the functionally graded material structures with different component distributions can be implemented by the microelement method.

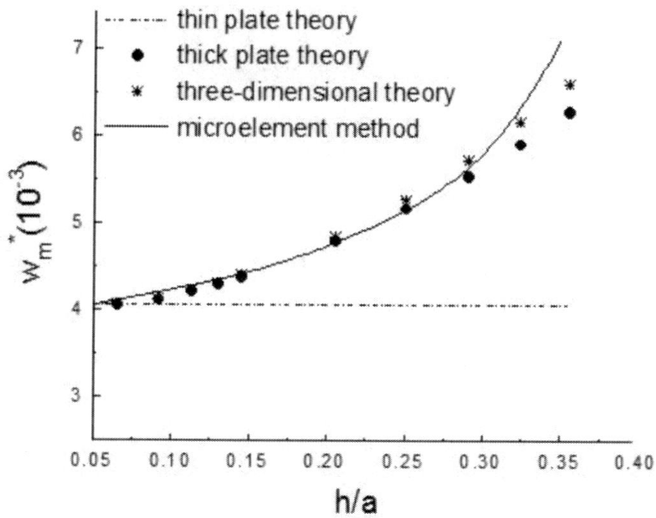

Figure1. Dimensionless deflection versus thickness-span ratio.

4.2. Different Component Functions Composed of Two Materials

Considering a simply-supported rectangular plate, the length of side $a = 1m$ and $b = 0.6m$, the thickness of plate $h = 0.1m$, the uniform load on the above surface of plate $q = 1000N/m^2$。 The plate is composed of the follow materials with the graded component: the material parameters of the metal (A) are $E_A = 150GPa$ and $\mu_A = 0.3$; the material parameters of the ceramics (B) are $E_B = 350GPa$ and $\mu_B = 0.33$, and the material parameters of the gap (O) are zero. Considering the following several component functions (η is the component, namely, it is the percentage of this material): $\eta_B(z) = 0.2(\frac{z}{h})^{\frac{1}{2}}$, $\eta_B(z) = 0.2(\frac{z}{h})$, $\eta_B(z) = 0.2(\frac{z}{h})^3$ and $\eta_o(z) = 0.001$, $\eta_A(z) = 1-(\eta_B + \eta_o)$, the plates with the above component functions are calculated by the microelement method respectively, and the distribution curves along the thickness direction of the dimensionless maximum of u, w, σ_x, σ_{xy}, σ_{xz} and σ_z are given in Figure 2 to Figure 7 respectively. The results indicate that the macro-response analysis of structures may be immediately calculated by the given material component distribution obtained from the production of functionally graded material. The difficulties and the errors of the determination and identification of functionally graded function [20] can be avoided by this microelement method.

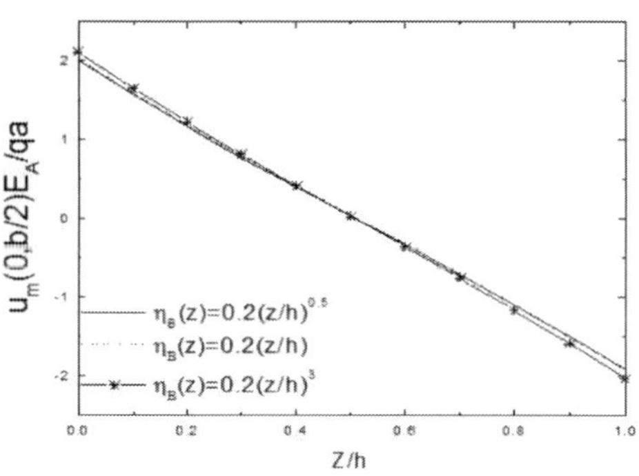

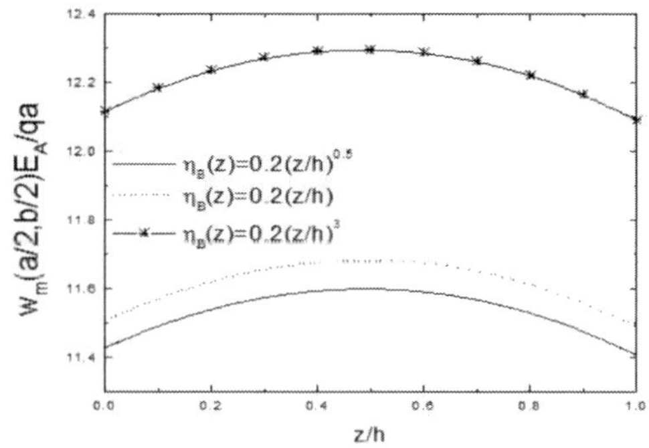

Figure 2. u_m distribution along the thickness direction.

Figure 3. w_m distribution along the thickness direction.

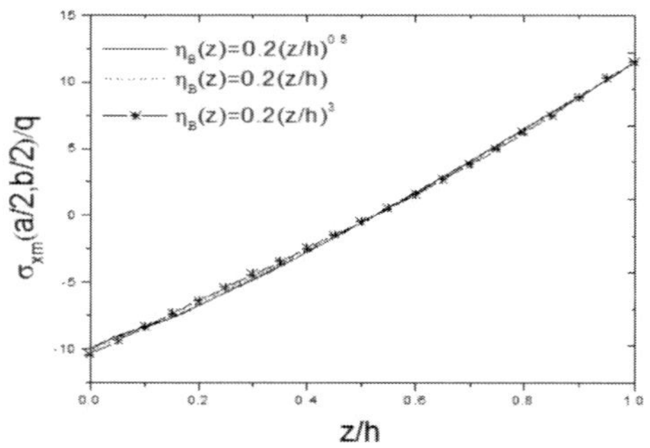

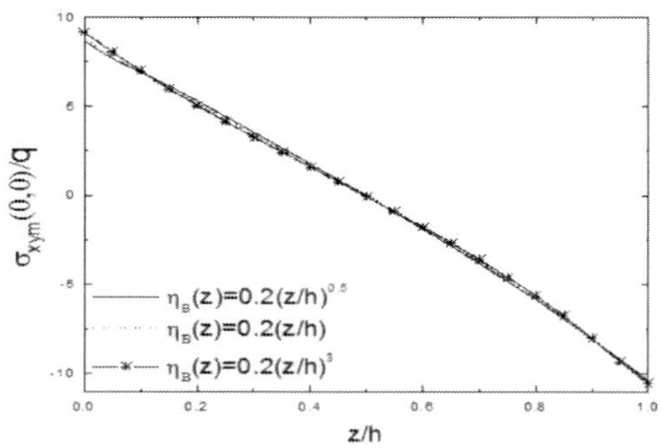

Figure 4. σ_{xm} distribution along the thickness direction.

Figure 5. σ_{xym} distribution along the thickness direction.

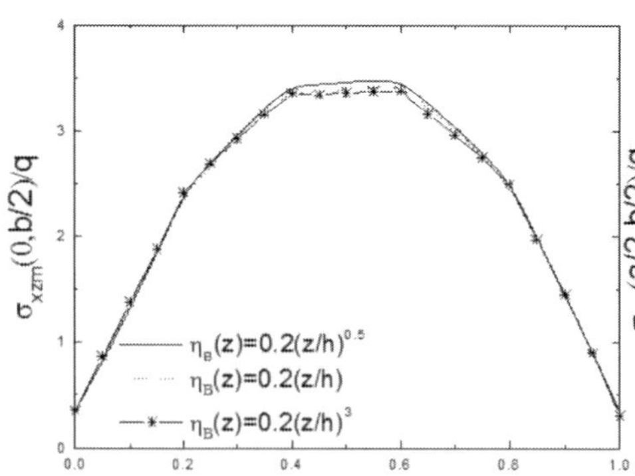

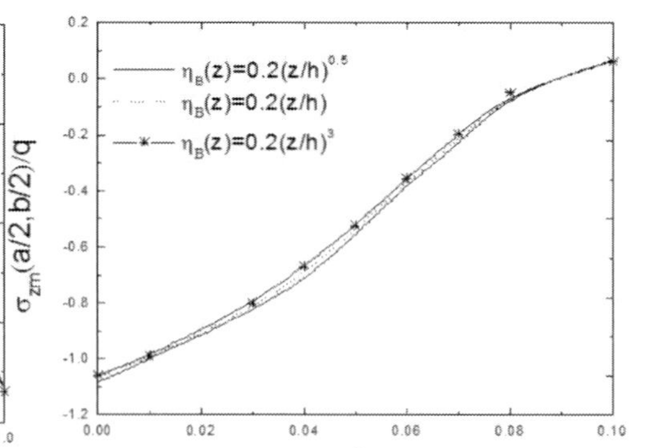

Figure 6. σ_{xzm} distribution along the thickness direction.

Figure 7. σ_{zm} distribution along the thickness direction.

Besides, the distribution shapes of mechanical quantities of the functionally graded plates can be influenced by the different component functions with same value on two surfaces of plate.

4.3. Different Component Values Composed of Two Materials

The plate and the material parameters are the same as example 4.2, while the component functions have the same distribution function but different component values as follows: $\eta_B(z)=0.4(\frac{z}{h})$, $\eta_B(z)=0.6(\frac{z}{h})$ and $\eta_B(z)=0.999(\frac{z}{h})$. The plates with the above component values are calculated by the microelement method respectively, and the distribution curves along thickness direction of the dimensionless maximum of u, w, σ_x, σ_{xy}, σ_{xz} and σ_z are given in Figure 8 to Figure 13 respectively.

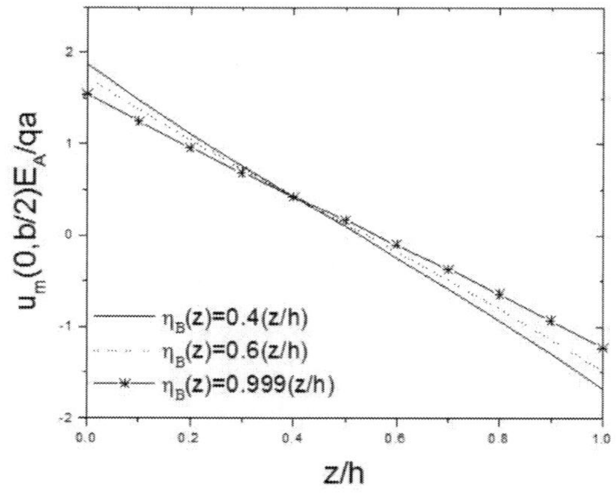

Figure 8. u_m distribution along the thickness direction.

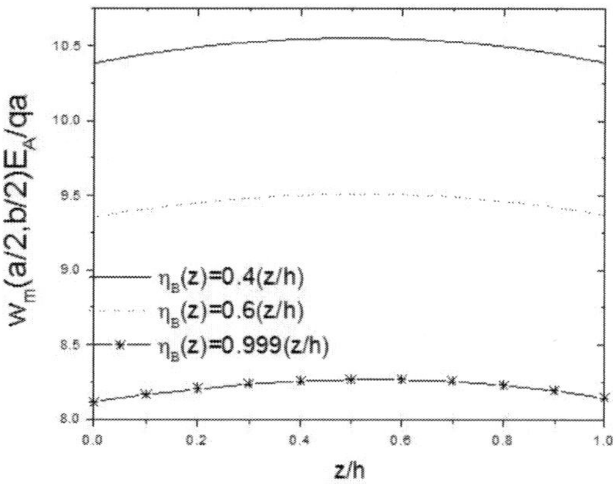

Figure 9. w_m distribution along the thickness direction.

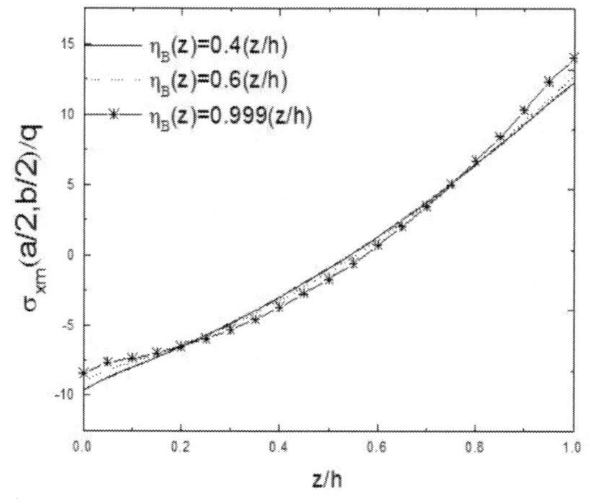

Figure10. σ_{xm} distribution along the thickness direction.

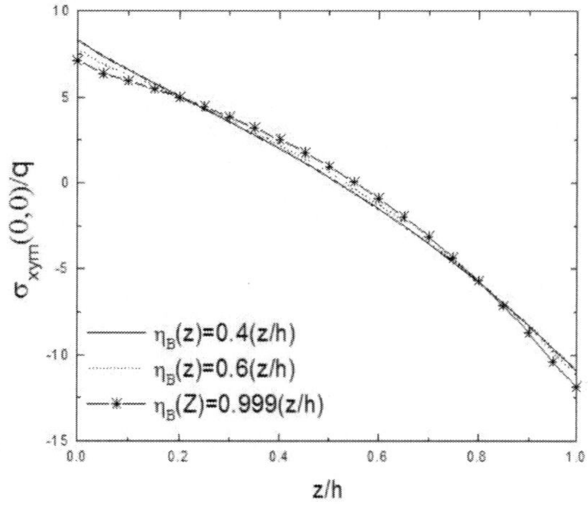

Figure 11. σ_{xym} distribution along the thickness direction.

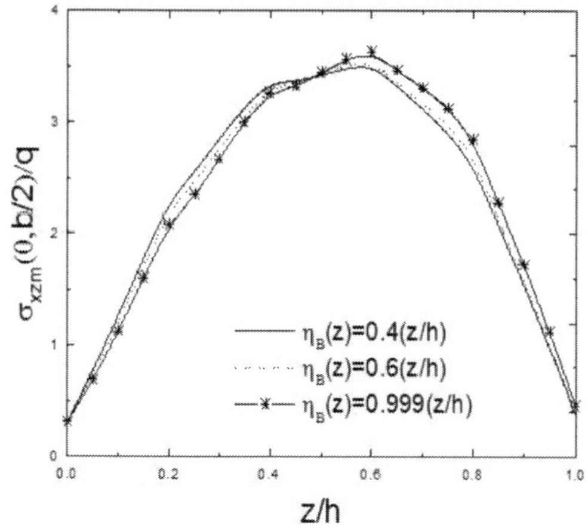

Figure12. σ_{xzm} distribution along the thickness direction.

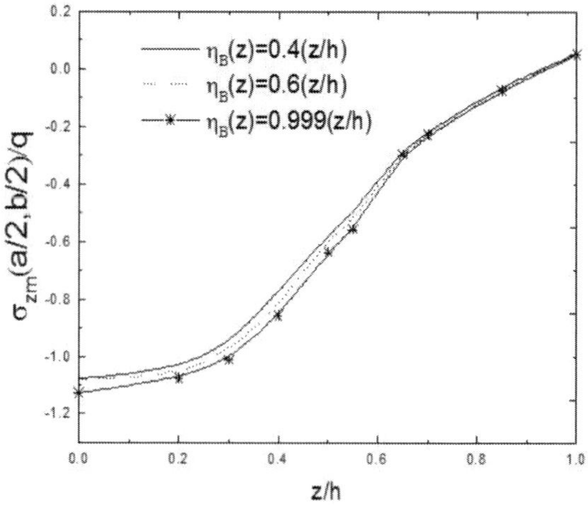

Figure 13. σ_{zm} distribution along the thickness direction .

The distribution shapes of mechanical quantities of the functionally graded plates can be greatly influenced by the different component values. With the increasing of the component of another ceramic material, the displacement u is never again antisymmetrical relative to the middle plane of plate; the normal deflection w still keeps being uniform along the thickness direction. the bending stress σ_x and the shear stress σ_{xy} don't maintain linear distribution along the thickness direction, the σ_{xz} along the thickness direction is also not symmetrical relative to the middle plane of plate; The stress σ_z can be described by cubic curve yet, but the component of cubic curve is slightly changed. The three-dimensional characteristic of the functionally graded plate [24] can be fully reflected, which deviates from the homogeneous thin plate theory.

4.4. Different Component Values Composed of Three Materials

The plate and the material parameters are the same as the above, but the parametersof another carbide (C) are $E_c = 300GPa$ and $\mu_c = 0.32$. The functionally graded plate is composed of three materials, where $\eta_B(z) = 0.2(\frac{z}{h})$, $\eta_C(z) = 0.1(\frac{z}{h})$, $0.2(\frac{z}{h})$, $0.3(\frac{z}{h})$, $\eta_A(z) = 1-(\eta_B+\eta_C)$. The plates with the above component values are calculated by the microelement method respectively, and the distribution curves along thickness direction of the dimensionless maximum of u, w, σ_x, σ_{xy}, σ_{xz} and σ_z re given in Figure 14 to Figure 19 respectively.

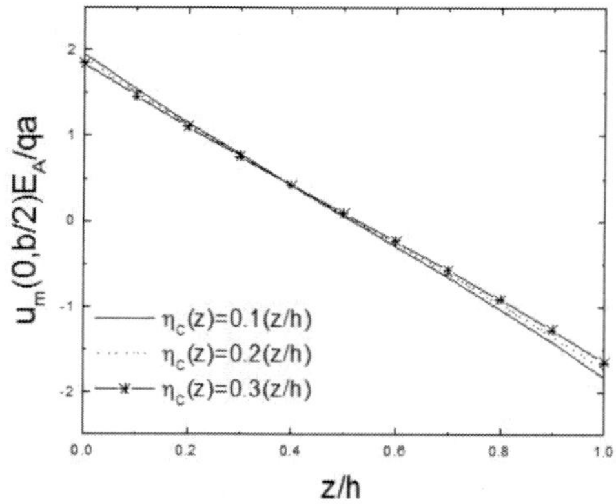

Figure14. u_m distribution along the thickness direction.

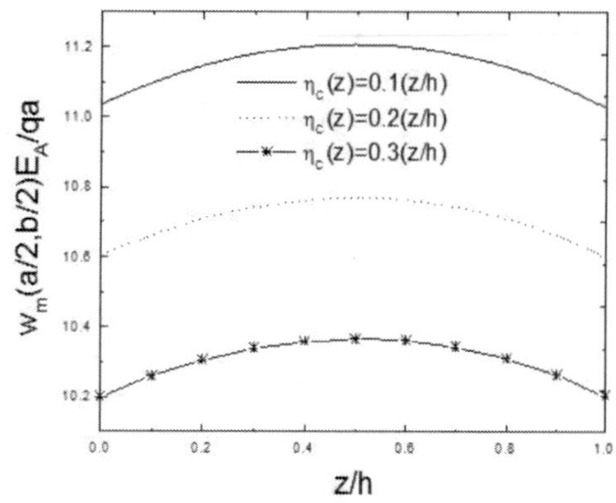

Figure 15. w_m distribution along the thickness direction.

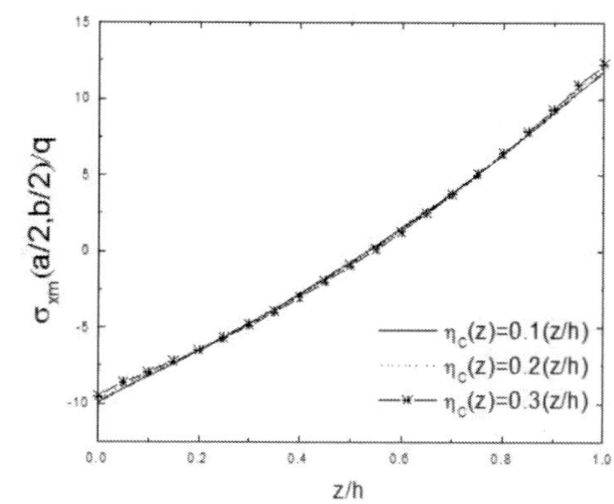

Figure16. σ_{xm} distribution along the thickness direction .

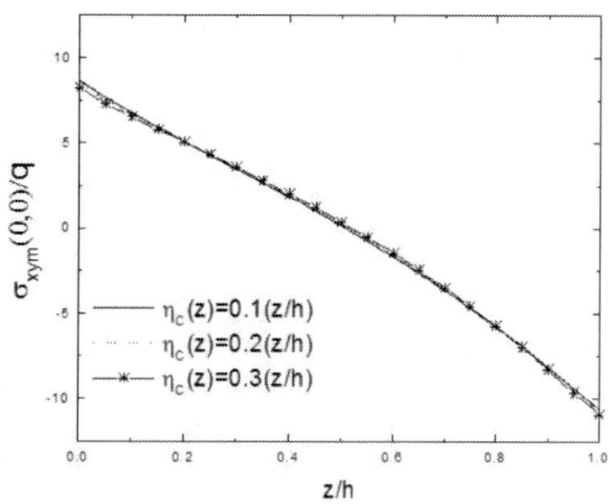

Figure 17. σ_{xym} distribution along the thickness direction.

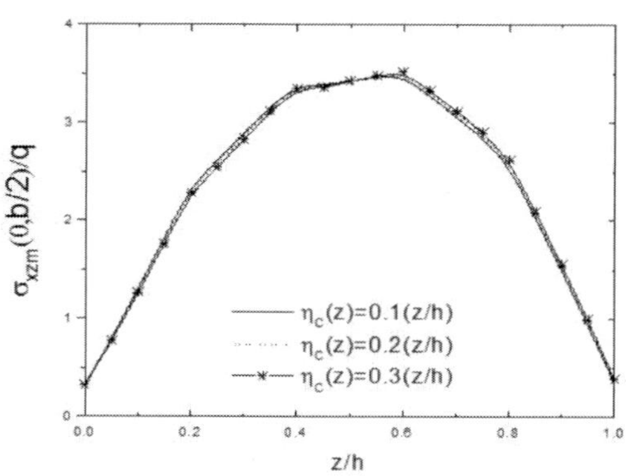

Figure18. σ_{xzm} distribution along the thickness direction.

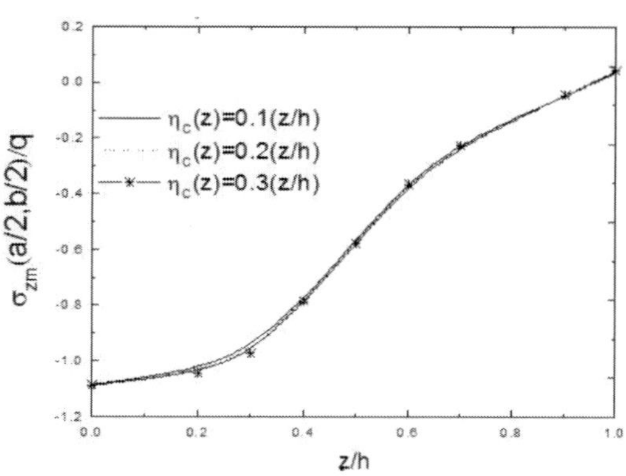

Figure 19. σ_{zm} distribution along the thickness direction.

For the functionally graded plates composed of three materials with the different component values, the macro-responses are similar as the example 4.3 , but the amplitude differences are very small, because of very limited components of the third material (carbide).

5. Three-Dimensional Analysis for Functionally Graded Plate with Complex Shape

5.1. Functionally Graded Plates with Complex Shape

The simply-supported functionally graded plates with complex shape include: (1) triangular functionally graded plate (Figure 20), (2) parallelogram functionally graded plate (Figure 21), (3) trapezoidal functionally graded plate (Figure 22), (4) semicircular functionally graded plate (Figure23). The thickness of plate $h = 0.1m$, the distributions of the different material components are $\eta_B(z) = 0.2(\frac{z}{h})$ and $\eta_A(z) = 1 - \eta_B$. The uniform load on the above surface of plate is $q = 1000N/m^2$, the material parameters of the metal (A) are $E_A = 150GPa$ and $\mu_A = 0.3$; the material parameters of the ceramics (B) are $E_B = 350GPa$ and $\mu_B = 0.33$.

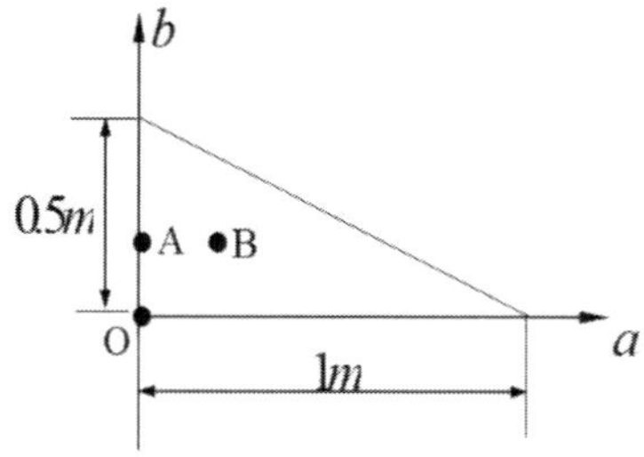

Figure 20. triangular functionally graded plate.

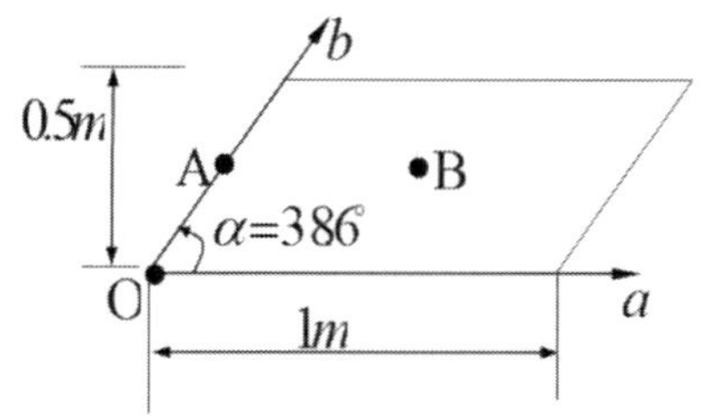

Figure 21. parallelogram functionally graded plate.

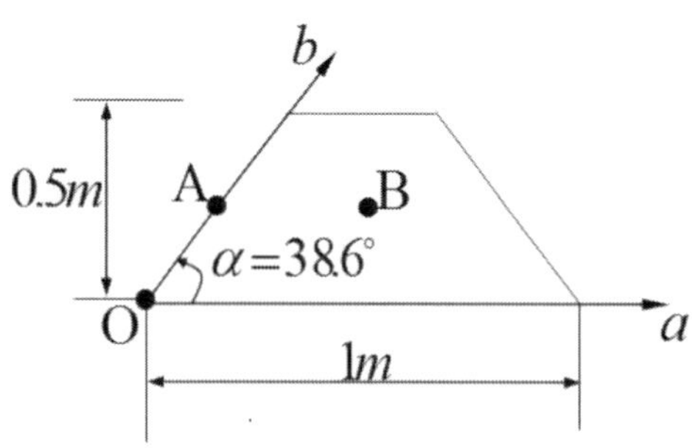

Figure 22. trapezoidal functionally graded plate.

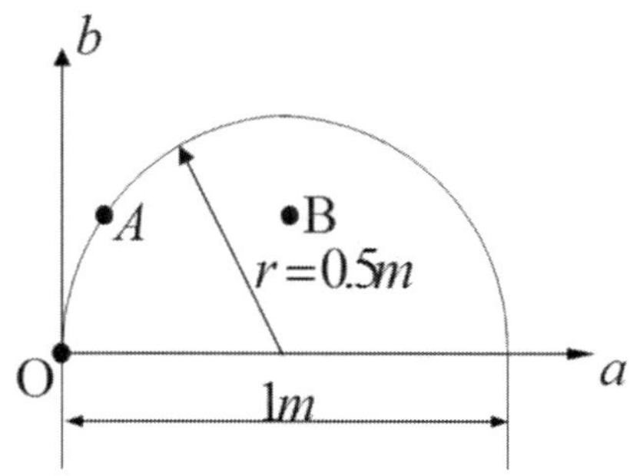

Figure 23. semicircular functionally graded plate.

5.2. Distribution Law of Mechanical Responses along the Thickness Direction of Triangular Functionally Graded Plate

The triangular functionally graded plate with the above component functions is calculated by the microelement method, and the distribution curves along the thickness direction of the dimensionless values of u, w, σ_x, σ_{xy}, σ_{xz} and σ_z are given in Figure24 to Figure29 respectively.

The law of the flat plate theory can be fully reflected by the macro-response of functionally graded plate, which calculated by the microelement method. The normal deflection w keeps being uniform along the thickness direction; the u, σ_x, and σ_{xy} along the thickness direction are antisymmetrical relative to the middle plane of plate ; The σ_{xz} is zero in each surface of plate, but its value is maximum in the middle plane of plate; the σ_z is external load in the loading surface, and is zero in another surface. These conclusions indicate that the microelement method is reliable and effective to solve the problems of functionally graded plate with complex shape.

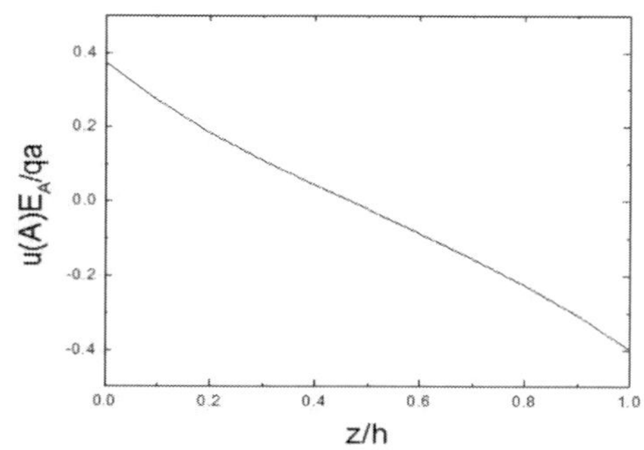

Figure 24. u distribution along the thickness direction.

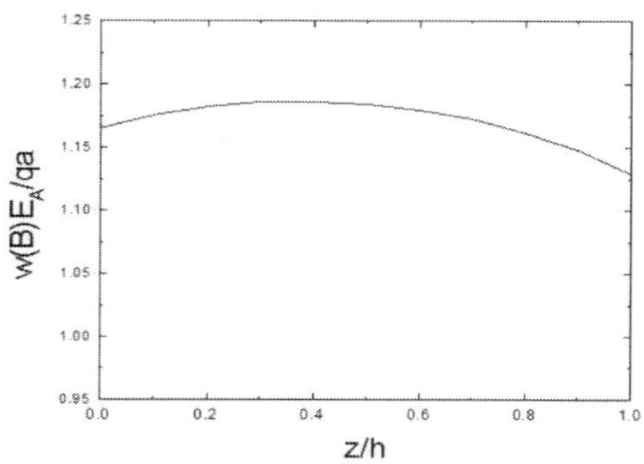

Figure 25. w distribution along the thickness direction.

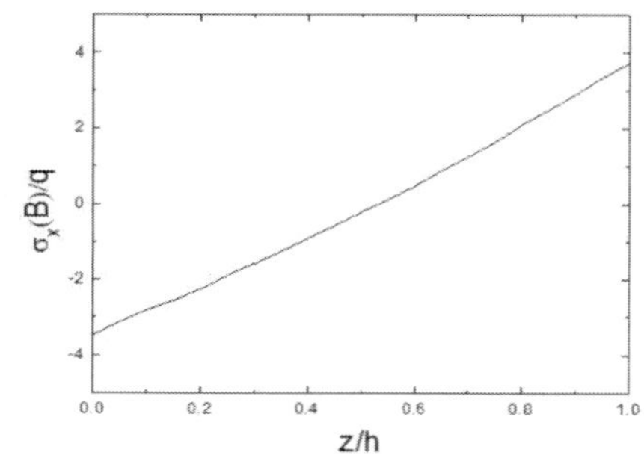

Figure 26. σ_x distribution along the thickness direction.

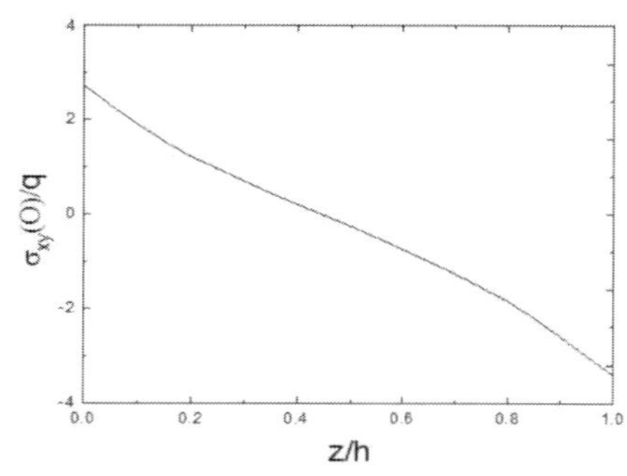

Figure 27. σ_{xy} distribution along the thickness direction.

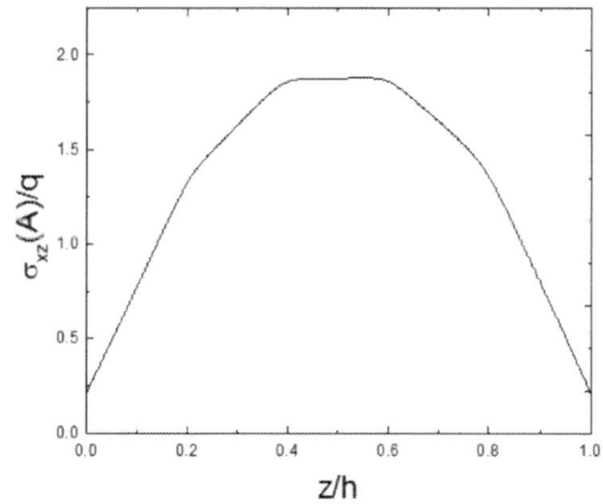

Figure 28. σ_{xz} distribution along the thickness direction.

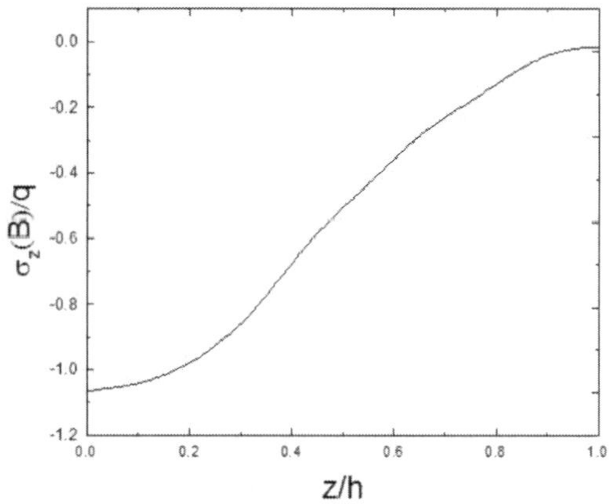

Figure 29. σ_z distribution along the thickness direction.

5.3. Mechanical Responses of Functionally Graded Plate with Different Shape

The plates and the material parameters are the same as the above, The plates with four complex shapes are calculated by the microelement method respectively, and the dimensionless values of u, w, σ_x, σ_{xy}, σ_{xz} and σ_z are given in Tab.1 to Tab.6 respectively.

Table 1. u distribution along the thickness direction

$u(A)E_A/qa$	$z/h=0$	$z/h=0.5$	$z/h=1$
Triangular plate（Figure20)	0.3710	-0.0225	-0.3969
Parallelogram plate（Figure21)	0.2346	0.0005	-0.2137
Trapezoidal plate（Figure22)	0.1790	-0.0019	-0.1628
Semicircular plate（Figure23)	0.1344	-0.0092	-0.1348

Table 2. w distribution along the thickness direction

$w(B)E_A/qa$	$z/h=0$	$z/h=0.5$	$z/h=1$
Triangular plate（Figure20)	1.1653	1.1843	1.1297
Parallelogram plate（Figure21)	4.4404	4.5287	4.4164
Trapezoidal plate（Figure22)	1.3419	1.3615	1.3089
Semicircular plate（Figure23)	1.4485	1.4671	1.4161

Table 3. σ_x distribution along the thickness direction

$\sigma_x(B)/q$	$z/h=0$	$z/h=0.5$	$z/h=1$
Triangular plate (Figure20)	-3.4854	-0.2263	3.7631
Parallelogram plate (Figure21)	-6.6050	-0.3982	7.3191
Trapezoidal plate (Figure22)	-3.8624	-0.2877	4.1164
Semicircular plate (Figure23)	-2.9009	-0.2842	2.9537

Table 4. σ_{xy} distribution along the thickness direction

$\sigma_{xy}(O)/q$	$z/h=0$	$z/h=0.5$	$z/h=1$
Triangular plate (Figure20)	2.7400	-0.2550	-3.4126
Parallelogram plate (Figure21)	0.3942	-0.1252	-0.3842
Trapezoidal plate (Figure22)	0.3934	-0.1201	-0.3375
Semicircular plate (Figure23)	0.8915	-0.2646	-1.1596

Table 5. σ_{xz} distribution along the thickness direction

$\sigma_{xz}(A)/q$	$z/h=0$	$z/h=0.5$	$z/h=1$
Triangular plate (Figure20)	0.2161	1.8754	0.2095
Parallelogram plate (Figure21)	0.0534	1.1595	-0.0086
Trapezoidal plate (Figure22)	0.0503	0.84127	-0.0095
Semicircular plate (Figure23)	-0.0277	0.6480	-0.0905

Table 6. σ_z distribution along the thickness direction

$\sigma_z(B)/q$	$z/h=0$	$z/h=0.5$	$z/h=1$
Triangular plate (Figure20)	-1.0645	-0.5109	0.0190
Parallelogram plate (Figure21)	-1.0684	-0.4614	0.0851
Trapezoidal plate (Figure22)	-1.0470	-0.4699	0.1222
Semicircular plate (Figure23)	-1.0478	-0.4697	0.0372

The difference of the macro-responses of functionally graded plates with different shapes is obvious. The three-dimensional characteristic [12] of the functionally graded plates with different shapes can be fully reflected, which deviates from the homogeneous thin plate theory.

6. Three-Dimensional Macro-Response Analyses of Functionally Graded Plates with Different Positions and Distributions of Gap

6.1. Realizing the Existence of Gap with Arbitrary Position and Distribution in the Functionally Graded Plate by the Microelement Method

Mixing air may occurr in the course of the production of functionally graded materials. The air content at some position may increase on the special condition. In order to further study the influence of internal gap of functionally graded materials on the macro-response of whole plate, the microelement method will be used to calculate this influence in this paper. The corresponding materials determined by the component are taken in different microelements. The existence and distribution of gap can be reflected by the disappearance of microelement material at the position of gap, and the three-dimensional analyses of functionally graded structures with gap can be implemented by the microelement method.

6.2. Uniform Gap with Different Content

Considering a simply-supported rectangular plate, the length of side $a = 1m$ and $b = 0.6m$ respectively, the thickness of the plate $h = 0.1m$, the uniform load on the above surface of the plate is $q = 1000N/m^2$. The plate was composed by the following materials: the material parameters of the metal (A) are $E_A = 150GPa$ and $\mu_A = 0.3$; the material parameters of the ceramics (B) are $E_B = 350GPa$ and $\mu_B = 0.33$. Consider the following values of the gap(O) (η is the component, namely, it is the percentage content of this material): $\eta_0(z) = 0$, 0.02 and 0.04 respectively, $\eta_B(z) = 0.2(\frac{z}{h})$, $\eta_A(z) = 1-(\eta_B + \eta_0)$. The plates with the above three gap values were calculated by the microelement method respectively, and the dimensionless maximums of u, w, σ_x, σ_{xy}, σ_{xz} and σ_z are given in Tab.7 respectively. The results indicate that the macro-responses of structures may be immediately calculated by the given material component distribution and the gap content from the production of functionally graded materials. The difficulties and the errors of the determination and identification of functionally graded function [20] can be avoided by this microelement method. The influence of existence and content of the gap on the macro-response of functionally graded structures can be discovered in Tab.7. The microelement method is an effective means to identify this influence, even if the influence is very tiny.

Table.7 Macro-responses of functionally graded plates with different contents of uniform gap

$\eta_0(z)$	$\frac{u_m(0,\frac{b}{2},0)E_A}{qa}$	$\frac{w_m(\frac{a}{2},\frac{b}{2},\frac{h}{2})E_A}{qa}$	$\frac{\sigma_{xm}(\frac{a}{2},\frac{b}{2},h)}{q}$	$\frac{\sigma_{xym}(0,0,h)}{q}$	$\frac{\sigma_{xzm}(0,\frac{b}{2},\frac{h}{2})}{q}$	$\frac{\sigma_{zm}(\frac{a}{2},\frac{b}{2},0)}{q}$
0	2.0098	11.677	11.506	10.431	3.413	-1.086
0.02	2.064	11.970	11.452	10.566	3.429	-1.101
0.04	2.117	12.255	11.382	10.681	3.415	-1.043

6.3. Non-Uniform Gap with Different Distribution Function

The plate and the material parameters are the same as example 6.2, but the gap is non-uniform, and its distribution functions are $\eta_0(z) = 0.04$, $\eta_0(z) = 0.04(\frac{z}{h})$ and $\eta_0(z) = 0.04(\frac{z}{h})^2$ respectively. The plates with the above three distribution values were calculated by the microelement method respectively, and the dimensionless maximums of u, w, σ_x, σ_{xy}, σ_{xz} and σ_z are given in Tab.8.

Table 8. Macro-responses of functionally graded plates with different distribution functions of non-uniform gap

$\eta_0(z)$	$\frac{u_m(0,\frac{b}{2},0)E_A}{qa}$	$\frac{w_m(\frac{a}{2},\frac{b}{2},\frac{h}{2})E_A}{qa}$	$\frac{\sigma_{xm}(\frac{a}{2},\frac{b}{2},h)}{q}$	$\frac{\sigma_{xym}(0,0,h)}{q}$	$\frac{\sigma_{xzm}(0,\frac{b}{2},\frac{h}{2})}{q}$	$\frac{\sigma_{zm}(\frac{a}{2},\frac{b}{2},0)}{q}$
0.04	2.117	12.255	11.382	10.681	3.415	-1.043
$0.04(z/h)$	2.052	11.996	11.265	10.554	3.421	-1.059
$0.04(z/h)^2$	2.033	11.878	11.236	10.481	3.418	-1.076

Owing to the increasing of values of non-uniform gap from the zero on the above surface of plate to the 0.04 on the below surface, the total amount of gap is smaller than the amount of the plate with uniform gap 0.04, therefore, the displacement decreasing and the stress redistribution of non-uniform gap plates can be discovered in Tab.8.

6.4. Non-Uniform Gap Centralized in Thin Layer

The plate and the material parameters are the same as example 6.3.The origional value of the gap is uniform and $\eta_0(z) = 0.04$, but the value of the gap in the thin layer $z = 0.04 \sim 0.06(\mathrm{m})$is abruptly increased by 0.1 due to the production accident, namely the gap distribution is $\eta_0(z) = 0.04 + 0.1\big|_{z=0.04\sim0.06(m)}$. The plate with the above gap value was calculated by the microelement method, and the curves along thickness direction of the dimensionless maximums of u, w, σ_x, σ_{xy}, σ_{xz} and σ_z are given in Figure30 to Figure35 respectively. In order to compare, the results of the uniform gap $\eta_0(z) = 0.04$ plate are also listed here.

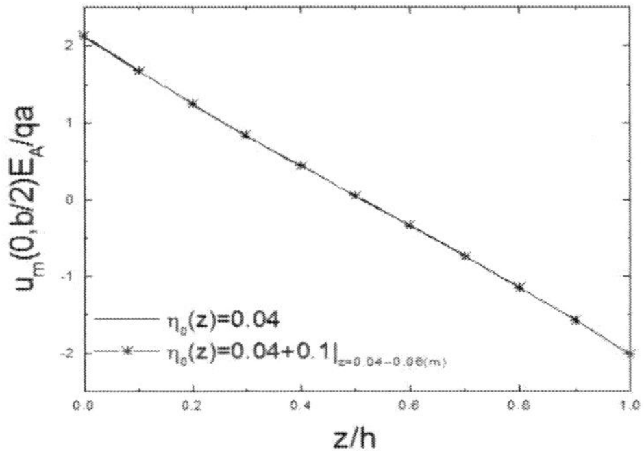

Figure 30. u distribution along the thickness direction.

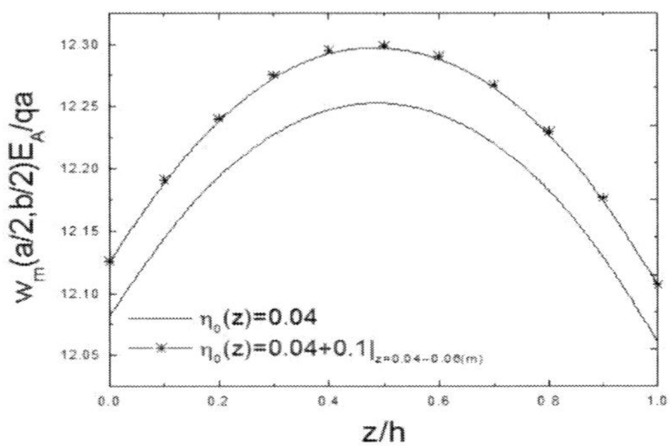

Figure 31. w distribution along the thickness direction.

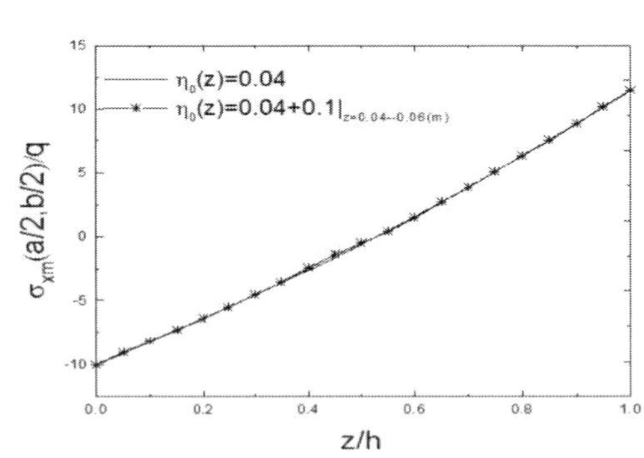

Figure 32. σ_x distribution along the thickness direction.

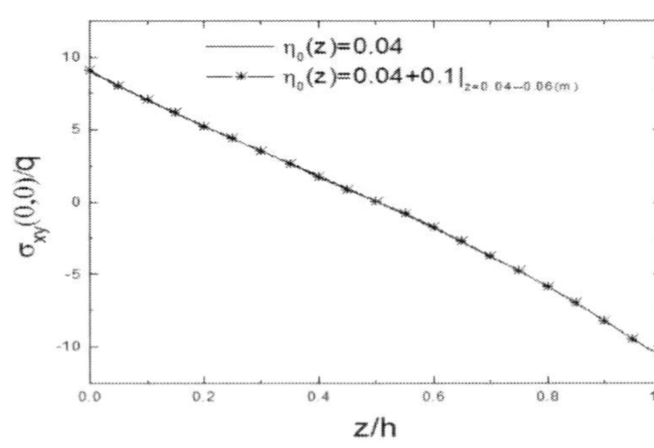

Figure 33. σ_{xy} distribution along the thickness direction.

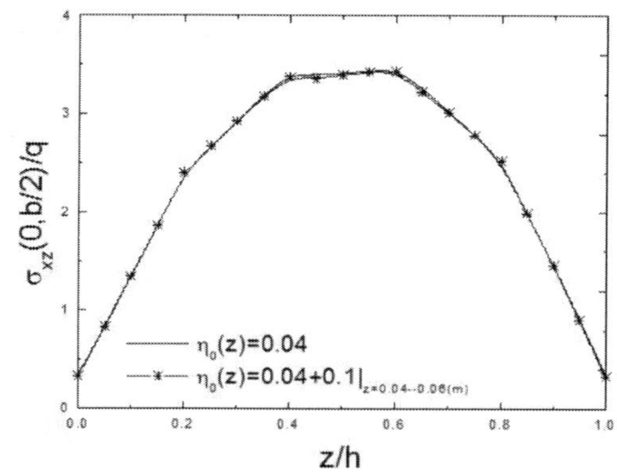

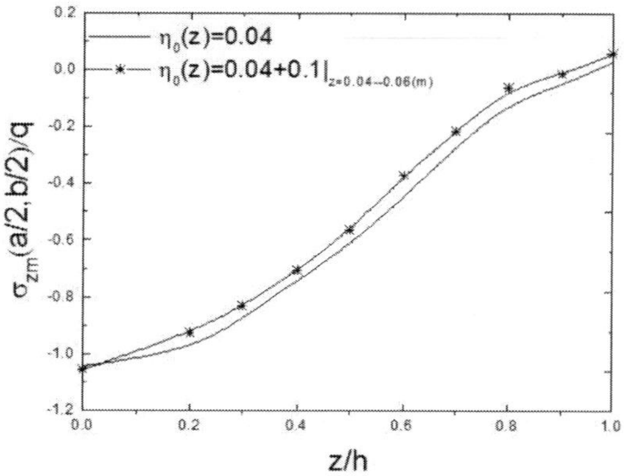

Figure 34. σ_{xz} distribution along the thickness direction. Figure 35. σ_z distribution along the thickness direction.

As a result from the above, the tiny change of micro-constitution can be fully reflected by the macro-responses calculated by the microelement method. The accuracy of microelement method can be further demonstrated.

6.5. Non-Uniform Gap Centralized in Local Zone

The plate and the material parameters are the same as example 6.2. The value of gap generally is $\eta_0 = 0.04$, but the gap component is $\eta_0(z) = 0.04 + 0.1 |_{\substack{x=0.4\sim0.6(m) \\ y=0.4\sim0.6(m)}}$ in the middle zone $0.2m \times 0.2m$ of functionally graded plate. In order to compare, The plates with uniform gap $\eta_0(z) = 0.04$ and with non-uniform gap centralized in local zone were calculated by the microelement method respectively, and the dimensionless maximums of u, w, σ_x, σ_{xy}, σ_{xz} and σ_z are given in Tab.9.

Table.9. Macro-responses of functionally graded plate with non-uniform gap centralized in local zone

$\eta_0(z)$	$\frac{u_m(0,\frac{b}{2},0)E_A}{qa}$	$\frac{w_m(\frac{a}{2},\frac{b}{2},\frac{h}{2})E_A}{qa}$	$\frac{\sigma_{xm}(\frac{a}{2},\frac{b}{2},h)}{q}$	$\frac{\sigma_{xym}(0,0,h)}{q}$	$\frac{\sigma_{xzm}(0,\frac{b}{2},\frac{h}{2})}{q}$	$\frac{\sigma_{zm}(\frac{a}{2},\frac{b}{2},0)}{q}$
0.04	2.117	12.255	11.382	10.681	3.415	-1.043
$\eta_0(z) = 0.04 + 0.1 \|_{\substack{x=0.4\sim0.6(m) \\ y=0.4\sim0.6(m)}}$	2.138	12.650	10.925	10.743	3.400	-0.988

From the above table, there is the apparent influence of the gap centralized in local zone of the plate on the macro-responses of functionally graded structure. There is further increase of the displacements relative to the original plate, the redistributions of stresses are also greater than the above examples.

7. Static Scale-Span Analysis of Functionally Graded Plate based on Metallographic Diagrams

In the development of solid mechanics of about two centuries, the mechanical responses of a structure are usually obtained under given geometry, material properties and external action. With the appearance of different new materials with microstructure and the development of computer science, the research work on the solution of mechanical macro response under the input based on metallographic diagrams of the material, as one of the development goals of 21st century by The American Society of Mechanics[25], has received attention from international academic circles.

7.1. Realization of FGM Metallographic Diagrams as Input Data by Microelement Method

As mentioned in Introduction, for FGM plate structure, the material components and correspondent micostructure vary in graded way across thickness (Figure 36(a)). One material near the two surfaces of a plate is in dominant position, while the other is relatively less and is suspends in graininess in the former part. The microstructure is shown in Figure 36(b) and Figure 36(d). Near the middle plane of the plate the two materials are close in component, with the microstructure similar to net structure (see Fig 36(c))[16]. Whatever microstructure, suspension, net or complicated one in Figure 36(a) can always be expressed and input by the microelement method. The process is as follows: the macroelement metallographic diagrams are first divided into a set of concentrated 3-D solid microelements, the scale of which equals or is less than the least scale of microstructure. The property of the microelement is determined by the property of the material at the location of the microelement after the metallographic structure is divided. Inputting the material properties, geometry and coordinate parameters of node degrees of freedom of each microelement by ordinary 3-D FEM completes the input of FGM metallographic diagrams. Finally microelement method program is used to finish the numerical computation for macro-mechanical responses of members.

Of course, highly concentrated microelements are necessary to express the microstructure of metallographic diagrams. However, as mentioned above, the increase in microelements does not increase the macroelement's degrees of freedom and the computation work. Therefore microelement method is an efficient approach to the scale-span analysis based on input of metallographic diagrams.

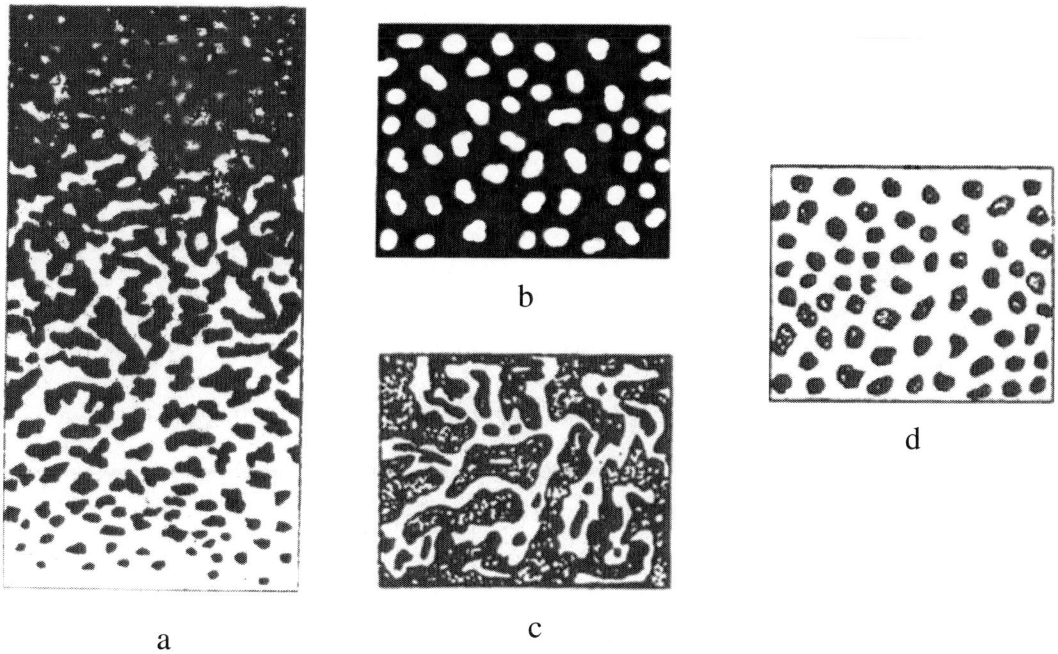

Figure 36. Metallograph of functionally graded plane. (a) graph of cross section, (b) longitudinal profile in upper surfacem, (c) longitudinal profile in middle plane, (d) longitudinal profile in lower surface.

7.2. Spatial Distribution of Macro-Mechanical Responses of Plate

Consider a simply supported rectangular plate with length $a = 1.0\,m$ and $b = 0.6\,m$, thickness $h = 0.1\,m$ under uniform loading $q = 1000\,N/m^2$. The metallographic diagrams of the plate are shown in Figure 36, where the black parts are metal material with modulus of elasticity $E_A = 150.0\,GPA$ and Poisson's ratio $\mu_A = 0.3$, and the white parts are ceramic material with modulus of elasticity $E_B = 350.0\,GPA$ and Poisson's ratio $\mu_B = 0.33$. The macro responses of the functionally graded plate are calculated according to the metallographic diagrams information input by microelement method. The main results are presented in the following.

For metal-ceramic functionally graded plate with microstructure, the distributions of dimensionless values of six main macro-mechanical variables across thickness are shown in Figure37(1)~Figure37(6), including horizontal displacement u, vertical deflection w, bending stress σ_x, horizontal shear stress σ_{xy}, vertical shear stress σ_{xz} and compressive stress σ_z. As comparison, the correspondent distributions of pure metal plate and pure ceramic plate are also presented.

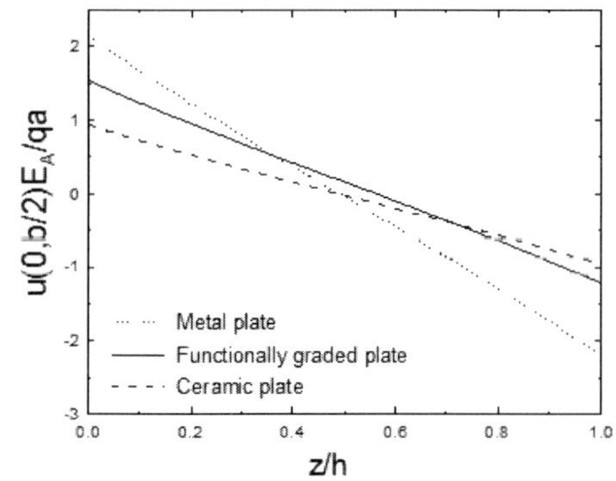

(1) u distribution along the thickness direction

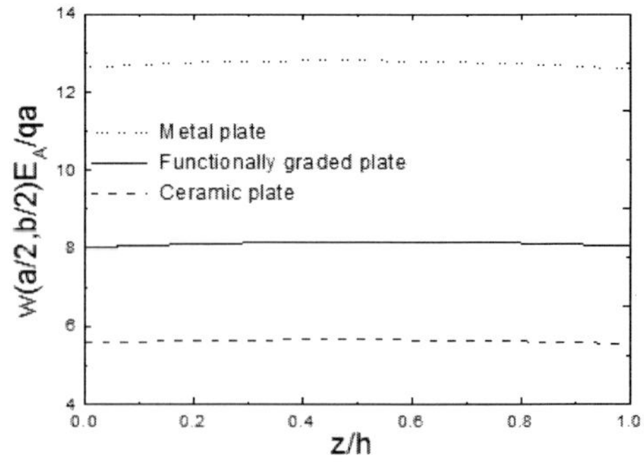

(2) w distribution along the thickness direction

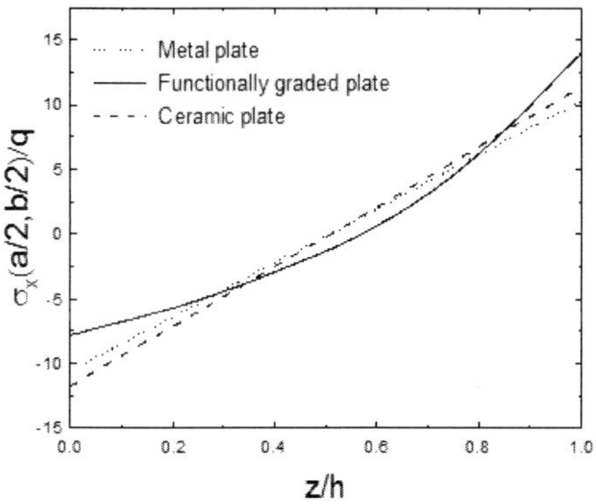

(3) σ_x distribution along the thickness direction

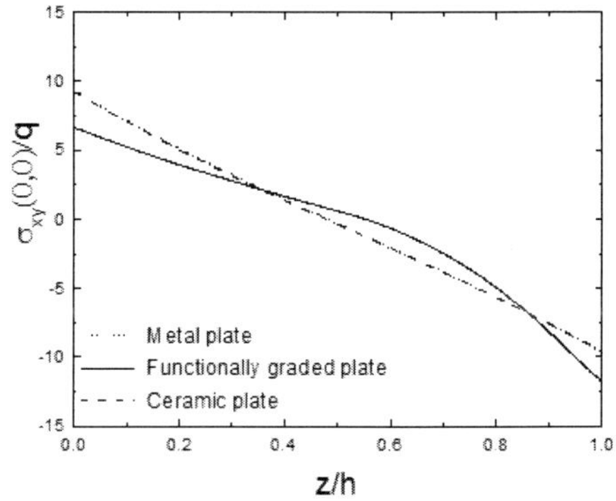

(4) σ_{xy} distribution along the thickness direction

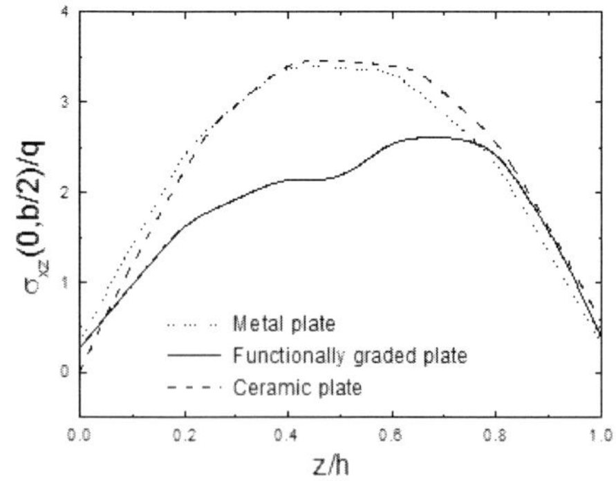

(5) σ_{xz} distribution along the thickness direction

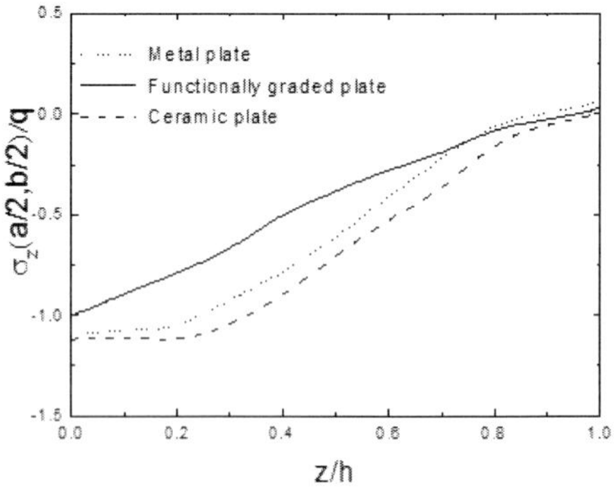

(6) σ_z distribution along the thicknessdirection

Figure 37. The dimensionless value of mechanical quantity distributions along the thickness direction.

Figure37(1)~ Figure37(6) show that the solutions by microelement method basically match that of the plate theory: w is uniformly distributed; u, σ_x and σ_{xy} are anti-symmetrically distributed; σ_{xz} is symmetrically distributed and is zero at both surfaces of the plate; σ_z is cubic curve distributed and equals external load beneath the upper surface and zero on the bottom surface. This method is reliable because the macroelement used here is a 3-D element and no artificial restrictions or assumptions are needed for mechanical variables along thickness.

That its numerical solutions comply with the law of well-known plate theory also proves it correct. It can also be observed from the Figures that the non-uniformity of functionally graded plate has slightly different mechanical quantities from those of the classical theory of uniform plate (i.e., the distributions of solutions across the thickness for pure metal and pure ceramic uniform plates) due to change in material properties along thickness, which agrees well with the present 3-D theoretical solutions[24] of functionally graded plates.

The numerical solutions in this section show that the macro-mechanical responses can be obtained by microelement method with the input of metallographic diagrams, and the reliability of solutions has been proved.

7.3. Micro-Stress Distribution on the Net Microstructure

The contour-lines of σ_x/q on microstructure of plane $z = 3h/5$ (net structure part) are given in Figure 38, which thoroughly expresses the tendency of identical stresses along net microstructure and gives the results of micromechanics distribution.

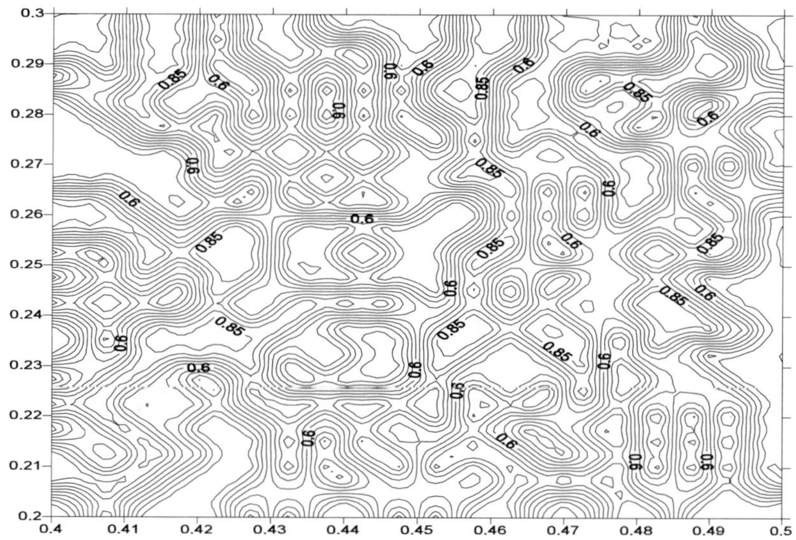

Figure38. The contour chart of σ_x/q in microstructure.

Therefore, microelement method can give not only the macro-mechanical responses, but the mechanical distribution on microelement and microstructure as well, which has been discussed in basic theoretic formulae (9a)~(9c) and is further proved by the present numerical analysis. Microelement method can be used for solutions of micromechanics, which proposes a very effective approach for failure analysis of material-micro structure and boundary stress analysis, and their elements and calculation work amount are sill the same as macroelement method.

8. Dynamic Property Analysis of Functionally Graded Plate based on Metallographic Diagrams

Consider a simply supported rectangular plate with length $a = 1.0\,m$, $b = 0.6\,m$, and thickness $h = 0.1\,m$. The metallographic diagrams of the plate are shown in Figure 36 where the black part means metal material with modulus of elasticity $E_A = 150.0\,GPA$ and Poisson's ratio $\mu_A = 0.3$, mass density $\rho_A = 9000 kg/m^3$ and the white part means ceramic material with modulus of elasticity $E_B = 350.0\,GPA$ and Poisson's ratio $\mu_B = 0.33$, mass density $\rho_B = 4500 kg/m^3$. The dynamic properties of the functionally graded plate are calculated according to the metallographic diagrams information inputted by use of microelement method. The main results are presented in the following sections , where the dimensionless natural frequency $\omega_i^* = \omega_i h\sqrt{\rho_A / E_A}$.

8.1. Natural Frequencies of Functionally Graded Plate

The fist six dimensionless natural frequencies of a metal-ceramic functionally graded plate with microstructure are shown in Table 10. As a comparison, the corresponding solutions of pure metal plate and pure ceramic plate are also presented in it.

Table 10 shows that the solutions of functionally graded plate are between the corresponding solutions of pure metal plate and pure ceramic plate, which is reasonable as the former plate material is composed of metal and ceramic materials in a certain ratio. The numerical example also shows that the microelement method is workable for dynamic property analysis under the input based on arbitrary metallographic diagrams of the composite.

Table10. Dimensionless natural frequencies of functionally graded plate and uniform plates

Type of plate	Order i	1	2	3	4	5	6
Pure metal plate	Frequency ω_i^*	0.1079	0.1837	0.1959	0.2996	0.3074	0.3253
Functionally graded plate	Frequency ω_i^*	0.1548	0.2661	0.2875	0.4357	0.4470	0.4782
Pure ceramic plate	Frequency ω_i^*	0.2297	0.3969	0.4156	0.6496	0.6665	0.6933

8.2. Basic Mode Shape Distribution Along Thickness of Functionally Graded Plate

As a part of basic mode shape results obtained via microelement method computation which takes metallographic diagrams as input, the distribution curves of displacement mode shape u, w along thickness of the functionally graded plate are presented in Figure39(a) and Figure39(b) , respectively. The correspondent distribution curves of stress mode shape

σ_x and σ_{xz} are given in Figure39(c) and Figure39(d) which show that σ_x is anti-symmetrical distributive; σ_{xz} is symmetrical distributive and vanishes at two surfaces of the plate. As the macroelement used here is a kind of 3-D element, no additional restrictions and assumptions are needed for mechanical variables across the thickness. The present numerical solutions agree quite well with the law of well-known plate theory, which has proved the correctness of the microelement method. On the other side, it can be found from the figures that the non-uniformity of functionally graded plate, due to the change of material properties across the thickness, leads to a slight deviation of mechanical quantities compared with the classical theory of uniform plate (i.e., the distributions of solutions across the thickness for pure metal and pure ceramic uniform plates). And their solutions are same as those obtained by present 3-D theory of functionally graded plates[24], which shows again the reliability of the present method.

The numerical solutions in this section show that the macro dynamic property law can be obtained by microelement method with the input of metallographic diagrams, and the reliability of solutions has been proved.

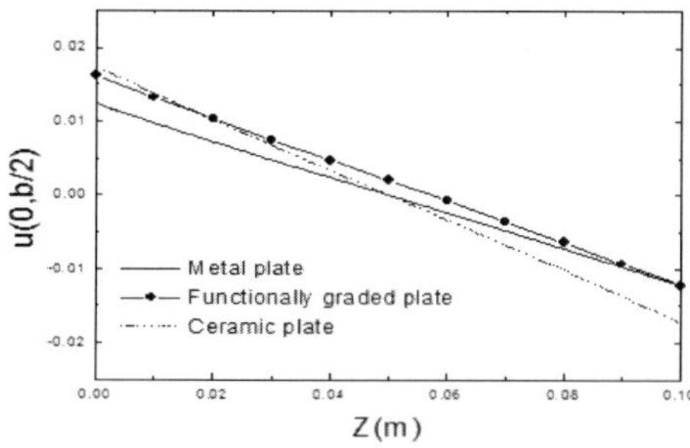

(a) u distribution along the thickness

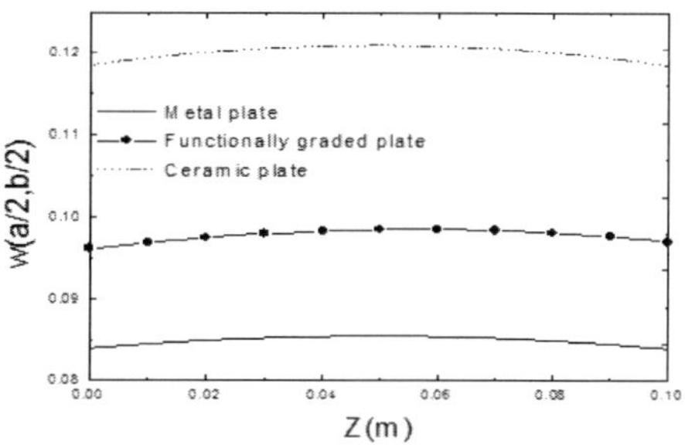

(b) w distribution along the thickness

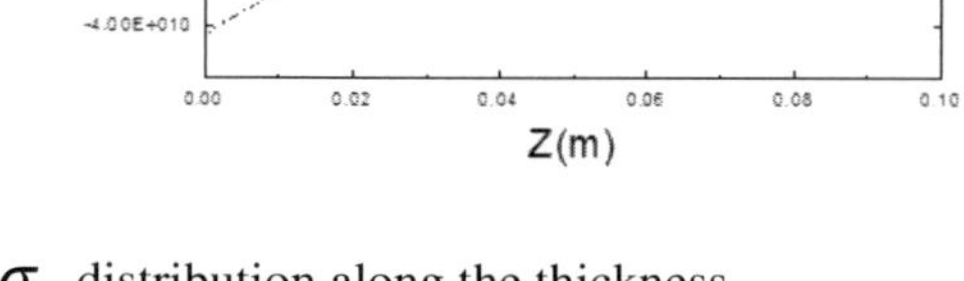

(c) σ_x distribution along the thickness

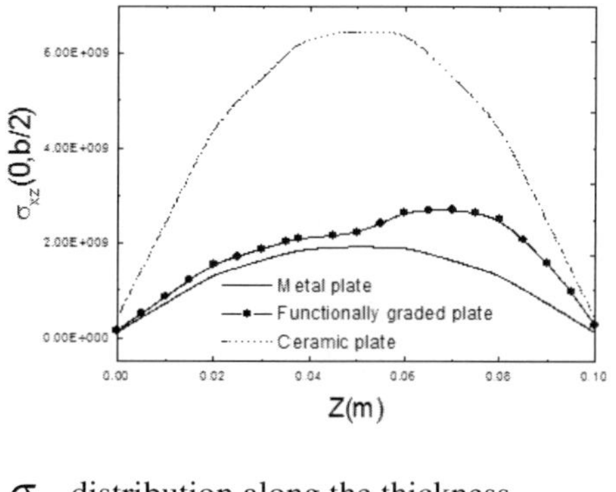

(d) σ_{xz} distribution along the thickness

Figure 39. Basic mode shape distributions along the thickness.

8.3. Micro-Stress Mode Shape Distribution on the Net Microstructure

The contour-lines of σ_x mode shape on microstructure of plane $z = 3h/5$ (net structure part) are given in Figure40, which thoroughly expresses the tendency of identical stresses along net microstructure and gives the results of micromechanics distribution.

Therefore, the microelement method can give not only the macro-mechanical responses, but the mechanical distribution on microelement and microstructure as well, which has been discussed in basic theoretic formulae (9a)~(9c) and is further proved by the present numerical analysis. The microelement method can be used for the solutions of micromechanics, which proposes a very effective approach to failure analysis of material-micro structure and boundary strcss analysis, and thcir clcmcnts and calculation work amount arc still thc samc as macroelement method.

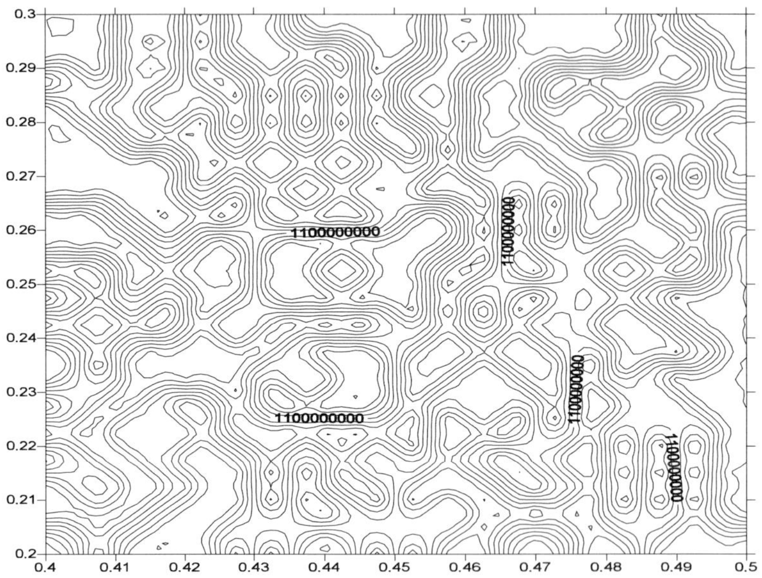

Figure 40. The contour chart of σ_x mode shape on microstructure.

CONCLUSION

Using the microelement method, the macro-response of structures can be immediately calculated by the material components given in the course of production, so the determination and identification of the functionally graded function can be avoided. The micro-constitution of the material can be fully reflected in the macro-response of structure, and the scale-span analyses of the functionally graded materials and structures can be also implemented.

There are many advantages with the higher accuracy and lower work amount when the microelement method is applied to the three-dimensional analyses of functionally graded structures.

The method can be applied to the three-dimensional analyses of functionally graded plates with various material components.

The three-dimensional mechanical distributions will deviate from the solution of the classical plate-shell theory when the material components of functionally graded plates and shells are changed violently.

The method can be applied to the analyses of functionally graded plates with complex shape.

The method can be applied to the three-dimensional analyses of functionally graded plates with the gap of arbitrary shape, position and distribution.

The method can be suitable for the materials with non-periodic micro-constitution (example 6.5).

The macro response of a structure can be calculated directly by use of microstructure of material metallographic diagrams, which changes traditional material-statistics-property input schedule. And hence, the microelement method realizes the exact macro-micro scale-span analysis.

The present method employs macro-computation to realize the micromechanical analysis.

The characteristics of general adaptability and program unity of ordinary FEM for engineering problems are reserved by the proposed method.

REFERENCES

Koizumi, M., The concept of FGM Ceramic Trans, Functionally Gradient Materials 1993 Vol 34, 3-10.

Hirai, T., Chen L., Recent and prospective development of functionally graded materials in Japan Mater Sci. Forum 1999 Vol 308-311, 509-514.

Obata, Y., Noda, N., Tsuji, T., Steady thermal stresses in a FGM plate Trans of the Japan Society of Mechanical Engineers Series A 1992 Vol 58,1689-95.

Obata, Y. Noda, N., Steady thermal stresses in a hollow circular cylinder and a hollow sphere of a FGM J. of Thermal Stresses 1994 Vol 17, 471-87

Cheng, Z.Q., Batra, R.C., 3-D thermoelastic deformations of a FGM elliptic plate Composites, Part B 2000 Vol 31, 97-106.

Praveen, G. N., Reddy, J. N., Nonlinear transient thermoelastic analysis of FGM ceramic-metal plates Int. J. of Solids and Structures 1998 Vol 35, 4457-76.

Tanigawa, Y., Matsumoto, M., Akat, T., Optimization problem of material composition for non-homogeneous plate to minimize thermal stress Trans. of the Japan Society of Mechanical Engineers, Series A 1996, Vol 62, 115-122.

Erdogan, F., Fracture mechanics Int. *J. of Solids and Structures* 2000 Vol 37, 171-83

Tanigawa, Y., Some basic thermoelastic problems for non-homogeneous structural materials, *Applied Mechanics Reviews* 1995 Vol 48, 287-300.

Chen, W. Q., Ding, H. J., Bending of functionally graded piezoelectric rectangular plates Acta Mech Solida Sinica 2000 Vol 13, 312-19.

Cheng, Z.Q., Lim, C.W., Kitipornchai, S., 3-D asymptotic approach to inhomogeneous and laminated piezoelectric plates Int. *J. of Solids and Structures* 2000 Vol 37, 3153-75.

Lim, C.W., He, L.H., Exact solution of a compositionally graded piezoelectric layer under uniform stretch, bending and twisting Int. *J. of Mechanical Sciences* 2001 Vol 43, 2479-92.

Yang, J., Chen, H.S., Dynamic response of initially stressed functionally graded rectangular thin plates Composite Structure 2001 Vol 54, 497-508.

Williamson, R. L., Rabin, B. H., Drake, J. T., Finite element analysis of thermal residual stresses at graded ceramic metal interfaces *J. Appl. Phy.* 1993 Vol 74, 1310-26.

Yang, S.Y. Zhang, Q.J., Zhai, P.C., Computational micromechanics analysis of the functionally graded materials under thermal shock loading, Materials Science Forum 2003, Vol 423-425, 687-92.

Markworth, A.J., Ramesh, K.S., Review modeling studies applied to functionally graded materials *J. of Materials Science* 1995 Vol 30, 2183-93.

Pindera, M.J., Aboudi, J., Arnold, S.M., Limitations of uncoupled, RVE-based micromechanical approaches in the analysis of FGM Mech. Mater. 1995 Vol 20, 77-94.

Zuiker, J. R., FGM: Choice of micromechanics model and limitations in property variation Composites Engineering 1995 Vol 5, 807-19.

Reiter T., Dvorak G.J., Tvergaard V. Micromechanical models for FGM *J. Mech. Phys. Solids* 1997 Vol 46, 1655-73.

Shen H.S., Bending, buckling and vibration of functionally graded composite material plates and shells Advances in Mechanics 2004 Vol 34, 53-60. (in Chinese)

Wang, X.C., Finite Element Method. Beijing: Tsinghua University Press, 2003. (in Chinese)

Timoshenko, S., Woinowsky S., Theory of plates and shell. London: McGraw–Hill, 1959.

Institute of Mechanics, Chinese Academy of Sciences, Bending, stability and vibration of sandwich plates and shells. Beijing: Science Press, 1977.

Zhong, Z., Sheng E.T., 3-D exact analysis of a simply supported functionally gradient piezoelectric plate. Inter. *J. of Solids and Structures* 2003 Vol 40, 5335-52.

Oden, J.T., Belytschko, T., Babuska, I.,et al. Research directions in computational mechanics: Report Prepared by the US National Committee on Theoretical and Applied Mechanics. *Computer Methods in Appl. Mech and Eng* 2003 Vol 192, 913-22.

About the Authors

Cao Zhiyuan received the bachelor's degree from Tsinghua University, China, 1963. Before retried in 2007, he is a Full Professor of the School of Aerospace Engineering and Applied Mechanics, Shanghai, China, and now, he is an honorary Professor of the School. His research interests include the solid mechanics, smart structures, thick plate theory, semi-analytic method, numerical method, construction mechanics.

In: Mechanics of Functionally Graded Materials and Structures ISBN 978-1-61668-920-9
Editors: Z. Zhong, L. Wu and W. Chen

Chapter 4

ANALYTICAL SOLUTIONS FOR ANISOTROPIC MAGNETO-ELECTRO-ELASTIC FUNCTIONALLY GRADED BEAMS

***Dejin Huang*[1] *and Haojiang Ding*[2]**
[1]Department of Engineering Mechanics, Ningbo University,
Ningbo,. China
[2] Department of Civil Engineering, Zhejiang University, Zijingang Campus,
Hangzhou, P.R. China

ABSTRACT

Methods are described in this chapter to acquire analytical solutions of anisotropic functionally graded magneto-electro-elastic (FGMEE) beams under polynomial and arbitrary loads. The material properties are functions of the thickness coordinate. The partial differential equations governing stress function, electric displacement function and magnetic induction function are derived for plane problems of anisotropic FGMEE materials. These functions are assumed in terms of unknown functions of thickness coordinate only, which are determined subsequently. The analytical expressions of axial force, bending moment, shear force, average electric displacement, average magnetic induction, displacements, electric potential and magnetic potential are then deduced. The integral constants are determinable completely from boundary conditions. A series of analytical solutions are thus obtained. The methods can also be used to find analytical solutions for FG piezoelastic beams and FG elastic beams under polynomial and arbitrary loads.

1. INTRODUCTION

Beams are important structures in engineering applications. For example, actuators and sensors widely applied in smart structures are often considered as beams of different kinds.

[1] e-mail: 1huangdejin@nbu.edu.cn.
[2] dinghj@zju.edu.cn

These smart structures are often made from piezoelectric materials or magneto-electro-elastic materials, due to their coupling of mechanical, electrical and magnetic properties. It has been a hot topic to investigate the problems of coupling among mechanical deformation, electric field and magnetic field.

Conventional actuators are often in forms of bimorphs and multi-morph stacks, which are bonded together with adhesive epoxy resin [1]. In actual application, the bonding layer between piezoelectric or magneto-electro-elastic sheets may crack at low temperature and creep or peel off at high temperature. The bonding of different piezoelectric materials or identical piezoelectric materials with different poling directions will cause severe interfacial stress concentration that may lead to the failure of devices. These disadvantages reduce the reliability and lifetime, and limit their applications [2,3].

To solve these problems, the concept of functionally graded materials (FGMs) has been introduced into designing piezoelectric and magneto-electro-elastic devices. The devices made of functionally graded materials have no discernible internal boundaries and do not produce stress concentration while they are loaded. Zhu et al. [2] and Wu et al. [3] individually prepared functionally graded piezoelectric materials (FGPMs) successfully and developed new kinds of FGP actuators. The experimental study on FGP cantilever by Wu et al. [3] showed that the cantilever produces relatively uniform stress distribution. In recent decades, great deals of works have been carried out on problems of FGP beams and plates by virtue of laminated plate theories. Saravanos and Heyliger [4] and Tauchert et al. [5] presented excellent reviews on the related topics. A state-space solution based on the three-dimensional equations of piezoelasticity, in conjugation with the approximate laminate model, was proposed to analyze simply-supported FGP plates by Chen and Ding [6]. Kruusing [7] reviewed the design and modeling of cantilever beam microactuators loaded normally at the tip, and presented a new simple and straightforward analytical model of the mechanical response. Hauke et al. [8] presented a new kind of multi-morph piezoelectric cantilever actuator, which was assumed to consist of *N*-layers with stepwise linear piezoelectric coefficient d_{31} in different layers, while other material parameters were assumed constant for all layers. Shi [9] investigated orthotropic density FGP cantilever by the stress function approach. He assumed the body force function is a cubic polynomial, and presented several analytical solutions under different loading conditions. Shi and Chen [10] presented solutions for an orthotropic FGP beam subjected to different polynomial loads, in which the material parameter s_{33} was assumed to be a quadratic polynomial and the body force function to be a cubic polynomial along the height of the beam. Liu and Shi [11] investigated the problem of an orthotropic FGP beam whose piezoelectric parameter g_{31} changes linearly along the beam height. Chen et al. [12,13] investigated the bending and free vibration problems of a transversely isotropic FGMEE plate using a particular state-space formulation along with the approximate laminate model. Pan and Han [14] gave an exact solution for functionally graded and layered MEE plates with all material constants varying in the same exponential way along the thickness direction. Bhangale and Ganesan [15] carried out static analysis of FGMEE plates using a semi-analytical finite element method. Bhangale and Ganesan [16,17] reported the finite element investigations on free vibration of FGMEE plates and cylindrical shells.

In this chapter, we present analytical solutions for anisotropic FGMEE beams under polynomial and arbitrary loads. The material properties vary arbitrarily through the thickness, and various boundary conditions at the two ends can be tackled. The basic equations for plane problem are given in Section 2. Analytical solutions and numerical examples are presented in

Sections 3 and 4 for FGMEE beams under polynomial and arbitrary loads respectively. Finally, a summary is given in Section 5.

2. BASIC EQUATIONS

The basic equations for an anisotropic FGMEE beam in the plane stress state include the equilibrium equations, geometrical equations and constitutive equations as follows.

The equilibrium equations

$$\frac{\partial \sigma_x}{\partial x}+\frac{\partial \tau_{xz}}{\partial z}=0, \qquad \frac{\partial \tau_{xz}}{\partial x}+\frac{\partial \sigma_z}{\partial z}=0, \tag{1}$$

$$\frac{\partial D_x}{\partial x}+\frac{\partial D_z}{\partial z}=0, \tag{2}$$

$$\frac{\partial B_x}{\partial x}+\frac{\partial B_z}{\partial z}=0, \tag{3}$$

where σ_x, σ_z and τ_{xz} denote stress components, D_x and D_z are electric displacement components, and B_x and B_z are magnetic induction components.

The geometrical equations

$$\varepsilon_x=\frac{\partial u}{\partial x}, \qquad \varepsilon_z=\frac{\partial w}{\partial z}, \qquad \gamma_{xz}=\frac{\partial w}{\partial x}+\frac{\partial u}{\partial z}, \tag{4}$$

$$E_x=-\frac{\partial \Phi}{\partial x}, \qquad E_z=-\frac{\partial \Phi}{\partial z}, \tag{5}$$

$$H_x=-\frac{\partial \Psi}{\partial x}, \qquad H_z=-\frac{\partial \Psi}{\partial z}, \tag{6}$$

where ε_x, ε_z and γ_{xz} denote strain components, u and w displacement components, E_x and E_z electric field components, Φ electrical potential, H_x and H_z magnetic field components, and Ψ magnetic potential.

The constitutive equations

$$\begin{aligned}
\varepsilon_x &= s_{11}\sigma_x + s_{13}\sigma_z + s_{15}\tau_{xz} + d_{11}D_x + d_{31}D_z + b_{11}B_x + b_{31}B_z, \\
\varepsilon_z &= s_{13}\sigma_x + s_{33}\sigma_z + s_{35}\tau_{xz} + d_{13}D_x + d_{33}D_z + b_{13}B_x + b_{33}B_z, \\
\gamma_{xz} &= s_{15}\sigma_x + s_{35}\sigma_z + s_{55}\tau_{xz} + d_{15}D_x + d_{35}D_z + b_{15}B_x + b_{35}B_z,
\end{aligned} \tag{7}$$

$$E_x = -d_{11}\sigma_x - d_{13}\sigma_z - d_{15}\tau_{xz} + \gamma_{11}D_x + \gamma_{13}D_z + \beta_{11}B_x + \beta_{13}B_z,$$

$$E_z = -d_{31}\sigma_x - d_{33}\sigma_z - d_{35}\tau_{xz} + \gamma_{13}D_x + \gamma_{33}D_z + \beta_{13}B_x + \beta_{33}B_z, \quad (8)$$

$$H_x = -b_{11}\sigma_x - b_{13}\sigma_z - b_{15}\tau_{xz} + \beta_{11}D_x + \beta_{13}D_z + \nu_{11}B_x + \nu_{13}B_z,$$

$$H_z = -b_{31}\sigma_x - b_{33}\sigma_z - b_{35}\tau_{xz} + \beta_{13}D_x + \beta_{33}D_z + \nu_{13}B_x + \nu_{33}B_z, \quad (9)$$

where s_{ij}, d_{ij}, b_{ij}, γ_{ij}, β_{ij} and ν_{ij} denote coefficients of elastic compliance, piezoelectricity, piezomagnetism, dielectric impermeability, magneto-electricity and magnetic permeability, respectively. For FGMEE beams, we generally have

$$s_{ij} = s_{ij}(z),\quad d_{ij} = d_{ij}(z),\quad b_{ij} = b_{ij}(z),\quad \gamma_{ij} = \gamma_{ij}(z),\quad \beta_{ij} = \beta_{ij}(z),\quad \nu_{ij} = \nu_{ij}(z).$$

From Eqs. (4), (5) and (6), we obtain the following compatibility equations, respectively,

$$\frac{\partial^2 \varepsilon_x}{\partial z^2} + \frac{\partial^2 \varepsilon_z}{\partial x^2} - \frac{\partial^2 \gamma_{xz}}{\partial x \partial z} = 0, \quad (10)$$

$$\frac{\partial E_x}{\partial z} - \frac{\partial E_z}{\partial x} = 0, \quad (11)$$

$$\frac{\partial H_x}{\partial z} - \frac{\partial H_z}{\partial x} = 0. \quad (12)$$

3. FGMEE BEAMS UNDER POLYNOMIAL LOADS

3.1. Stress Function, Electric Displacement Function and Magnetic Induction Function

Consider an FGMEE beam of unit thickness with height h and span l, referred to a Cartesian system x-z as shown in Figure 1.

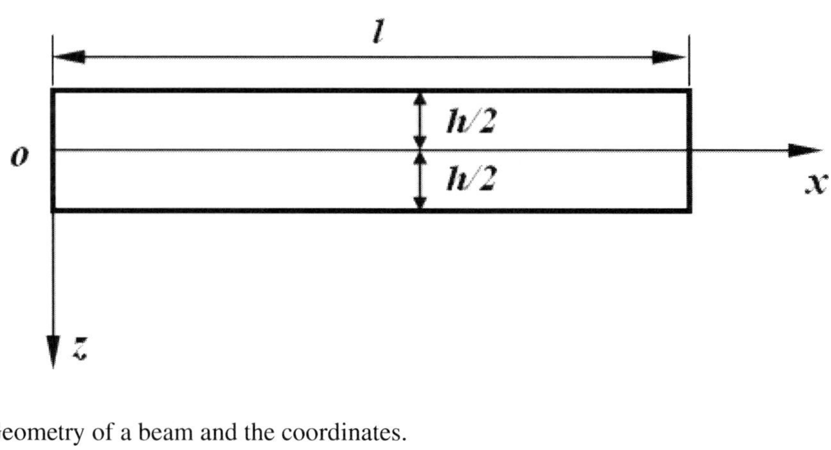

Figure 1. Geometry of a beam and the coordinates.

Introducing stress function φ, electric displacement function ψ and magnetic induction function χ, we obtain the general solutions of Eqs.(1), (2) and (3) as

$$\sigma_x = \frac{\partial^2 \varphi}{\partial z^2}, \quad \sigma_z = \frac{\partial^2 \varphi}{\partial x^2}, \quad \tau_{xz} = -\frac{\partial^2 \varphi}{\partial x \partial z}, \tag{13}$$

$$D_x = \frac{\partial \psi}{\partial z}, \quad D_z = -\frac{\partial \psi}{\partial x}, \tag{14}$$

$$B_x = \frac{\partial \chi}{\partial z}, \quad B_z = -\frac{\partial \chi}{\partial x}. \tag{15}$$

Further, we assume that φ, ψ and χ take the following polynomial forms

$$\begin{aligned} \varphi &= \varphi_0(z) + x\varphi_1(z) + x^2\varphi_2(z), \\ \psi &= \psi_0(z) + x\psi_1(z) + x^2\psi_2(z), \\ \chi &= \chi_0(z) + x\chi_1(z) + x^2\chi_2(z). \end{aligned} \tag{16}$$

where φ_i, ψ_i and χ_i (i=1,2,3) are unknown functions to be determined. Substituting Eq. (16) into Eqs. (13)-(15), we have

$$\sigma_x = \varphi_0''(z) + x\varphi_1''(z) + x^2\varphi_2''(z), \quad \sigma_z = 2\varphi_2(z), \quad \tau_{xz} = -\varphi_1'(z) - 2x\varphi_2'(z), \tag{17}$$

$$D_x = \psi_0'(z) + x\psi_1'(z) + x^2\psi_2'(z), \quad D_z = -\psi_1(z) - 2x\psi_2(z), \tag{18}$$

$$B_x = \chi_0'(z) + x\chi_1'(z) + x^2\chi_2'(z), \quad B_z = -\chi_1(z) - 2x\chi_2(z), \tag{19}$$

where $(.)'$ and $(.)''$ denote the first and the second derivatives with respect to z, respectively.

Substitution Eqs. (17)-(19) into Eqs. (7)-(9) gives

$$\begin{aligned} \varepsilon_x &= s_{11}\varphi_0'' + 2s_{13}\varphi_2 - s_{15}\varphi_1' + d_{11}\psi_0' - d_{31}\psi_1 + b_{11}\chi_0' - b_{31}\chi_1 \\ &\quad + x\left(s_{11}\varphi_1'' - 2s_{15}\varphi_2' + d_{11}\psi_1' - 2d_{31}\psi_2 + b_{11}\chi_1' - 2b_{31}\chi_2\right) \\ &\quad + x^2\left(s_{11}\varphi_2'' + d_{11}\psi_2' + b_{11}\chi_2'\right), \\ \varepsilon_z &= s_{13}\varphi_0'' + 2s_{33}\varphi_2 - s_{35}\varphi_1' + d_{13}\psi_0' - d_{33}\psi_1 + b_{13}\chi_0' - b_{33}\chi_1 \\ &\quad + x\left(s_{13}\varphi_1'' - 2s_{35}\varphi_2' + d_{13}\psi_1' - 2d_{33}\psi_2 + b_{13}\chi_1' - 2b_{33}\chi_2\right) \\ &\quad + x^2\left(s_{13}\varphi_2'' + d_{13}\psi_2' + b_{13}\chi_2'\right), \end{aligned}$$

$$\begin{aligned}\gamma_{xz} &= s_{15}\varphi_0'' + 2s_{35}\varphi_2 - s_{55}\varphi_1' + d_{15}\psi_0' - d_{35}\psi_1 + b_{15}\chi_0' - b_{35}\chi_1 \\ &+ x\left(s_{15}\varphi_1'' - 2s_{55}\varphi_2' + d_{15}\psi_1' - 2d_{35}\psi_2 + b_{15}\chi_1' - 2b_{35}\chi_2\right) \\ &+ x^2\left(s_{15}\varphi_2'' + d_{15}\psi_2' + b_{15}\chi_2'\right);\end{aligned} \tag{20}$$

$$\begin{aligned}E_x &= -d_{11}\varphi_0'' - 2d_{13}\varphi_2 + d_{15}\varphi_1' + \gamma_{11}\psi_0' - \gamma_{13}\psi_1 + \beta_{11}\chi_0' - \beta_{13}\chi_1 \\ &+ x\left(-d_{11}\varphi_1'' + 2d_{15}\varphi_2' + \gamma_{11}\psi_1' - 2\gamma_{13}\psi_2 + \beta_{11}\chi_1' - 2\beta_{13}\chi_2\right) \\ &+ x^2\left(-d_{11}\varphi_2'' + \gamma_{11}\psi_2' + \beta_{11}\chi_2'\right), \\ E_z &= -d_{31}\varphi_0'' - 2d_{33}\varphi_2 + d_{35}\varphi_1' + \gamma_{13}\psi_0' - \gamma_{33}\psi_1 + \beta_{13}\chi_0' - \beta_{33}\chi_1 \\ &+ x\left(-d_{31}\varphi_1'' + 2d_{35}\varphi_2' + \gamma_{13}\psi_1' - 2\gamma_{33}\psi_2 + \beta_{13}\chi_1' - 2\beta_{33}\chi_2\right) \\ &+ x^2\left(-d_{31}\varphi_2'' + \gamma_{13}\psi_2' + \beta_{13}\chi_2'\right);\end{aligned} \tag{21}$$

$$\begin{aligned}H_x &= -b_{11}\varphi_0'' - 2b_{13}\varphi_2 + b_{15}\varphi_1' + \beta_{11}\psi_0' - \beta_{13}\psi_1 + \nu_{11}\chi_0' - \nu_{13}\chi_1 \\ &+ x\left(-b_{11}\varphi_1'' + 2b_{15}\varphi_2' + \beta_{11}\psi_1' - 2\beta_{13}\psi_2 + \nu_{11}\chi_1' - 2\nu_{13}\chi_2\right) \\ &+ x^2\left(-b_{11}\varphi_2'' + \beta_{11}\psi_2' + \nu_{11}\chi_2'\right), \\ H_z &= -b_{31}\varphi_0'' - 2b_{33}\varphi_2 + b_{35}\varphi_1' + \beta_{13}\psi_0' - \beta_{33}\psi_1 + \nu_{13}\chi_0' - \nu_{33}\chi_1 \\ &+ x\left(-b_{31}\varphi_1'' + 2b_{35}\varphi_2' + \beta_{13}\psi_1' - 2\beta_{33}\psi_2 + \nu_{13}\chi_1' - 2\nu_{33}\chi_2\right) \\ &+ x^2\left(-b_{31}\varphi_2'' + \beta_{13}\psi_2' + \nu_{13}\chi_2'\right).\end{aligned} \tag{22}$$

Substitutions of Eqs. (20)-(22) into Eqs. (10)-(12) educe the following equations

$$\left(-d_{11}\varphi_2'' + \gamma_{11}\psi_2' + \beta_{11}\chi_2'\right)' = 0, \tag{23}$$

$$\left(-b_{11}\varphi_2'' + \beta_{11}\psi_2' + \nu_{11}\chi_2'\right)' = 0, \tag{24}$$

$$\left(s_{11}\varphi_2'' + d_{11}\psi_2' + b_{11}\chi_2'\right)'' = 0, \tag{25}$$

$$\begin{aligned}&\left(-d_{11}\varphi_1'' + 2d_{15}\varphi_2' + \gamma_{11}\psi_1' - 2\gamma_{13}\psi_2 + \beta_{11}\chi_1' - 2\beta_{13}\chi_2\right)' \\ &\quad + 2\left(d_{31}\varphi_2'' - \gamma_{13}\psi_2' - \beta_{13}\chi_2'\right) = 0,\end{aligned} \tag{26}$$

$$\begin{aligned}&\left(-b_{11}\varphi_1'' + 2b_{15}\varphi_2' + \beta_{11}\psi_1' - 2\beta_{13}\psi_2 + \nu_{11}\chi_1' - 2\nu_{13}\chi_2\right)' \\ &\quad + 2\left(b_{31}\varphi_2'' - \beta_{13}\psi_2' - \nu_{13}\chi_2'\right) = 0,\end{aligned} \tag{27}$$

$$\left(s_{11}\varphi_1'' - 2s_{15}\varphi_2' + d_{11}\psi_1' - 2d_{31}\psi_2 + b_{11}\chi_1' - 2b_{31}\chi_2\right)'' - 2\left(s_{15}\varphi_2'' + d_{15}\psi_2' + b_{15}\chi_2'\right)' = 0, \tag{28}$$

$$\left(-d_{11}\varphi_0'' - 2d_{13}\varphi_2 + d_{15}\varphi_1' + \gamma_{11}\psi_0' - \gamma_{13}\psi_1 + \beta_{11}\chi_0' - \beta_{13}\chi_1\right)' + d_{31}\varphi_1'' - 2d_{35}\varphi_2' - \gamma_{13}\psi_1' + 2\gamma_{33}\psi_2 - \beta_{13}\chi_1' + 2\beta_{33}\chi_2 = 0, \tag{29}$$

$$\left(-b_{11}\varphi_0'' - 2b_{13}\varphi_2 + b_{15}\varphi_1' + \beta_{11}\psi_0' - \beta_{13}\psi_1 + \nu_{11}\chi_0' - \nu_{13}\chi_1\right)' + b_{31}\varphi_1'' - 2b_{35}\varphi_2' - \beta_{13}\psi_1' + 2\beta_{33}\psi_2 - \nu_{13}\chi_1' + 2\nu_{33}\chi_2 = 0, \tag{30}$$

$$\left(s_{11}\varphi_0'' - s_{15}\varphi_1' + 2s_{13}\varphi_2 + d_{11}\psi_0' - d_{31}\psi_1 + b_{11}\chi_0' - b_{31}\chi_1\right)'' - \left(s_{15}\varphi_1'' - 2s_{55}\varphi_2' + d_{15}\psi_1' - 2d_{35}\psi_2 + b_{15}\chi_1' - 2b_{35}\chi_2\right)' + 2\left(s_{13}\varphi_2'' + d_{13}\psi_2' + b_{13}\chi_2'\right) = 0. \tag{31}$$

Integrating Eqs. (23)-(25) yields

$$-d_{11}\varphi_2'' + \gamma_{11}\psi_2' + \beta_{11}\chi_2' = a_1, \qquad -b_{11}\varphi_2'' + \beta_{11}\psi_2' + \nu_{11}\chi_2' = a_2, \qquad s_{11}\varphi_2'' + d_{11}\psi_2' + b_{11}\chi_2' = a_3 z + a_4, \tag{32}$$

where and hereafter a_i $(i = 1, 2, ..., 24)$ are integral constants. From Eq. (32), we obtain

$$\varphi_2'' = a_1 A_{1,1} + a_2 A_{2,1} + a_3 A_{3,1} + a_4 A_{4,1} = \sum_{i=1}^{19} a_i A_{i,1}, \tag{33}$$

$$\psi_2' = a_1 A_{1,2} + a_2 A_{2,2} + a_3 A_{3,2} + a_4 A_{4,2} = \sum_{i=1}^{19} a_i A_{i,2}, \tag{34}$$

$$\chi_2' = a_1 A_{1,3} + a_2 A_{2,3} + a_3 A_{3,3} + a_4 A_{4,3} = \sum_{i=1}^{19} a_i A_{i,3}, \tag{35}$$

where

$$\begin{aligned} &A_{i,j}(z) = \Delta_{ij}/\Delta, \qquad (i = 1, 2, \ j = 1, 2, 3) \\ &A_{3,j}(z) = z\Delta_{3j}/\Delta, \qquad A_{4,j}(z) = \Delta_{3j}/\Delta, \qquad (j = 1, 2, 3), \\ &A_{i,j} = 0, \quad (i = 5, 6, ..., 19, \quad j = 1, 2, 3.) \end{aligned} \tag{36}$$

$$\begin{aligned}
&\Delta = -2d_{11}\beta_{11}b_{11} + s_{11}\gamma_{11}\nu_{11} + \nu_{11}d_{11}^2 + \gamma_{11}b_{11}^2 - s_{11}\beta_{11}^2, \\
&\Delta_{11} = b_{11}\beta_{11} - d_{11}\nu_{11}, \quad \Delta_{12} = b_{11}^2 + s_{11}\nu_{11}, \quad \Delta_{13} = -(d_{11}b_{11} + \beta_{11}s_{11}), \\
&\Delta_{21} = -(\gamma_{11}b_{11} - d_{11}\beta_{11}), \quad \Delta_{22} = -(d_{11}b_{11} + \beta_{11}s_{11}), \quad \Delta_{23} = d_{11}^2 + \gamma_{11}s_{11}, \\
&\Delta_{31} = \gamma_{11}\nu_{11} - \beta_{11}^2, \quad \Delta_{32} = -(b_{11}\beta_{11} - d_{11}\nu_{11}), \quad \Delta_{33} = \gamma_{11}b_{11} - d_{11}\beta_{11}.
\end{aligned} \tag{37}$$

Integration of Eqs. (33)-(35) yields

$$\varphi_2' = a_1A_{1,1}^0 + a_2A_{2,1}^0 + a_3A_{3,1}^0 + a_4A_{4,1}^0 + a_5 = \sum_{i=1}^{19} a_iA_{i,1}^0, \tag{38}$$

$$\varphi_2 = a_1A_{1,1}^1 + a_2A_{2,1}^1 + a_3A_{3,1}^1 + a_4A_{4,1}^1 + a_5z + a_6 = \sum_{i=1}^{19} a_iA_{i,1}^1. \tag{39}$$

$$\psi_2 = a_1A_{1,2}^0 + a_2A_{2,2}^0 + a_3A_{3,2}^0 + a_4A_{4,2}^0 + a_7 = \sum_{i=1}^{19} a_iA_{i,2}^0. \tag{40}$$

$$\chi_2 = a_1A_{1,3}^0 + a_2A_{2,3}^0 + a_3A_{3,3}^0 + a_4A_{4,3}^0 + a_8 = \sum_{i=1}^{19} a_iA_{i,3}^0, \tag{41}$$

Where

$$\begin{aligned}
&A_{i,j}^n = \int_{-h/2}^{z} (z-\xi)^n A_{i,j}(\xi)\, d\xi, \qquad (i=1,2,3,4,\ j=1,2,3,\ n=0,1), \\
&A_{5,1}^0 = 1, \quad A_{i,1}^0 = 0, \quad (i=6,7,\dots,19), \\
&A_{5,1}^1 = z, \quad A_{6,1}^1 = 1, \quad A_{i,1}^1 = 0, \quad (i=7,8,\dots,19), \\
&A_{5,2}^0 = A_{6,2}^0 = 0, \quad A_{7,2}^0 = 1, \quad A_{i,2}^0 = 0, \quad (i=8,9,\dots,19) \\
&A_{5,3}^0 = A_{6,3}^0 = A_{7,3}^0 = 0, \quad A_{8,3}^0 = 1, \quad A_{i,3}^0 = 0, \quad (i=9,10,\dots,19),
\end{aligned} \tag{42}$$

Substituting Eqs. (33)-(35), (38), (40) and (41) into Eqs. (26)-(28), and then integrating, we obtain

$$\begin{aligned}
&-d_{11}\varphi_1'' + \gamma_{11}\psi_1' + \beta_{11}\chi_1' \\
&\quad = a_1C_{1,1} + a_2C_{2,1} + a_3C_{3,1} + a_4C_{4,1} - 2d_{15}a_5 + 2\gamma_{13}a_7 + 2\beta_{13}a_8 + a_9,
\end{aligned} \tag{43a}$$

$$\begin{aligned}
&-b_{11}\varphi_1'' + \beta_{11}\psi_1' + \nu_{11}\chi_1' \\
&\quad = a_1C_{1,2} + a_2C_{2,2} + a_3C_{3,2} + a_4C_{4,2} - 2b_{15}a_5 + 2\beta_{13}a_7 + 2\nu_{13}a_8 + a_{10},
\end{aligned} \tag{43b}$$

$$s_{11}\varphi_1'' + d_{11}\psi_1' + b_{11}\chi_1' \\ = a_1 C_{1,3} + a_2 C_{2,3} + a_3 C_{3,3} + a_4 C_{4,3} + 2s_{15}a_5 + 2d_{31}a_7 + 2b_{31}a_8 + a_{11}z + a_{12}, \tag{43c}$$

where

$$C_{i,1} = B_{i,1}^0 - 2d_{15}A_{i,1}^0 + 2\gamma_{13}A_{i,2}^0 + 2\beta_{13}A_{i,3}^0, \\ C_{i,2} = B_{i,2}^0 - 2b_{15}A_{i,1}^0 + 2\beta_{13}A_{i,2}^0 + 2\nu_{13}A_{i,3}^0, \\ C_{i,3} = B_{i,3}^0 + 2s_{15}A_{i,1}^0 + 2d_{31}A_{i,2}^0 + 2b_{31}A_{i,3}^0, \qquad (i = 1,2,3,4), \tag{44}$$

$$B_{i,j}^0 = \int_{-h/2}^{z} B_{i,j}(\xi)\, d\xi, \qquad (\ i = 1,2,3,4; \quad j = 1,2,3\ ), \tag{45}$$

$$B_{i,1} = -2d_{31}A_{i,1} + 2\gamma_{13}A_{i,2} + 2\beta_{13}A_{i,3}, \\ B_{i,2} = -2b_{31}A_{i,1} + 2\beta_{13}A_{i,2} + 2\nu_{13}A_{i,3}, \\ B_{i,3} = 2s_{15}A_{i,1} + 2d_{15}A_{i,2} + 2b_{15}A_{i,3}, \qquad (i = 1,2,3,4). \tag{46}$$

From Eq. (43), we have

$$\varphi_1'' = \sum_{i=1}^{19} a_i D_{i,1}, \tag{47}$$

$$\psi_1' = \sum_{i=1}^{19} a_i D_{i,2}, \tag{48}$$

$$\chi_1' = \sum_{i=1}^{19} a_i D_{i,3}, \tag{49}$$

where

$$D_{i,j} = C_{i,j}\,\Delta_{1j}/\Delta + C_{i,2}\,\Delta_{2j}/\Delta + C_{i,3}\,\Delta_{3j}/\Delta, \quad (i = 1,2,3,4, \quad j = 1,2,3), \\ D_{5,j} = -2d_{15}\,\Delta_{1j}/\Delta - 2b_{15}\,\Delta_{2j}/\Delta + 2s_{15}\,\Delta_{3j}/\Delta, \qquad D_{6,j} = 0, \\ D_{7,j} = 2\gamma_{13}\,\Delta_{1j}/\Delta + 2\beta_{13}\,\Delta_{2j}/\Delta + 2d_{31}\,\Delta_{3j}/\Delta, \\ D_{8,j} = 2\beta_{13}\,\Delta_{1j}/\Delta + 2\nu_{13}\,\Delta_{2j}/\Delta + 2b_{31}\,\Delta_{3j}/\Delta, \\ D_{9,j} = \Delta_{1j}/\Delta, \qquad D_{10,j} = \Delta_{2j}/\Delta, \qquad D_{11,j} = z\,\Delta_{3j}/\Delta, \\ D_{12,j} = \Delta_{3j}/\Delta, \quad (j = 1,2,3), \\ D_{i,j} = 0, \quad (i = 13,14,\ldots,19,\ j = 1,2,3). \tag{50}$$

Integration of Eqs. (47)-(49) yields

$$\varphi_1' = \sum_{i=1}^{19} a_i D_{i,1}^0, \tag{51}$$

$$\varphi_1 = \sum_{i=1}^{19} a_i D_{i,1}^1 + a_{20}, \tag{52}$$

$$\psi_1 = \sum_{i=1}^{19} a_i D_{i,2}^0, \tag{53}$$

$$\chi_1 = \sum_{i=1}^{19} D_{i,3}^0, \tag{54}$$

where

$$\begin{aligned}
&D_{i,j}^n = \int_{-h/2}^{z} (z-\xi)^n D_{i,j}(\xi)\, d\xi, \qquad (i=1,2,...12, \quad j=1,2,3, \quad n=0,1),\\
&D_{13,1}^0 = 1, \qquad D_{i,1}^0 = 0, \qquad (i=14,15,...,19),\\
&D_{13,1}^1 = z, \qquad D_{i,1}^1 = 0, \qquad (i=14,15,...,19),\\
&D_{13,2}^0 = 0, \qquad D_{14,2}^0 = 1, \quad D_{i,2}^0 = 0, \quad (i=15,16,...,19),\\
&D_{13,3}^0 = 0, \quad D_{14,3}^0 = 0, \quad D_{15,3}^0 = 1, \qquad D_{i,3}^0 = 0, \quad (i=16,17,18,19).
\end{aligned} \tag{55}$$

Substituting Eqs. (33-35), (38)-(41), (47)-(49), (51), (53) and (54) into Eqs. (29)-(31), and then integrating, gives rise to

$$-d_{11}\varphi_0'' + \gamma_{11}\psi_0' + \beta_{11}\chi_0' = \sum_{i=1}^{19} a_i F_{i,1}, \tag{56a}$$

$$-b_{11}\varphi_0'' + \beta_{11}\psi_0' + \nu_{11}\chi_0' = \sum_{i=1}^{19} a_i F_{i,2}^0, \tag{56b}$$

$$s_{11}\varphi_0'' + d_{11}\psi_0' + b_{11}\chi_0' = \sum_{i=1}^{19} a_i F_{i,3}^0, \tag{56c}$$

where

$$\begin{aligned}
&F_{i,1} = E_{i,1}^0 + 2d_{13}A_{i,1}^1 - d_{15}D_{i,1}^0 + \gamma_{13}D_{i,2}^0 + \beta_{13}D_{i,3}^0, \; (i=1,2,...,16), \; F_{i,1} = 0, \; (i=17,18,19),\\
&F_{i,2}^0 = E_{i,2}^0 + 2b_{13}A_{i,1}^1 - b_{15}D_{i,1}^0 + \beta_{13}D_{i,2}^0 + \nu_{13}D_{i,3}^0, \; (i=1,2,...,17), \quad F_{i,2} = 0, \; (i=18,19),\\
&F_{i,3}^0 = E_{i,3}^0 + s_{15}D_{i,1}^0 - 2s_{13}A_{i,1}^1 + d_{31}D_{i,2}^0 + b_{31}D_{i,3}^0, \quad (i=1,2,...,19),
\end{aligned}$$

$$E_{i,1}^0 = \int_{-h/2}^{z} E_{i,1}\, d\xi, \quad (i=1,2,...,15,) \quad E_{16,1}^0 = 1, \quad E_{i,1}^0 = 0, \quad (i=17,18,19),$$
$$E_{i,2}^0 = \int_{-h/2}^{z} E_{i,2}\, d\xi, \quad (i=1,2,...,15,) \quad E_{17,2}^0 = 1, \quad E_{i,2}^0 = 0, \quad (i=16,18,19),$$
$$E_{i,3}^0 = \int_{-h/2}^{z} E_{i,3}\, d\xi, \quad (i=1,2,...,18), \quad E_{19,3}^0 = 1,$$
$$E_{i,1} = 2d_{35}A_{i,1}^0 - 2\gamma_{33}A_{i,2}^0 - 2\beta_{33}A_{i,3}^0 - d_{31}D_{i,1} + \gamma_{13}D_{i,2} + \beta_{13}D_{i,3}, \quad (i=1,2,...,19),$$
$$E_{i,2} = -b_{31}D_{i,1} + 2b_{35}A_{i,1}^0 + \beta_{13}D_{i,2} - 2\beta_{33}A_{i,2}^0 + \nu_{13}D_{i,3} - 2\nu_{33}A_{i,3}^0, \quad (i=1,2,...,19),$$
$$E_{i,3} = s_{15}D_{i,1} - 2s_{55}A_{i,1}^0 + d_{15}D_{i,2} - 2d_{35}A_{i,2}^0 + b_{15}D_{i,3} - 2b_{35}A_{i,3}^0 + E_i^0, \quad (i=1,2,...,17),$$
$$E_{18,3} = 1, \quad E_{19,3} = 0,$$
$$E_i^0 = \int_{-h/2}^{z} E_i\, d\xi, \quad (i=1,2,...,17), \quad E_{18}^0 = 1, \quad E_{19}^0 = 0,$$
$$E_i = -2s_{13}A_{i,1} - 2d_{13}A_{i,2} - 2b_{13}A_{i,3}, \quad (i=1,2,...,17), \quad E_{18} = E_{19} = 0. \tag{57}$$

From Eq. (56), we obtain

$$\varphi_0'' = \sum_{i=1}^{19} a_i G_{i,1}, \tag{58}$$

$$\psi_0' = \sum_{i=1}^{19} a_i G_{i,2}, \tag{59}$$

$$\chi_0' = \sum_{i=1}^{19} a_i G_{i,3}, \tag{60}$$

where

$$G_{i,j} = F_{i,1}\Delta_{1j}/\Delta + F_{i,2}\Delta_{2j}/\Delta + F_{i,3}^0\Delta_{3j}/\Delta, \quad (i=1,2,...,19, \quad j=1,2,3), \tag{61}$$

Integrating Eqs. (58)-(60) gives

$$\varphi_0' = \sum_{i=1}^{19} a_i G_{i,1}^0 + a_{21}, \tag{62}$$

$$\varphi_0 = \sum_{i=1}^{19} a_i G_{i,1}^1 + a_{21}z + a_{22}. \tag{63}$$

$$\psi_0 = \sum_{i=1}^{19} a_i G_{i,2}^0 + a_{23}, \tag{64}$$

$$\chi_0 = \sum_{i=1}^{19} a_i G_{i,3}^0 + a_{24}. \tag{65}$$

where

$$G_{i,j}^n = \int_{-h/2}^{z} (z-\xi)^n G_{i,j}(\xi)\, d\xi, \qquad (i=1,2,...,19, \quad j=1,2,3, \quad n=0,1). \tag{66}$$

The integral constants, a_{20} in Eq. (52), a_{21} and a_{22} in Eqs. (62) and (63), a_{23} in Eq. (64), and a_{24} in Eq. (65), make no contribution to the stress, electric displacement and magnetic induction fields, and hence we shall omit these terms in the following expressions.

3.2. Internal Forces, Average Electric Displacement, Average Magnetic Induction Displacements, Electrical Potential and Magnetic Potential

By virtue of Eqs. $(17)_{1,3}$, $(18)_1$ and $(19)_1$, the axial force T, bending moment M, shear force Q, average electric displacement D^* and average magnetic induction B^* can be obtained as follows

$$T = \int_{-h/2}^{h/2} \sigma_x\, dz = \sum_{k=0}^{2} x^k \left[\varphi_k'(h/2) - \varphi_k'(-h/2)\right], \tag{67}$$

$$M = \int_{-h/2}^{h/2} \sigma_x z\, dz = \sum_{k=0}^{2} x^k \left\{\left[\varphi_k'(h/2) + \varphi_k'(-h/2)\right] h/2 - \varphi_k(h/2) + \varphi_k(-h/2)\right\}, \tag{68}$$

$$Q = \int_{-h/2}^{h/2} \tau_{xz}\, dz = \sum_{k=1}^{2} k x^{k-1} \left[\varphi_k(-h/2) - \varphi_k(h/2)\right], \tag{69}$$

$$D^* = \int_{-h/2}^{h/2} D_x\, d\xi = \sum_{k=0}^{2} x^k \left[\psi_k(h/2) - \psi_k(-h/2)\right], \tag{70}$$

$$B^* = \int_{-h/2}^{h/2} B_x\, d\xi = \sum_{k=0}^{2} x^k \left[\chi_k(h/2) - \chi_k(-h/2)\right]. \tag{71}$$

Substituting Eq. (4) into Eq. (20) and then integrating, by virtue of Eqs. (32c), (43*c*) and (56*c*), we obtain the following expressions for displacements

$$u = \left(s_{11}\varphi_0'' + 2s_{13}\varphi_2 - s_{15}\varphi_1' + d_{11}\psi_0' - d_{31}\psi_1 + b_{11}\chi_0' - b_{31}\chi_1\right)x$$
$$+\frac{x^2}{2}\left(s_{11}\varphi_1'' - 2s_{15}\varphi_2' + d_{11}\psi_1' - 2d_{31}\psi_2 + b_{11}\chi_1' - 2b_{31}\chi_2\right)$$
$$+\frac{x^3}{3}\left(s_{11}\varphi_2'' + d_{11}\psi_2' + b_{11}\chi_2'\right) + \int_{-h/2}^{z} F(\xi)\,d\xi + \omega z + u_0, \tag{72}$$

$$w = \int_{-h/2}^{z}\left(s_{13}\varphi_0'' + 2s_{33}\varphi_2 - s_{35}\varphi_1' + g_{13}\psi_0' - g_{33}\psi_1 + h_{13}\chi_0' - h_{33}\chi_1\right)d\xi$$
$$+x\int_{-h/2}^{z}\left(s_{13}\varphi_1'' - 2s_{35}\varphi_2' + g_{13}\psi_1' - 2g_{33}\psi_2 + h_{13}\chi_1' - 2h_{33}\chi_2\right)d\xi$$
$$+x^2\int_{-h/2}^{z}\left(s_{13}\varphi_2'' + g_{13}\psi_2' + h_{13}\chi_2'\right)d\xi - \frac{x^4}{12}a_3 - \frac{x^3}{6}a_{11} - \frac{x^2}{2}a_{19} - \omega x + w_0, \tag{73}$$

where u_0, w_0 and ω are arbitrary constants, and

$$F(z) = -\int_{-h/2}^{z}\left(s_{13}\varphi_1'' - 2s_{35}\varphi_2' + d_{13}\psi_1' - 2d_{33}\psi_2 + b_{13}\chi_1' - 2b_{33}\chi_2\right)d\xi$$
$$+ s_{15}\varphi_0'' + 2s_{35}\varphi_2 - s_{55}\varphi_1' + d_{15}\psi_0' - d_{35}\psi_1 + b_{15}\chi_0' - b_{35}\chi_1. \tag{74}$$

By virtue of Eq. (5), integrating the first equation in Eq. (21), and substituting it into the second, noticing Eqs. (32*a*), (43*a*) and (56*a*), we have

$$\Phi = x\left(d_{11}\varphi_0'' + 2d_{13}\varphi_2 - d_{15}\varphi_1' - \gamma_{11}\psi_0' + \gamma_{13}\psi_1 - \beta_{11}\chi_0' + \beta_{13}\chi_1\right)$$
$$+\frac{x^2}{2}\left(d_{11}\varphi_1'' - 2d_{15}\varphi_2' - \gamma_{11}\psi_1' + 2\gamma_{13}\psi_2 - \beta_{11}\chi_1' + 2\beta_{13}\chi_2\right) - \frac{x^3}{3}a_1$$
$$+\int_{-h/2}^{z}\left(d_{31}\varphi_0'' + 2d_{33}\varphi_2 - d_{35}\varphi_1' - \gamma_{13}\psi_0' + \gamma_{33}\psi_1 - \beta_{13}\chi_0' + \beta_{33}\chi_1\right)d\xi + \Phi_0, \tag{75}$$

where Φ_0 is an integral constant.

By virtue of Eq. (6), integrating the first equation in Eq. (22), and substituting it into the second, noticing Eqs. (32*b*), (43*b*) and (56*b*), we have

$$\Psi = x\left(b_{11}\varphi_0'' + 2b_{13}\varphi_2 - b_{15}\varphi_1' - \gamma_{11}\psi_0' + \gamma_{13}\psi_1 - \nu_{11}\chi_0' + \nu_{13}\chi_1\right)$$
$$+\frac{x^2}{2}\left(b_{11}\varphi_1'' - 2b_{15}\varphi_2' - \beta_{11}\psi_1' + 2\beta_{13}\psi_2 - \nu_{11}\chi_1' + 2\nu_{13}\chi_2\right) - \frac{x^3}{3}a_2$$
$$+\int_{-h/2}^{z}\left(b_{31}\varphi_0'' + 2b_{33}\varphi_2 - b_{35}\varphi_1' - \beta_{13}\psi_0' + \beta_{33}\psi_1 - \nu_{13}\chi_0' + \nu_{33}\chi_1\right)d\xi + \Psi_0, \tag{76}$$

where Ψ_0 is an integral constant.

3.3. Boundary Conditions

3.3.1. Boundary Conditions on the Upper and Lower Surfaces

Boundary conditions on the upper and lower surfaces of the beam are presented according the load applied on the beam. For example, when the beam is only subjected to a uniform load on its upper surfaces, the boundary conditions are

$$\sigma_z(x,-h/2)=-q,\ \tau_{xz}(x,-h/2)=0,\ D_z(x,-h/2)=0,\ B_z(x,-h/2)=0,$$

$$\sigma_z(x,h/2)=0,\quad \tau_{xz}(x,h/2)=0,\quad D_z(x,h/2)=0,\quad B_z(x,h/2)=0,$$

3.3.2. Boundary Conditions at the Two Ends

The electric boundary conditions at the two ends are

$$D^* = const.,\qquad \text{at } x=0 \text{ or } x=l,$$

or

$$\Phi = const.,\quad \text{at } (0,0) \text{ or } (l,0),$$

The magnetic boundary conditions at the two ends are

$$B^* = const.,\qquad \text{at } x=0 \text{ or } x=l,$$

or

$$\Psi = const.,\quad \text{at } (0,0) \text{ or } (l,0),$$

The mechanic boundary conditions at the two ends can be expressed as follows

Free

$$T=\bar{T},\quad M=\bar{M},\quad Q=\bar{Q},\qquad \text{at } x=0 \text{ or } x=l,$$

where $\bar{T}$, $\bar{M}$ and $\bar{Q}$ are the prescribed values.

(2) Fixed (or clamped)

$$u=w=0,\ \partial w/\partial x=0 \text{ or } \partial u/\partial z=0,\ \text{at } (0,0) \text{ or } (l,0),$$

(3) Hinged

$$T=0,\ M=0,\quad \text{at } x=0 \text{ or } x=l,$$

$$w=0,\ \text{at } (0,0) \text{ or } (l,0),$$

(4) The boundary conditions for a simply supported beam

$$u=0,\quad w=0,\quad \text{at } (0,0),\quad M=0,\quad \text{at } x=0,$$

$$w=0,\quad \text{at } (l,0),\quad T=0,\quad M=0,\quad \text{at } x=l,$$

3.4. Applications

3.4.1 $\varphi=\varphi_0(z)$, $\psi=\psi_0(z)$, $\chi=\chi_0(z)$.

Setting $a_1 \sim a_{15}$ to zero gives rise to $\varphi_1=0$, $\varphi_2=0$, $\psi_1=0$, $\psi_2=0$, $\chi_1=0$ and $\chi_2=0$. Thus from Eqs. (16) and (63)-(65), we obtain the stress function, electric displacement function and magnetic induction function as follows

$$\varphi=\varphi_0=a_{16}G^1_{16,1}+a_{17}G^1_{17,1}+a_{18}G^1_{18,1}+a_{19}G^1_{19,1}, \tag{77}$$

$$\psi=\psi_0=a_{16}G^0_{16,2}+a_{17}G^0_{17,2}+a_{18}G^0_{18,2}+a_{19}G^0_{19,2}, \tag{78}$$

$$\chi=\chi_0=a_{16}G^0_{16,3}+a_{17}G^0_{17,3}+a_{18}G^0_{18,3}+a_{19}G^0_{19,3}. \tag{79}$$

Substituting Eqs. (77)-(79) into Eqs. (17)-(19) gives

$$\sigma_x=\sum_{i=16}^{19}a_iG_{i,1},\quad \sigma_z=0,\quad \tau_{xz}=0; \tag{80}$$

$$D_x=\sum_{i=16}^{19}a_iG_{i,2},\quad D_z=0, \tag{81}$$

$$B_x=\sum_{i=16}^{19}a_iG_{i,3},\quad B_z=0. \tag{82}$$

By virtue of Eqs. (67)-(71) and $\varphi_0(-h/2)=0$, $\varphi_0'(-h/2)=0$, $\psi_0(-h/2)=0$ and $\chi_0(-h/2)=0$, we obtain the axial force, bending moment, average electric displacement D^* and average magnetic induction B^* as

$$T = \varphi_0'\left(\frac{h}{2}\right) = \sum_{i=16}^{19} a_i G_{i,1}^0\left(\frac{h}{2}\right), \tag{83}$$

$$M = \varphi_0'\left(\frac{h}{2}\right)\frac{h}{2} - \varphi_0\left(\frac{h}{2}\right) = \frac{h}{2}\sum_{i=16}^{19} a_i G_{i,1}^0\left(\frac{h}{2}\right) - \sum_{i=16}^{19} a_i G_{i,1}^1\left(\frac{h}{2}\right), \tag{84}$$

$$D^* = \psi_0\left(\frac{h}{2}\right) = \sum_{i=16}^{19} a_i G_{i,2}^0\left(\frac{h}{2}\right), \tag{85}$$

$$B^* = \chi_0\left(\frac{h}{2}\right) = \sum_{i=16}^{19} a_i G_{i,3}^0\left(\frac{h}{2}\right). \tag{86}$$

Example 3.1: A free beam subjected to axial force $\bar{T}$, bending moment $\bar{M}$, average electric displacement $\bar{D}$ and average magnetic induction $\bar{B}$ at the two ends.

Since $T_0 = T_l = \bar{T}$, $M_0 = M_l = \bar{M}$, $D_0^* = D_l^* = \bar{D}$ and $B_0^* = B_l^* = \bar{B}$ at x=0 and x=l, we have

$$\sum_{i=16}^{19} a_i G_{i,1}^0\left(\frac{h}{2}\right) = \bar{T}, \qquad \frac{h}{2}\sum_{i=16}^{19} a_i G_{i,1}^0\left(\frac{h}{2}\right) - \sum_{i=16}^{19} a_i G_{i,1}^1\left(\frac{h}{2}\right) = \bar{M},$$
$$\sum_{i=16}^{19} a_i G_{i,2}^0\left(\frac{h}{2}\right) = \bar{D}, \qquad \sum_{i=16}^{19} a_i G_{i,3}^0\left(\frac{h}{2}\right) = \bar{B}. \tag{87}$$

From Eq. (87), we obtain

$$a_{16} = \frac{1}{\Delta_H}\left[\bar{T}\Delta_H^{11} + \bar{D}\Delta_H^{21} + \bar{B}\Delta_H^{31} + \left(\frac{h}{2}\bar{T} - \bar{M}\right)\Delta_H^{41}\right],$$

$$a_{17} = \frac{1}{\Delta_H}\left[\bar{T}\Delta_H^{12} + \bar{D}\Delta_H^{22} + \bar{B}\Delta_H^{32} + \left(\frac{h}{2}\bar{T} - \bar{M}\right)\Delta_H^{42}\right],$$

$$a_{18} = \frac{1}{\Delta_H}\left[\bar{T}\Delta_H^{13} + \bar{D}\Delta_H^{23} + \bar{B}\Delta_H^{33} + \left(\frac{h}{2}\bar{T} - \bar{M}\right)\Delta_H^{43}\right],$$

$$a_{19} = \frac{1}{\Delta_H}\left[\bar{T}\Delta_H^{14} + \bar{D}\Delta_H^{24} + \bar{B}\Delta_H^{34} + \left(\frac{h}{2}\bar{T} - \bar{M}\right)\Delta_H^{44}\right], \tag{88}$$

where

$$H=\begin{bmatrix} G_{17,1}^{0}\left(\frac{h}{2}\right) & G_{18,1}^{0}\left(\frac{h}{2}\right) & G_{19,1}^{0}\left(\frac{h}{2}\right) & G_{20,1}^{0}\left(\frac{h}{2}\right) \\ G_{17,2}^{0}\left(\frac{h}{2}\right) & G_{18,2}^{0}\left(\frac{h}{2}\right) & G_{19,2}^{0}\left(\frac{h}{2}\right) & G_{20,2}^{0}\left(\frac{h}{2}\right) \\ G_{17,3}^{0}\left(\frac{h}{2}\right) & G_{18,3}^{0}\left(\frac{h}{2}\right) & G_{19,3}^{0}\left(\frac{h}{2}\right) & G_{20,3}^{0}\left(\frac{h}{2}\right) \\ G_{17,1}^{1}\left(\frac{h}{2}\right) & G_{18,1}^{1}\left(\frac{h}{2}\right) & G_{19,1}^{1}\left(\frac{h}{2}\right) & G_{20,1}^{1}\left(\frac{h}{2}\right) \end{bmatrix}. \tag{89}$$

and $\Delta_H = \det H$, Δ_H^{ij} is the algebraic complement for element (i, j) of the matrix H.

Substitution of Eq. (88) into Eqs. (80)-(82) gives

$$\begin{aligned} \sigma_x = & \frac{\bar{T}}{\Delta_H}\left(\sum_{i=16}^{19} G_{i,1}\Delta_H^{1(i-15)} + \frac{h}{2}\sum_{i=16}^{19} G_{i,1}\Delta_H^{4(i-16)}\right) - \frac{\bar{M}}{\Delta_H}\sum_{i=16}^{19} G_{i,1}\Delta_H^{4(i-15)} \\ & + \frac{\bar{D}}{\Delta_H}\sum_{i=16}^{19} G_{i,1}\Delta_H^{2(i-15)} + \frac{\bar{B}}{\Delta_H}\sum_{i=16}^{19} G_{i,1}\Delta_H^{3(i-15)}, \\ \sigma_z = & \ 0, \qquad \tau_{xz} = 0; \end{aligned} \tag{90}$$

$$\begin{aligned} D_x = & \frac{\bar{T}}{\Delta_H}\left(\sum_{i=16}^{19} G_{i,2}\Delta_H^{1(i-15)} + \frac{h}{2}\sum_{i=16}^{19} G_{i,2}\Delta_H^{4(i-15)}\right) - \frac{\bar{M}}{\Delta_H}\sum_{i=16}^{19} G_{i,2}\Delta_H^{4(i-15)} \\ & + \frac{\bar{D}}{\Delta_H}\sum_{i=16}^{19} G_{i,2}\Delta_H^{2(i-15)} + \frac{\bar{B}}{\Delta_H}\sum_{i=16}^{19} G_{i,2}\Delta_H^{3(i-15)}, \\ D_z = & \ 0; \end{aligned} \tag{91}$$

$$\begin{aligned} B_x = & \frac{\bar{T}}{\Delta_H}\left(\sum_{i=16}^{19} G_{i,3}\Delta_H^{1(i-15)} + \frac{h}{2}\sum_{i=16}^{19} G_{i,3}\Delta_H^{4(i-15)}\right) - \frac{\bar{M}}{\Delta_H}\sum_{i=16}^{19} G_{i,3}\Delta_H^{4(i-15)} \\ & + \frac{\bar{D}}{\Delta_H}\sum_{i=16}^{19} G_{i,3}\Delta_H^{2(i-15)} + \frac{\bar{B}}{\Delta_H}\sum_{i=16}^{19} G_{i,3}\Delta_H^{3(i-15)}, \\ B_z = & \ 0. \end{aligned} \tag{92}$$

From the above general expressions, we can obtain the results for several particular cases: (1) when $\bar{T} \neq 0,\ \bar{M} = 0,\ \bar{D} = 0,\ \bar{B} = 0$, they correspond to a simple tension/compression problem; (2) when $\bar{T} = 0,\ \bar{M} \neq 0,\ \bar{D} = 0,\ \bar{B} = 0$, they are for a bending problem; (3) when $\bar{T} = 0, \bar{M} = 0, \bar{D} \neq 0, \bar{B} = 0$, they become the solution for the problem of a beam subjected to electric displacement in the x direction; and (4) when

$\bar{T}=0, \bar{M}=0, \bar{D}=0, \bar{B}\neq 0$, we obtain the ones for a beam subjected to magnetic induction in the x direction.

For homogeneous piezoelectric materials, the coefficients s_{ij}, d_{ij}, b_{ij}, γ_{ij}, β_{ij} and ν_{ij} keep constant. Thus we have

$$\sigma_x=\frac{\bar{M}z}{h^3/12}+\frac{\bar{T}}{h},\qquad \sigma_z=\tau_{xz}=0; \tag{93}$$

$$D_x=\frac{\bar{M}z}{h^3/12}\cdot\frac{b_{11}\beta_{11}-d_{11}\nu_{11}}{\beta_{11}^2-\gamma_{11}\nu_{11}}+\frac{\bar{D}}{h},\quad D_z=0; \tag{94}$$

$$B_x=\frac{\bar{M}z}{h^3/12}\cdot\frac{d_{11}\beta_{11}-b_{11}\gamma_{11}}{\beta_{11}^2-\gamma_{11}\nu_{11}}+\frac{\bar{B}}{h},\qquad B_z=0. \tag{95}$$

Eq. (93) coincides with the results of a homogeneous isotropic beam, as presented in Ref. [18], the stresses being independent of material constants. From the first equation in Eqs. (93)-(95), we find the axial force $\bar{T}$ will not influence the electric displacements and magnetic inductions, the average electric displacement $\bar{D}$ will not influence the stress σ_x and magnetic induction B_x, and the average magnetic induction will not influence the stress σ_x and electric displacement D_x. While for FGMEEs, from Eqs. (90)-(92), we find $\bar{T}$, $\bar{M}$, $\bar{D}$ and $\bar{B}$ will have effects on σ_x, D_x and B_x in a general way.

3.4.2. $\varphi=\varphi_0(z)+x\varphi_1(z),\quad \psi=\psi_0(z)+x\psi_1(z),\quad \chi=\chi_0(z)+x\chi_1(z).$

On setting $a_1\sim a_8$ to zero, we have $\varphi_2=0,\ \psi_2=0,\ \chi_2=0$. Thus from Eqs. (16), (52)-(54) and (63)-(65), we obtain the stress function, electric displacement function and magnetic induction function as

$$\varphi=\sum_{i=9}^{19}a_iG_{i,1}^1+x\left(a_9D_{9,1}^1+a_{10}D_{10,1}^1+a_{11}D_{11,1}^1+a_{12}D_{12,1}^1+a_{13}z\right), \tag{96}$$

$$\psi=\sum_{i=9}^{19}a_iG_{i,2}^0+x\left(a_9D_{9,2}^0+a_{10}D_{10,2}^0+a_{11}D_{11,2}^0+a_{12}D_{12,2}^0\right), \tag{97}$$

$$\chi=\sum_{i=9}^{19}a_iG_{i,3}^0+x\left(a_9D_{9,3}^0+a_{10}D_{10,3}^0+a_{11}D_{11,3}^0+a_{12}D_{12,3}^0+a_{14}\right). \tag{98}$$

Substitution of Eqs. (96)-(98) into Eqs. (17)-(19) gives

$$\sigma_x = \sum_{i=9}^{19} a_i G_{i,1} + x\left(a_9 D_{9,1} + a_{10} D_{10,1} + a_{11} D_{11,1} + a_{12} D_{12,1}\right),$$
$$\sigma_z = 0,$$
$$\tau_{xz} = -\left(a_9 D_{9,1}^0 + a_{10} D_{10,1}^0 + a_{11} D_{11,1}^0 + a_{12} D_{12,1}^0 + a_{13}\right); \quad (99)$$

$$D_x = \sum_{i=9}^{19} a_i G_{i,2} + x\left(a_9 D_{9,2} + a_{10} D_{10,2} + a_{11} D_{11,2} + a_{12} D_{12,2}\right),$$
$$D_z = -\left(a_9 D_{9,2}^0 + a_{10} D_{10,2}^0 + a_{11} D_{11,2}^0 + a_{12} D_{12,2}^0 + a_{14}\right); \quad (100)$$

$$B_x = \sum_{i=9}^{19} a_i G_{i,3} + x\left(a_9 D_{9,3} + a_{10} D_{10,3} + a_{11} D_{11,3} + a_{12} D_{12,3}\right),$$
$$B_z = -\left(a_9 D_{9,3}^0 + a_{10} D_{10,3}^0 + a_{11} D_{11,3}^0 + a_{12} D_{12,3}^0 + a_{15}\right). \quad (101)$$

From Eqs. (67)-(71) and $\varphi_k(-h/2) = 0$, $\varphi_k'(-h/2) = 0$, $\psi_k(-h/2) = 0$, $\chi_k(-h/2) = 0$, $(k = 0,1)$, we obtain the axial force *T*, bending moment *M*, average electric displacement D^* and average magnetic induction B^* as follows

$$T = \sum_{i=9}^{19} a_i G_{i,1}^0\left(\frac{h}{2}\right) + x\left[\sum_{i=9}^{12} a_i D_{i,1}^0\left(\frac{h}{2}\right) + a_{13}\right], \quad (102)$$

$$Q = -\sum_{i=1}^{12} a_i D_{i,1}^1\left(\frac{h}{2}\right), \quad (103)$$

$$M = \frac{h}{2}\sum_{i=9}^{19} a_i G_{i,1}^0\left(\frac{h}{2}\right) - \sum_{i=9}^{20} a_i G_{i,1}^1\left(\frac{h}{2}\right) + x\left[\left(\sum_{i=9}^{12} a_i D_{i,1}^0 + a_{13}\right)\frac{h}{2} - \left(\sum_{i=9}^{12} a_i D_{i,1}^1 + a_{13} z\right)\right], \quad (104)$$

$$D^* = \sum_{i=9}^{19} a_i G_{i,2}^0\left(\frac{h}{2}\right) + x\left[\sum_{i=9}^{12} a_i D_{i,2}^0\left(\frac{h}{2}\right) + a_{14}\right], \quad (105)$$

$$B^* = \sum_{i=9}^{19} a_i G_{i,3}^0\left(\frac{h}{2}\right) + x\left[\sum_{i=9}^{12} a_i D_{i,3}^0\left(\frac{h}{2}\right) + a_{15}\right]. \quad (106)$$

Example 3.2: A cantilever beam subjected to a transverse force *P* at the free end. The boundary conditions are

$$z = \pm h/2: \quad \sigma_z = 0, \quad \tau_{xz} = 0, \quad D_z = 0, \quad B_z = 0, \quad (107)$$

$$x = 0: \quad T = 0, \quad M = 0, \quad Q = -P, \quad D^* = 0, \quad B^* = 0. \tag{108}$$

From the boundary conditions $\tau_{xz} = 0$, $D_z = 0$, $B_z = 0$ at $z = -h/2$ and Eqs. $(99)_3$, $(100)_2$ and $(101)_2$, we find

$$a_{13} = 0, \quad a_{14} = 0, \quad a_{15} = 0. \tag{109}$$

From the boundary conditions $\tau_{xz} = 0$, $D_z = 0$, $B_z = 0$ at $z = h/2$ and Eqs. (99)-(101), we have

$$a_9 D_{9,1}^0\left(\frac{h}{2}\right) + a_{10} D_{10,1}^0\left(\frac{h}{2}\right) + a_{11} D_{11,1}^0\left(\frac{h}{2}\right) + a_{12} D_{12,1}^0\left(\frac{h}{2}\right) = 0, \tag{110}$$

$$a_9 D_{9,2}^0\left(\frac{h}{2}\right) + a_{10} D_{10,2}^0\left(\frac{h}{2}\right) + a_{11} D_{11,2}^0\left(\frac{h}{2}\right) + a_{12} D_{12,2}^0\left(\frac{h}{2}\right) = 0, \tag{111}$$

$$a_9 D_{9,3}^0\left(\frac{h}{2}\right) + a_{10} D_{10,3}^0\left(\frac{h}{2}\right) + a_{11} D_{11,3}^0\left(\frac{h}{2}\right) + a_{12} D_{12,3}^0\left(\frac{h}{2}\right) = 0. \tag{112}$$

The boundary conditions in Eq. (108) along with Eqs. (102)-(106) yield

$$\sum_{i=9}^{19} a_i G_{i,1}^0\left(\frac{h}{2}\right) = 0, \tag{113}$$

$$\frac{h}{2}\sum_{i=9}^{19} a_i G_{i,1}^0\left(\frac{h}{2}\right) - \sum_{i=9}^{19} a_i G_{i,1}^1\left(\frac{h}{2}\right) = 0, \tag{114}$$

$$\sum_{i=9}^{12} a_i D_{i,1}^1\left(\frac{h}{2}\right) = P, \tag{115}$$

$$\sum_{i=9}^{19} a_i G_{i,2}^0\left(\frac{h}{2}\right) = 0, \tag{116}$$

$$\sum_{i=9}^{19} a_i G_{i,3}^0\left(\frac{h}{2}\right) = 0. \tag{117}$$

From Eqs. (110)-(117), we can obtain the eight constants $a_9 \sim a_{12}$ and $a_{16} \sim a_{19}$, and hence completely determine the expressions in Eqs. (99)-(101).

3.4.3. $\varphi=\varphi_0(z)+x\varphi_1(z)+x^2\varphi_2(z), \quad \psi=\psi_0(z)+x\psi_1(z)+x^2\psi_2(z),$
$\chi=\chi_0(z)+x\chi_1(z)+x^2\chi_2(z).$

When the stress function, electric displacement function and magnetic induction function take the forms of Eq. (16), we obtain from Eqs. (17)-(19)

$$\sigma_x=\sum_{i=1}^{19}a_iG_{i,1}+x\sum_{i=1}^{12}a_iD_{i,1}+x^2\sum_{i=1}^{4}a_iA_{i,1},$$
$$\sigma_z=2\left(\sum_{i=1}^{4}a_iA_{i,1}^1+a_5z+a_6\right),$$
$$\tau_{xz}=-\left(\sum_{i=1}^{5}a_iD_{i,1}^0+\sum_{i=7}^{12}a_iD_{i,1}^0+a_{13}\right)-2x\left(\sum_{i=1}^{4}a_iA_{i,1}^0+a_5\right); \tag{118}$$

$$D_x=\sum_{i=9}^{19}a_iG_{i,2}+x\left(\sum_{i=1}^{5}a_iD_{i,2}+\sum_{i=7}^{12}a_iD_{i,2}\right)+x^2\sum_{i=1}^{4}a_iA_{i,2},$$
$$D_z=-\left(\sum_{i=1}^{5}a_iD_{i,2}^0+\sum_{i=7}^{12}a_iD_{i,2}^0+a_{14}\right)-2x\left(\sum_{i=1}^{4}a_iA_{i,2}^0+a_7\right); \tag{119}$$

$$B_x=\sum_{i=9}^{19}a_iG_{i,3}+x\left(\sum_{i=1}^{5}a_iD_{i,3}+\sum_{i=7}^{12}a_iD_{i,3}\right)+x^2\sum_{i=1}^{4}a_iA_{i,3},$$
$$B_z=-\left(\sum_{i=1}^{5}a_iD_{i,3}^0+\sum_{i=7}^{12}a_iD_{i,3}^0+a_{15}\right)-2x\left(\sum_{i=1}^{4}a_iA_{i,3}^0+a_8\right). \tag{120}$$

From Eqs. (67)-(71), and by virtue of $\varphi_k(-h/2)=0, \quad \varphi_k'(-h/2)=0, \ \psi_k(-h/2)=0$ and $\chi_k(-h/2)=0, \ (k=0,1,2)$, the axial force *T*, bending moment *M*, average electric displacement D^* and average magnetic induction B^* at the free end (x=0) can be obtained as

$$T=\sum_{i=1}^{19}a_iG_{i,1}^0\left(\frac{h}{2}\right), \tag{121}$$
$$M=\frac{h}{2}\sum_{i=1}^{19}a_iG_{i,1}^0\left(\frac{h}{2}\right)-\sum_{i=1}^{19}a_iG_{i,1}^1\left(\frac{h}{2}\right), \tag{122}$$

$$Q=-\sum_{i=1}^{12}a_iD_{i,1}^1\left(\frac{h}{2}\right), \tag{123}$$

$$D^* = \sum_{i=1}^{19} a_i G_{i,2}^0 \left(\frac{h}{2}\right), \tag{124}$$

$$B^* = \sum_{i=1}^{19} a_i G_{i,3}^0 \left(\frac{h}{2}\right). \tag{125}$$

Example 3.3: A cantilever beam subjected to a uniform mechanical load q.
The boundary conditions are

$$z = -h/2: \quad \sigma_z = -q, \quad \tau_{xz} = 0, \quad D_z = 0, \quad B_z = 0, \tag{126}$$

$$z = h/2: \quad \sigma_z = 0, \quad \tau_{xz} = 0, \quad D_z = 0, \quad B_z = 0, \tag{127}$$

$$x = 0: \quad T = 0, \quad M = 0, \quad Q = 0, \quad D^* = 0, \quad B^* = 0. \tag{128}$$

From Eq. (126) and Eqs. (118)-(120), we have $a_5 = 0$, $a_6 = -\frac{q}{2}$, $a_7 = 0$, $a_8 = 0$, $a_{13} = 0$, $a_{14} = 0$, $a_{15} = 0$. From Eqs. (127) and (118)-(120), we get

$$\sum_{i=1}^{4} a_i A_{i,1}^1 \left(\frac{h}{2}\right) = \frac{q}{2}, \tag{129}$$

$$\sum_{i=1}^{4} a_i A_{i,1}^0 \left(\frac{h}{2}\right) = 0, \tag{130}$$

$$\sum_{i=1}^{4} a_i A_{i,2}^0 \left(\frac{h}{2}\right) = 0, \tag{131}$$

$$\sum_{i=1}^{4} a_i A_{i,3}^0 \left(\frac{h}{2}\right) = 0, \tag{132}$$

$$\sum_{i=1}^{12} a_i D_{i,1}^0 \left(\frac{h}{2}\right) = 0, \tag{133}$$

$$\sum_{i=1}^{12} a_i D_{i,2}^0 \left(\frac{h}{2} \right) = 0, \tag{134}$$

$$\sum_{i=1}^{12} a_i D_{i,3}^0 \left(\frac{h}{2} \right) = 0. \tag{135}$$

From Eqs. (128) and (121)-(125), we obtain

$$\sum_{i=1}^{19} a_i G_{i,1}^0 \left(\frac{h}{2} \right) = 0, \tag{136}$$

$$\sum_{i=1}^{19} a_i G_{i,1}^1 \left(\frac{h}{2} \right) = 0, \tag{137}$$

$$\sum_{i=1}^{12} a_i D_{i,1}^1 \left(\frac{h}{2} \right) = 0, \tag{138}$$

$$\sum_{i=1}^{19} a_i G_{i,2}^0 \left(\frac{h}{2} \right) = 0, \tag{139}$$

$$\sum_{i=1}^{19} a_i G_{i,3}^0 \left(\frac{h}{2} \right) = 0. \tag{140}$$

Thus, Eqs. (129)-(132) determine $a_1 \sim a_4$, Eqs. (133)-(135) and (138) give rise to $a_9 \sim a_{12}$, while Eqs. (136), (137), (139) and (140) fix $a_{16} \sim a_{19}$. Thus the expressions in Eqs.(118)-(120) are completely determined.

3.5. A Numerical Example

For numerical calculations, we consider an FGMEE cantilever beam subjected to a concentrated force P at its free end. The parameters are as follows: the span of the beam is $l = 0.6$ m, the height $h = 0.02$ m, the width $t = 0.002$ m, and the load $P = 1000$ N/m. Boundary conditions for the fixed-end are: $u = w = 0$, $\partial w/\partial x = 0$, $\Phi = 0$, $\Psi = 0$, at point $(l, 0)$. The material parameters s_{11}, γ_{11}, β_{11} and ν_{11} are exponential functions of z, i.e.

$$s_{11}(z) = s_{11}^0 \exp\left(\lambda \frac{z + h/2}{h} \right), \qquad \gamma_{11}(z) = \gamma_{11}^0 \exp\left(\lambda \frac{z + h/2}{h} \right),$$

$$\beta_{11}(z) = \beta_{11}^0 \exp\left(\lambda \frac{z + h/2}{h}\right), \qquad \nu_{11}(z) = \nu_{11}^0 \exp\left(\lambda \frac{z + h/2}{h}\right), \tag{141}$$

where λ is an arbitrary constant, s_{11}^0, γ_{11}^0, β_{11}^0 and ν_{11}^0 are coefficients of material parameters at $z = -h/2$. All other material parameters are constant. The material properties are listed in Table 1.

Table 1. Material properties

(s_{ij}^0 *in* $10^{-12}\, m^2/N$, d_{ij}^0 *in* m^2/C, b_{ij}^0 *in* $10^{-3}\, m/NA$, γ_{ij}^0 *in* $10^9\, m/F$, β_{ij}^0 *in* $10^6\, C^2/Ns^2$, ν_{ij}^0 *in* $10^6\, CV/Ns$)

s_{11}^0	s_{13}^0	s_{15}^0	s_{33}^0	s_{35}^0	s_{55}^0	d_{11}^0	d_{13}^0	d_{15}^0
9.235	-4.736	0	14.113	0	-33.883	0	0	0.0351
d_{31}^0	d_{33}^0	d_{35}^0	b_{11}^0	b_{13}^0	b_{15}^0	b_{31}^0	b_{33}^0	b_{35}^0
0.0102	-0.0225	0	0	0	3.727	-0.205	-0.713	0
γ_{11}^0	γ_{13}^0	γ_{13}^0	β_{11}^0	β_{13}^0	β_{33}^0	ν_{11}^0	ν_{13}^0	ν_{33}^0
0.0529	0	0.116	-3.860	0	0.981	-0.210	0	0.162

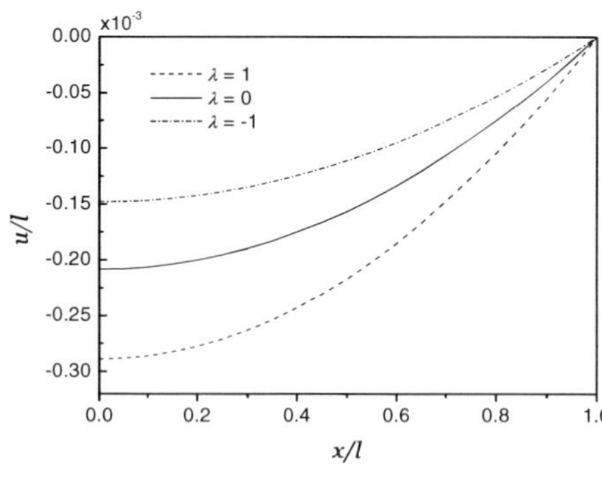

Figure 2. u/l versus x/l $(z = -h/2)$

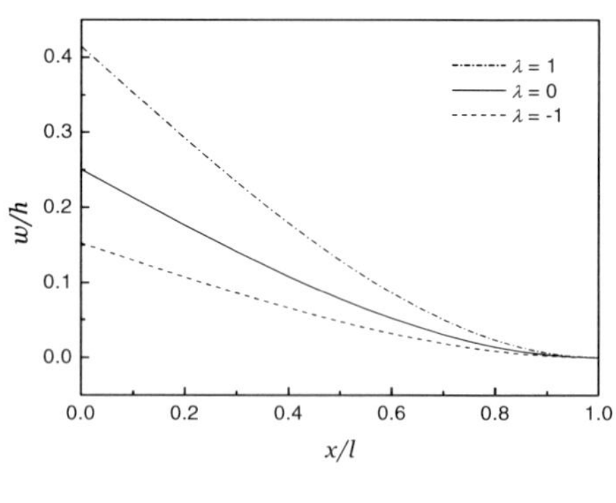

Figure 3. w/h versus x/l.

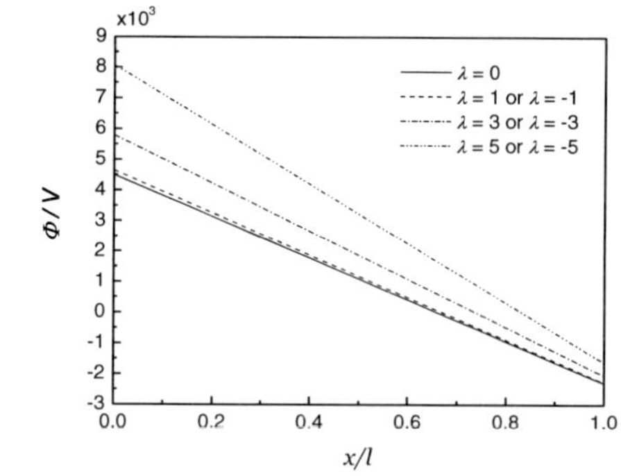

Figure 4. Φ versus x/l $(z = h/2)$

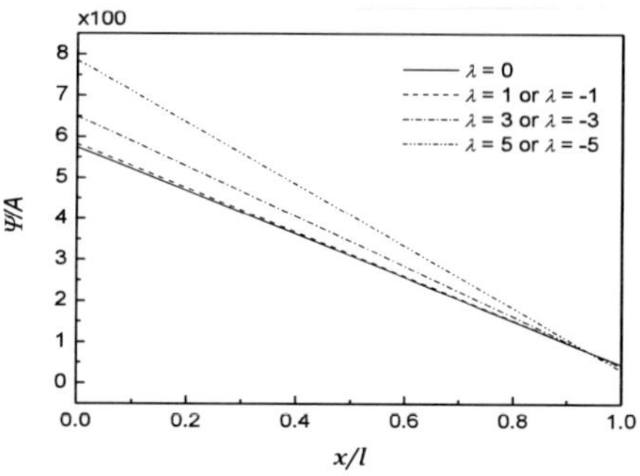

Figure 5. Ψ versus x/l $(z = h/2)$.

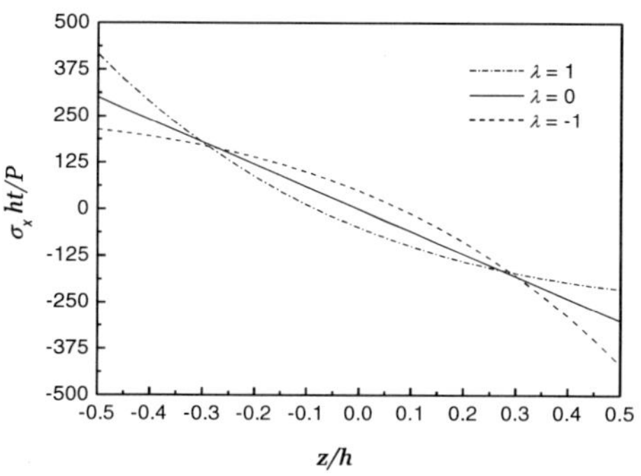

Figure 6. $\sigma_x ht/P$ versus z/h $(x = l/6)$

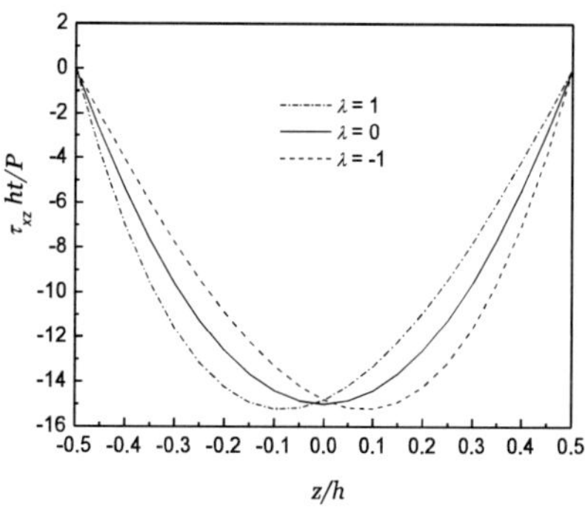

Figure 7. $\tau_{xz} ht/P$ versus z/h $(x = l/6)$.

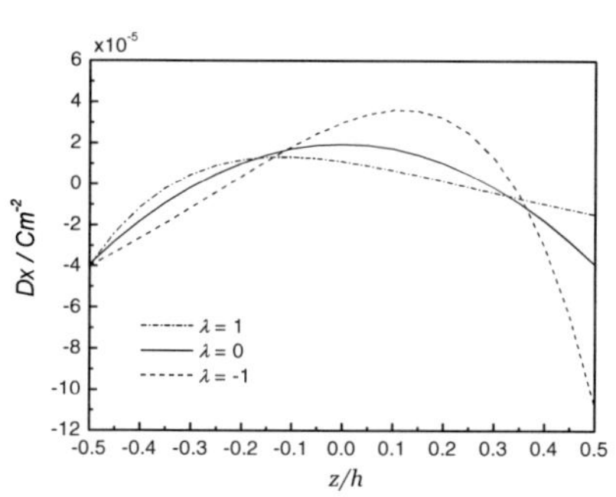

Figure 8. D_x versus z/h $(x = l/6)$

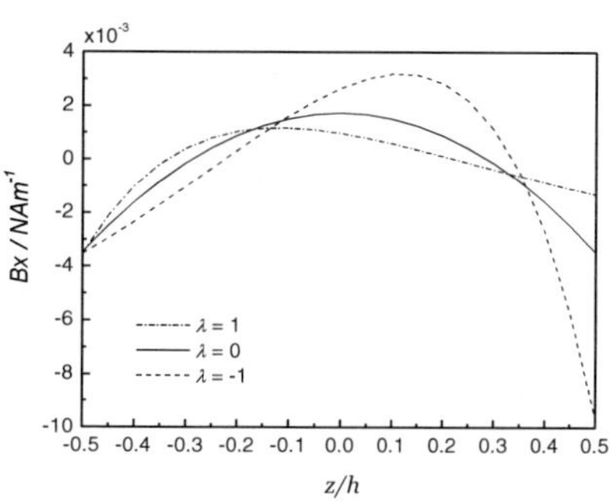

Figure 9. B_x versus z/h $(x = l/6)$.

The numerical results can be carried out with Eqs. (72), (73), (75), (76) and (99)-(101). For the problem, we obtain $D_z = 0$ and $B_z = 0$, and some other results are given in the form of figures. The curves for displacements, electric potential, magnetic potential, stresses, electric displacements and magnetic inductions are depicted in Figures 2-9, clearly showing the effect of the inhomogeneity parameter λ. As seen from Figures 2 and 3, both the displacement components, u at $z = -h/2$ and w at $z = 0$, increase with λ. From Figures 4 and 5, we find that the electric potential curve and magnetic potential curve for $\lambda = 1, 3, 5$ coincide with those for $\lambda = -1, -3, -5$, which means that the electric potential and magnetic potential are dependent only on the absolute value of λ, irrelative to the sign of λ. The two potentials increase with the absolute value of λ. The curves for the dimensionless stress $\sigma_x ht/P$ versus z/h at $x = l/6$ are shown in Figure 6. The curve for $\lambda = 1$ is concave downwards, it becomes a straight line for $\lambda = 0$, while it is concave upwards for $\lambda = -1$. The curves for the electric displacement component D_x and magnetic induction component B_x are shown in Figures 8 and 9, indicating similar distributions of the two quantities, which is reasonable since the electric field and magnetic field almost play an identical role in the current problem.

4. FGMEE Beams Subjected to Arbitrary Loading

4.1. Stress Function, Electric Displacement Function and Magnetic Induction Function

Consider an FGMEE beam subjected to an arbitrary distributed loading $q(x)$, as shown in Figure 10, with $q(0) = q(l) = 0$. The loading $q(x)$ can be expanded into a sinusoidal series as

$$q(x) = \sum_{n=1}^{\infty} q_n \sin \alpha_n x, \tag{142}$$

where $\alpha_n = n\pi/l$, n=1,2,3…, and

$$q_n = \frac{2}{l}\int_0^l q(x)\sin(\alpha_n x)\,dx. \tag{143}$$

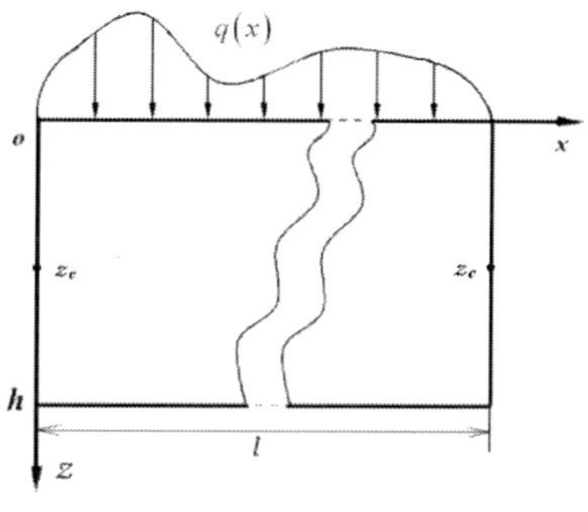

Figure 10. An FGMEE beam and the coordinate system.

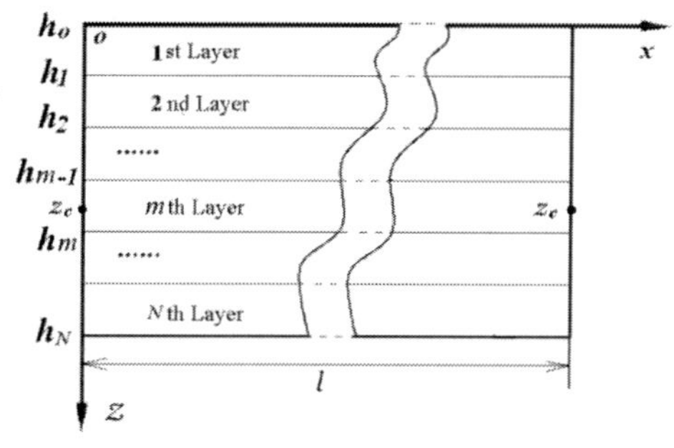

Figure 11. The sub-layers along the thickness.

By the superposition theorem for linear problems, we can only consider the representative term in Eq. (142), $q_n \sin(\alpha_n x)$.

We further assume that

$$\begin{aligned} \varphi &= \sin(\alpha_n x)F(z)+\cos(\alpha_n x)G(z)+\varphi_0(z)+x\varphi_1(z), \\ \psi &= \sin(\alpha_n x)R(z)+\cos(\alpha_n x)S(z)+\psi_0(z)+x\psi_1(z), \\ \chi &= \sin(\alpha_n x)U(z)+\cos(\alpha_n x)V(z)+\chi_0(z)+x\chi_1(z). \end{aligned} \tag{144}$$

where $F(z)$, $G(z)$, $R(z)$, $S(z)$, $U(z)$, $V(z)$, $\varphi_0(z)$, $\varphi_1(z)$, $\psi_0(z)$, $\psi_1(z)$, $\chi_0(z)$ and $\chi_1(z)$ are unknown functions to be determined. Substitution of Eq. (144) into Eqs. (13)-(15), respectively, yields

$$\begin{aligned} \sigma_x &= \left[\sin(\alpha_n x)F''+\cos(\alpha_n x)G''\right]+\varphi_0''+x\varphi_1'', \\ \sigma_z &= -\alpha_n^2\left[\sin(\alpha_n x)F+\cos(\alpha_n x)G\right], \\ \tau_{xz} &= -\alpha_n\left[\cos(\alpha_n x)F'-\sin(\alpha_n x)G'\right]-\varphi_1'; \end{aligned} \tag{145}$$

$$\begin{aligned} D_x &= \left[\sin(\alpha_n x)R'+\cos(\alpha_n x)S'\right]+\psi_0'+x\psi_1', \\ D_z &= -\alpha_n\left[\cos(\alpha_n x)R-\sin(\alpha_n x)S\right]-\psi_1; \end{aligned} \tag{146}$$

$$\begin{aligned} B_x &= \left[\sin(\alpha_n x)U'+\cos(\alpha_n x)V'\right]+\chi_0'+x\chi_1', \\ B_z &= -\alpha_n\left[\cos(\alpha_n x)U-\sin(\alpha_n x)V\right]-\chi_1. \end{aligned} \tag{147}$$

where a prime denotes derivative with respect to z. Substitutions of Eqs. (145)-(147) into Eqs. (7)-(9), and then into Eqs. (10)-(12), give rise to

$$\begin{aligned}
&\left(s_{11}F''-\alpha_n^2 s_{13}F+\alpha_n s_{15}G'+d_{11}R'+\alpha_n d_{31}S+b_{11}U'+\alpha_n b_{31}V\right)''\\
&\quad -\alpha_n^2\left(s_{13}F''-\alpha_n^2 s_{33}F+\alpha_n s_{35}G'+d_{13}R'+\alpha_n d_{33}S+b_{13}U'+\alpha_n b_{33}V\right)\\
&\quad +\alpha_n\left(s_{15}G''-\alpha_n^2 s_{35}G-\alpha_n s_{55}F'+d_{15}S'-\alpha_n d_{35}R+b_{15}V'-\alpha_n b_{35}U\right)'=0,\\
&\left(s_{11}G''-\alpha_n^2 s_{13}G-\alpha_n s_{15}F'+d_{11}S'-\alpha_n d_{31}R+b_{11}V'-\alpha_n b_{31}U\right)''\\
&\quad -\alpha_n^2\left(s_{13}G''-\alpha_n^2 s_{33}G-\alpha_n s_{35}F'+d_{13}S'-\alpha_n d_{33}R+b_{13}V'-\alpha_n b_{33}U\right)\\
&\quad -\alpha_n\left(s_{15}F''-\alpha_n^2 s_{35}F+\alpha_n s_{55}G'+d_{15}R'+\alpha_n d_{35}S+b_{15}U'+\alpha_n b_{35}V\right)'=0,\\
&\left(d_{11}F''-\alpha_n^2 d_{13}F+\alpha_n d_{15}G'-\gamma_{11}R'-\alpha_n\gamma_{13}S-\beta_{11}U'-\alpha_n\beta_{13}V\right)'\\
&\quad +\alpha_n\left(d_{31}G''-\alpha_n^2 d_{33}G-\alpha_n d_{35}F'-\gamma_{13}S'+\alpha_n\gamma_{33}R-\beta_{13}V'+\alpha_n\beta_{33}U\right)=0,\\
&\left(d_{11}G''-\alpha_n^2 d_{13}G-\alpha_n d_{15}F'-\gamma_{11}S'+\alpha_n\gamma_{13}R-\beta_{11}V'+\alpha_n\beta_{13}U\right)'\\
&\quad -\alpha_n\left(d_{31}F''-\alpha_n^2 d_{33}F+\alpha_n d_{35}G'-\gamma_{13}R'-\alpha_n\gamma_{33}S-\beta_{13}U'-\alpha_n\beta_{33}V\right)=0,\\
&\left(b_{11}F''-\alpha_n^2 b_{13}F+\alpha_n b_{15}G'-\beta_{11}R'-\alpha_n\beta_{13}S-\nu_{11}U'-\alpha_n\nu_{13}V\right)'\\
&\quad +\alpha_n\left(b_{31}G''-\alpha_n^2 b_{33}G-\alpha_n b_{35}F'-\beta_{13}S'+\alpha_n\beta_{33}R-\nu_{13}V'+\alpha_n\nu_{33}U\right)=0,\\
&\left(b_{11}G''-\alpha_n^2 b_{13}G-\alpha_n b_{15}F'-\beta_{11}S'+\alpha_n\beta_{13}R-\nu_{11}V'+\alpha_n\nu_{13}U\right)'\\
&\quad -\alpha_n\left(b_{31}F''-\alpha_n^2 b_{33}F+\alpha_n b_{35}G'-\beta_{13}R'-\alpha_n\beta_{33}S-\nu_{13}U'-\alpha_n\nu_{33}V\right)=0,
\end{aligned} \tag{148}$$

$$\begin{aligned}
&\left(s_{11}\varphi_1''+d_{11}\psi_1'+b_{11}\chi_1'\right)''=0, \qquad \left(-d_{11}\varphi_1''+\gamma_{11}\psi_1'+\beta_{11}\chi_1'\right)'=0,\\
&\left(-b_{11}\varphi_1''+\beta_{11}\psi_1'+\nu_{11}\chi_1'\right)'=0,\\
&\left(s_{11}\varphi_0''-s_{15}\varphi_1'+d_{11}\psi_0'-d_{31}\psi_1+b_{11}\chi_0'-b_{31}\chi_1\right)''-\left(s_{15}\varphi_1''+d_{15}\psi_1'+b_{15}\chi_1'\right)'=0,\\
&\left(-d_{11}\varphi_0''+d_{15}\varphi_1'+\gamma_{11}\psi_0'-\gamma_{13}\psi_1+\beta_{11}\chi_0'-\beta_{13}\chi_1\right)'+\left(d_{31}\varphi_1''-\gamma_{13}\psi_1'-\beta_{13}\chi_1'\right)=0,\\
&\left(-b_{11}\varphi_0''+b_{15}\varphi_1'+\beta_{11}\psi_0'-\beta_{13}\psi_1+\nu_{11}\chi_0'-\nu_{13}\chi_1\right)'+\left(b_{31}\varphi_1''-\beta_{13}\psi_1'-\nu_{13}\chi_1'\right)=0.
\end{aligned} \tag{149}$$

The unknown functions F, G, R, S, U, V, φ_1, φ_0, ψ_1, ψ_0, χ_1 and χ_0 can be obtained by integrating Eqs. (148) and (149), where the integral constants can be fixed with the boundary conditions. However, when the material properties M_{ij} vary in an arbitrary continuous fashion (where M_{ij} stand for s_{jk}, d_{jk}, b_{jk}, ν_{jk}, γ_{jk} or β_{jk}), although the analytical solutions can be obtained from Eq. (149), it is difficult to obtain those of Eq. (148) expect when M_{ij} vary in certain special forms (for instance, in form of exponential function). Hence, for the general case, the approach of sub-layer approximation will be used to acquire the semi-analytical solutions. With this approach the beam is divided into N layers virtually, as shown in Figure 11, with the height h_m-h_{m-1} (m = 1,2, …, N) of the mth layer, where $h_0 = 0,\ h_N = h$. When the height of each sub-layer is small enough, the material parameters within the sub-layer can be considered constant, and the material properties in the mth layer can be taken as

$$M_{jk}^{(m)} = M_{jk}\left(h_{m-1}/2 + h_m/2\right).$$

4.2. Boundary Conditions

4.2.1 Boundary Conditions on the Upper and Lower Surfaces

Boundary conditions on the upper and lower surfaces of the beam are

$$\sigma_z(x,0) = -q_n \sin\alpha_n x, \quad \tau_{xz}(x,0) = 0, \quad D_z(x,0) = 0, \quad B_z(x,0) = 0, \tag{150}$$

$$\sigma_z(x,h) = 0, \quad \tau_{xz}(x,h) = 0, \quad D_z(x,h) = 0, \quad B_z(x,h) = 0, \tag{151}$$

Substitutions of Eqs. $(145)_{2,3}$, $(146)_2$ and $(147)_2$ into Eqs. (150) and (151) give

$$F(0) = q_n/\alpha_n^2, \qquad F'(0) = 0, \qquad F(h) = 0, \qquad F'(h) = 0, \tag{152}$$

$$G(0) = 0, \qquad G'(0) = 0, \qquad G(h) = 0, \qquad G'(h) = 0, \tag{153}$$

$$R(0) = 0, \qquad R(h) = 0, \qquad S(0) = 0, \qquad S(h) = 0, \tag{154}$$

$$U(0) = 0, \qquad U(h) = 0, \qquad V(0) = 0, \qquad V(h) = 0, \tag{155}$$

$$\varphi_1'(0) = 0, \quad \varphi_1'(h) = 0, \quad \psi_1(0) = 0, \quad \psi_1(h) = 0, \quad \chi_1(0) = 0, \quad \chi_1(h) = 0, \tag{156}$$

When the sub-layer approximation approach is used, the boundary conditions in Eqs. (152)-(155) should be rewritten accordingly, i.e.,

$$F_1(0) = q_n/\alpha_n^2, \quad F_N(h) = 0, \quad F_1'(0) = 0, \quad F_N'(h) = 0, \tag{157}$$

$$G_1(0) = 0, \quad G_N(h) = 0, \quad G_1'(0) = 0, \quad G_N'(h) = 0, \tag{158}$$

$$R_1(0) = 0, \quad R_N(h) = 0, \quad S_1(0) = 0, \quad S_N(h) = 0, \tag{159}$$

$$U_1(0)=0, \quad U_N(h)=0, \quad V_1(0)=0, \quad V_N(h)=0, \tag{160}$$

where the subscripts 1 and N represent the 1st and Nth layers, respectively.

4.2.2 .Boundary Conditions at the Two Ends

The electric boundary conditions at the two ends are

$$D^* = const., \qquad \text{at } x=0 \text{ or } x=l, \tag{161}$$

or

$$\Phi = const., \quad \text{at } (0,h/2) \text{ or } (l,h/2), \tag{162}$$

where Φ is the electric potential, and $D^*(x)$ is the average electric displacement,

$$D^*(x) = \int_0^h D_x dz. \tag{163}$$

The magnetic boundary conditions at the two ends are

$$B^* = const., \qquad \text{at } x=0 \text{ or } x=l, \tag{164}$$

or

$$\Psi = const., \quad \text{at } (0,h/2) \text{ or } (l,h/2), \tag{165}$$

where Ψ is the magnetic potential, $B^*(x)$ is the average magnetic induction,

$$B^*(x) = \int_0^h B_x dz. \tag{166}$$

The mechanic boundary conditions at the two ends can be expressed as follows

(1) Free

$$T=\bar{T}, \quad M=\bar{M}, \quad Q=\bar{Q}, \qquad \text{at } x=0 \text{ or } x=l, \tag{167}$$

where $\bar{T}$, $\bar{M}$ and $\bar{Q}$ are the prescribed values.

(2) Fixed (or clamped)

$$u=w=0, \partial w/\partial x=0 \; or \; \partial u/\partial z=0, \text{ at } (0,h/2) \text{ or } (l,h/2), \tag{168}$$

(3) Hinged

$$T = 0, \ M = 0, \quad \text{at} \ x = 0 \ \text{or} \ x = l, \tag{169}$$

$$w = 0, \quad \text{at} \ (0, h/2) \ \text{or} \ (l, h/2), \tag{170}$$

(4) The boundary conditions for a simply supported beam

$$u = 0, \quad w = 0, \quad \text{at} \ (0, h/2), \quad M = 0, \quad \text{at} \ x = 0, \tag{171}$$

$$w = 0, \quad \text{at} \ (l, h/2), \quad T = 0, \quad M = 0, \quad \text{at} \ x = l, \tag{172}$$

4.3. Exact Solutions For Eq. (149)

Integrating Eq. $(149)_{1,2,3}$, we obtain

$$[A]\{\zeta_1\} = [b_1, b_2, b_3 z + b_4]^T, \tag{173}$$

where and hereafter b_i (i=1,2,...,8) are integral constants , the superscript T denotes the transpose operator, $\{\zeta_1\} = [\varphi_1'', \psi_1', \chi_1']^T$, and

$$[A] = \begin{bmatrix} -d_{11} & \gamma_{11} & \beta_{11} \\ -b_{11} & \beta_{11} & v_{11} \\ s_{11} & d_{11} & b_{11} \end{bmatrix}. \tag{174}$$

From Eq. (173), we obtain

$$\varphi_1'' = \sum_{k=1}^{4} b_k A_{k1}, \qquad \psi_1' = \sum_{k=1}^{4} b_k A_{k2}, \qquad \chi_1' = \sum_{k=1}^{4} b_k A_{k3}, \tag{175}$$

where

$$A_{ij} = \Delta_{ij}/\Delta, \quad A_{3j} = z\Delta_{3j}/\Delta, \quad A_{4j} = \Delta_{3j}/\Delta, \quad (i = 1, 2, \quad j = 1, 2, 3),$$
$$\Delta = \det[A], \tag{176}$$

in which Δ_{ij} is algebraic complement for element (i, j) of matrix A. Integrating Eq.(175), and by virtue of boundary conditions Eq. $(156)_{1,3,5}$, we obtain

$$\varphi_1' = \sum_{k=1}^{4} b_k A_{k1}^0, \qquad \varphi_1 = \sum_{k=1}^{4} b_k A_{k1}^1, \qquad \psi_1 = \sum_{k=1}^{4} b_k A_{k2}^0, \qquad \chi_1 = \sum_{k=1}^{4} b_k A_{k3}^0, \tag{177}$$

where the constant term in φ_1 is omitted, because it contributes nothing to the stress field, and

$$A_{ki}^{j} = \int_0^z (z-\xi)^j A_{ki} d\xi, \qquad (j=0,1,\ k=1,2,3,4,\ i=1,2,3), \tag{178}$$

Substituting Eqs. (175) and $(177)_{1,3,4}$ into Eq. $(149)_{4,5,6}$, and then integrating, we obtain

$$[A]\{\zeta_0\} = [B_1,\ B_2,\ B_3]^T, \tag{179}$$

where

$$\begin{aligned}
&\{\zeta_0\} = [\varphi_0'',\ \psi_0',\ \chi_0']^T, \\
&B_1 = b_1 B_{11} + b_2 B_{21} + b_3 B_{31} + b_4 B_{41} + b_5, \quad B_2 = b_1 B_{12} + b_2 B_{22} + b_3 B_{32} + b_4 B_{42} + b_6, \\
&B_3 = b_1 B_{13} + b_2 B_{23} + b_3 B_{33} + b_4 B_{43} + b_7 z + b_8, \\
&B_{k1} = -d_{15} A_{k1}^0 + \gamma_{13} A_{k2}^0 + \beta_{13} A_{k3}^0 - \int_0^z (d_{31} A_{k1} - \gamma_{13} A_{k2} - \beta_{13} A_{k3}) d\xi, (k=1,2,3,4), \\
&B_{k2} = -b_{15} A_{k1}^0 + \beta_{13} A_{k2}^0 + \nu_{13} A_{k3}^0 - \int_0^z (b_{31} A_{k1} - \beta_{13} A_{k2} - \nu_{13} A_{k3}) d\xi, (k=1,2,3,4), \\
&B_{k3} = s_{15} A_{k1}^0 + d_{31} A_{k2}^0 + b_{31} A_{k3}^0 + \int_0^z (s_{15} A_{k1} + d_{15} A_{k2} + b_{15} A_{k3}) d\xi, (k=1,2,3,4),
\end{aligned} \tag{180}$$

From Eq. (179) we obtain

$$\varphi_0'' = \sum_{k=1}^{8} b_k C_{k1}, \qquad \psi_0' = \sum_{k=1}^{8} b_k C_{k2}, \qquad \chi_0' = \sum_{k=1}^{8} b_k C_{k3}, \tag{181}$$

where

$$\begin{aligned}
&C_{ij} = \sum_{k=1}^{3} B_{ik} \frac{\Delta_{kj}}{\Delta}, \qquad (i=1,2,3,4, \quad j=1,2,3) \\
&C_{5k} = \frac{\Delta_{1k}}{\Delta}, \quad C_{6k} = \frac{\Delta_{2k}}{\Delta}, \quad C_{7k} = \frac{z\Delta_{3k}}{\Delta}, \quad C_{8k} = \frac{\Delta_{3k}}{\Delta}, \quad (k=1,2,3).
\end{aligned} \tag{182}$$

Integrating Eq. (181), we have

$$\varphi_0' = \sum_{k=1}^{8} b_k C_{k1}^0, \qquad \varphi_0 = \sum_{k=1}^{8} b_k C_{k1}^1, \qquad \psi_0 = \sum_{k=1}^{8} b_k C_{k2}^0, \qquad \chi_0 = \sum_{k=1}^{8} b_k C_{k3}^0, \tag{183}$$

where

$$C_{ki}^{j} = \int_{0}^{z} (z-\xi)^{j} C_{ki} d\xi, \qquad (j=0,1,\ k=1,2,...,8,\ i=1,2,3), \tag{184}$$

Substituting Eqs. $(177)_{1,3,4}$ into Eqs. $(156)_{2,4,6}$, we obtain

$$\sum_{k=1}^{4} b_k A_{k1}^{0}(h) = 0, \qquad \sum_{k=1}^{4} b_k A_{k2}^{0}(h) = 0, \qquad \sum_{k=1}^{4} b_k A_{k3}^{0}(h) = 0. \tag{185}$$

4.4. Resultant Force, Average Electric Displacement, Average Magnetic Induction, Displacement, Electric Potential and Magnetic Potential

By using of Eqs. $(145)_{1,3}$, $(146)_1$, $(147)_1$ and (152)-(156), we can obtain the axial force T, bending moment M, shear force Q, average electric displacement D^* and average magnetic induction B^*

$$T = \int_{0}^{h} \sigma_x dz = \sum_{k=1}^{8} b_k C_{k1}^{0}(h), \tag{186}$$

$$M = \int_{0}^{h} \sigma_x z dz$$
$$= \frac{q_n}{\alpha_n^2} \sin \alpha_n x + h \sum_{k=1}^{8} b_k C_{k1}^{0}(h) - \sum_{k=1}^{8} b_k C_{k1}^{1}(h) - x \sum_{k=1}^{4} b_k A_{k1}^{1}(h), \tag{187}$$

$$Q = \int_{0}^{h} \tau_{xz}\, dz = \frac{q_n}{\alpha_n} \cos \alpha_n x - \sum_{k=1}^{4} b_k A_{k1}^{1}(h), \tag{188}$$

$$D^* = \int_{0}^{h} D_x\, dz = \sum_{k=1}^{8} b_k C_{k2}^{0}(h), \tag{189}$$

$$B^* = \int_{0}^{h} B_x\, dz = \sum_{k=1}^{8} b_k (h) C_{k3}^{0}. \tag{190}$$

Substituting Eqs. (145)-(147) into Eqs. (7)-(9), and then into Eqs. (4)-(6), integrating them and by virtue of Eqs. (10)-(12), we obtain the displacement, electric potential and magnetic potential expressions,

$$u = \frac{1}{\alpha_n}\left(s_{11}G'' - \alpha_n^2 s_{13}G - \alpha_n s_{15}F' + d_{11}S' - \alpha_n d_{31}R + b_{11}V' - \alpha_n b_{31}U\right)\sin(\alpha_n x)$$
$$-\frac{1}{\alpha_n}\left(s_{11}F'' - \alpha_n^2 s_{13}F + \alpha_n s_{15}G' + d_{11}R' + \alpha_n d_{31}S + b_{11}U' + \alpha_n b_{31}V\right)\cos(\alpha_n x)$$
$$+x\left(s_{11}\varphi_0'' - s_{15}\varphi_1' + d_{11}\psi_0' - d_{31}\psi_1 + b_{11}\chi_0' - b_{31}\chi_1\right) + \frac{x^2}{2}\left(s_{11}\varphi_1'' + d_{11}\psi_1' + b_{11}\chi_1'\right)$$
$$+\int_0^z\left(s_{15}\varphi_0'' - s_{55}\varphi_1' + d_{15}\psi_0' - d_{35}\psi_1 + b_{15}\chi_0' - b_{35}\chi_1\right)d\xi$$
$$-\int_0^z(z-\xi)\left(s_{13}\varphi_1'' + d_{13}\psi_1' + b_{13}\chi_1'\right)d\xi - \omega z + u_0, \tag{191}$$

$$w = \left[\left(s_{11}F'' - \alpha_n^2 s_{13}F + \alpha_n s_{15}G' + d_{11}R' + \alpha_n d_{31}S + b_{11}U' + \alpha_n b_{31}V\right)' + \alpha_n\right.$$
$$\left.\times\left(s_{15}G'' - \alpha_n^2 s_{35}G - \alpha_n s_{55}F' + d_{15}S' - \alpha_n d_{35}R + b_{15}V' - \alpha_n b_{35}U\right)\right]\frac{\sin(\alpha_n x)}{\alpha_n^2}$$
$$+\left[\left(s_{11}G'' - \alpha_n^2 s_{13}G - \alpha_n s_{15}F' + d_{11}S' - \alpha_n d_{31}R + b_{11}V' - \alpha_n b_{31}U\right)' - \alpha_n\right.$$
$$\left.\times\left(s_{15}F'' - \alpha_n^2 s_{35}F + \alpha_n s_{55}G' + d_{15}R' + \alpha_n d_{35}S + b_{15}U' + \alpha_n b_{35}V\right)\right]\frac{\cos(\alpha_n x)}{\alpha_n^2}$$
$$+\int_0^z\left(s_{13}\varphi_0'' - s_{35}\varphi_1' + d_{13}\psi_0' - d_{33}\psi_1 + b_{13}\chi_0' - b_{33}\chi_1\right)d\xi$$
$$+x\int_0^z\left(s_{13}\varphi_1'' + d_{13}\psi_1' + b_{13}\chi_1'\right)d\xi - \frac{b_7}{2}x^2 - \frac{b_3}{6}x^3 + \omega x + w_0, \tag{192}$$

$$\Phi = \left(d_{11}G'' - \alpha_n^2 d_{13}G - \alpha_n d_{15}F' - \gamma_{11}S' + \alpha_n\gamma_{13}R - \beta_{11}V' + \alpha_n\beta_{13}U\right)\frac{\sin(\alpha_n x)}{\alpha_n}$$
$$+\left(-d_{11}F'' + \alpha_n^2 d_{13}F - d_{15}\alpha_n G' + \gamma_{11}R' + \alpha_n\gamma_{13}S + \beta_{11}U' + \alpha_n\beta_{13}V\right)\frac{\cos(\alpha_n x)}{\alpha_n}$$
$$+x\left(d_{11}\varphi_0'' + \alpha_n d_{15}\varphi_1' + \gamma_{11}\psi_0' - \gamma_{13}\psi_1 + \beta_{11}\chi_0' - \beta_{13}\chi_1\right) + \frac{x^2}{2}\left(d_{11}\varphi_1'' - \gamma_{11}\psi_1' - \beta_{11}\chi_1'\right)$$
$$-\int_0^z\left(d_{31}\varphi_0'' - d_{35}\varphi_1' - \gamma_{13}\psi_0' + \gamma_{33}\psi_1 - \beta_{13}\chi_0' + \beta_{33}\chi_1\right)d\xi + \Phi_0, \tag{193}$$

$$\Psi = \left(b_{11}G'' - \alpha_n^2 b_{13}G - \alpha_n b_{15}F' - \beta_{11}S' + \alpha_n\beta_{13}R - \nu_{11}V' + \alpha_n\nu_{13}U\right)\frac{\sin(\alpha_n x)}{\alpha_n}$$
$$+\left(-b_{11}F'' + \alpha_n^2 b_{13}F - \alpha_n b_{15}G' + \beta_{11}R' + \alpha_n\beta_{13}S + \nu_{11}U' + \alpha_n\nu_{13}V\right)\frac{\cos(\alpha_n x)}{\alpha_n}$$
$$+x\left(b_{11}\varphi_0'' - b_{15}\varphi_1' - \beta_{11}\psi_0' + \beta_{13}\psi_1 - \nu_{11}\chi_0' + \nu_{13}\chi_1\right) + \frac{x^2}{2}\left(b_{11}\varphi_1'' - \beta_{11}\psi_1' - \nu_{11}\chi_1'\right)$$
$$-\int_0^z\left(b_{31}\varphi_0'' - b_{35}\varphi_1' - \beta_{13}\psi_0' + \beta_{33}\psi_1 - \nu_{13}\chi_0' + \nu_{33}\chi_1\right)d\xi + \Psi_0, \tag{194}$$

where u_0, w_0, ω, Φ_0 and Ψ_0 are integral constants.

4.5. Analytical Solutions of Eq. (148)

When the material properties M_{ij} vary in the following way, Eq. (148) can be simplified into ordinary differential equations whose solutions can be obtained analytically,

$$M_{jk} = M_{jk}^{0} \cdot e^{\lambda z/h}, \tag{195}$$

where λ is a gradient index, M_{jk}^{0} are values of s_{jk}^{0}, d_{jk}^{0}, β_{jk}^{0}, b_{jk}^{0}, γ_{jk}^{0} and ν_{jk}^{0} at z=0. Let

$$\begin{aligned} &F = \tilde{f}\exp(\eta z), \quad G = \tilde{g}\exp(\eta z), \quad R = \tilde{r}\exp(\eta z), \\ &S = \tilde{s}\exp(\eta z), \quad U = \tilde{u}\exp(\eta z), \quad V = \tilde{v}\exp(\eta z), \end{aligned} \tag{196}$$

where $\tilde{f}$, $\tilde{g}$, $\tilde{r}$, $\tilde{s}$, $\tilde{u}$, $\tilde{v}$, and η are undetermined constants. Substitution of Eq. (196) in to (148) gives rise to

$$[H]\{\zeta_2\} = 0, \tag{197}$$

where $\{\zeta_2\} = \left[\tilde{f},\ \tilde{g},\ \tilde{r},\ \tilde{s},\ \tilde{u},\ \tilde{v}\right]^T$, and

$$[H] = \begin{bmatrix} H_{11} & H_{12} & \cdots & H_{16} \\ H_{21} & H_{22} & \cdots & \cdots \\ \cdots & \cdots & \cdots & \cdots \\ H_{61} & \cdots & \cdots & H_{66} \end{bmatrix},$$

$$\begin{aligned}
H_{11} &= s_{11}^0\left(\eta^4 + 2\lambda/h\eta^3 + \lambda^2/h^2\eta^2\right) - \alpha_n^2\left(2\eta^2 + 2\lambda/h\eta + \lambda^2/h^2\right)s_{13}^0 \\
&\quad + \alpha_n^4 s_{33}^0 - \alpha_n^2\left(\eta + \lambda/h\right)s_{55}^0\eta, \\
H_{12} &= \alpha_n\left(2\eta^2 + 3\lambda/h\eta + \lambda^2/h^2\right)s_{15}^0\eta - \alpha_n^3\left(2\eta + \lambda/h\right)s_{35}^0, \\
H_{13} &= \eta\left(\eta^2 + 2\lambda/h\eta + \lambda^2/h^2\right)d_{11}^0 - \alpha_n^2\eta d_{13}^0 - \alpha_n^2\left(\eta + \lambda/h\right)d_{35}^0, \\
H_{14} &= \alpha_n\eta\left(\eta + \lambda/h\right)d_{15}^0 + \alpha_n\left(\eta^2 + 2\lambda/h\eta + \lambda^2/h^2\right)d_{31}^0 - \alpha_n^3 d_{33}^0, \\
H_{15} &= \eta\left(\eta^2 + 2\lambda/h\eta + \lambda^2/h^2\right)b_{11}^0 - \alpha_n^2 b_{13}^0\eta - \alpha_n^2\left(\eta + \lambda/h\right)b_{35}^0, \\
H_{16} &= \alpha_n\eta\left(\eta + \lambda/h\right)b_{15}^0 + \alpha_n\left(\eta^2 + 2\lambda/h\eta + \lambda^2/h^2\right)b_{31}^0 - \alpha_n^3 b_{33}^0, \\
H_{21} &= -H_{12}, \quad H_{22} = H_{11}, \quad H_{23} = -H_{14}, \quad H_{24} = H_{13}, \quad H_{25} = -H_{16}, \quad H_{26} = H_{15}, \\
H_{31} &= \eta^2\left(\eta + \lambda/h\right)d_{11}^0 - \alpha_n^2\left(\eta + \lambda/h\right)d_{13}^0 - \alpha_n^2 d_{35}^0\eta, \\
H_{32} &= \alpha_n\eta\left(\eta + \lambda/h\right)d_{15}^0 + \alpha_n d_{31}^0\eta^2 - \alpha_n^3 d_{33}^0,
\end{aligned}$$

$$H_{33}=-\eta(\eta+\lambda/h)\gamma_{11}^0+\alpha_n^2\gamma_{33}^0,\qquad H_{34}=-\alpha_n(2\eta+\lambda/h)\gamma_{13}^0,$$
$$H_{35}=-\eta(\eta+\lambda/h)\beta_{11}^0+\alpha_n^2\beta_{33}^0,\qquad H_{36}=-\alpha_n(2\eta+\lambda/h)\beta_{13}^0,$$
$$H_{41}=-H_{32},\quad H_{42}=H_{31},\quad H_{43}=-H_{34},\quad H_{44}=H_{33},\quad H_{45}=-H_{36},\quad H_{46}=H_{35},$$
$$H_{51}=\eta^2(\eta+\lambda/h)b_{11}^0-\alpha_n^2b_{13}^0\eta-\alpha_n^2\lambda/hb_{13}^0-\alpha_n^2b_{35}^0\eta,$$
$$H_{52}=\alpha_n\eta(\eta+\lambda/h)b_{15}^0+\alpha_nb_{31}^0\eta^2-\alpha_n^3b_{33}^0,$$
$$H_{53}=-\eta(\eta+\lambda/h)\beta_{11}^0+\alpha_n^2\beta_{33}^0,\qquad H_{54}=-\alpha_n(2\eta+\lambda/h)\beta_{13}^0,$$
$$H_{55}=-\eta(\eta+\lambda/h)\nu_{11}^0+\alpha_n^2\nu_{33}^0,\qquad H_{56}=-\alpha_n(2\eta+\lambda/h)\nu_{13}^0,$$
$$H_{61}=-H_{52},\ H_{62}=H_{51},\ H_{63}=-H_{54},\ H_{64}=H_{53},\ H_{65}=-H_{56},\ H_{66}=H_{55},\tag{198}$$

For non-trivial solutions to exist, the determinant of coefficient matrix of Eq. (197) should vanish,

$$\det[H]=0,\tag{199}$$

From Eq. (199), we can obtain the characteristic roots η_k (k=1,2,…,16) of Eq. (197). Hence, the analytical solutions of Eq. (148) are

$$F=\sum_{k=1}^{16}\tilde{f}_k\exp(\eta_k\cdot z),\quad G=\sum_{k=1}^{16}\tilde{g}_k\exp(\eta_k\cdot z),\quad R=\sum_{k=1}^{16}\tilde{r}_k\exp(\eta_k\cdot z),$$
$$S=\sum_{k=1}^{16}\tilde{s}_k\exp(\eta_k\cdot z),\quad U=\sum_{k=1}^{16}\tilde{u}_k\exp(\eta_k\cdot z),\quad V=\sum_{k=1}^{16}\tilde{v}_k\exp(\eta_k\cdot z).\tag{200}$$

There 96 unknown constants in Eq. (200) to be determined altogether, $\tilde{f}_k$, $\tilde{g}_k$, $\tilde{r}_k$, $\tilde{s}_k$, $\tilde{u}_k$ and $\tilde{v}_k$ (k=1,2,…,16). Substitution of Eq. (200) into Eqs. (152)-(155) yields 16 equations. From Eq. (197) we can obtain 5 independent equations, hence when k=1,2,…,16, we obtain 80 equations. Thus we have 96 equations altogether to fix 96 unknown constants $\tilde{f}_k$, $\tilde{g}_k$, $\tilde{r}_k$, $\tilde{s}_k$, $\tilde{u}_k$ and $\tilde{v}_k$,(k=1,2,…, 16).

So far, we still have 8 undetermined constants b_1~b_8 appearing in stress function, electric displacement function and magnetic induction function, 5 undetermined constants u_0, w_0, ω , Φ_0 and Ψ_0 in the expressions of displacements, electric potential and magnetic potential. Hence we have 13 undetermined constants to be fixed. Since there are 10 equations corresponding to the boundary conditions at the two ends, plus the already derived 3 relations in Eq. (185), we have totally 13 equations to solve for the 13 undetermined constants.

4.6. Semi-Analytical Solution of Eq. (148)

When the material properties s_{ij}, d_{ij}, b_{ij}, γ_{ij}, β_{ij} and ν_{ij} vary in an arbitrary continuous fashion, it is very difficult to obtain the analytical solution of the ordinary differential equations (148). Hence the approach of sub-layer approximation will be employed to seek for a semi-analytical solution. The expressions for stresses, electric displacements and magnetic induction in the *m*th sub-layer are:

$$\begin{aligned}\sigma_x^{(m)} &= \sin(\alpha_n x)F_m'' + \cos(\alpha_n x)G_m'' + \varphi_0'' + x\varphi_1'',\\ \sigma_z^{(m)} &= -\alpha_n^2\left[\sin(\alpha_n x)F_m + \cos(\alpha_n x)G_m\right],\\ \tau_{xz}^{(m)} &= -\alpha_n\left[\cos(\alpha_n x)F_m' - \sin(\alpha_n x)G_m'\right] - \varphi_1'.\end{aligned} \tag{201}$$

$$\begin{aligned}D_x^{(m)} &= \sin(\alpha_n x)R_m' + \cos(\alpha_n x)S_m' + \psi_0' + x\psi_1',\\ D_z^{(m)} &= -\alpha_n\left[\cos(\alpha_n x)R_m - \sin(\alpha_n x)S_m\right] - \psi_1,\end{aligned} \tag{202}$$

$$\begin{aligned}B_x^{(m)} &= \left[\sin(\alpha_n x)U_m' + \cos(\alpha_n x)V_m'\right] + \chi_0' + x\chi_1',\\ B_z^{(m)} &= -\alpha_n\left[\cos(\alpha_n x)U_m - \sin(\alpha_n x)V_m\right] - \chi_1.\end{aligned} \tag{203}$$

where the superscript (m) and the subscript m represent the *m*th sub-layer. The constitutive equations are:

$$\begin{aligned}\varepsilon_x^{(m)} &= s_{11}^{(m)}\sigma_x^{(m)} + s_{13}^{(m)}\sigma_z^{(m)} + s_{15}^{(m)}\tau_{xz}^{(m)} + d_{11}^{(m)}D_x^{(m)} + d_{31}^{(m)}D_z^{(m)} + b_{11}^{(m)}B_x^{(m)} + b_{31}^{(m)}B_z^{(m)},\\ \varepsilon_z^{(m)} &= s_{13}^{(m)}\sigma_x^{(m)} + s_{33}^{(m)}\sigma_z^{(m)} + s_{35}^{(m)}\tau_{xz}^{(m)} + d_{13}^{(m)}D_x^{(m)} + d_{33}^{(m)}D_z^{(m)} + b_{13}^{(m)}B_x^{(m)} + b_{33}^{(m)}B_z^{(m)},\\ \gamma_{xz}^{(m)} &= s_{15}^{(m)}\sigma_x^{(m)} + s_{35}^{(m)}\sigma_z^{(m)} + s_{55}^{(m)}\tau_{xz}^{(m)} + d_{15}^{(m)}D_x^{(m)} + d_{35}^{(m)}D_z^{(m)} + b_{15}^{(m)}B_x^{(m)} + b_{35}^{(m)}B_z^{(m)}.\end{aligned} \tag{204}$$

$$\begin{aligned}E_x^{(m)} &= -d_{11}^{(m)}\sigma_x^{(m)} - d_{13}^{(m)}\sigma_z^{(m)} - d_{15}\tau_{xz}^{(m)} + \gamma_{11}^{(m)}D_x^{(m)} + \gamma_{13}^{(m)}D_z^{(m)} + \beta_{11}^{(m)}B_x^{(m)} + \beta_{13}^{(m)}B_z^{(m)},\\ E_z^{(m)} &= -d_{31}^{(m)}\sigma_x^{(m)} - d_{33}^{(m)}\sigma_z^{(m)} - d_{35}^{(m)}\tau_{xz}^{(m)} + \gamma_{13}D_x^{(m)} + \gamma_{33}^{(m)}D_z^{(m)} + \beta_{13}B_x^{(m)} + \beta_{33}^{(m)}B_z^{(m)}.\end{aligned} \tag{205}$$

$$\begin{aligned}H_x^{(m)} &= -b_{11}^{(m)}\sigma_x^{(m)} - b_{13}^{(m)}\sigma_z^{(m)} - b_{15}\tau_{xz}^{(m)} + \beta_{11}^{(m)}D_x^{(m)} + \beta_{13}^{(m)}D_z^{(m)} + \nu_{11}^{(m)}B_x^{(m)} + \nu_{13}^{(m)}B_z^{(m)},\\ H_z^{(m)} &= -b_{31}^{(m)}\sigma_x^{(m)} - b_{33}^{(m)}\sigma_z^{(m)} - b_{35}^{(m)}\tau_{xz}^{(m)} + \beta_{13}D_x^{(m)} + \beta_{33}^{(m)}D_z^{(m)} + \nu_{13}B_x^{(m)} + \nu_{33}^{(m)}B_z^{(m)}.\end{aligned} \tag{206}$$

where $s_{jk}^{(m)}$, $g_{jk}^{(m)}$ and $\beta_{jk}^{(m)}$ are the material properties in the *m*th sub-layer, $M_{jk}^{(m)} = M_{jk}\left(h_{m-1}/2 + h_m/2\right)$.

Hence Eqs. (148) in *m*th layer are rewritten as

$$\begin{aligned}
&\left(s_{11}^{(m)}F_m'' - \alpha_n^2 s_{13}^{(m)}F_m + \alpha_n s_{15}^{(m)}G_m' + d_{11}^{(m)}R_m' + \alpha_n d_{31}^{(m)}S_m + b_{11}^{(m)}U_m' + \alpha_n b_{31}^{(m)}V_m\right)'' \\
&\quad - \alpha_n^2\left(s_{13}^{(m)}F_m'' - \alpha_n^2 s_{33}^{(m)}F_m + \alpha_n s_{35}^{(m)}G_m' + d_{13}^{(m)}R_m' + \alpha_n d_{33}^{(m)}S_m + b_{13}^{(m)}U_m' + \alpha_n b_{33}^{(m)}V_m\right) \\
&\quad + \alpha_n\left(s_{15}^{(m)}G_m'' - \alpha_n^2 s_{35}^{(m)}G_m - \alpha_n s_{55}^{(m)}F_m' + d_{15}^{(m)}S_m' - \alpha_n d_{35}^{(m)}R_m + b_{15}^{(m)}V_m' - \alpha_n b_{35}^{(m)}U_m\right)' = 0, \\
&\left(s_{11}^{(m)}G_m'' - \alpha_n^2 s_{13}^{(m)}G_m - \alpha_n s_{15}^{(m)}F_m' + d_{11}^{(m)}S_m' - \alpha_n d_{31}^{(m)}R_m + b_{11}^{(m)}V_m' - \alpha_n b_{31}^{(m)}U_m\right)'' \\
&\quad - \alpha_n^2\left(s_{13}^{(m)}G_m'' - \alpha_n^2 s_{33}^{(m)}G_m - \alpha_n s_{35}^{(m)}F_m' + d_{13}^{(m)}S_m' - \alpha_n d_{33}^{(m)}R_m + b_{13}^{(m)}V_m' - \alpha_n b_{33}^{(m)}U_m\right) \\
&\quad - \alpha_n\left(s_{15}^{(m)}F_m'' - \alpha_n^2 s_{35}^{(m)}F_m + \alpha_n s_{55}^{(m)}G_m' + d_{15}^{(m)}R_m' + \alpha_n d_{35}^{(m)}S_m + b_{15}^{(m)}U_m' + \alpha_n b_{35}^{(m)}V_m\right)' = 0, \\
&\left(d_{11}^{(m)}F_m'' - \alpha_n^2 d_{13}^{(m)}F_m + \alpha_n d_{15}^{(m)}G_m' - \gamma_{11}^{(m)}R_m' - \alpha_n\gamma_{13}^{(m)}S_m - \beta_{11}^{(m)}U_m' - \alpha_n\beta_{13}^{(m)}V_m\right)' \\
&\quad + \alpha_n\left(d_{31}^{(m)}G_m'' - \alpha_n^2 d_{33}^{(m)}G_m - \alpha_n d_{35}^{(m)}F_m' - \gamma_{13}^{(m)}S_m' + \alpha_n\gamma_{33}^{(m)}R_m - \beta_{13}^{(m)}V_m' + \alpha_n\beta_{33}^{(m)}U_m\right) = 0, \\
&\left(d_{11}^{(m)}G_m'' - \alpha_n^2 d_{13}^{(m)}G_m - \alpha_n d_{15}^{(m)}F_m' - \gamma_{11}^{(m)}S_m' + \alpha_n\gamma_{13}^{(m)}R_m - \beta_{11}^{(m)}V_m' + \alpha_n\beta_{13}^{(m)}U_m\right)' \\
&\quad - \alpha_n\left(d_{31}^{(m)}F_m'' - \alpha_n^2 d_{33}^{(m)}F_m + \alpha_n d_{35}^{(m)}G_m' - \gamma_{13}^{(m)}R_m' - \alpha_n\gamma_{33}^{(m)}S_m - \beta_{13}^{(m)}U_m' - \alpha_n\beta_{33}^{(m)}V_m\right) = 0, \\
&\left(b_{11}^{(m)}F_m'' - \alpha_n^2 b_{13}^{(m)}F_m + \alpha_n b_{15}^{(m)}G_m' - \beta_{11}^{(m)}R_m' - \alpha_n\beta_{13}^{(m)}S_m - \nu_{11}^{(m)}U_m' - \alpha_n\nu_{13}^{(m)}V_m\right)' \\
&\quad + \alpha_n\left(b_{31}^{(m)}G_m'' - \alpha_n^2 b_{33}^{(m)}G_m - \alpha_n b_{35}^{(m)}F_m' - \beta_{13}^{(m)}S_m' + \alpha_n\beta_{33}^{(m)}R_m - \nu_{13}^{(m)}V_m' + \alpha_n\nu_{33}^{(m)}U_m\right) = 0, \\
&\left(b_{11}^{(m)}G_m'' - \alpha_n^2 b_{13}^{(m)}G_m - \alpha_n b_{15}^{(m)}F_m' - \beta_{11}^{(m)}S_m' + \alpha_n\beta_{13}^{(m)}R_m - \nu_{11}^{(m)}V_m' + \alpha_n\nu_{13}^{(m)}U_m\right)' \\
&\quad - \alpha_n\left(b_{31}^{(m)}F_m'' - \alpha_n^2 b_{33}^{(m)}F_m + \alpha_n b_{35}^{(m)}G_m' - \beta_{13}^{(m)}R_m' - \alpha_n\beta_{33}^{(m)}S_m - \nu_{13}^{(m)}U_m' - \alpha_n\nu_{33}^{(m)}V_m\right) = 0,
\end{aligned} \tag{207}$$

Assume that

$$\begin{aligned}
&F_m = \tilde{f}^{(m)}\exp\left(\eta^{(m)}z\right), \quad G_m = \tilde{g}^{(m)}\exp\left(\eta^{(m)}z\right), \quad R_m = \tilde{r}^{(m)}\exp\left(\eta^{(m)}z\right), \\
&S_m = \tilde{s}^{(m)}\exp\left(\eta^{(m)}z\right), \quad U_m = \tilde{u}^{(m)}\exp\left(\eta^{(m)}z\right), \quad V_m = \tilde{v}^{(m)}\exp\left(\eta^{(m)}z\right).
\end{aligned} \tag{208}$$

Substitution of Eq. (208) into Eq. (207) yields

$$\left[H^{(m)}\right]\left\{\varsigma_2^{(m)}\right\} = 0, \tag{209}$$

where $\left\{\varsigma_2^{(m)}\right\} = \left[\tilde{f}^{(m)},\ \tilde{g}^{(m)},\ \tilde{r}^{(m)},\ \tilde{s}^{(m)},\ \tilde{u}^{(m)},\ \tilde{v}^{(m)}\right]^T$, and

$$\left[H^{(m)}\right]=\begin{bmatrix} H_{11}^{(m)} & H_{12}^{(m)} & \dots & H_{16}^{(m)} \\ H_{21}^{(m)} & H_{22}^{(m)} & \dots & \dots \\ \dots & \dots & \dots & \dots \\ H_{61}^{(m)} & \dots & \dots & H_{66}^{(m)} \end{bmatrix},$$

$$H_{11}^{(m)}=\left(\eta^{(m)}\right)^4 s_{11}^{(m)}-2\alpha_n^2\left(\eta^{(m)}\right)^2 s_{13}^{(m)}+\alpha_n^4 s_{33}^{(m)}-\alpha_n^2\left(\eta^{(m)}\right)^2 s_{55}^{(m)},$$

$$H_{12}^{(m)}=2\alpha_n\left(\eta^{(m)}\right)^3 s_{15}^{(m)}-2\alpha_n^3\eta^{(m)}s_{35}^{(m)},\quad H_{13}^{(m)}=\left(\eta^{(m)}\right)^3 d_{11}^{(m)}-\alpha_n^2\eta^{(m)}\left(d_{13}^{(m)}+d_{35}^{(m)}\right),$$

$$H_{14}^{(m)}=\alpha_n\left(\eta^{(m)}\right)^2\left(d_{15}^{(m)}+d_{31}^{(m)}\right)-d_{33}^{(m)}\alpha_n^3,\quad H_{15}^{(m)}=b_{11}^{(m)}\left(\eta^{(m)}\right)^3-\alpha_n^2\left(b_{13}^{(m)}+b_{35}^{(m)}\right)\eta^{(m)},$$

$$H_{16}^{(m)}=\alpha_n\left(b_{15}^{(m)}+b_{31}^{(m)}\right)\left(\eta^{(m)}\right)^2-\alpha_n^3 b_{33}^{(m)},$$

$$H_{21}^{(m)}=-H_{12}^{(m)},\ H_{22}^{(m)}=H_{11}^{(m)},\ H_{23}^{(m)}=-H_{14}^{(m)},\ H_{24}^{(m)}=H_{13}^{(m)},\ H_{25}^{(m)}=-H_{26}^{(m)},\ H_{26}^{(m)}=H_{15}^{(m)},$$

$$H_{31}^{(m)}=\left(\eta^{(m)}\right)^3 d_{11}^{(m)}-\alpha_n^2\eta^{(m)}\left(d_{13}^{(m)}+d_{35}^{(m)}\right),\quad H_{32}^{(m)}=\alpha_n\left(\eta^{(m)}\right)^2\left(d_{15}^{(m)}+d_{31}^{(m)}\right)-\alpha_n^3 d_{33}^{(m)},$$

$$H_{33}^{(m)}=-\gamma_{11}^{(m)}\left(\eta^{(m)}\right)^2+\alpha_n^2\gamma_{33}^{(m)},\qquad H_{34}^{(m)}=-2\alpha_n\gamma_{13}^{(m)}\eta^{(m)},$$

$$H_{35}^{(m)}=-\beta_{11}^{(m)}\left(\eta^{(m)}\right)^2+\alpha_n^2\beta_{33}^{(m)},\qquad H_{36}^{(m)}=-2\alpha_n\beta_{13}^{(m)}\eta^{(m)}$$

$$H_{41}^{(m)}=-H_{32}^{(m)},\ H_{42}^{(m)}=H_{31}^{(m)},\ H_{43}^{(m)}=-H_{34}^{(m)},\ H_{44}^{(m)}=H_{33}^{(m)},\ H_{45}^{(m)}=-H_{36}^{(m)},\ H_{46}^{(m)}=H_{35}^{(m)},$$

$$H_{51}^{(m)}=b_{11}^{(m)}\left(\eta^{(m)}\right)^3-\alpha_n^2\left(b_{13}^{(m)}+b_{35}^{(m)}\right)\eta^{(m)},\quad H_{52}^{(m)}=\alpha_n\left(b_{15}^{(m)}+b_{31}^{(m)}\right)\left(\eta^{(m)}\right)^2-\alpha_n^3 b_{33}^{(m)},$$

$$H_{53}^{(m)}=-\beta_{11}^{(m)}\left(\eta^{(m)}\right)^2+\alpha_n^2\beta_{33}^{(m)},\qquad H_{54}^{(m)}=-2\alpha_n\beta_{13}^{(m)}\left(\eta^{(m)}\right),$$

$$H_{55}^{(m)}=-\nu_{11}^{(m)}\left(\eta^{(m)}\right)^2+\alpha_n^2\nu_{33}^{(m)},\quad H_{56}^{(m)}=-2\alpha_n\nu_{13}^{(m)}\eta^{(m)},\quad H_{61}^{(m)}=-H_{52}^{(m)},\quad H_{62}^{(m)}=H_{51}^{(m)},$$

$$H_{63}^{(m)}=-H_{54}^{(m)},\ H_{64}^{(m)}=H_{53}^{(m)},\ H_{65}^{(m)}=-H_{56}^{(m)},\ H_{66}^{(m)}=H_{55}^{(m)}, \tag{210}$$

The condition of non-trivial solution of Eq. (209) demands

$$\det\left[H^{(m)}\right]=0. \tag{211}$$

From Eq. (211) we obtain the characteristic roots $\eta_k^{(m)}$ (k=1,2,…,16) of Eq. (209). Thus the solutions of Eq. (207) are

$$F_m=\sum_{k=1}^{16}\tilde{f}_k^{(m)}\exp\left(\eta_k^{(m)}\cdot z\right),\quad G_m=\sum_{k=1}^{16}\tilde{g}_k^{(m)}\exp\left(\eta_k^{(m)}\cdot z\right),\quad R_m=\sum_{k=1}^{16}\tilde{r}_k^{(m)}\exp\left(\eta_k^{(m)}\cdot z\right),$$

$$S_m=\sum_{k=1}^{16}\tilde{s}_k^{(m)}\exp\left(\eta_k^{(m)}\cdot z\right),\ U_m=\sum_{k=1}^{16}\tilde{u}_k^{(m)}\exp\left(\eta_k^{(m)}\cdot z\right),\ V_m=\sum_{k=1}^{16}\tilde{v}_k^{(m)}\exp\left(\eta_k^{(m)}\cdot z\right). \tag{212}$$

Substituting Eq. (212) into Eqs. (201)-(203), and then into Eqs. (204)-(206), integrating them and by virtue of Eqs. (10)-(12), we obtain the displacement, electric potential and magnetic potential expressions for mth layer,

$$
\begin{aligned}
u_m &= \left(s_{11}^{(m)}G_m'' - \alpha_n^2 s_{13}^{(m)}G_m - \alpha_n s_{15}^{(m)}F_m' + d_{11}^{(m)}S_m' - \alpha_n d_{31}^{(m)}R_m + b_{11}^{(m)}V_m' - \alpha_n b_{31}^{(m)}U_m\right)\frac{\sin(\alpha_n x)}{\alpha_n} \\
&\quad - \left(s_{11}^{(m)}F_m'' - \alpha_n^2 s_{13}^{(m)}F_m + \alpha_n s_{15}^{(m)}G_m' + d_{11}^{(m)}R_m' + \alpha_n d_{31}^{(m)}S_m + b_{11}^{(m)}U_m' + \alpha_n b_{31}^{(m)}V_m\right)\frac{\cos(\alpha_n x)}{\alpha_n} \\
&\quad + x\left(s_{11}\varphi_0'' - s_{15}\varphi_1' + d_{11}\psi_0' - d_{31}\psi_1 + b_{11}\chi_0' - b_{31}\chi_1\right) + \frac{x^2}{2}\left(s_{11}\varphi_1'' + d_{11}\psi_1' + b_{11}\chi_1'\right) \\
&\quad + \int_0^z\left(s_{15}\varphi_0'' - s_{55}\varphi_1' + d_{15}\psi_0' - d_{35}\psi_1 + b_{15}\chi_0' - b_{35}\chi_1\right)d\xi \\
&\quad - \int_0^z(z-\xi)\left(s_{13}\varphi_1'' + d_{13}\psi_1' + b_{13}\chi_1'\right)d\xi - \omega^{(m)}z + u_0^{(m)}, \\
w_m &= \Bigg[\left(s_{11}^{(m)}F_m'' - \alpha_n^2 s_{13}^{(m)}F_m + \alpha_n s_{15}^{(m)}G_m' + d_{11}^{(m)}R_m' + \alpha_n d_{31}^{(m)}S_m + b_{11}^{(m)}U_m' + \alpha_n b_{31}^{(m)}V_m\right)' \\
&\quad + \alpha_n\left(s_{15}^{(m)}G_m'' - \alpha_n^2 s_{35}^{(m)}G_m - \alpha_n s_{55}^{(m)}F_m' + d_{15}^{(m)}S_m' - \alpha_n d_{35}^{(m)}R_m + b_{15}^{(m)}V_m' - \alpha_n b_{35}^{(m)}U_m\right)\Bigg]\frac{\sin(\alpha_n x)}{\alpha_n^2} \\
&\quad + \Bigg[\left(s_{11}^{(m)}G_m'' - \alpha_n^2 s_{13}^{(m)}G_m - \alpha_n s_{15}^{(m)}F_m' + d_{11}^{(m)}S_m' - \alpha_n d_{31}^{(m)}R_m + b_{11}^{(m)}V_m' - \alpha_n b_{31}^{(m)}U_m\right)' \\
&\quad - \alpha_n\left(s_{15}^{(m)}F_m'' - \alpha_n^2 s_{35}^{(m)}F_m + \alpha_n s_{55}^{(m)}G_m' + d_{15}^{(m)}R_m' + \alpha_n d_{35}^{(m)}S_m + b_{15}^{(m)}U_m' + \alpha_n b_{35}^{(m)}V_m\right)\Bigg]\frac{\cos(\alpha_n x)}{\alpha_n^2} \\
&\quad + \int_0^z\left(s_{13}\varphi_0'' - s_{35}\varphi_1' + d_{13}\psi_0' - d_{33}\psi_1 + b_{13}\chi_0' - b_{33}\chi_1\right)d\xi + x\int_0^z\left(s_{13}\varphi_1'' + d_{13}\psi_1' + b_{13}\chi_1'\right)d\xi \\
&\quad - \frac{b_7}{2}x^2 - \frac{b_3}{6}x^3 + \omega^{(m)}x + w_0^{(m)},
\end{aligned}
\tag{213}
$$

$$
\begin{aligned}
\Phi^{(m)} &= \Big(d_{11}^{(m)}G_m'' - \alpha_n^2 d_{13}^{(m)}G_m - \alpha_n d_{15}^{(m)}F_m' - \gamma_{11}^{(m)}S_m' + \alpha_n\gamma_{13}^{(m)}R_m \\
&\quad - \beta_{11}^{(m)}V_m' + \alpha_n\beta_{13}^{(m)}U_m\Big)\frac{\sin(\alpha_n x)}{\alpha_n} \\
&\quad + \Big(-d_{11}^{(m)}F_m'' + \alpha_n^2 d_{13}^{(m)}F_m - \alpha_n d_{15}^{(m)}G_m' + \gamma_{11}^{(m)}R_m' + \alpha_n\gamma_{13}^{(m)}S_m \\
&\quad + \beta_{11}^{(m)}U_m' + \alpha_n\beta_{13}^{(m)}V_m\Big)\frac{\cos(\alpha_n x)}{\alpha_n} \\
&\quad + x\left(d_{11}\varphi_0'' + \alpha_n d_{15}\varphi_1' + \gamma_{11}\psi_0' - \gamma_{13}\psi_1 + \beta_{11}\chi_0' - \beta_{13}\chi_1\right) \\
&\quad + \frac{x^2}{2}\left(d_{11}\varphi_1'' - \gamma_{11}\psi_1' - \beta_{11}\chi_1'\right) \\
&\quad - \int_0^z\left(d_{31}\varphi_0'' - d_{35}\varphi_1' - \gamma_{13}\psi_0' + \gamma_{33}\psi_1 - \beta_{13}\chi_0' + \beta_{33}\chi_1\right)d\xi + \Phi_0^{(m)},
\end{aligned}
\tag{214}
$$

$$\begin{aligned}\Psi^{(m)} &= \Big(b_{11}^{(m)}G_m'' - \alpha_n^2 b_{13}^{(m)}G_m - \alpha_n b_{15}^{(m)}F_m' - \beta_{11}^{(m)}S_m' + \alpha_n\beta_{13}^{(m)}R_m \\ &\quad -\nu_{11}^{(m)}V_m' + \alpha_n\nu_{13}^{(m)}U_m\Big)\frac{\sin(\alpha_n x)}{\alpha_n} \\ &\quad +\Big(-b_{11}^{(m)}F_m'' + \alpha_n^2 b_{13}^{(m)}F_m - \alpha_n b_{15}^{(m)}G_m' + \beta_{11}^{(m)}R_m' + \alpha_n\beta_{13}^{(m)}S_m \\ &\quad +\nu_{11}^{(m)}U_m' + \alpha_n\nu_{13}^{(m)}V_m\Big)\frac{\cos(\alpha_n x)}{\alpha_n} \\ &\quad + x\left(b_{11}\varphi_0'' - b_{15}\varphi_1' - \beta_{11}\psi_0' + \beta_{13}\psi_1 - \nu_{11}\chi_0' + \nu_{13}\chi_1\right) \\ &\quad + \frac{x^2}{2}\left(b_{11}\varphi_1'' - \beta_{11}\psi_1' - \nu_{11}\chi_1'\right) \\ &\quad - \int_0^z\left(b_{31}\varphi_0'' - b_{35}\varphi_1' - \beta_{13}\psi_0' + \beta_{33}\psi_1 - \nu_{13}\chi_0' + \nu_{33}\chi_1\right)d\xi + \Psi_0^{(m)}.\end{aligned} \tag{215}$$

where $u_0^{(m)}$, $w_0^{(m)}$, $\omega^{(m)}$, $\Phi_0^{(m)}$ and $\Psi_0^{(m)}$ are integral constants.

The continuity of stresses σ_z and τ_{xz}, electric displacements D_z, magnetic induction B_z, displacements u and w, electric potential Φ and magnetic potential Ψ at the interface $z=h_m$ requires

$$\sigma_z^{(m)}(x,h_m) = \sigma_z^{(m+1)}(x,h_m), \qquad \tau_{xz}^{(m)}(x,h_m) = \tau_{xz}^{(m+1)}(x,h_m), \tag{216}$$

$$D_z^{(m)}(x,h_m) = D_z^{(m+1)}(x,h_m), \qquad B_z^{(m)}(x,h_m) = B_z^{(m+1)}(x,h_m), \tag{217}$$

$$u^{(m)}(x,h_m) = u^{(m+1)}(x,h_m), \quad w^{(m)}(x,h_m) = w^{(m+1)}(x,h_m), \tag{218}$$

$$\Phi^{(m)}(x,h_m) = \Phi^{(m+1)}(x,h_m), \qquad \Psi^{(m)}(x,h_m) = \Psi^{(m+1)}(x,h_m), \tag{219}$$

where m=1,2,…, N-1. Substitution of Eqs. $(201)_{2,3}$ into Eq. (216), Eq. $(202)_2$ into Eq. $(217)_1$, Eq. $(203)_2$ into $(217)_2$, Eq. (213) into Eq. (218), Eq. (214) into Eq. $(219)_1$, and Eq. (215) into Eq.$(219)_2$, yields

$$F_m(h_m) = F_{m+1}(h_m), \qquad G_m(h_m) = G_{m+1}(h_m), \tag{220}$$

$$F_m'(h_m) = F_{m+1}'(h_m), \qquad G_m'(h_m) = G_{m+1}'(h_m), \tag{221}$$

$$R_m(h_m) = R_{m+1}(h_m), \qquad S_m(h_m) = S_{m+1}(h_m), \tag{222}$$

$$U_m(h_m) = U_{m+1}(h_m), \qquad V_m(h_m) = V_{m+1}(h_m), \tag{223}$$

$$\left[s_{11}^{(m)}G_m''-\alpha_n^2 s_{13}^{(m)}G_m-\alpha_n s_{15}^{(m)}F_m'+d_{11}^{(m)}S_m'-\alpha_n d_{31}^{(m)}R_m+b_{11}^{(m)}V_m'-\alpha_n b_{31}^{(m)}U_m\right]_{z=h_m}$$
$$=\left[s_{11}^{(m+1)}G_{m+1}''-\alpha_n^2 s_{13}^{(m+1)}G_{m+1}-\alpha_n s_{15}^{(m+1)}F_{m+1}'+d_{11}^{(m+1)}S_{m+1}'-\alpha_n d_{31}^{(m+1)}R_{m+1}\right.$$
$$\left.+b_{11}^{(m+1)}V_{m+1}'-\alpha_n b_{31}^{(m+1)}U_{m+1}\right]_{z=h_m}, \tag{224}$$

$$\left[s_{11}^{(m)}F_m''-\alpha_n^2 s_{13}^{(m)}F_m+\alpha_n s_{15}^{(m)}G_m'+d_{11}^{(m)}R_m'+\alpha_n d_{31}^{(m)}S_m+b_{11}^{(m)}U_m'+\alpha_n b_{31}^{(m)}V_m\right]_{z=h_m}$$
$$=\left[s_{11}^{(m+1)}F_{m+1}''-\alpha_n^2 s_{13}^{(m+1)}F_{m+1}+\alpha_n s_{15}^{(m+1)}G_{m+1}'+d_{11}^{(m+1)}R_{m+1}'+\alpha_n d_{31}^{(m+1)}S_{m+1}\right.$$
$$\left.+b_{11}^{(m+1)}U_{m+1}'+\alpha_n b_{31}^{(m+1)}V_{m+1}\right]_{z=h_m}, \tag{225}$$

$$\left[\left(s_{11}^{(m)}F_m''-\alpha_n^2 s_{13}^{(m)}F_m+\alpha_n s_{15}^{(m)}G_m'+d_{11}^{(m)}R_m'+\alpha_n d_{31}^{(m)}S_m+b_{11}^{(m)}U'+\alpha_n b_{31}^{(m)}V_m\right)'\right.$$
$$\left.+\alpha_n\left(s_{15}^{(m)}G_m''-\alpha_n^2 s_{35}^{(m)}G_m-\alpha_n s_{55}^{(m)}F_m'+d_{15}^{(m)}S_m'-\alpha_n d_{35}^{(m)}R_m+b_{15}^{(m)}V_m'-\alpha_n b_{35}^{(m)}U_m\right)\right]_{z=h_m}$$
$$=\left[\left(s_{11}^{(m+1)}F_{m+1}''-\alpha_n^2 s_{13}^{(m+1)}F_{m+1}+\alpha_n s_{15}^{(m+1)}G_{m+1}'+d_{11}^{(m+1)}R_{m+1}'+\alpha_n d_{31}^{(m+1)}S_{m+1}+b_{11}^{(m+1)}U_{m+1}'\right.\right.$$
$$\left.+\alpha_n b_{31}^{(m+1)}V_{m+1}\right)'+\alpha_n\left(s_{15}^{(m+1)}G_{m+1}''-\alpha_n^2 s_{35}^{(m+1)}G_{m+1}-\alpha_n s_{55}^{(m+1)}F_{m+1}'+d_{15}^{(m+1)}S_{m+1}'\right.$$
$$\left.\left.-\alpha_n d_{35}^{(m+1)}R_{m+1}+b_{15}^{(m+1)}V_{m+1}'-\alpha_n b_{35}^{(m+1)}U_{m+1}\right]_{z=h_m} \tag{226}$$

$$\left[\left(s_{11}^{(m)}G_m''-\alpha_n^2 s_{13}^{(m)}G_m-\alpha_n s_{15}^{(m)}F_m'+d_{11}^{(m)}S_m'-\alpha_n d_{31}^{(m)}R_m+b_{11}^{(m)}V_m'-\alpha_n b_{31}^{(m)}U_m\right)'\right.$$
$$\left.-\alpha_n\left(s_{15}^{(m)}F_m''-\alpha_n^2 s_{35}^{(m)}F_m+\alpha_n s_{55}^{(m)}G_m'+d_{15}^{(m)}R_m'+\alpha_n d_{35}^{(m)}S_m+b_{15}^{(m)}U_m'+\alpha_n b_{35}^{(m)}V_m\right)\right]_{z=h_m}$$
$$=\left[\left(s_{11}^{(m+1)}G_{m+1}''-\alpha_n^2 s_{13}^{(m+1)}G_{m+1}-\alpha_n s_{15}^{(m+1)}F_{m+1}'+d_{11}^{(m+1)}S_{m+1}'-\alpha_n d_{31}^{(m+1)}R_{m+1}\right.\right.$$
$$\left.+b_{11}^{(m+1)}V_{m+1}'-\alpha_n b_{31}^{(m+1)}U_{m+1}\right)'-\alpha_n\left(s_{15}^{(m+1)}F_{m+1}''-\alpha_n^2 s_{35}^{(m+1)}F_{m+1}+\alpha_n s_{55}^{(m+1)}G_{m+1}'\right.$$
$$\left.\left.+d_{15}^{(m+1)}R_{m+1}'+\alpha_n d_{35}^{(m+1)}S_{m+1}+b_{15}^{(m+1)}U_{m+1}'+\alpha_n b_{35}^{(m+1)}V_{m+1}\right)\right]_{z=h_m}, \tag{227}$$

$$\left[d_{11}^{(m)}G_m''-\alpha_n^2 d_{13}^{(m)}G_m-\alpha_n d_{15}^{(m)}F_m'-\gamma_{11}^{(m)}S_m'+\alpha_n\gamma_{13}^{(m)}R_m-\beta_{11}^{(m)}V_m'+\alpha_n\beta_{13}^{(m)}U_m\right]_{z=h_m}$$
$$=\left[d_{11}^{(m+1)}G_{m+1}''-\alpha_n^2 d_{13}^{(m+1)}G_{m+1}-\alpha_n d_{15}^{(m+1)}F_{m+1}'-\gamma_{11}^{(m+1)}S_{m+1}'+\alpha_n\gamma_{13}^{(m+1)}R_{m+1}\right.$$
$$\left.-\beta_{11}^{(m+1)}V_{m+1}'+\alpha_n\beta_{13}^{(m+1)}U_{m+1}\right]_{z=h_m}, \tag{228}$$

$$\left[-d_{11}^{(m)}F_m''+\alpha_n^2 d_{13}^{(m)}F_m-\alpha_n d_{15}^{(m)}G_m'+\gamma_{11}^{(m)}R_m'+\alpha_n\gamma_{13}^{(m)}S_m+\beta_{11}^{(m)}U_m'+\alpha_n\beta_{13}^{(m)}V_m\right]_{z=h_m}$$
$$=\left[-d_{11}^{(m+1)}F_{m+1}''+\alpha_n^2 d_{13}^{(m+1)}F_{m+1}-\alpha_n d_{15}^{(m+1)}G_{m+1}'+\gamma_{11}^{(m+1)}R_{m+1}'+\alpha_n\gamma_{13}^{(m+1)}S_{m+1}\right.$$
$$\left.+\beta_{11}^{(m+1)}U_{m+1}'+\alpha_n\beta_{13}^{(m+1)}V_{m+1}\right]_{z=h_m}, \quad (229)$$

$$\left[b_{11}^{(m)}G_m''-\alpha_n^2 b_{13}^{(m)}G_m-\alpha_n b_{15}^{(m)}F_m'-\beta_{11}^{(m)}S_m'+\alpha_n\beta_{13}^{(m)}R_m-\nu_{11}^{(m)}V_m'+\alpha_n\nu_{13}^{(m)}U_m\right]_{z=h_m}$$
$$=\left[b_{11}^{(m+1)}G_{m+1}''-\alpha_n^2 b_{13}^{(m+1)}G_{m+1}-\alpha_n b_{15}^{(m+1)}F_{m+1}'-\beta_{11}^{(m+1)}S_{m+1}'+\alpha_n\beta_{13}^{(m+1)}R_{m+1}\right.$$
$$\left.-\nu_{11}^{(m+1)}V_{m+1}'+\alpha_n\nu_{13}^{(m+1)}U_{m+1}\right]_{z=h_m}, \quad (230)$$

$$\left[-b_{11}^{(m)}F_m''+\alpha_n^2 b_{13}^{(m)}F_m-\alpha_n b_{15}^{(m)}G_m'+\beta_{11}^{(m)}R_m'+\alpha_n\beta_{13}^{(m)}S_m+\nu_{11}^{(m)}U_m'+\alpha_n\nu_{13}^{(m)}V_m\right]_{z=h_m}$$
$$=\left[-b_{11}^{(m+1)}F_{m+1}''+\alpha_n^2 b_{13}^{(m+1)}F_{m+1}-\alpha_n b_{15}^{(m+1)}G_{m+1}'+\beta_{11}^{(m+1)}R_{m+1}'+\alpha_n\beta_{13}^{(m+1)}S_{m+1}\right.$$
$$\left.+\nu_{11}^{(m+1)}U_{m+1}'+\alpha_n\nu_{13}^{(m+1)}V_{m+1}\right]_{z=h_m}, \quad (231)$$

$$-\omega^{(m)}h_m+u_0^{(m)}=-\omega^{(m+1)}h_m+u_0^{(m+1)},\quad \omega^{(m)}=\omega^{(m+1)},\quad w_0^{(m)}=w_0^{(m+1)}, \quad (232)$$

$$\Phi_0^{(m)}=\Phi_0^{(m+1)},\qquad \Psi_0^{(m)}=\Psi_0^{(m+1)}, \quad (233)$$

By using of Eqs. $(201)_{1,3}$, $(202)_1$, $(203)_1$, (157)-(160) and (220)-(223), we obtain the same expressions of the axial force, bending moment, shear force and average electric displacement as Eqs. (186)-(190).

So far, the $16\cdot(N-1)$ equations in Eqs. (220)-(231), together with 16 equations in Eqs. (157)-(160) completely determine the $16N$ integral constants $\tilde{f}_k^{(m)}$, $\tilde{g}_k^{(m)}$, $\tilde{r}_k^{(m)}$, $\tilde{s}_k^{(m)}$, $\tilde{u}_k^{(m)}$, $\tilde{v}_k^{(m)}$ (m=1,2,…,N, k=1,2,…,16). The $5\cdot(N-1)$ equations in Eqs. (232) and (233) together with 10 equations of boundary conditions at the two ends and another 3 equations in Eq. (185) also completely determine the $5N$+8 unknown constants, $u_0^{(m)}$, $w_0^{(m)}$, $\omega^{(m)}$, $\Phi_0^{(m)}$, $\Psi_0^{(m)}$ (m=1, 2,…, N) and b_k (k= 1, 2,…,8).

4.7. Numerical Examples

Example 4.1: A simply supported beam with material properties varying exponentially through the thickness

Consider a simply supported FGMEE beam with span l = 1m, height h = 0.1 m, subjected to a distributed load $q=q_0\sin(\pi x/l)$. The boundary conditions at the two ends are: $D^*=0$, $B^*=0$, $M=0$, (at x = 0) $u=0$, $w=0$, (at $(0,0)$); $M=0$, $T=0$, (at x = 1), $\Phi=0$, $\Psi=0$, $w=0$, (at $(l,0)$). Every material property varies through the thickness according to Eq. (195). The material properties at z = 0 are listed in Table 2. In this example, the analytical solution can be obtained.

Table 2. Material properties

(s_{ij}^0 *in* $10^{-12}\,m^2/N$, d_{ij}^0 *in* $10^{-3}\,m^2/C$, b_{ij}^0 *in* $10^{-3}\,m/NA$, γ_{ij}^0 *in* $10^9\,m/F$, β_{ij}^0 *in* $10^6\,C^2/Ns^2$, ν_{ij}^0 *in* $10^6\,CV/Ns$)

s_{11}^0	s_{13}^0	s_{15}^0	s_{33}^0	s_{35}^0	s_{55}^0	d_{11}^0	d_{13}^0	d_{15}^0
6.81	-3.24	0.51	5.77	-0.58	8.58	0.80	-0.60	9.00
d_{31}^0	d_{33}^0	d_{35}^0	b_{11}^0	b_{13}^0	b_{15}^0	b_{31}^0	b_{33}^0	b_{35}^0
-7.00	9.50	0.50	0.14	-0.12	0.94	0.16	0.22	-0.063
γ_{11}^0	γ_{13}^0	γ_{13}^0	β_{11}^0	β_{13}^0	β_{33}^0	ν_{11}^0	ν_{13}^0	ν_{33}^0
0.080	-0.0012	0.063	-0.98	0.051	-0.26	0.095	-0.0014	0.075

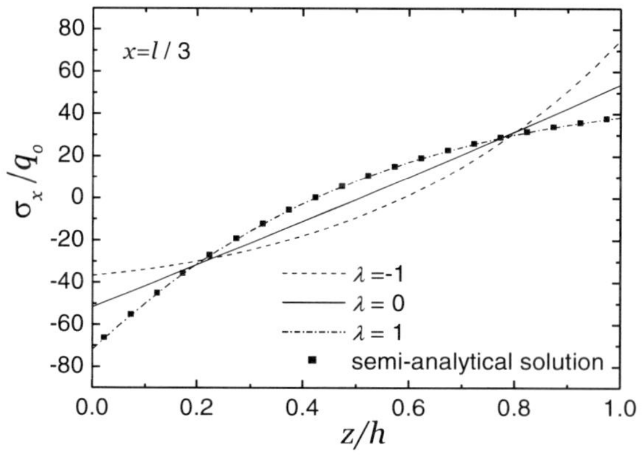

Figure 12. σ_x/q_0 vs. z/h at x=l/3

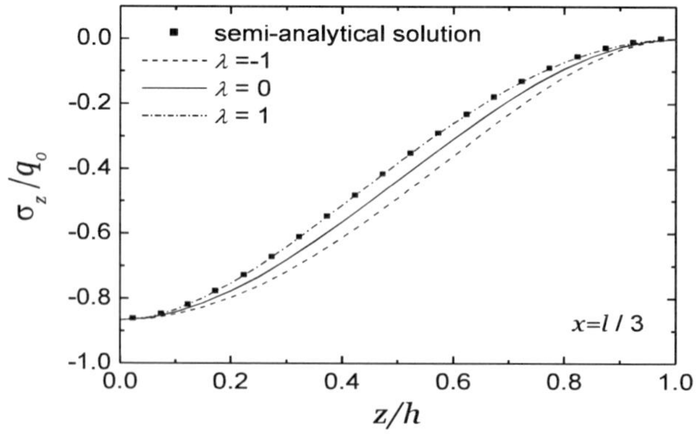

Figure 13. σ_z/q_0 vs. *z/h* at *x=l/3*.

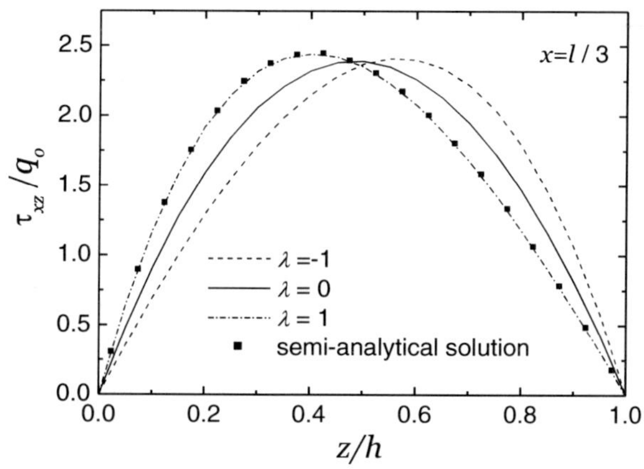

Figure 14. τ_{xz}/q_0 vs. z/h at $x=l/3$

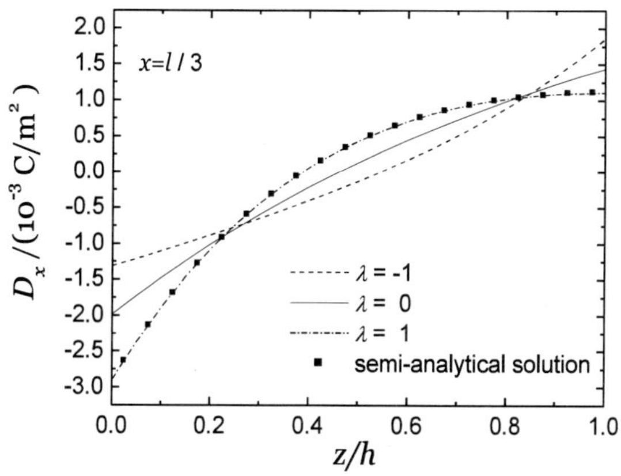

Figure 15. D_x vs. z/h at $x=l/3$.

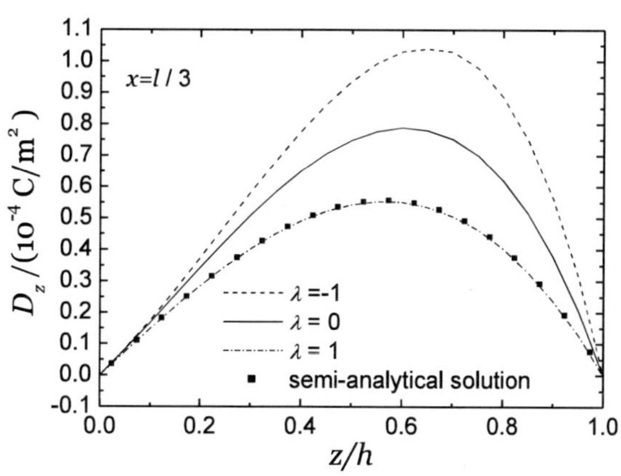

Figure 16. D_z vs. z/h at x=l/3.

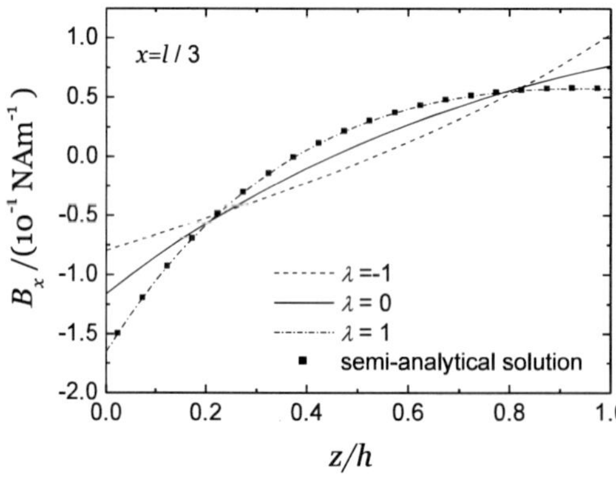

Figure 17. B_x vs. z/h at x=l/3

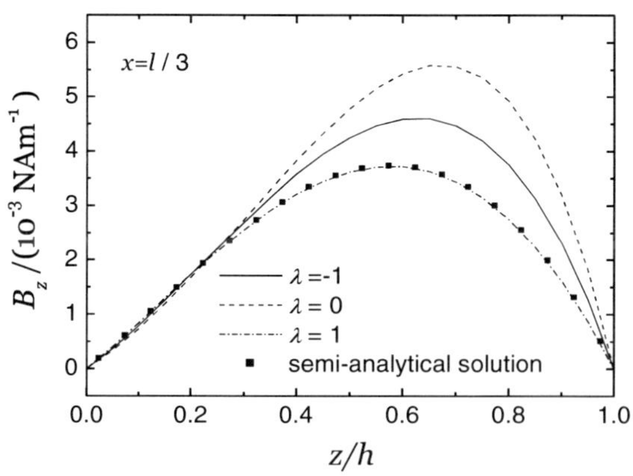

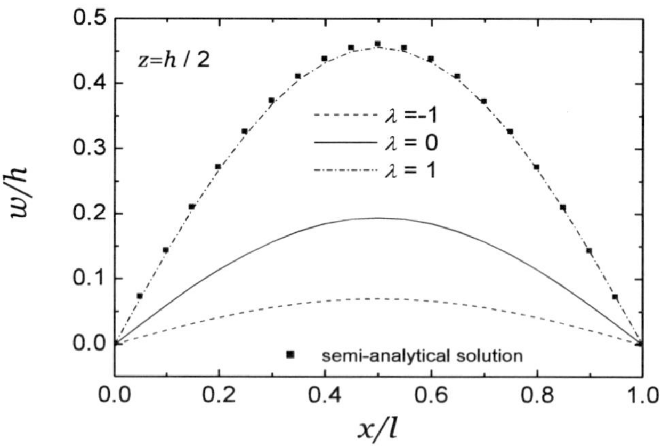

Figure 18. B_z vs. z/h at x=l/3 Figure 19. The deflection curve of the beam

From Figures 12-18, we find that the gradient index λ will influence the distribution of stresses, electric displacements, displacements and electric potential to different degrees. The value of λ has little influence on the maximum σ_z and τ_{xz}, however it influences the maximal absolute value of σ_x. The maximal absolute value of σ_x of $\lambda = \pm 1$ is 39.3% larger than that of $\lambda = 0$.

Comparisons between Figures 15 and 17 and between Figures 16 and 18 show that the similar distributions between D_x and B_x, and between D_z and B_z. which is reasonable since the electric field and magnetic field almost play an identical role in the problem. The value of λ influences the distribution of D_x, D_z, B_x and B_z as well as their maximum. The maximal absolute value of D_x of $\lambda = \pm 1$ is 46.5% larger than that of $\lambda = 0$, and the maximal absolute value of B_x of $\lambda = \pm 1$ is 42.2% larger than that of $\lambda = 0$. With the increase of λ, the maximal absolute values of D_z and B_z decrease. The maximal absolute value of D_z of $\lambda = -1$ is as 1.88 times larger as that of $\lambda = 1$. The maximal absolute value of B_z of $\lambda = -1$ is as 1.50 times large as that of $\lambda = 1$.

The deflection curve of the central line of the beam is shown in Figure 19. The parameter λ has great effect on the deflection curve. With the increase of λ, the maximal absolute value of the deflection decreases. The maximal deflection of $\lambda = -1$ is as 6.53 times large as that of $\lambda = 1$.

The semi-analytical solution is acquired by the approach of sub-layer approximation in which the beam is divided into 20 layers virtually. A quite good agreement from these figures can be observed between the analytical and semi-analytical results.

Example 4.2: A cantilever beam with material properties varying in a power law.

In this example, both the geometry of the beam and the external load are the same as those of Example 4.1. Equations (157)-(160) are taken for the boundary conditions on the upper and lower surfaces, Eqs. (161), (164) and (167) (where $\bar{T} = 0$, $\bar{M} = 0$, $\bar{Q} = 0$, $D^* = 0$, $B^* = 0$) for the boundary conditions at the left end (free, $x = 0$), and Eqs (162), (165) and $(168)_{1,2}$ for that at the right end (fixed, $x = l$). The variation of material properties through the thickness obeys the following rule

$$M_{ij} = M_{ij}^{I}\left(1 - z/h\right)^{\kappa} + M_{ij}^{II}\left[1 - \left(1 - z/h\right)^{\kappa}\right], \tag{239}$$

where M_{ij} denote the properties at z, M_{ij}^{I} and M_{ij}^{II} stand for materials I and II, κ is another gradient index. We assume that , $M_{ij}^{I} = M_{ij}^{0}$, $M_{ij}^{II} = M_{ij}^{0}\exp(1)$. M_{ij}^{0} are listed in Table 2. In this example, the sub-layer approximation is employed, in which the beam is divided virtually into 20 uniform sub-layers along the thickness.

The stress, electric displacement and deflection curves are shown in Figures 20-26. When the material properties are designated at the upper and lower surfaces, the gradient index κ influences distribution of stress, electric displacement and magnetic induction, but just to a limited extent. Figure 27 shows the deflection curve of the central line, indicating that κ influences the deflection greatly. With the increase of κ, the deflection at the free end increases. The deflection for $\kappa = 5$ is as 1.35 times large as that for $\kappa = 0.5$, and as 2.54 times large as that for $\kappa = 0$.

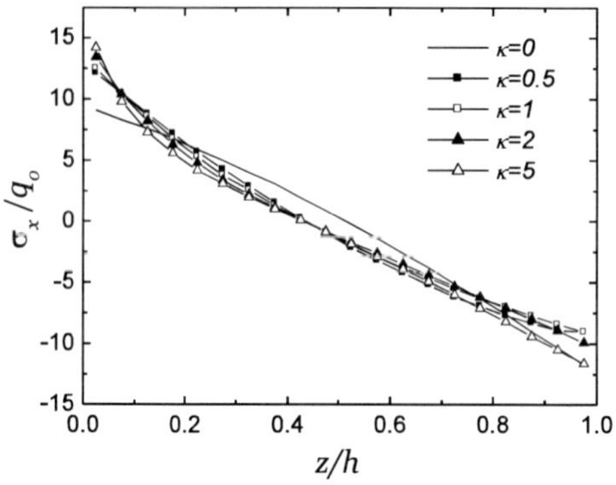

Figure 20. σ_x / q_0 vs. z/h at $x=l/3$.

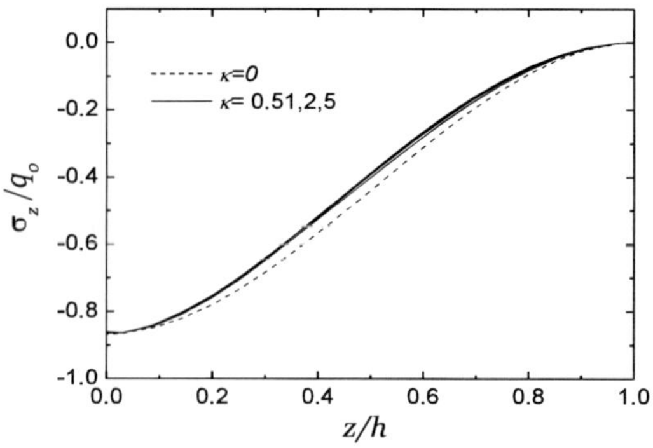

Figure 21. σ_z / q_0 vs. z/h at $x=l/3$.

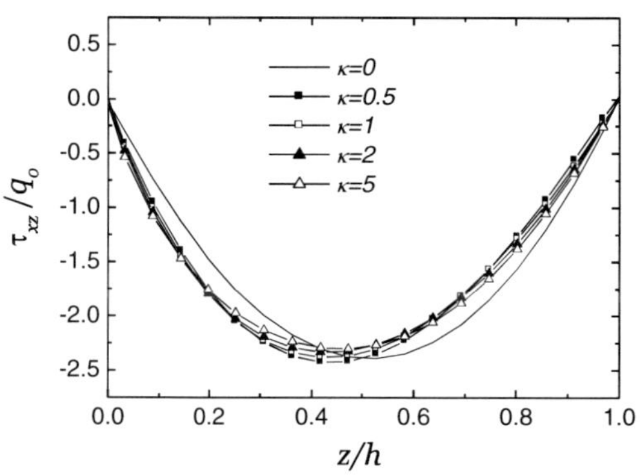

Figure 22. τ_{xz} / q_0 vs. z/h at $x=l/3$.

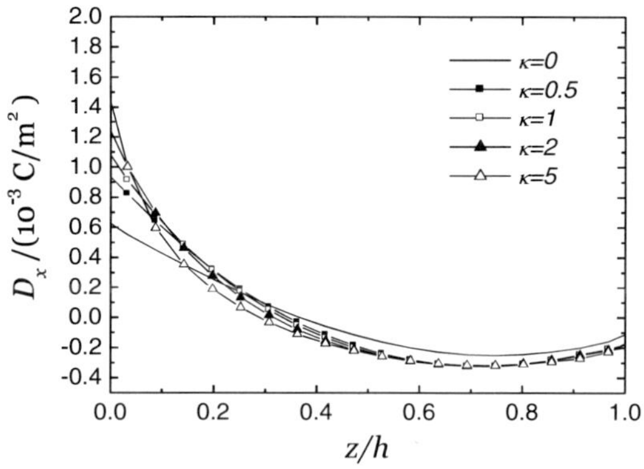

Figure 23. D_x vs. z/h at $x=l/3$.

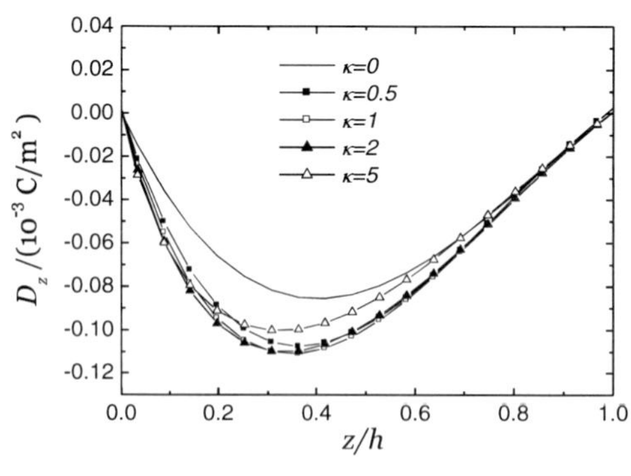

Figure 24. D_z vs. z/h at $x=l/3$.

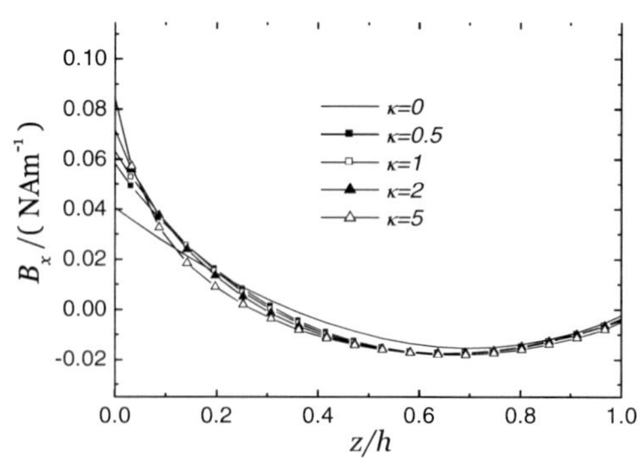

Figure 25. B_x vs. z/h at $x=l/3$.

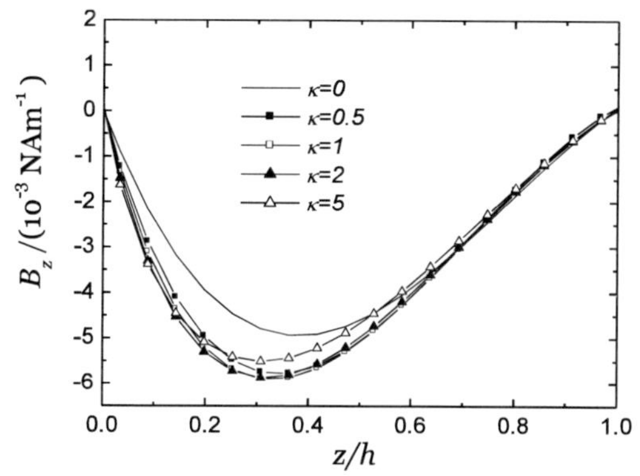

Figure 26. B_z vs. z/h at $x=l/3$.

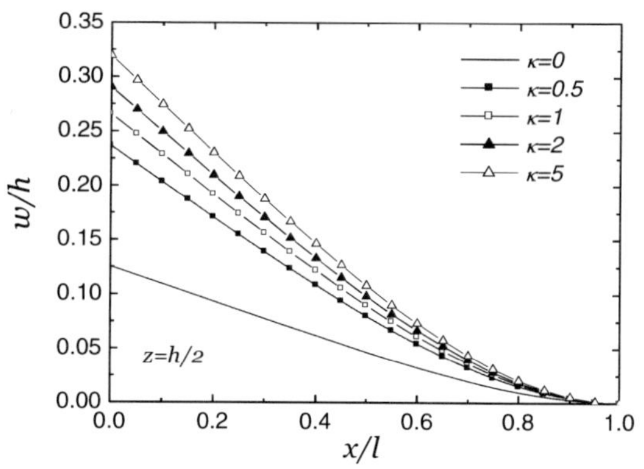

Figure 27. The deflection curve of the beam.

CONCLUSION

Analytical solutions are presented for anisotropic FGMEE beams subjected to polynomial loads, whose material properties vary in arbitrary fashions along the thickness. Various kinds of boundary conditions at the two ends of the beam can be considered. If the stress function, electric displacement function and magnetic induction function are taken to be a polynomial of higher order of x, the method can be used to solve problems of beams subjected to distributed loads of higher orders.

Analytical solution and semi-analytical solution are also presented for anisotropic FGMEE beams subjected to distributed sinusoidal loads. The material properties are assumed to vary in exponential function or be arbitrary functions of the thickness coordinate. Various kinds of boundary conditions at the two ends of the beam are considered. When the arbitrarily distributed load vanishes at the two ends of the beam, it can be expanded in terms of sinusoidal series according to Eq. (142). The solutions for beam subjected to sinusoidal loading with different α_n are first obtained one by one. Then we can obtain the complete solution of the problem by virtue of the theorem of superposition. It is noted that when the load is not zero at the two ends, it can be decomposed into a load represented by a sinusoidal series and a linearly distributed load $q(0)+\left[q(l)-q(0)\right]\cdot x/l$. The solution for the latter load case can be acquired by the method proposed in Section 3.

Besides the analytical solutions for FGMEE beams under polynomial and arbitrary loads, the methods described in this chapter can be conveniently used to find analytical solutions for FG piezoelastic beams and FG elastic beams under polynomial and arbitrary loads. The reader is referred to Refs. [19] and [20] for the details.

Acknowledgments

Our work on functionally graded structures has been supported by the National Natural Science Foundation of China (Nos. 10432030, 10725210 and 10832009), and the National Basic Research Program of China (No. 2009CB623200).

References

Uchino, K. Piezoelectric Actuators and Ultrasonic Motors; Kluwer: Boston, 1997.

Zhu, X. H., Wang, Q., Meng, Z. Y. A functionally graded piezoelectric actuator prepared by powder metallurgical process in PNN-PZ-PT system. *J. Mate. Sci. Lett.* 1995, Vol 14, 516-518.

Wu, C. C. M., Kahn, M., Moy, W. Piezoelectric ceramics with functional gradients: a new application in material design. *J. Am. Ceram. Soc.* 1996, Vol 79, 809-812.

Saravanos, D. A., Heyliger, P. R. Mechanics and computational models for laminated piezoelectric beams, plates, and shells. *Appl. Mech. Rev.* 1999, Vol 52, 305-320.

Tauchert, T. R., Ashida, F., Noda, N., Adali, S., Verijenko, V. Developments in thermopiezoelasticity with relevance to smart composite structures. Compos. Struct. 2000, Vol 48, 31-38.

Chen, W. Q., Ding, H. J. Bending of functionally graded piezoelectric rectangular plates. *Acta Mech. Solida Sinica* 2000, Vol 13, 312-319.

Kruusing, A. Analysis and optimization of loaded cantilever beam microactuators. Smart Mater. Struct. 2000, Vol 9, 186-196.

Hauke, T., Kouvatov, A., Steinhausen, R., Seifert, W., Beige, H., Hanghammer H. T., Abicht, H. Bending behavior of functionally graded materials. *Ferroelect.* 2001, Vol 238, 195-202.

Shi, Z. F. General solution of a density functionally gradient piezoelectric cantilever and its applications. *Smart Mater. Struct.* 2002, Vol 11, 122-129.

Shi, Z. F., Chen, Y. Functionally graded piezoelectric cantilever beam under load. *Arch. Appl. Mech.* 2004, Vol 74, 237-247.

Liu, T. T., Shi, Z. F. Bending behavior of functionally gradient piezoelectric cantilever. *Ferroelect.* 2004, Vol 308, 43-51.

Chen, W. Q., Lee, K. Y. Alternative state space formulations for magnetoelectric thermoelasticity with transverse isotropy and the application to bending analysis of non-homogeneous plates. *Int. J. Solids Struct.* 2003, Vol 40, 5689-5705.

Chen, W.Q., Lee, K.Y., Ding, H.J. On free vibration of non-homogeneous transversely isotropic magneto-electro-elastic plates. *J. Sound Vib.* 2005, Vol 279 237-251.

Pan, E., Han, F. Exact solution for functionally graded and layered magneto- electro-elastic plates. *Int. J. Eng. Sci.* 2005, Vol 43, 321-339.

Bhangale, R. K., Ganesan, N. Static analysis of simply supported functionally graded and layered magneto-electro-elastic plates. *Int. J. Solids Struct.* 2006, Vol 43, 3230- 3253.

Bhangale, R. K., Ganesan, N. Free vibration studies of simply supported non-homogeneous functionally graded magneto-electro-elastic finite cylindrical shells. *J. Sound Vib.* 2005, Vol 288, 412-422.

Bhangale, R. K., Ganesan, N. Free vibration of simply supported functionally graded and layered magneto-electro-elastic plates by finite element method. *J. Sound Vib.* 2005, 294, 1016-1038.

Gere, J. M., Timoshenko, S. P. Mechanics of Materials, 2nd edition, PWS-KENT Publishing Company: Boston, 1984.

Huang, D. J., Ding, H. J., Chen, W. Q. Piezoelasticity solutions for functionally graded piezoelectric beams. *Smart Mater. Struct.* 2007, Vol 16, 687- 695.

Ding, H. J., Huang, D. J., Chen W. Q. Elasticity solutions for plane anisotropic functionally graded beams. *Int. J. Solids Struct.* 2007, Vol 44, 176-196.

ABOUT THE AUTHORS

Haojiang Ding received his B.S. degree in Applied Mathematics and Mechanics in 1957 from Peking University, Beijing, China. After his graduation, he joined the Department of Engineering Mechanics, Zhejiang University, Hangzhou, China, and was promoted to be an associate Professor in 1981 and a full Professor in 1985. He has been the dean of the Department of Engineering Mechanics, Zhejiang University from 1984 to 1992. In 1995, he moved to the Department of Civil Engineering, and worked as a senior Professor until his retirement in 2006. Through years, he has been the supervisor of more than 30 Ph. D. candidates. His research interests include elasticity of transversely isotropic materials, piezoelasticity, mechanics of functionally graded materials and structures, analytical and numerical methods in applied mechanics, as well as mechanics of porous media. He is the author of seven monographs and textbooks as well as 300+ journal papers. In 2000, he was awarded the Citation Classic Award from ISI, USA.

Dejin Huang received the Ph.D. degree in 2010 from Zhejiang University, Hangzhou, China. He is an associate Professor of the Faculty of Mechanical Engineering and Mechanics, Ningbo university, Ningbo, China. His research interests include solid mechanics, smart structures, microsensor, actuator, structural health monitoring and impact engineering.

In: Mechanics of Functionally Graded Materials and Structures ISBN 978-1-61668-920-9
Editors: Z. Zhong, L. Wu and W. Chen

Chapter 5

THE STATE-SPACE METHOD AND ITS APPLICATION IN ANALYSES OF FGM STRUCTURES

***Weiqiu Chen*[1] *and Haojiang Ding*[2]**
Department of Engineering Mechanics,
Zhejiang University, Yuquan Campus, Hangzhou, P.R. China
Department of Civil Engineering,
Zhejiang University, Zijingang Campus, Hangzhou, P.R. China

ABSTRACT

The state-space description has been widely adopted in the modern control theory, while it is only occasionally employed for seeking the solutions of elasticity problems until recently. This chapter presents a brief overview of the state-space method for elasticity, with particular emphasis on its applications in analyses of functionally graded materials plates and structures. Because of the material inhomogeneity, it is generally very difficult to obtain exact static/dynamic solutions for these structures directly based on three-dimensional (or even two-dimensional) elasticity equations. Except for some particular cases, a layerwise scheme thus has been incorporated into the analysis, facilitating us to derive analytical solutions. These though approximate solutions can be made as accurately as possible by involving more layers in the approximation, and hence can be benchmarks for clarifying other approximate and numerical solutions. A newly developed semi-analytical method, which combines the state-space method and the differential quadrature method, is also presented.

1. INTRODUCTION

In modern control theory, the system equations are usually expressed as a set of first order differential equations, called the state equation of the system, with respect to the essential variables, called the state (or primary) variables which constitute a particular state

[1] e-mail: chenwq@zju.edu.cn.
[2] e-mail: dinghj@zju.edu.cn.

space. Other variables of the system, known as the output (or secondary) variables, can be determined completely if the state variables are known. Both the state and output variables are functions of time t only. Since only first-order differential equations are involved, the matrix theory and particularly the fast-growing computer facilities can be fully utilized. Along with other superiorities, the state-space based method has exhibited a more powerful ability than other approaches and the state-space description has become a standard description in modern control theory.

On the contrary, the elasticity of material bodies is governed by partial differential equations, rather than the ordinary differential ones. For a three-dimensional dynamic problem, all field variables, including displacements, stresses and strains, are functions of three coordinate variables and the time variable. Even in the static case, they are still functions of three coordinate variables. This makes it very difficult to derive the exact or analytical solutions. Traditionally, the three sets of elasticity equations, namely the equations of motion, the kinematic equations and the constitutive equations, are transformed to a set of higher-order partial differential equations, expressed purely in terms of displacements, strains or stresses. When the material is isotropic or transversely isotropic, displacement functions or stress functions have been introduced to further simplify the resulting equations into harmonic or bi-harmonic equations, whose solutions, except for some particular cases, are still difficult to obtain.

When structural theories (e.g. the rod, beam, plate, and shell theories) are employed, the situations become simpler because of the reduced dimensions compared to the original three-dimensional elasticity theory. As early in 1952, Thomson [1] formulated the basic equations of the bending of a beam with non-uniform section as a set of first-order ordinary differential equations, in contrast to the well-known fourth-order differential equation that governs the beam deflection. The state variables include the deflection, rotation, shear force and bending moment. The independent variable is the coordinate variable along the beam axis, while that in control theory is time. Thus, he established a similarity between the structural theory and control theory. To deal with the non-uniformity of beam section, he introduced the piecewise approximation by dividing the beam into many segments so that the section of each segment is taken as constant. The continuity conditions at the interfaces of two adjacent segments and the boundary conditions at two ends of the beam are easily taken into consideration because they are all expressed directly in terms of state variables. Moreover, since a transfer relation between the state variables at two ends of a segment can be established, a global transfer relation between the state variables at the two ends of the beam can be established finally by eliminating those at the interfaces through the use of continuity conditions. This makes the scale of the final system algebraic equations as small as the homogeneous beam no matter how many beam segments are divided. Thus, this state-space method is very efficient when dealing with structures consisting of a large number of components that are connected in serial. It is noted that almost simultaneously, Thomson recognized that this approach is also applicable to wave propagation in infinite stratified medium [2]. The key is that, by appropriate transforms the original problem is reduced to a likely one-dimensional problem with the resulting governing equations, expressed in the transformed wave-number and frequency domains, being ordinary differential ones. The only independent variable is the coordinate variable corresponding to the direction perpendicular to the layer. Similar ideas can be used to solve problems of finite bodies (e.g. plates and shells) with the dimension larger than one, as will be illustrated in this chapter. Since Thomson's pioneer works, the state-space method, also called the transfer matrix method, has been widely adopted in structural engineering as well as in geophysics. The reader is referred to Pestel and Leckie [3] and Kennett [4] for its applications in these two areas, respectively.

The applications of state-space method in general elasticity problems are relatively few before 1990 [5-8]. The recent two decades, however, have witnessed the burst of its applications especially in laminated composite structures [9-13]. Along with the introduction of displacement and stress functions, Ding et al. [14,15] developed new state-space formulations for transversely isotropic materials. Chen and Ding [16,17] used a similar idea to derive the new state-space formulations for spherically isotropic materials. Parallel works on piezoelectric materials can be found in Refs. [18-21]. Further extension to magnetoelectric thermoelasticity was also accomplished by Chen et al. [22]. Another noteworthy advance is that Zhong [23,24] established a symplectic framework for elasticity based on the state equation.

The inhomogeneous characteristic of functionally graded materials (FGMs) imposes a great difficulty on obtaining exact and even analytical solutions for FGM structures. Exact solutions have been obtained only for a few special cases [25-30], see also Chapters 1, 4 and 7 of this book for the related topics, and the references therein. Generally, for an FGM structure with an arbitrary variation of material properties along the thickness direction, a feasible way for seeking analytical elasticity solution is to divide the structure into many layers [25], i.e. an FGM plate or shell is approximated by a laminated plate or shell. It is known that with the increase of layer number, the solution of the approximate laminated structure approaches gradually to the exact solution of the original FGM structure. However, the conventional displacement method becomes very inefficient when the layer number is large. Thus, combining the layerwise approximation with the state-space method, which can lead to a final system equation with the same small scale as that of a single-layer structure, becomes a natural choice [31-38].

In this chapter, we will first introduce the basic concepts and formulations of the state-space method. The state equation for a general anisotropic elastic body is then presented. By adopting the layerwise approximation, a general solution procedure is suggested for FGM plates and shells based on the three-dimensional elasticity equations. These solutions, although not strictly exact, can be as accurate as possible by selecting a proper number of divided layers. To overcome the difficulty associated with the conventional state-space method for plates or shells with arbitrary boundary conditions, a semi-analytical technique with the idea of discretizing the problem domain along the in-plane direction only via the differential quadrature method [39-44] is also put forward for the analysis of FGM structures.

2. FUNDAMENTALS OF THE STATE-SPACE METHOD

Consider a time-dependent linear physical system with a certain amount of variables. The state variables (the elements of the state vector $\mathbf{x}$) are defined as the smallest possible subset of system variables that can represent the entire state of the system at any given time. The other variables are called the output variables, which constitute the output vector $\mathbf{y}$. The state-space description of the system is given by

$$\dot{\mathbf{x}}(t) = \mathbf{A}(t)\mathbf{x}(t) + \mathbf{B}(t)\mathbf{u}(t) \tag{1}$$

$$\mathbf{y}(t) = \mathbf{C}(t)\mathbf{x}(t) + \mathbf{D}(t)\mathbf{u}(t) \tag{2}$$

where the first is the state equation of the system, a set of linear first-order ordinary differential equations, and the second the output equation, usually a set of linear algebraic equations. In Eqs. (1) and (2), $\mathbf{A}$, $\mathbf{B}$, $\mathbf{C}$ and $\mathbf{D}$ are the state matrix, input matrix, output matrix, and feedthrough (or feedforward) matrix, respectively, $\mathbf{u}$ is the input (or control) vector, and a dot over any quantity signifies the differentiation with respect to time t. All the matrices and vectors can be time-dependent as indicated above.

It is seen that once the state vector $\mathbf{x}$ is determined from Eq. (1) and the appropriate initial conditions, the output vector $\mathbf{y}$ can be easily calculated by Eq. (2). The solution of Eq. (1) can be obtained via various analytical or numerical methods [45]. For simplicity, we consider here the case when $\mathbf{A}$ is time-independent, for which the solution can be written as

$$\mathbf{x}(t) = \mathrm{e}^{\mathbf{A}(t-t_0)}\mathbf{x}(t_0) + \int_{t_0}^{t} \mathrm{e}^{\mathbf{A}(t-\tau)}\mathbf{B}(\tau)\mathbf{u}(\tau)\,\mathrm{d}\tau \tag{3}$$

where $\mathrm{e}^{\mathbf{A}(t-t_0)}$ is the matrix exponential function, which can be calculated by various methods [46]. It is seen that the above state-space solution expresses a transfer relation between the state vector at any instance and its initial value, and hence $\mathrm{e}^{\mathbf{A}(t-t_0)}$ is also called the transfer matrix. This transfer characteristic will lead to significant numerical efficiency for elastic problems with repetitive components or alike structures, as to be shown later. Moreover, since all the operations are in matrix form, the computer facilities and their abilities can be fully explored. It is also noted that the mathematical symbolic simulation codes such as Matlab, Mathmatica and Maple all have embedded functions for automatic computation of the matrix exponential function.

3. State-Space Formulations for Anisotropic Elasticity

In this section, we will show the derivation of state equation for general anisotropic elasticity in Cartesian coordinates (x, y, z) or (x_1, x_2, x_3) [11,15]. The equations of motion are

$$\sigma_{ji,j} + f_i = \ddot{u}_i \tag{4}$$

where σ_{ij} are the stress components, u_i the displacement components, f_i the body force components, and the subscript j following a comma indicates partial derivative with respect to x_j. The geometric or kinematic equations are

$$\varepsilon_{ij} = \frac{1}{2}(u_{i,j} + u_{j,i}) \tag{5}$$

where ε_{ij} are the strain components. The constitutive equations read as

$$\sigma_{ij} = c_{ijkl}\varepsilon_{ij} \tag{6}$$

where c_{ijkl} are the elastic stiffness constants. Equations (4) through (6) can be rewritten in matrix form as

$$\frac{\partial}{\partial z}\boldsymbol{\sigma}_2 + \mathbf{L}_2^{\mathrm{T}}\boldsymbol{\sigma}_2 + \mathbf{L}_1^{\mathrm{T}}\boldsymbol{\sigma}_1 + \mathbf{f} = \rho\ddot{\mathbf{u}} \tag{7}$$

$$\boldsymbol{\varepsilon}_1 = \mathbf{L}_1\mathbf{u},\ \boldsymbol{\varepsilon}_2 = \frac{\partial}{\partial z}\mathbf{u} + \mathbf{L}_2\mathbf{u} \tag{8}$$

$$\boldsymbol{\sigma}_1 = \mathbf{C}_1\boldsymbol{\varepsilon}_1 + \mathbf{C}_2\boldsymbol{\varepsilon}_2,\quad \boldsymbol{\sigma}_2 = \mathbf{C}_2^{\mathrm{T}}\boldsymbol{\varepsilon}_1 + \mathbf{C}_4\boldsymbol{\varepsilon}_2 \tag{9}$$

where

$$\boldsymbol{\sigma}_1 = [\sigma_x, \sigma_y, \tau_{xy}]^{\mathrm{T}},\quad \boldsymbol{\sigma}_2 = [\sigma_z, \tau_{zx}, \tau_{yz}]^{\mathrm{T}}$$

$$\boldsymbol{\varepsilon}_1 = [\varepsilon_x, \varepsilon_y, \gamma_{xy}]^{\mathrm{T}},\quad \boldsymbol{\varepsilon}_2 = [\varepsilon_z, \gamma_{zx}, \gamma_{yz}]^{\mathrm{T}}$$

$$\mathbf{u} = [u_z, u_x, u_y]^{\mathrm{T}},\quad \mathbf{f} = [f_z, f_x, f_y]^{\mathrm{T}}$$

$$\mathbf{L}_1 = \begin{bmatrix} 0 & \frac{\partial}{\partial x} & 0 \\ 0 & 0 & \frac{\partial}{\partial y} \\ 0 & \frac{\partial}{\partial y} & \frac{\partial}{\partial x} \end{bmatrix},\quad \mathbf{L}_2 = \begin{bmatrix} 0 & 0 & 0 \\ \frac{\partial}{\partial x} & 0 & 0 \\ \frac{\partial}{\partial y} & 0 & 0 \end{bmatrix}$$

$$\mathbf{C}_1 = \begin{bmatrix} c_{11} & c_{12} & c_{16} \\ c_{12} & c_{22} & c_{26} \\ c_{16} & c_{26} & c_{66} \end{bmatrix},\quad \mathbf{C}_2 = \begin{bmatrix} c_{13} & c_{15} & c_{14} \\ c_{23} & c_{25} & c_{24} \\ c_{36} & c_{56} & c_{46} \end{bmatrix},\quad \mathbf{C}_4 = \begin{bmatrix} c_{33} & c_{35} & c_{34} \\ c_{35} & c_{55} & c_{45} \\ c_{34} & c_{45} & c_{44} \end{bmatrix}$$

Note that the double superscripts (*ij* and *kl*) of the stiffness constants c_{ijkl} have been replaced by a single superscript (Voigt notation) by the rule 11→1, 22→2, 33→3, 32 or 23→4, 13 or 31 →5, 12 or 21 →6. From Eqs. (8) and (9), we get

$$\boldsymbol{\varepsilon}_2 = \mathbf{C}_4^{-1}(\boldsymbol{\sigma}_2 - \mathbf{C}_2^{\mathrm{T}}\boldsymbol{\varepsilon}_1) \tag{10}$$

$$\boldsymbol{\sigma}_1 = \mathbf{B}_1\boldsymbol{\sigma}_2 + \mathbf{B}_2\mathbf{u} \tag{11}$$

$$\frac{\partial}{\partial z}\mathbf{u} = \mathbf{A}_3\boldsymbol{\sigma}_2 + \mathbf{A}_4\mathbf{u} \tag{12}$$

where

$$\mathbf{B}_1 = \mathbf{C}_2\mathbf{C}_4^{-1},\ \mathbf{B}_2 = (\mathbf{C}_1 - \mathbf{B}_1\mathbf{C}_2^{\mathrm{T}})\mathbf{L}_1,\ \mathbf{A}_3 = \mathbf{C}_4^{-1},\ \mathbf{A}_4 = -\mathbf{L}_2 - \mathbf{B}_1^{\mathrm{T}}\mathbf{L}_1$$

Substituting Eq. (11) into Eq. (7) gives

$$\frac{\partial}{\partial z}\boldsymbol{\sigma}_2 = \mathbf{A}_1\boldsymbol{\sigma}_2 + \mathbf{A}_2\mathbf{u} - \mathbf{f} \tag{13}$$

where

$$\mathbf{A}_1 = \mathbf{A}_4^{\mathrm{T}}, \quad \mathbf{A}_2 = \rho\mathbf{I}\frac{\partial^2}{\partial t^2} - \mathbf{L}_1^{\mathrm{T}}\mathbf{B}_2$$

with $\mathbf{I}$ being an identity matrix.

Now combining Eq. (12) with Eq. (13), leads to

$$\frac{\partial}{\partial z}\mathbf{V} = \mathbf{A}\mathbf{V} + \mathbf{F} \tag{14}$$

which is the state equation for general anisotropic elastic body, where

$$\mathbf{V} = \begin{Bmatrix} \boldsymbol{\sigma}_2 \\ \mathbf{u} \end{Bmatrix}, \quad \mathbf{A} = \begin{bmatrix} \mathbf{A}_1 & \mathbf{A}_2 \\ \mathbf{A}_3 & \mathbf{A}_4 \end{bmatrix}, \quad \mathbf{F} = -\begin{Bmatrix} \mathbf{f} \\ \mathbf{0} \end{Bmatrix} \tag{15}$$

$\mathbf{V}$ is the state vector, which includes both displacement components and stress components (on the surface of $z = \text{const.}$). It may be found that, if we deal with a layered medium with the z coordinate perpendicular to the interfaces, then the interfacial continuity conditions are identical to the continuity of state vector. Furthermore, since the state matrix $\mathbf{A}$ contains both operator matrices $\mathbf{L}_1$ and $\mathbf{L}_2$, the state equation (14) is more complex than the one in control theory. To obtain analytical solutions, certain constraints should be imposed on material symmetry and boundary conditions for a finite elastic body. Finally, it is pointed out that, the above choice of z direction is not preferable; it simply depends on the problem geometry and the coordinates orientated.

4. Application to FGM Plates with the Layerwise Approximation

4.1. Orthotropic FGM Rectangular Plates

For a rectangular plate shown in Figure 1, we are able to derive three-dimensional analytical solutions when the material is orthotropic and the four edges are simply supported or rigidly slipperily supported [15]. For orthotropic materials, the sub-matrices of the state matrix $\mathbf{A}$ read as

$$\mathbf{A}_3 = \begin{bmatrix} \frac{1}{c_{33}} & 0 & 0 \\ 0 & \frac{1}{c_{55}} & 0 \\ 0 & 0 & \frac{1}{c_{44}} \end{bmatrix}, \ \mathbf{A}_4 = \begin{bmatrix} 0 & -\frac{c_{13}}{c_{33}}\frac{\partial}{\partial x} & -\frac{c_{23}}{c_{33}}\frac{\partial}{\partial y} \\ -\frac{\partial}{\partial x} & 0 & 0 \\ -\frac{\partial}{\partial y} & 0 & 0 \end{bmatrix} = \mathbf{A}_1^{\mathrm{T}}$$

$$\mathbf{A}_2 = \begin{bmatrix} \rho\frac{\partial^2}{\partial t^2} & 0 & 0 \\ 0 & \rho\frac{\partial^2}{\partial t^2} - \left(c_{11} - \frac{c_{13}^2}{c_{33}}\right)\frac{\partial^2}{\partial x^2} - c_{66}\frac{\partial^2}{\partial y^2} & -\left(c_{12} + c_{66} - \frac{c_{13}c_{23}}{c_{33}}\right)\frac{\partial^2}{\partial x \partial y} \\ 0 & -\left(c_{12} + c_{66} - \frac{c_{13}c_{23}}{c_{33}}\right)\frac{\partial^2}{\partial x \partial y} & \rho\frac{\partial^2}{\partial t^2} - \left(c_{22} - \frac{c_{23}^2}{c_{33}}\right)\frac{\partial^2}{\partial y^2} - c_{66}\frac{\partial^2}{\partial x^2} \end{bmatrix}$$

The elastic constants c_{ij} as well as the density ρ are all assumed to be functions of the coordinate variable z, i.e. the plate is functionally graded in the thickness direction.

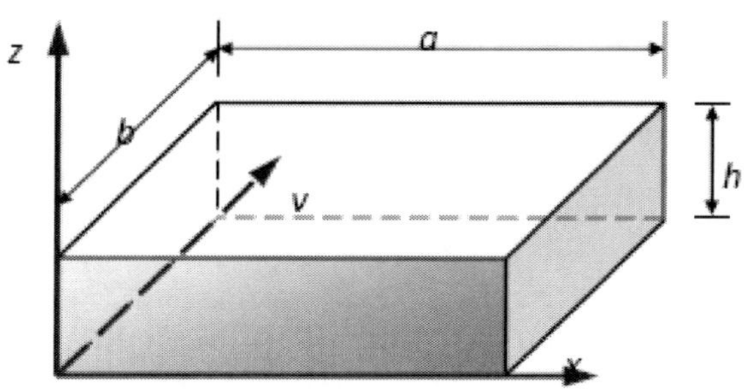

Figure 1. The geometry of a rectangular plate.

Suppose that the plate is simply supported at all four edges. We have the following boundary conditions:

$$x = 0, a: \ u_z = u_y = \sigma_x = 0 \tag{16a}$$

$$y = 0, b: \ u_z = u_x = \sigma_y = 0 \tag{16b}$$

To satisfy these conditions, we assume

$$\mathbf{V}=\begin{Bmatrix} c_{44}^{0}\bar{\sigma}_z(\zeta)\sin(m\pi\xi)\sin(n\pi\eta) \\ c_{44}^{0}\bar{\tau}_{zx}(\zeta)\cos(m\pi\xi)\sin(n\pi\eta) \\ c_{44}^{0}\bar{\tau}_{yz}(\zeta)\sin(m\pi\xi)\cos(n\pi\eta) \\ h\bar{u}_z(\zeta)\sin(m\pi\xi)\sin(n\pi\eta) \\ h\bar{u}_x(\zeta)\cos(m\pi\xi)\sin(n\pi\eta) \\ h\bar{u}_y(\zeta)\sin(m\pi\xi)\cos(n\pi\eta) \end{Bmatrix}\mathrm{e}^{\mathrm{i}\omega t} \tag{17}$$

where an overbar denotes the dimensionless quantity, $c_{44}^{0}=c_{44}\big|_{z=0}$, m and n are half-wave numbers (integers) in the x- and y-directions, $\xi=x/a$, $\eta=y/b$ and $\zeta=z/h$ are dimensionless coordinates, and ω is the circular frequency of vibration.

Substituting into the state equation and omitting the body force, we obtain

$$\frac{\mathrm{d}}{\mathrm{d}\zeta}\bar{\mathbf{V}}=\bar{\mathbf{A}}\bar{\mathbf{V}} \tag{18}$$

where $\bar{\mathbf{V}}=[\bar{\sigma}_z,\bar{\tau}_{zx},\bar{\tau}_{yz},\bar{u}_z,\bar{u}_x,\bar{u}_y]^{\mathrm{T}}$ is the dimensionless state vector, and

$$\bar{\mathbf{A}}=\begin{bmatrix} 0 & s_1 & s_2 & -\frac{\rho}{\rho_0}\Omega^2 & 0 & 0 \\ -\frac{c_{13}}{c_{33}}s_1 & 0 & 0 & 0 & -\frac{\rho}{\rho_0}\Omega^2+\frac{c_{11}c_{33}-c_{13}^2}{c_{33}c_{44}^0}s_1^2+\frac{c_{66}}{c_{44}^0}s_2^2 & \left(\frac{c_{12}+c_{66}}{c_{44}^0}-\frac{c_{13}c_{23}}{c_{33}c_{44}^0}\right)s_1s_2 \\ -\frac{c_{23}}{c_{33}}s_2 & 0 & 0 & 0 & \left(\frac{c_{12}+c_{66}}{c_{44}^0}-\frac{c_{13}c_{23}}{c_{33}c_{44}^0}\right)s_1s_2 & -\frac{\rho}{\rho_0}\Omega^2+\frac{c_{22}c_{33}-c_{23}^2}{c_{33}c_{44}^0}s_2^2+\frac{c_{66}}{c_{44}^0}s_1^2 \\ \frac{c_{44}^0}{c_{33}} & 0 & 0 & 0 & \frac{c_{13}}{c_{33}}s_1 & \frac{c_{23}}{c_{33}}s_2 \\ 0 & \frac{c_{44}^0}{c_{55}} & 0 & -s_1 & 0 & 0 \\ 0 & 0 & \frac{c_{44}^0}{c_{44}} & -s_2 & 0 & 0 \end{bmatrix} \tag{19}$$

where $\Omega=\omega h\sqrt{\rho^0/c_{44}^0}$ is the dimensionless frequency, $s_1=m\pi h/a$ and $s_2=n\pi h/b$. It is seen that, with the assumption (17), not only are the boundary conditions satisfied, but also the operator matrix $\mathbf{A}$ has been transformed to an algebraic matrix $\bar{\mathbf{A}}$, which however still varies with z. Thus, exact solution to Eq. (18) is very difficult to obtain, except for the particular case when the material constants all vary with z in an identical exponential form [47]. In the following, we will follow the idea proposed by Chen et al. [31, 34] to present computationally efficient analytical solutions by combining the state space formulations with a layerwise approximation.

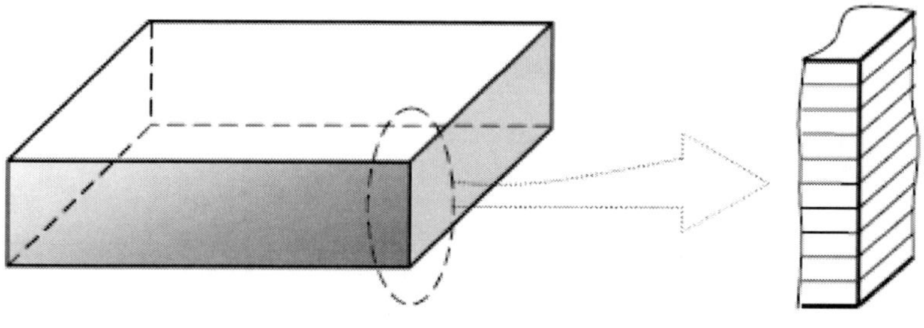

Figure 2. Layerwise approximation.

To this end, the original FGM plate is divided along the thickness direction into p thin layers, see Figure 2. p is usually taken to be large to assure the accuracy of the approximate solution. In such a way, the material constants in each thin layer are regarded as constant, approximately equal to the values at the middle surface for example. Thus, we have $c_{ij}^k = c_{ij}(h\zeta_m^k)$ and $\rho^k = \rho(h\zeta_m^k)$, with $\zeta_m^k = (\zeta_0^k + \zeta_1^k)/2$ being the dimensionless middle surface coordinate of the k-th layer, and $\zeta_0^k = (k-1)/p$ and $\zeta_1^k = k/p$ the coordinates of the lower and upper surfaces of the k-th layer, respectively. Consequently, in the k-th layer, we have $\bar{\mathbf{A}} \approx \bar{\mathbf{A}}^k$, and Eq. (18) becomes a state equation with constant coefficient matrix

$$\frac{\mathrm{d}}{\mathrm{d}\zeta}\bar{\mathbf{V}}^k = \bar{\mathbf{A}}^k\bar{\mathbf{V}}^k \quad (\zeta_0^k \le \zeta \le \zeta_1^k) \tag{20}$$

Its solution, according to Section 2, is

$$\bar{\mathbf{V}}^k(\zeta) = \mathrm{e}^{\bar{\mathbf{A}}^k(\zeta-\zeta_0^k)}\,\bar{\mathbf{V}}^k(\zeta_0^k) \tag{21}$$

Setting $\zeta = \zeta_1^k$ in the above equation gives a relation between the state vectors at the lower and upper surfaces as follows

$$\bar{\mathbf{V}}^k(\zeta_1^k) = \mathbf{T}^k\bar{\mathbf{V}}^k(\zeta_0^k) \tag{22}$$

where $\mathbf{T}^k = \mathrm{e}^{\bar{\mathbf{A}}^k(\zeta_1^k-\zeta_0^k)}$ is the transfer matrix. Making use of the interfacial continuity conditions, we finally get

$$\bar{\mathbf{V}}^p(1) = \mathbf{T}^G\bar{\mathbf{V}}^1(0) \tag{23}$$

where $\mathbf{T}^G = \prod_{k=p}^{1}\mathbf{T}^k$ is the global transfer matrix of the FGM plate. The fact of $\zeta_1^p = 1$ and $\zeta_0^1 = 0$ has been noticed above. Equation (23) relates the state variables at the top surface to those at the bottom surface, and contains only six linear algebraic equations, the number of which is much smaller than that of the usual displacement method [48] which involves $6p$ unknowns when p is large. By applying the boundary conditions at the top and bottom surfaces to Eq. (23), we can readily obtain the final linear algebraic equations for the solutions. For instance, let's consider the general steady-state response of the FGM plate subject to a harmonic dynamic pressure applied at the upper surface, $q = q_0(x, y)\mathrm{e}^{\mathrm{i}\omega t}$. The bottom surface is tractions-free. We have the following boundary conditions

$$\sigma_z = -q, \quad \tau_{zx} = \tau_{yz} = 0 \text{ at } z = h \tag{24a}$$

$$\sigma_z = \tau_{zx} = \tau_{yz} = 0 \text{ at } z = 0 \tag{24b}$$

The pressure can be expanded in terms of double sinusoidal series as

$$q = c_{44}^0 \sum_{m=1,n=1}^{\infty} q_{mn} \sin(m\pi\xi)\sin(n\pi\eta)\mathrm{e}^{\mathrm{i}\omega t} \tag{25}$$

with

$$q_{mn} = \frac{4}{c_{44}^0} \int_0^1 \int_0^1 q_0 \sin(m\pi\xi)\sin(n\pi\eta)\,\mathrm{d}\xi\,\mathrm{d}\eta$$

Substituting Eq. (24) into Eq. (23), and making use of Eq. (25), yields

$$\begin{Bmatrix} -q_{mn} \\ 0 \\ 0 \\ \bar{u}_z(1) \\ \bar{u}_x(1) \\ \bar{u}_y(1) \end{Bmatrix} = \mathbf{T}^G \begin{Bmatrix} 0 \\ 0 \\ 0 \\ \bar{u}_z(0) \\ \bar{u}_x(0) \\ \bar{u}_y(0) \end{Bmatrix} \tag{26}$$

Then, from the first three equations we get

$$\begin{bmatrix} T_{14}^G & T_{15}^G & T_{16}^G \\ T_{24}^G & T_{25}^G & T_{26}^G \\ T_{34}^G & T_{35}^G & T_{36}^G \end{bmatrix} \begin{Bmatrix} \bar{u}_z(0) \\ \bar{u}_x(0) \\ \bar{u}_y(0) \end{Bmatrix} = \begin{Bmatrix} -q_{mn} \\ 0 \\ 0 \end{Bmatrix} \tag{27}$$

where T_{ij}^G are the elements of $\mathbf{T}^G$. We can see that the scale of the equations is further reduced by half. From Eq. (27) we can solve for the unknown displacements at the bottom surface $\zeta = 0$. With these, we can obtain those at the top surface $\zeta = 1$ from Eq. (26). The state vector at any position then can be determined from Eq. (21) and the appropriate continuity conditions. Other physical variables (the output variables) can be calculated from Eq. (11). It is noted that, when the forcing frequency becomes identical with one of the natural frequencies of the plate, the coefficient determinant of Eq. (27) vanishes, yielding infinite amplitudes of the displacements. This is known as resonance, and the vanishing determinant

$$\begin{vmatrix} T_{14}^G & T_{15}^G & T_{16}^G \\ T_{24}^G & T_{25}^G & T_{26}^G \\ T_{34}^G & T_{35}^G & T_{36}^G \end{vmatrix} = 0 \tag{28}$$

also gives the characteristic equation for determining the nature frequencies, which depend on the half-wave numbers m and n. Generally, either m or n can be taken zero, which corresponds to the in-plane vibration or deformation modes.

Numerical results will be given in Section 4.2 for a transversely isotropic FGM plate using alternative state-space formulations.

If we replace the sinusoidal function with the cosine function and vice versa in Eq. (17), we may see that the following boundary conditions are satisfied

$$x = 0, a:\ u_x = \tau_{xz} = \tau_{xy} = 0 \tag{29a}$$

$$y = 0, b:\ u_y = \tau_{yz} = \tau_{xy} = 0 \tag{29b}$$

This corresponds to the case when the plate is rigidly slipperily-supported at all four edges. For a plate having simple support at some edges and having rigidly slippery support at the other edges, analytical solutions are also derivable. For example, when the edge $x = a$ is simply supported, and the other three edges have rigidly slippery conditions, we can assume

$$\mathbf{V} = \begin{Bmatrix} c_{44}^0 \bar{\sigma}_z(\zeta)\sin[(m+1/2)\pi\xi]\cos(n\pi\eta) \\ c_{44}^0 \bar{\tau}_{zx}(\zeta)\cos[(m+1/2)\pi\xi]\cos(n\pi\eta) \\ c_{44}^0 \bar{\tau}_{yz}(\zeta)\sin[(m+1/2)\pi\xi]\sin(n\pi\eta) \\ h\bar{u}_z(\zeta)\sin[(m+1/2)\pi\xi]\cos(n\pi\eta) \\ h\bar{u}_x(\zeta)\cos[(m+1/2)\pi\xi]\cos(n\pi\eta) \\ h\bar{u}_y(\zeta)\sin[(m+1/2)\pi\xi]\sin(n\pi\eta) \end{Bmatrix} \mathrm{e}^{\mathrm{i}\omega t} \tag{30}$$

The succeeding analysis is the same as that for the simply supported conditions. Of course, the loads should also be expanded in a proper form.

It is also noted that the state equation can be rearranged by choosing the state vector as $\mathbf{V} = [\sigma_z, u_x, u_y, u_z, \tau_{zx}, \tau_{yz}]^{\mathrm{T}}$. In this case, the operator matrix $\mathbf{A}$ looks simpler and neater with

$$\mathbf{A}_1 = \mathbf{A}_4 = \mathbf{0},\ \mathbf{A}_2 = \begin{bmatrix} \rho\dfrac{\partial^2}{\partial t^2} & -\dfrac{\partial}{\partial x} & -\dfrac{\partial}{\partial y} \\ & \dfrac{1}{c_{55}} & 0 \\ \text{sym.} & & \dfrac{1}{c_{44}} \end{bmatrix}$$

$$\mathbf{A}_3 = \begin{bmatrix} \dfrac{1}{c_{33}} & -\dfrac{c_{13}}{c_{33}}\dfrac{\partial}{\partial x} & -\dfrac{c_{23}}{c_{33}}\dfrac{\partial}{\partial y} \\ & \rho\dfrac{\partial^2}{\partial t^2} - \left(c_{11} - \dfrac{c_{13}^2}{c_{33}}\right)\dfrac{\partial^2}{\partial x^2} - c_{66}\dfrac{\partial^2}{\partial y^2} & -\left(c_{12} + c_{66} - \dfrac{c_{13}c_{23}}{c_{33}}\right)\dfrac{\partial^2}{\partial x \partial y} \\ \text{sym.} & & \rho\dfrac{\partial^2}{\partial t^2} - \left(c_{22} - \dfrac{c_{23}^2}{c_{33}}\right)\dfrac{\partial^2}{\partial y^2} - c_{66}\dfrac{\partial^2}{\partial x^2} \end{bmatrix}$$

4.2. New State Space Formulations for Transversely Isotropic Materials

For transversely isotropic materials, alternative state-space formulations can be developed by introducing proper displacement functions (Ψ, G) and stress functions (τ_1, τ_2) as follows

$$u_x = \frac{\partial \Psi}{\partial y} - \frac{\partial G}{\partial x},\ u_y = -\frac{\partial \Psi}{\partial x} - \frac{\partial G}{\partial y} \tag{31a}$$

$$\tau_{zx} = \frac{\partial \tau_1}{\partial y} - \frac{\partial \tau_2}{\partial x},\ \tau_{yz} = -\frac{\partial \tau_1}{\partial x} - \frac{\partial \tau_2}{\partial y} \tag{31b}$$

The body forces are also expressed as

$$f_x = \frac{\partial f_1}{\partial y} - \frac{\partial f_2}{\partial x},\ f_y = -\frac{\partial f_1}{\partial x} - \frac{\partial f_2}{\partial y} \tag{31c}$$

Through some manipulations, we can arrive at [14,15]

$$\frac{\partial}{\partial z}\begin{Bmatrix} \Psi \\ \tau_1 \end{Bmatrix} = \begin{bmatrix} 0 & \dfrac{1}{c_{44}} \\ \rho\dfrac{\partial^2}{\partial t^2} - c_{66}\Lambda & 0 \end{bmatrix}\begin{Bmatrix} \Psi \\ \tau_1 \end{Bmatrix} + \begin{Bmatrix} 0 \\ -f_1 \end{Bmatrix} \tag{32}$$

$$\frac{\partial}{\partial z}\begin{Bmatrix} G \\ \sigma_z \\ \tau_2 \\ u_z \end{Bmatrix} = \begin{bmatrix} 0 & 0 & \frac{1}{c_{44}} & 1 \\ 0 & 0 & \Lambda & \rho\frac{\partial^2}{\partial t^2} \\ \rho\frac{\partial^2}{\partial t^2} - \left(c_{11} - \frac{c_{13}^2}{c_{33}}\right)\Lambda & \frac{c_{13}}{c_{33}} & 0 & 0 \\ \frac{c_{13}}{c_{33}}\Lambda & \frac{1}{c_{33}} & 0 & 0 \end{bmatrix} \begin{Bmatrix} G \\ \sigma_z \\ \tau_2 \\ u_z \end{Bmatrix} + \begin{Bmatrix} 0 \\ -f_z \\ -f_2 \\ 0 \end{Bmatrix} \tag{33}$$

where $\Lambda = \partial^2 / \partial x^2 + \partial^2 / \partial y^2$ is the two-dimensional Laplacian. For transversely isotropic materials, we have $c_{11} = c_{22}$, $c_{13} = c_{23}$, $c_{44} = c_{55}$ and $c_{66} = (c_{11} - c_{12})/2$. These are the material symmetry constraints on the elastic constants of an orthotropic material.

As shown in Eqs. (32) and (33), the six state variables Ψ, τ_1, G, σ_z, τ_2 and u_z are dropped into two groups: one is only related to Ψ and τ_1 and the other is about the remanent four state variables. It is obvious that either Eq. (32) or Eq. (33) has an order lower than Eq. (11), leading to a somehow higher numerical efficiency. Furthermore, the separation of state equations will clearly show some particular characteristics occupied by practical problems that cannot be revealed by the sixth-order state equation. For example, the first group characterized by Ψ and τ_1 relates to the elastic in-plane deformation only. Also, for the bending of a plate, it is only necessary to solve Eqs. (33), the order of which is four, less than that of Eq. (11).

Another superiority of the new formulations is that relative simpler operators are involved. As can be seen, the right-hand sides of Eqs. (32) and (33) include the two-dimensional Laplace operator Λ only, which enables us to write down directly the corresponding formulations in circular cylindrical coordinates (r, θ, z). In fact, Eqs. (32) and (33) are still valid except for

$$\Lambda = \frac{\partial^2}{\partial r^2} + r\frac{\partial}{\partial r} + \frac{1}{r^2}\frac{\partial^2}{\partial \theta^2}$$

while the separation formulae in Eq. (30) should be replaced by

$$\begin{aligned} u_r &= -\frac{1}{r}\frac{\partial \Psi}{\partial \theta} - \frac{\partial G}{\partial r}, \quad u_\theta = \frac{\partial \Psi}{\partial r} - \frac{1}{r}\frac{\partial G}{\partial \theta} \\ \tau_{rz} &= -\frac{1}{r}\frac{\partial \tau_1}{\partial \theta} - \frac{\partial \tau_2}{\partial r}, \quad \tau_{\theta z} = \frac{\partial \tau_1}{\partial r} - \frac{1}{r}\frac{\partial \tau_2}{\partial \theta} \\ f_r &= -\frac{1}{r}\frac{\partial f_1}{\partial \theta} - \frac{\partial f_2}{\partial r}, \quad f_\theta = \frac{\partial f_1}{\partial r} - \frac{1}{r}\frac{\partial f_2}{\partial \theta} \end{aligned} \tag{34}$$

Using the new state-space formulations, Ding et al. [14] performed the bending, buckling, and free vibration analysis of transversely isotropic laminated rectangular plates. With the layerwise approximation, the problems of transversely isotropic FGM rectangular plates can be tackled similarly.

As a numerical example, let's consider the bending of an FGM plate with $a/h=6$ and $b/h=4$, subject to a sinusoidal pressure $q=Q\sin(\pi x/a)\sin(\pi y/b)$, which is applied to the lower surface $z=0$. Thus only Eq. (33) is employed. The material property varies along the thickness direction according to following rule

$$M = M_I\left(\frac{h-z}{h}\right)^{\alpha} + M_{II}\left[1-\left(\frac{h-z}{h}\right)^{\alpha}\right] \tag{35}$$

where M represents any material constant of the functionally graded material, while M_I and M_{II} are the corresponding ones for two homogeneous materials (see Table 1), and α is the material inhomogeneity parameter or gradient index, which is taken to be 20 in the numerical calculation.

Table 1. Material constants of two homogeneous transversely isotropic materials (units: c_{ij} -10^{10} N/m^2; ρ-kg/m^3)

	c_{11}	c_{12}	c_{13}	c_{33}	c_{44}	ρ
Material I	13.90	7.80	1.40	33.64	16.25	7500
Material II	20.97	12.11	10.51	21.09	4.25	5676

Table 2. Convergence study

	$wc_{44}^0/(hQ)$ Center of lower surface Center of middle surface Center of upper surface	$\sigma_x/Q, \sigma_y/Q, \sigma_z/Q$ Center of lower surface Center of middle surface Center of upper surface
$p=20$	19.8893 20.1934 19.2383	-4.7496, -5.9168, -1.0000 -0.0857, -0.0278, -0.4993 4.2566, 5.7678, 0.0000
$p=22$	19.8961 20.2008 19.2456	-4.7204, -5.8754, -1.0000 -0.0863, -0.0286, -0.4994 4.2578, 5.7694, 0.0000

Table 2 gives the results when the plate is divided into $p=20$ layers and $p=22$ layers, respectively. It can be seen that the relative error between the two approximate models is almost neglectable, indicating that the calculation converges rapidly and the proposed approach is very efficient.

Figures 3 and 4 show the variations of dimensionless normal stress $\bar{\sigma} = \sigma_z / Q$ and deflection $\bar{w} = wc_{44}^0 / (hQ)$ at the center of the middle surface of the plate with the gradient index α. A significant effect of material inhomogeneity is seen from the figures. Thus, it is possible to optimize the structure by selecting a proper distribution of material properties along the thickness direction.

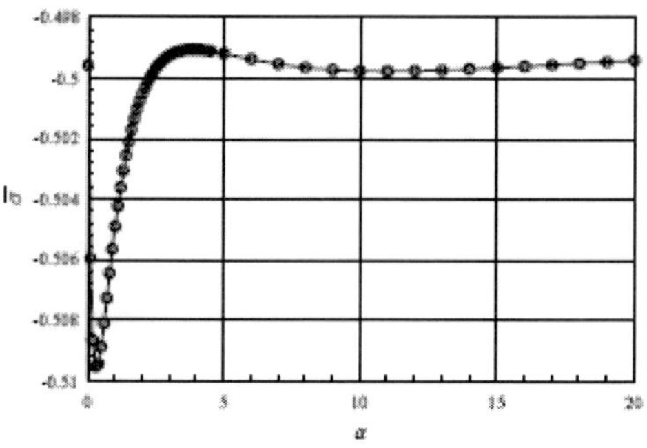

Figure 3. $\bar{\sigma}$ at the center of the plate

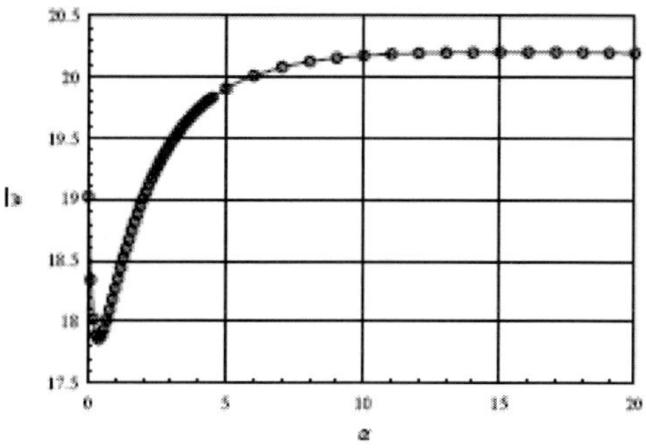

Figure 4. $\bar{w}$ at the center of the plate.

4.3. Transversely Isotropic Circular and Annular Plates

First we focus on the axisymmetric flexural deformation or vibration of an FGM circular plate of radius a, Figure 5. In cylindrical coordinates (r, θ, z), the equations of motion are

$$\begin{aligned} &\frac{\partial \sigma_r}{\partial r} + \frac{\partial \tau_{zr}}{\partial z} + \frac{\sigma_r - \sigma_\theta}{r} = \rho \frac{\partial^2 u_r}{\partial t^2} \\ &\frac{\partial \tau_{zr}}{\partial r} + \frac{\partial \sigma_z}{\partial z} + \frac{\tau_{zr}}{r} = \rho \frac{\partial^2 u_z}{\partial t^2} \end{aligned} \tag{36}$$

and the constitutive relations are

$$\begin{aligned} \sigma_r &= c_{11} \frac{\partial u_r}{\partial r} + c_{12} \frac{u_r}{r} + c_{13} \frac{\partial u_z}{\partial z} \\ \sigma_\theta &= c_{12} \frac{\partial u_r}{\partial r} + c_{11} \frac{u_r}{r} + c_{13} \frac{\partial u_z}{\partial z} \\ \sigma_z &= c_{13} \frac{\partial u_r}{\partial r} + c_{13} \frac{u_r}{r} + c_{33} \frac{\partial u_z}{\partial z} \\ \tau_{zr} &= c_{44} \left(\frac{\partial u_r}{\partial z} + \frac{\partial u_z}{\partial r} \right) \end{aligned} \tag{37}$$

From Eqs. (36) and (37), we may derive the following state equation

$$\frac{\partial}{\partial z}\begin{Bmatrix} u_r \\ \sigma_z \\ \tau_{zr} \\ u_z \end{Bmatrix} = \begin{bmatrix} 0 & 0 & \frac{1}{c_{44}} & -\frac{\partial}{\partial r} \\ 0 & 0 & -\left(\frac{\partial}{\partial r}+\frac{1}{r}\right) & \rho\frac{\partial^2}{\partial t^2} \\ \rho\frac{\partial^2}{\partial t^2}-\frac{c_{11}c_{33}-c_{13}^2}{c_{33}}\left(\frac{\partial^2}{\partial r^2}+\frac{1}{r}\frac{\partial}{\partial r}-\frac{1}{r^2}\right) & -\frac{c_{13}}{c_{33}}\frac{\partial}{\partial r} & 0 & 0 \\ -\frac{c_{13}}{c_{33}}\left(\frac{\partial}{\partial r}+\frac{1}{r}\right) & \frac{1}{c_{33}} & 0 & 0 \end{bmatrix}\begin{Bmatrix} u_r \\ \sigma_z \\ \tau_{zr} \\ u_z \end{Bmatrix} \tag{38}$$

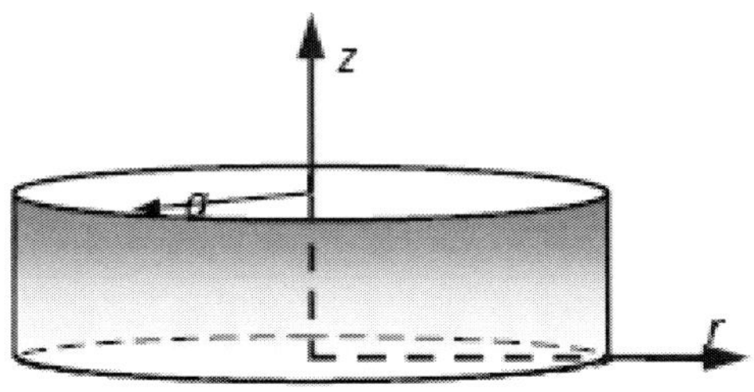

Figure 5. The geometry of a circular plate.

To eliminate the differential operators contained in the right-hand side of Eq. (38), we assume a common time factor $e^{i\omega t}$, and take the following finite Hankel transform

$$H_\mu[f(r)] = \int_0^a f(r)J_\mu(\kappa r)r\,\mathrm{d}r \tag{39}$$

where $J_\mu(x)$ is the Bessel function of the first kind of order μ, such that the following transformed variables are introduced

$$U(\kappa,z) = H_1[u_r],\ \sigma(\kappa,z) = H_0[\sigma_z],\ \tau(\kappa,z) = H_1[\tau_{zr}],\ W(\kappa,z) = H_0[u_z] \tag{40}$$

The parameter κ in Eq. (39) is to be determined from the boundary conditions at $r = a$. Equation (38) then reduces to

$$\frac{\mathrm{d}}{\mathrm{d}z}\mathbf{V}(\kappa,z) = \mathbf{A}\mathbf{V}(\kappa,z)+\mathbf{Q} \tag{41}$$

where $\mathbf{V} = [U(\kappa,z), \sigma(\kappa,z), \tau(\kappa,z), W(\kappa,z)]^{\mathrm{T}}$, and

$$\mathbf{A}=\begin{bmatrix} 0 & 0 & -\dfrac{c_{44}^{(1)}}{c_{44}} & \kappa \\ 0 & 0 & -\kappa & -\rho\omega^2 \\ -\rho\omega^2+\dfrac{c_{11}c_{33}-c_{13}^2}{c_{33}}\kappa^2 & \dfrac{c_{13}}{c_{33}}\kappa & 0 & 0 \\ -\dfrac{c_{13}}{c_{33}}\kappa & \dfrac{1}{c_{33}} & 0 & 0 \end{bmatrix}$$

$$\mathbf{Q}=\begin{Bmatrix} -u_z(a,z)J_1(\kappa a) \\ -\tau_{zr}(a,z)J_0(\kappa a) \\ -\left[(c_{11}-c_{12})u_r(a,z)+\sigma_r(a,z)\right]J_1(\kappa a)+\dfrac{c_{11}c_{33}-c_{13}^2}{c_{33}}ku_r(a,z)J_0(\kappa a) \\ -\dfrac{c_{13}}{c_{33}}u_r(a,z)J_0(\kappa a) \end{Bmatrix}$$

It is seen that the inclusion of $\mathbf{Q}$, which contains state variables, makes it very difficult to obtain the analytical solution to Eq. (41). However, for the following two special types of boundary conditions, $\mathbf{Q}$ disappears:

(*a*) Elastically supported

$$u_z=0,\ (c_{11}-c_{12})u_r+\sigma_r=0\ (r=a) \tag{42a}$$

$$J_0(\kappa a)=0 \tag{42b}$$

(*b*) Rigidly slipperily supported

$$u_r=0,\ \tau_{zr}=0\ (r=a) \tag{43a}$$

$$J_1(\kappa a)=0 \tag{43b}$$

Equations (42b) and (43b) determine the parameter κ for the two particular types of boundary conditions, respectively. Under these two types of boundary conditions, Eq. (41) becomes

$$\frac{\mathrm{d}}{\mathrm{d}z}\mathbf{V}(\kappa,z)=\mathbf{AV}(\kappa,z) \tag{44}$$

Its solution can be obtained by following the procedure for FGM rectangular plates presented in Section 4.1. Details for laminated circular plates can be found in Ref. [49], and the extension to FGM plates is very straightforward.

If an annular plate with inner radius b and outer radius a is considered, we need to use the following finite Hankel transform

$$H_\mu[f(r)] = \int_b^a f(r)[AJ_\mu(\kappa r) + BY_\mu(\kappa r)]r\,\mathrm{d}r \tag{45}$$

where A and B can be determined form the boundary conditions at $r = a$, b. Then adopting the transformed state variables similar to that in Eq. (40), we obtain a state equation with inhomogeneous terms, which disappear only under special types of boundary conditions. A detailed procedure can be found in Ding et al. [50] for laminated annular plates. With the layerwise approximation, we get easily get the analytical solution for FGM annular plates.

5. Application To FGM Cylindrical and Spherical Shells

5.1. Cylindrically Orthotropic Cylindrical Shells

Let's consider for example the free vibration of a cylindrically orthotropic hollow cylinder, Figure 6.

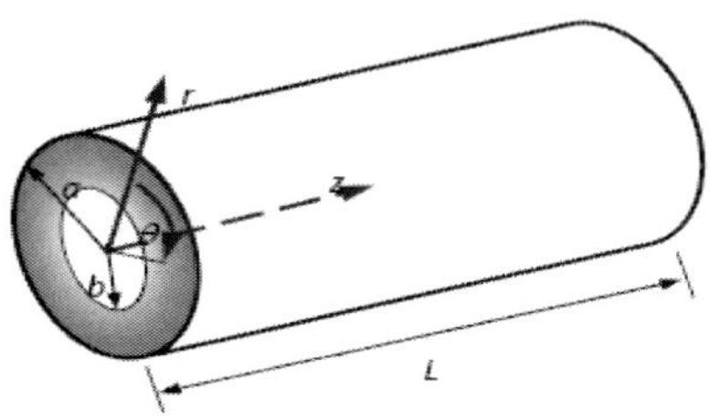

Figure 6. The geometry of a hollow cylinder.

In cylindrical coordinates (r, θ, z), the equations of motion are

$$\begin{aligned}
&\frac{\partial \sigma_r}{\partial r} + \frac{1}{r}\frac{\partial \tau_{r\theta}}{\partial \theta} + \frac{\partial \tau_{zr}}{\partial z} + \frac{\sigma_r - \sigma_\theta}{r} = \rho\frac{\partial^2 u_r}{\partial t^2} \\
&\frac{\partial \tau_{r\theta}}{\partial r} + \frac{1}{r}\frac{\partial \sigma_\theta}{\partial \theta} + \frac{\partial \tau_{\theta z}}{\partial z} + \frac{2\tau_{r\theta}}{r} = \rho\frac{\partial^2 u_\theta}{\partial t^2} \\
&\frac{\partial \tau_{zr}}{\partial r} + \frac{1}{r}\frac{\partial \tau_{\theta z}}{\partial \theta} + \frac{\partial \sigma_z}{\partial z} + \frac{\tau_{zr}}{r} = \rho\frac{\partial^2 u_z}{\partial t^2}
\end{aligned} \tag{46}$$

The constitutive relations are

$$\begin{Bmatrix} \sigma_z \\ \sigma_\theta \\ \sigma_r \\ \tau_{r\theta} \\ \tau_{zr} \\ \tau_{\theta z} \end{Bmatrix} = \begin{bmatrix} c_{11} & c_{12} & c_{13} & 0 & 0 & 0 \\ c_{12} & c_{22} & c_{23} & 0 & 0 & 0 \\ c_{13} & c_{23} & c_{33} & 0 & 0 & 0 \\ 0 & 0 & 0 & c_{44} & 0 & 0 \\ 0 & 0 & 0 & 0 & c_{55} & 0 \\ 0 & 0 & 0 & 0 & 0 & c_{66} \end{bmatrix} \begin{Bmatrix} \dfrac{\partial u_z}{\partial z} \\ \dfrac{1}{r}\dfrac{\partial u_r}{\partial \theta} + \dfrac{u_r}{r} \\ \dfrac{\partial u_r}{\partial r} \\ \dfrac{1}{r}\dfrac{\partial u_r}{\partial \theta} + \dfrac{\partial u_\theta}{\partial r} - \dfrac{u_\theta}{r} \\ \dfrac{\partial u_z}{\partial r} + \dfrac{\partial u_r}{\partial z} \\ \dfrac{\partial u_\theta}{\partial z} + \dfrac{1}{r}\dfrac{\partial u_z}{\partial \theta} \end{Bmatrix} \tag{47}$$

From Eqs. (46) and (47), we may get

$$\frac{\partial}{\partial r}\mathbf{V} = \mathbf{AV} \tag{48}$$

where $\mathbf{V} = [u_z, u_\theta, u_r, \sigma_r, \tau_{r\theta}, \tau_{zr}]^{\mathrm{T}}$, and

$$\mathbf{A} = \begin{bmatrix} 0 & 0 & -\dfrac{\partial}{\partial z} & 0 & 0 & \dfrac{1}{c_{55}} \\ 0 & \dfrac{1}{r} & -\dfrac{1}{r}\dfrac{\partial}{\partial \theta} & 0 & \dfrac{1}{c_{44}} & 0 \\ -\dfrac{c_{13}}{c_{33}}\dfrac{\partial}{\partial z} & -\dfrac{1}{r}\dfrac{c_{23}}{c_{33}}\dfrac{\partial}{\partial \theta} & -\dfrac{1}{r}\dfrac{c_{23}}{c_{33}} & \dfrac{1}{c_{33}} & 0 & 0 \\ \dfrac{c_2}{r}\dfrac{\partial}{\partial z} & \dfrac{c_3}{r^2}\dfrac{\partial}{\partial \theta} & \rho\dfrac{\partial^2}{\partial t^2} + \dfrac{c_3}{r^2} & \dfrac{1}{r}\left(\dfrac{c_{23}}{c_{33}} - 1\right) & -\dfrac{1}{r}\dfrac{\partial}{\partial \theta} & -\dfrac{\partial}{\partial z} \\ -\dfrac{(c_{66}+c_2)}{r}\dfrac{\partial^2}{\partial z \partial \theta} & \rho\dfrac{\partial^2}{\partial t^2} - c_{66}\dfrac{\partial^2}{\partial z^2} - \dfrac{c_3}{r^2}\dfrac{\partial^2}{\partial \theta^2} & -\dfrac{c_3}{r^2}\dfrac{\partial}{\partial \theta} & -\dfrac{1}{r}\dfrac{c_{23}}{c_{33}}\dfrac{\partial}{\partial \theta} & -\dfrac{2}{r} & 0 \\ \rho\dfrac{\partial^2}{\partial t^2} - c_1\dfrac{\partial^2}{\partial z^2} - \dfrac{c_{66}}{r^2}\dfrac{\partial^2}{\partial \theta^2} & -\dfrac{(c_{66}+c_2)}{r}\dfrac{\partial^2}{\partial z \partial \theta} & -\dfrac{c_2}{r}\dfrac{\partial}{\partial z} & -\dfrac{c_{13}}{c_{33}}\dfrac{\partial}{\partial z} & 0 & -\dfrac{1}{r} \end{bmatrix} \tag{49}$$

with $c_1 = c_{11} - c_{13}^2 / c_{33}$, $c_2 = c_{12} - c_{13}c_{23} / c_{33}$, and $c_3 = c_{22} - c_{23}^2 / c_{33}$. The state equation (48) was first derived by Fan and Ye [51] to study statics, dynamics and buckling of laminated cylindrical shells. It is noted here that Chen et al. [52] has developed a systematic derivation based on matrix notations to derive the state equation of arbitrarily anisotropic materials in cylindrical coordinates.

Similar to plate problems, analytical solutions to Eq. (48) can be obtained only for certain special types of boundary conditions imposed at the two ends of the hollow cylinder ($z = 0$ and $z = L$). One such type is simply supported at both ends, for which we have

$$u_\theta = u_r = \sigma_z = 0 \ (z = 0, L) \tag{50}$$

Another is rigidly slipperily supported at both ends

$$u_z = \tau_{\theta z} = \tau_{zr} = 0 \ (z = 0, L) \tag{51}$$

Analytical solution is also obtainable for any combination of Eqs. (50) and (51) at the two ends.

Consider a simply-supported hollow cylinder for example. We assume

$$\mathbf{V} = \begin{Bmatrix} b\bar{u}_z(\xi)\cos(\pi\pi\zeta)\cos(n\theta) \\ b\bar{u}_\theta(\xi)\sin(m\pi\zeta)\sin(n\theta) \\ b\bar{u}_r(\xi)\sin(m\pi\zeta)\cos(n\theta) \\ c_{44}^0\bar{\sigma}_r(\xi)\sin(m\pi\zeta)\cos(n\theta) \\ c_{44}^0\bar{\tau}_{r\theta}(\xi)\sin(m\pi\zeta)\sin(n\theta) \\ c_{44}^0\bar{\tau}_{zr}(\xi)\cos(m\pi\zeta)\cos(n\theta) \end{Bmatrix} \exp(i\omega t) \tag{52}$$

where $\xi = r/b$, $\zeta = z/L$, m and n are integers (wave numbers in axial and circumferential directions, respectively), c_{44}^0 represents the value of c_{44} at the inner surface $r = b$. It is immediately seen that the solution in Eq. (52) satisfies the simply supported conditions in Eq. (50). If the boundary conditions in Eq. (51) are considered, we can simply change $\cos(m\pi\zeta)$ with $\sin(m\pi\zeta)$ and vice versa in Eq. (52).

Substituting Eq. (52) into Eq. (48) gives

$$\frac{\mathrm{d}}{\mathrm{d}\xi}\bar{\mathbf{V}} = \bar{\mathbf{A}}\bar{\mathbf{V}} \tag{53}$$

where $\bar{\mathbf{V}} = [\bar{u}_z, \bar{u}_\theta, \bar{u}_r, \bar{\sigma}_r, \bar{\tau}_{r\theta}, \bar{\tau}_{zr}]^{\mathrm{T}}$, and the expression for $\bar{\mathbf{A}}$ can be easily obtained, see Ref. [52] for example. The state equation (53) needed to be solved analytically with the layerwise approximation. Based on such a methodology, Chen et al. [52] investigated the effect of inner fluid, which is compressible but nonviscous, on the free vibration characteristics of a fluid-filled FGM hollow cylinder.

It can be seen that when $n = 0$, $u_\theta = 0$. Thus, to study the pure torsional deformation of vibration, if it exists, we need to exchange the sinusoidal and cosine functions in Eq. (52). Also, if we assume

$$\mathbf{V} = \begin{Bmatrix} b\bar{u}_z(\xi)\cos(n\theta) \\ b\bar{u}_\theta(\xi)\sin(n\theta) \\ b\bar{u}_r(\xi)\cos(n\theta) \\ c_{44}^0\bar{\sigma}_r(\xi)\cos(n\theta) \\ c_{44}^0\bar{\tau}_{r\theta}(\xi)\sin(n\theta) \\ c_{44}^0\bar{\tau}_{zr}(\xi)\cos(n\theta) \end{Bmatrix} \exp[i(kz-\omega t)] \tag{54}$$

we may able to consider the wave propagation in an FGM hollow cylinder of infinite length, where k is the wave number in the axial direction.

The method can also be applied to cylindrical panels, Figure 7. If the panel has a central angle of α, then we should specify the boundary conditions at $\theta = 0$ and $\theta = \alpha$. Again, for two special types of boundary conditions mentioned earlier, we may obtain analytical solutions. For example, for the panel simply-supported at all edges, we can assume

$$\mathbf{V} = \begin{Bmatrix} b\bar{u}_z(\xi)\cos(\pi\pi\zeta)\sin(n\pi\theta/\alpha) \\ b\bar{u}_\theta(\xi)\sin(m\pi\zeta)\cos(n\pi\theta/\alpha) \\ b\bar{u}_r(\xi)\sin(m\pi\zeta)\sin(n\pi\theta/\alpha) \\ c_{44}^0\bar{\sigma}_r(\xi)\sin(m\pi\zeta)\sin(n\pi\theta/\alpha) \\ c_{44}^0\bar{\tau}_{r\theta}(\xi)\sin(m\pi\zeta)\cos(n\pi\theta/\alpha) \\ c_{44}^0\bar{\tau}_{zr}(\xi)\cos(m\pi\zeta)\sin(n\pi\theta/\alpha) \end{Bmatrix} \exp(i\omega t) \tag{55}$$

It may be easily verified that, when $a \to \infty$, the formulations for a cylindrical panel become those for a rectangular plate.

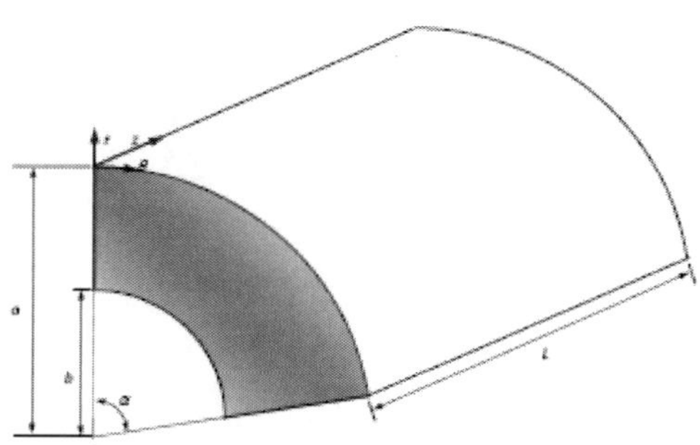

Figure 7. The geometry of a cylindrical panel.

It is also noted that, if we consider the cylindrical bending deformation (vibration) that the displacement components are independent of z, the material doesn't need to be orthotropic. It can be monoclinic, with the following constitutive relations:

$$\sigma_z = c_{12}\left(\frac{1}{r}\frac{\partial u_\theta}{\partial \theta} + \frac{u_r}{r}\right) + c_{13}\frac{\partial u_r}{\partial r} + c_{16}\frac{1}{r}\frac{\partial u_z}{\partial \theta}$$

$$\sigma_\theta = c_{22}\left(\frac{1}{r}\frac{\partial u_\theta}{\partial \theta} + \frac{u_r}{r}\right) + c_{23}\frac{\partial u_r}{\partial r} + c_{26}\frac{1}{r}\frac{\partial u_z}{\partial \theta}$$

$$\sigma_r = c_{23}\left(\frac{1}{r}\frac{\partial u_\theta}{\partial \theta} + \frac{u_r}{r}\right) + c_{33}\frac{\partial u_r}{\partial r} + c_{36}\frac{1}{r}\frac{\partial u_z}{\partial \theta}$$

$$\tau_{\theta z} = c_{26}\left(\frac{1}{r}\frac{\partial u_\theta}{\partial \theta} + \frac{u_r}{r}\right) + c_{36}\frac{\partial u_r}{\partial r} + c_{66}\frac{1}{r}\frac{\partial u_z}{\partial \theta}$$

$$\tau_{r\theta} = c_{44}\left(\frac{1}{r}\frac{\partial u_r}{\partial \theta} + \frac{\partial u_\theta}{\partial r} - \frac{u_\theta}{r}\right) + c_{45}\frac{\partial u_z}{\partial r}$$

The analysis procedure for such a problem can be referred to Chen and Lee [53], which is for a laminated cylindrical panel though. It is also true for a rectangular plate.

5.2. Spherically Isotropic Spherical Shells

The basic equations of a spherically isotropic elastic body are well described in the monograph of Lekhnitskii [54] or more recently in the book of Ding et al. [15]. For the sake of the analysis to be followed, we here follow Chen and Ding [16,17] to present these equations in a different way. Assuming the center of the spherical isotropy coincident with the origin of spherical coordinates (r, θ, ϕ), Figure 8, the linear constitutive relations can be rewritten as

$$\begin{aligned}
\Sigma_\theta &= r\sigma_\theta = c_{11}S_\theta + c_{12}S_\phi + c_{13}S_r \\
\Sigma_\phi &= r\sigma_\phi = c_{12}S_\theta + c_{11}S_\phi + c_{13}S_r \\
\Sigma_r &= r\sigma_r = c_{13}S_\theta + c_{13}S_\phi + c_{33}S_r \\
\Sigma_{r\theta} &= r\tau_{r\theta} = c_{44}S_{r\theta} \\
\Sigma_{\phi r} &= r\tau_{\phi r} = c_{44}S_{\phi r} \\
\Sigma_{\theta\phi} &= r\,\tau_{\theta\phi} = c_{66}S_{\theta\varphi}
\end{aligned} \tag{56}$$

where the relation $c_{11} = c_{12} + 2c_{66}$ holds for spherical isotropy, and

$$S_r = \nabla_2 u_r, \quad S_\theta = \frac{\partial u_\theta}{\partial \theta} + u_r, \quad S_\phi = \frac{1}{\sin\theta}\frac{\partial u_\phi}{\partial \phi} + u_r + u_\theta \cot\theta$$
$$S_{r\theta} = \frac{\partial u_r}{\partial \theta} + \nabla_2 u_\theta - u_\theta, \quad S_{\phi r} = \frac{1}{\sin\theta}\frac{\partial u_r}{\partial \phi} + \nabla_2 u_\phi - u_\phi \tag{57}$$
$$S_{\theta\phi} = \frac{1}{\sin\theta}\frac{\partial u_\theta}{\partial \phi} + \frac{\partial u_\phi}{\partial \theta} - u_\phi \cot\theta$$

where $\nabla_2 = r\partial / \partial r$. The equations of motion in terms of stresses can easily be transformed into the following form

$$\nabla_2 \Sigma_{r\theta} + \csc\theta \frac{\partial \Sigma_{\theta\phi}}{\partial \phi} + \frac{\partial \Sigma_\theta}{\partial \theta} + 2\Sigma_{r\theta} + (\Sigma_\theta - \Sigma_\phi)\cot\theta = \rho r^2 \frac{\partial^2 u_\theta}{\partial t^2}$$
$$\nabla_2 \Sigma_{\phi r} + \csc\theta \frac{\partial \Sigma_\phi}{\partial \phi} + \frac{\partial \Sigma_{\theta\phi}}{\partial \theta} + 2\Sigma_{\phi r} + 2\Sigma_{\theta\phi}\cot\theta = \rho r^2 \frac{\partial^2 u_\phi}{\partial t^2} \tag{58}$$
$$\nabla_2 \Sigma_{rr} + \csc\theta \frac{\partial \Sigma_{r\phi}}{\partial \phi} + \frac{\partial \Sigma_{r\theta}}{\partial \theta} + \Sigma_r - \Sigma_\theta - \Sigma_\phi + \Sigma_{r\theta}\cot\theta = \rho r^2 \frac{\partial^2 u_r}{\partial t^2}$$

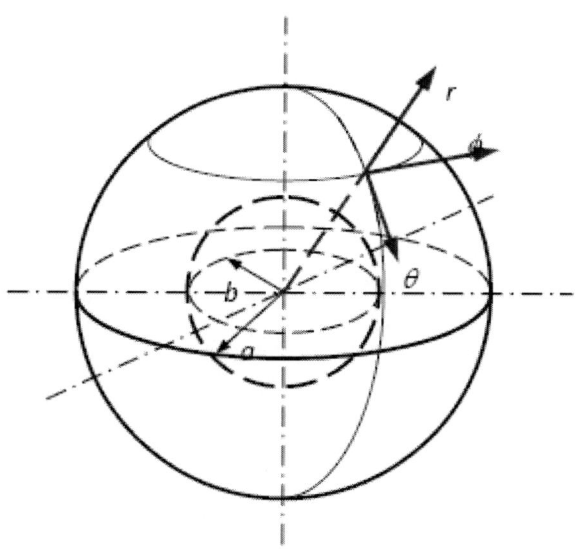

Figure 8. The geometry of a hollow sphere.

Just as for transversely isotropic rectangular plates, we shall employ the following separation formulae [55,56]

$$u_\theta = -\frac{1}{\sin\theta}\frac{\partial \Psi}{\partial \phi} - \frac{\partial G}{\partial \theta}, \quad u_\phi = \frac{\partial \Psi}{\partial \theta} - \frac{1}{\sin\theta}\frac{\partial G}{\partial \phi}, \quad u_r = W \tag{59}$$

$$\Sigma_{r\theta} = -\frac{1}{\sin\theta}\frac{\partial \Sigma_1}{\partial \phi} - \frac{\partial \Sigma_2}{\partial \theta}, \ \Sigma_{\phi r} = \frac{\partial \Sigma_1}{\partial \theta} - \frac{1}{\sin\theta}\frac{\partial \Sigma_2}{\partial \phi} \tag{60}$$

where W, G and Ψ are three displacement functions while Σ_1 and Σ_2 are two stress functions.

Utilizing Eqs. (59) and (60), through some lengthy manipulations, one can obtain the following equations from Eqs. (56)-(58)

$$\nabla_2 \begin{Bmatrix} \Sigma_1 \\ \Psi \end{Bmatrix} = \begin{bmatrix} -2 & -c_{66}(\nabla_1^2 + 2) + r^2\rho\frac{\partial^2}{\partial t^2} \\ \frac{1}{c_{44}} & 1 \end{bmatrix} \begin{Bmatrix} \Sigma_1 \\ \Psi \end{Bmatrix} \tag{61}$$

$$\nabla_2 \begin{Bmatrix} \Sigma_r \\ \Sigma_2 \\ G \\ W \end{Bmatrix} = \begin{bmatrix} 2\beta - 1 & \nabla_1^2 & k_1\nabla_1^2 & -2k_1 + r^2\rho\frac{\partial^2}{\partial t^2} \\ \beta & -2 & k_2\nabla_1^2 - 2c_{66} + r^2\rho\frac{\partial^2}{\partial t^2} & -k_1 \\ 0 & \frac{1}{c_{44}} & 1 & 1 \\ \frac{1}{c_{33}} & 0 & \beta\nabla_1^2 & -2\beta \end{bmatrix} \begin{Bmatrix} \Sigma_r \\ \Sigma_2 \\ G \\ W \end{Bmatrix} \tag{62}$$

where

$$\nabla_1^2 = \frac{\partial^2}{\partial\theta^2} + \cot\theta\frac{\partial}{\partial\theta} + \csc^2\theta\frac{\partial^2}{\partial\phi^2}$$

is the two-dimensional Laplacian in spherical coordinates and

$$\beta = c_{13}/c_{33}, k_1 = 2c_{13}\beta - (c_{11} + c_{12}), k_2 = k_1/2 - c_{66}$$

For the free vibration of an FGM hollow sphere (closed spherical shell), it is assumed that

$$\begin{Bmatrix} \Sigma_1 \\ \Psi \end{Bmatrix} = \begin{Bmatrix} \Sigma_{1n}(r) \\ \Psi_n(r) \end{Bmatrix} S_n^m(\theta,\phi)\mathrm{e}^{\mathrm{i}\omega t}$$
$$\begin{Bmatrix} \Sigma_r \\ \Sigma_2 \\ G \\ W \end{Bmatrix} = \begin{Bmatrix} \Sigma_{rn}(r) \\ \Sigma_{2n}(r) \\ G_n(r) \\ W_n(r) \end{Bmatrix} S_n^m(\theta,\phi)\mathrm{e}^{\mathrm{i}\omega t} \tag{63}$$

where $S_n^m(\theta,\phi)=P_n^m(\cos\theta)\mathrm{e}^{\mathrm{i}m\phi}$ are the spherical harmonic functions and $P_n^m(x)$ are the associated Legendre polynomials, n and m are integers, and ω is the circular frequency. It has been shown by Chen and Ding [16,17] that the integer m does not appear in the resulting ordinary differential equations about the unknown functions $\Sigma_{1n}(r)$ and $\Psi_n(r)$ etc., so that we needn't indicate it in the subscript of these functions in Eq. (63) and hereafter. Substituting Eq. (63) into Eqs. (62) and (61) gives

$$\frac{\mathrm{d}}{\mathrm{d}r}\begin{Bmatrix}\Sigma_{1n}\\ \Psi_n\end{Bmatrix}=\frac{1}{r}\begin{bmatrix}-2 & c_{66}(l-2)-r^2\rho\omega^2\\ \dfrac{1}{c_{44}} & 1\end{bmatrix}\begin{Bmatrix}\Sigma_{1n}\\ \Psi_n\end{Bmatrix} \tag{64}$$

$$\frac{\mathrm{d}}{\mathrm{d}r}\begin{Bmatrix}\Sigma_{rn}\\ \Sigma_{2n}\\ G_n\\ W_n\end{Bmatrix}=\frac{1}{r}\begin{bmatrix}2\beta-1 & -l & -k_1 l & -2k_1-r^2\rho\omega^2\\ \beta & -2 & -k_2 l-2c_{66}-r^2\rho\omega^2 & -k_1\\ 0 & \dfrac{1}{c_{44}} & 1 & 1\\ \dfrac{1}{c_{33}} & 0 & -\beta l & -2\beta\end{bmatrix}\begin{Bmatrix}\Sigma_{rn}\\ \Sigma_{2n}\\ G_n\\ W_n\end{Bmatrix} \tag{65}$$

where $l=n(n+1)$. Equations (64) and (65) are state equations with variable coefficients. Again, we can adopt the layerwise approximation to obtain the analytical solutions to these equations. For static problems of laminated hollow spheres, Chen and Ding [16] suggested an exact analysis by appropriate variable substitution, while for dynamic problems, Chen and Ding [17] put forward a method based on Taylor's expansion. These techniques can also be incorporated into the analysis of FGM hollow spheres along with the layerwise approximation.

Chen and Ding [17] showed that two independent classes of vibrations exist for laminated spherical shells. The first class, governed by Eq. (64) and for $n\geq 1$, corresponds to an equivoluminal motion of the shell without radial displacement. The second class, specified by Eq. (65) and for $n\geq 0$, possesses in general both transverse and radial components except the case of $n=0$. This is also true for a spherically isotropic FGM hollow sphere.

Consider the free vibration of a spherically isotropic FGM hollow sphere with $b/a=0.5$. The material properties follow Eq. (35), where however $z=r-b$. The material properties of the two homogeneous materials are still given in Table 1.

Figures 9 and 10 depict the lowest dimensionless natural frequencies ($\Omega=a\omega\sqrt{\rho^0/c_{44}^0}$) of the first and second classes of vibrations versus the material gradient index. It is seen that the effect of material inhomogeneity on the natural frequency is different for different mode n.

In the end of this section, we may notice that, in addition to the cylindrical and spherical shells, state-space formulations have also been developed for general doubly curved shells [57]. Adopting the layerwise approximation, it is easy to carry out the corresponding analysis of doubly curved FGM shells.

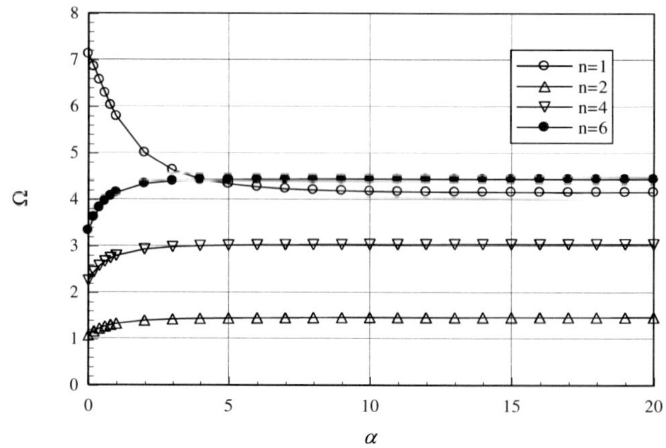

Figure 9. Variation of Ω versus α for the first class.

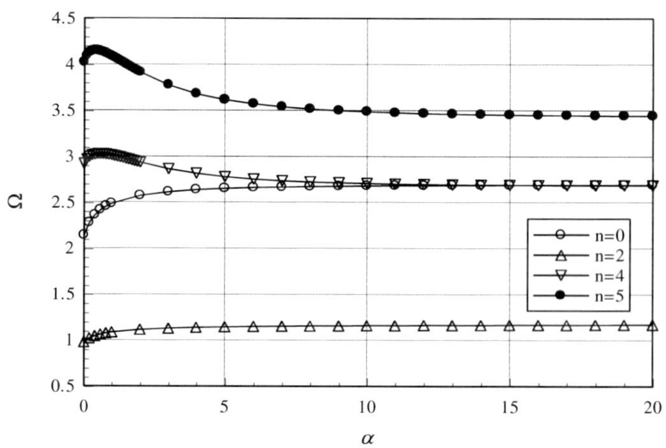

Figure 10. Variation of Ω versus α for the second class.

6. MULTI-FIELD COUPLED MEDIA

The methods described in Sections 4 and 5 for elastic FGM plates and shells can be extended to structures of advanced materials in which more than two different physical fields (elastic, electric, magnetic, temperature, etc.) are coupled together. For illustration, we consider here the bending of a multiferroic (or magneto-electro-elastic) FGM rectangular plates. The constitutive relations with thermal effect are [58]

$$\begin{aligned}
\sigma_{ij} &= c_{ijkl}\varepsilon_{ij} - e_{kij}E_k - q_{kij}H_k - \beta_{ij}T \\
D_i &= e_{ikl}\varepsilon_{kl} + \kappa_{ik}E_k + d_{ki}H_k - p_iT \\
B_i &= q_{ikl}\varepsilon_{kl} + d_{ik}E_k + \mu_{ik}H_k - m_iT
\end{aligned} \tag{66}$$

where E_i, H_i, D_i, B_i and T are the electric field, magnetic flux, electric displacement, magnetic induction and temperature difference relative to a referenced temperature at the stress free state, respectively, κ_{ik}, e_{ikl}, q_{ikl}, d_{ik}, μ_{ik}, β_{ij}, p_i and m_i are the dielectric, piezoelectric, piezomagnetic, electromagnetic, magnetic induction, thermal elastic, pyroelectric and pyromagnetic constants, respectively. There are different forms of constitutive relations, see the discussion of Soh and Liu [59].

In addition to the equations of motion in Eq. (4), the governing equations of the electric and magnetic fields are

$$D_{i,i} = f_e \tag{67}$$

$$B_{i,i} = f_m \tag{68}$$

where f_e and f_m the free charge and current density, respectively. The electric field and magnetic flux are related to the electric and magnetic potentials as follows

$$E_i = -\phi_{,i} \tag{69}$$

$$H_i = -\psi_{,i} \tag{70}$$

where ϕ and ψ are the electric and magnetic potentials, respectively.

The state equation for orthotropic multiferroic materials was derived by Wang et al. [60]. Chen et al. [22] presented separate state equations for transversely isotropic multiferroic materials by introducing appropriate displacement and stress functions in a manner similar to that described in Section 4.2. Here, we present the state equation for general anisotropic magneto-electro-elastic materials with thermal effect; the basic equations have been summarized above. As for elastic materials [11,15] and piezoelectric materials [61], the following formulations are readily derived for magneto-electro- elastic materials [62]

$$\frac{\partial}{\partial z}\begin{Bmatrix}\mathbf{U}\\ \boldsymbol{\sigma}_1\end{Bmatrix} = \begin{bmatrix}\mathbf{D}_{11} & \mathbf{C}_{11}^{-1}\\ \mathbf{D}_{21} & \mathbf{D}_{11}^{\mathrm{T}}\end{bmatrix}\begin{Bmatrix}\mathbf{U}\\ \boldsymbol{\sigma}_1\end{Bmatrix} + \begin{Bmatrix}\mathbf{C}_{11}^{-1}\mathbf{B}_1\\ \mathbf{L}_2^{\mathrm{T}}\tilde{\mathbf{B}}_1\end{Bmatrix}T - \begin{Bmatrix}\mathbf{0}\\ \mathbf{G}\end{Bmatrix} + \rho\frac{\partial^2}{\partial t^2}\begin{Bmatrix}\mathbf{0}\\ \mathbf{KU}\end{Bmatrix} \tag{71}$$

$$\boldsymbol{\sigma}_2 = [\tilde{\mathbf{C}}_{11}\mathbf{L}_2 \quad \mathbf{C}_{12}^{\mathrm{T}}\mathbf{C}_{11}^{-1}]\begin{Bmatrix}\mathbf{U}\\ \boldsymbol{\sigma}_1\end{Bmatrix} - \tilde{\mathbf{B}}_1 T \tag{72}$$

where

$$\mathbf{D}_{11} = -\mathbf{L}_1 - \mathbf{C}_{11}^{-1}\mathbf{C}_{12}\mathbf{L}_2, \quad \mathbf{D}_{21} = -\mathbf{L}_2^{\mathrm{T}}\tilde{\mathbf{C}}_{11}\mathbf{L}_2$$

$$\tilde{\mathbf{C}}_{11} = \mathbf{C}_{22} - \mathbf{C}_{12}^{\mathrm{T}}\mathbf{C}_{11}^{-1}\mathbf{C}_{12}, \quad \tilde{\mathbf{B}}_1 = \mathbf{B}_2 - \mathbf{C}_{12}^{\mathrm{T}}\mathbf{C}_{11}^{-1}\mathbf{B}_1$$

$$\mathbf{C}_{11} = \begin{bmatrix} c_{33} & c_{34} & c_{35} & e_{33} & q_{33} \\ & c_{44} & c_{45} & e_{34} & q_{34} \\ & & c_{55} & e_{35} & q_{35} \\ & & & -\kappa_{33} & -d_{33} \\ \text{sym.} & & & & -\mu_{33} \end{bmatrix}, \quad \mathbf{B}_1 = \begin{Bmatrix} \beta_3 \\ \beta_4 \\ \beta_5 \\ -p_3 \\ -m_3 \end{Bmatrix}$$

$$\mathbf{C}_{12} = \begin{bmatrix} c_{13} & c_{23} & c_{36} & e_{13} & e_{23} & q_{13} & q_{23} \\ c_{14} & c_{24} & c_{46} & e_{14} & e_{24} & q_{14} & q_{24} \\ c_{15} & c_{25} & c_{56} & e_{15} & e_{25} & q_{15} & q_{25} \\ e_{31} & e_{32} & e_{36} & -\kappa_{13} & -\kappa_{23} & -d_{13} & -d_{23} \\ q_{31} & q_{32} & q_{36} & -d_{13} & -d_{23} & -\mu_{13} & -\mu_{23} \end{bmatrix}$$

$$\mathbf{C}_{22} = \begin{bmatrix} c_{11} & c_{12} & c_{16} & e_{11} & e_{21} & q_{11} & q_{21} \\ & c_{22} & c_{26} & e_{12} & e_{22} & q_{12} & q_{22} \\ & & c_{66} & e_{16} & e_{26} & q_{16} & q_{26} \\ & & & -\kappa_{11} & -\kappa_{12} & -d_{11} & -d_{12} \\ & & & & -\kappa_{22} & -d_{12} & -d_{22} \\ & & & & & -\mu_{11} & -\mu_{12} \\ \text{sym.} & & & & & & -\mu_{22} \end{bmatrix}, \quad \mathbf{B}_2 = \begin{Bmatrix} \beta_1 \\ \beta_2 \\ \beta_6 \\ -p_1 \\ -p_2 \\ -m_1 \\ -m_2 \end{Bmatrix}$$

$$\mathbf{L}_1 = \begin{bmatrix} 0 & 0 & 0 & 0 & 0 \\ \partial/\partial y & 0 & 0 & 0 & 0 \\ \partial/\partial x & 0 & 0 & 0 & 0 \\ 0 & 0 & 0 & 0 & 0 \\ 0 & 0 & 0 & 0 & 0 \end{bmatrix}, \quad \mathbf{L}_2 = \begin{bmatrix} 0 & \partial/\partial x & 0 & 0 & 0 \\ 0 & 0 & \partial/\partial y & 0 & 0 \\ 0 & 0 & \partial/\partial x & 0 & 0 \\ 0 & \partial/\partial y & 0 & \partial/\partial x & 0 \\ 0 & 0 & 0 & \partial/\partial y & 0 \\ 0 & 0 & 0 & 0 & \partial/\partial x \\ 0 & 0 & 0 & 0 & \partial/\partial y \end{bmatrix}$$

$$\mathbf{K} = \begin{bmatrix} 1 & 0 & 0 & 0 & 0 \\ 0 & 1 & 0 & 0 & 0 \\ 0 & 0 & 1 & 0 & 0 \\ 0 & 0 & 0 & 0 & 0 \\ 0 & 0 & 0 & 0 & 0 \end{bmatrix}, \quad \mathbf{U} = \begin{Bmatrix} u_z \\ u_x \\ u_y \\ \phi \\ \psi \end{Bmatrix}, \quad \boldsymbol{\sigma}_1 = \begin{Bmatrix} \sigma_z \\ \tau_{yz} \\ \tau_{zx} \\ D_z \\ B_z \end{Bmatrix}$$

$$\boldsymbol{\sigma}_2 = \begin{Bmatrix} \sigma_x \\ \sigma_y \\ \tau_{xy} \\ D_x \\ D_y \\ B_x \\ B_y \end{Bmatrix}, \quad \mathbf{G} = \begin{Bmatrix} f_z \\ f_x \\ f_y \\ -f_e \\ -f_m \end{Bmatrix}$$

Equation (71) is the state equation, while Eq. (72) is the output equation.

For the static problem of orthotropic materials without body forces, free charge, current density and temperature effect, Eq. (71) reduces to

$$\frac{\partial}{\partial z}\mathbf{V} = \mathbf{A}\mathbf{V} = \begin{bmatrix} \mathbf{0} & \mathbf{A}_1 \\ \mathbf{A}_2 & \mathbf{0} \end{bmatrix}\mathbf{V} \tag{73}$$

where $\mathbf{V} = [u_x, u_y, \sigma_z, D_z, B_z, \tau_{zx}, \tau_{yz}, u_z, \phi, \psi]^{\mathrm{T}}$ is the state vector, and $\mathbf{A}_1$ and $\mathbf{A}_2$ are expressed as follows

$$\mathbf{A}_1 = \begin{bmatrix} \frac{1}{c_{55}} & 0 & -\frac{\partial}{\partial x} & -\frac{e_{15}}{c_{55}}\frac{\partial}{\partial x} & -\frac{q_{15}}{c_{55}}\frac{\partial}{\partial x} \\ & \frac{1}{c_{44}} & -\frac{\partial}{\partial y} & -\frac{e_{24}}{c_{44}}\frac{\partial}{\partial y} & -\frac{q_{24}}{c_{44}}\frac{\partial}{\partial y} \\ & & 0 & 0 & 0 \\ & & & k_1\frac{\partial^2}{\partial x^2} + k_3\frac{\partial^2}{\partial y^2} & k_2\frac{\partial^2}{\partial x^2} + k_4\frac{\partial^2}{\partial y^2} \\ \text{sym.} & & & & k_5\frac{\partial^2}{\partial x^2} + k_6\frac{\partial^2}{\partial y^2} \end{bmatrix}$$

$$\mathbf{A}_2 = \begin{bmatrix} -\gamma_1\frac{\partial^2}{\partial x^2} - c_{66}\frac{\partial^2}{\partial y^2} & -(\gamma_2 + c_{66})\frac{\partial^2}{\partial x \partial y} & -g_{11}\frac{\partial}{\partial x} & -g_{12}\frac{\partial}{\partial x} & -g_{13}\frac{\partial}{\partial x} \\ & -c_{66}\frac{\partial^2}{\partial x^2} - \gamma_3\frac{\partial^2}{\partial y^2} & -g_{21}\frac{\partial}{\partial y} & -g_{22}\frac{\partial}{\partial y} & -g_{23}\frac{\partial}{\partial y} \\ & & \frac{\alpha_{11}}{\alpha} & \frac{\alpha_{12}}{\alpha} & \frac{\alpha_{13}}{\alpha} \\ & & & \frac{\alpha_{22}}{\alpha} & \frac{\alpha_{23}}{\alpha} \\ \text{sym.} & & & & \frac{\alpha_{33}}{\alpha} \end{bmatrix} \tag{74}$$

The induced variables are obtained as

$$\begin{Bmatrix} \sigma_x \\ \sigma_y \\ \tau_{xy} \\ D_x \\ D_y \\ B_x \\ B_y \end{Bmatrix} = \begin{bmatrix} \gamma_1 \frac{\partial}{\partial x} & \gamma_2 \frac{\partial}{\partial y} & g_{11} & g_{12} & g_{13} & 0 & 0 & 0 & 0 & 0 \\ \gamma_2 \frac{\partial}{\partial x} & \gamma_3 \frac{\partial}{\partial y} & g_{21} & g_{22} & g_{23} & 0 & 0 & 0 & 0 & 0 \\ c_{66} \frac{\partial}{\partial y} & c_{66} \frac{\partial}{\partial x} & 0 & 0 & 0 & 0 & 0 & 0 & 0 & 0 \\ 0 & 0 & 0 & 0 & 0 & \frac{e_{15}}{c_{55}} & 0 & 0 & -k_1 \frac{\partial}{\partial x} & -k_2 \frac{\partial}{\partial x} \\ 0 & 0 & 0 & 0 & 0 & \frac{e_{24}}{c_{44}} & 0 & 0 & -k_3 \frac{\partial}{\partial y} & -k_4 \frac{\partial}{\partial y} \\ 0 & 0 & 0 & 0 & 0 & \frac{q_{15}}{c_{55}} & 0 & 0 & -k_2 \frac{\partial}{\partial x} & -k_5 \frac{\partial}{\partial x} \\ 0 & 0 & 0 & 0 & 0 & \frac{q_{24}}{c_{44}} & 0 & 0 & -k_4 \frac{\partial}{\partial y} & -k_6 \frac{\partial}{\partial y} \end{bmatrix} \mathbf{V} \tag{75}$$

In Eqs. (74) and (75), various coefficients are defined by

$$k_1 = \kappa_{11} + e_{15}^2 / c_{55}, \quad k_2 = d_{11} + e_{15} q_{15} / c_{55}, \quad k_3 = \kappa_{22} + e_{24}^2 / c_{44}$$

$$k_4 = d_{22} + e_{24} q_{24} / c_{44}, \quad k_5 = \mu_{11} + q_{15}^2 / c_{55}, \quad k_6 = \mu_{22} + q_{24}^2 / c_{44}$$

$$g_{1i} = (\alpha_{1i} c_{13} + \alpha_{2i} e_{31} + \alpha_{3i} q_{31}) / \alpha, \quad g_{2i} = (\alpha_{1i} c_{23} + \alpha_{2i} e_{32} + \alpha_{3i} q_{32}) / \alpha$$

$$\gamma_1 = c_{11} - c_{13} g_{11} - e_{31} g_{12} - q_{31} g_{13}, \quad \gamma_2 = c_{12} - c_{13} g_{21} - e_{31} g_{22} - q_{31} g_{23}$$

$$\gamma_3 = c_{22} - c_{23} g_{21} - e_{32} g_{22} - q_{32} g_{23}, \quad \alpha = \begin{vmatrix} c_{33} & e_{33} & q_{33} \\ e_{33} & -\kappa_{33} & -d_{33} \\ q_{33} & -d_{33} & -\mu_{33} \end{vmatrix}$$

where α_{ij} are the algebraic cofactors of α. The above equations are essentially the same as those in Wang et al. [60], but arranged in a more concise way, with the symmetry of matrices $\mathbf{A}_1$ and $\mathbf{A}_2$ clearly shown in Eq. (74).

Chen et al. [62] presented an exact solution of the bending of a multiferroic laminated rectangular plate featuring with interfacial bonding imperfections. The boundary conditions are assumed to be simply-supported in a generalized sense that

$$x = 0, a: u_z = \sigma_x = u_y = \phi = \psi = 0 \tag{76a}$$

$$y = 0, b:\ u_z = \sigma_y = u_x = \phi = \psi = 0 \tag{76b}$$

The analysis is also applicable to the generalized rigidly slipperily supported conditions

$$x = 0, a:\ \tau_{zx} = \tau_{xy} = u_x = D_x = B_x = 0 \tag{77a}$$

$$y = 0, b:\ \tau_{yz} = \tau_{xy} = u_y = D_y = B_y = 0 \tag{77b}$$

or for a combination of boundary conditions in Eq. (76) and (77).

It is obvious that, by the layerwise approximation, the analysis in Chen et al. [62] can be readily extended to multiferroic FGM rectangular plates.

7. A State-Space-based Semi-Analytical Method

It has been seen that, even with the layerwise approximation, analytical solutions of beams, plates and shells can be obtained only for some special types of boundary conditions. The frequently encountered boundary conditions such as free and clamped ones can not be tackled within the classical state-space framework. Fan and Ding [63] has suggested a particular methodology, which makes use of special functions and their Fourier expansions, to derive analytical solutions based on the state equation. However, this method is very complex, and the convergence of solution is relatively slow.

On the other hand, Chen et al. [64-67] suggested an alternative semi-analytical approach by combining the state-space formulations with the efficient differential quadrature method (DQM). The state-space formulations are still established with the transfer direction along the thickness, but the DQM, instead of the Fourier series expansion, is used to discretize the in-plane domain so as to eliminate the partial differential operators in the operator matrix **A** in Eq. (14) for example. Since DQM is a very efficient and accurate numerical method [68], the proposed semi-analytical is also very efficient and accurate, especially for laminated structures.

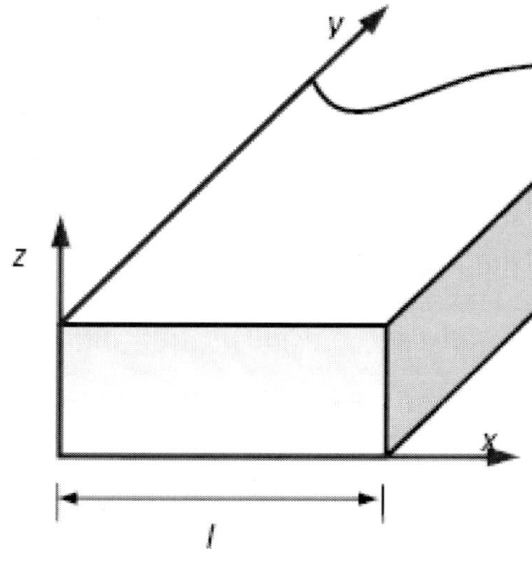

Figure 7. A plate in cylindrical bending.

As an example, let's consider an orthotropic FGM plate under the assumption of cylindrical bending, Figure 7. In this case, we have only two non-zero displacements u_x and u_z, in the x- and z- directions, respectively, which are independent of the coordinate y. The constitutive relations for a cross-ply laminate for cylindrical bending are

$$\begin{aligned}\sigma_x &= c_{11}\frac{\partial u_x}{\partial x}+c_{13}\frac{\partial u_z}{\partial z}, \quad \sigma_y = c_{12}\frac{\partial u_x}{\partial x}+c_{23}\frac{\partial u_z}{\partial z} \\ \sigma_z &= c_{13}\frac{\partial u_x}{\partial x}+c_{33}\frac{\partial u_z}{\partial z}, \quad \tau_{zx} = c_{55}\left(\frac{\partial u_x}{\partial z}+\frac{\partial u_z}{\partial x}\right)\end{aligned} \tag{78}$$

and $\tau_{xy} = \tau_{yz} = 0$. The equations of motion are

$$\frac{\partial \sigma_x}{\partial x}+\frac{\partial \tau_{zx}}{\partial z} = \rho\frac{\partial^2 u_x}{\partial t^2}, \quad \frac{\partial \tau_{zx}}{\partial x}+\frac{\partial \sigma_z}{\partial z} = \rho\frac{\partial^2 u_z}{\partial t^2} \tag{79}$$

The following state equation can be established from Eqs. (78) and (79)

$$\frac{\partial}{\partial z}\begin{Bmatrix}\sigma_z \\ u_x \\ u_z \\ \tau_{zx}\end{Bmatrix} = \begin{bmatrix} 0 & 0 & \rho\frac{\partial^2}{\partial t^2} & -\frac{\partial}{\partial x} \\ 0 & 0 & -\frac{\partial}{\partial x} & \frac{1}{c_{55}} \\ \frac{1}{c_{33}} & -\frac{c_{13}}{c_{33}}\frac{\partial}{\partial x} & 0 & 0 \\ -\frac{c_{13}}{c_{33}}\frac{\partial}{\partial x} & \rho\frac{\partial^2}{\partial t^2}+\alpha\frac{\partial^2}{\partial x^2} & 0 & 0 \end{bmatrix}\begin{Bmatrix}\sigma_z \\ u_x \\ u_z \\ \tau_{zx}\end{Bmatrix} \tag{80}$$

where $\alpha = -c_{11}+c_{13}^2/c_{33}$. The two induced variables can be determined as

$$\sigma_x = \frac{c_{13}}{c_{33}}\sigma_z - \alpha\frac{\partial u_x}{\partial x}, \quad \sigma_y = \frac{c_{23}}{c_{33}}\sigma_z + \left(c_{12}-\frac{c_{23}c_{13}}{c_{33}}\right)\frac{\partial u_x}{\partial x} \tag{81}$$

In the regime of three-dimensional elasticity, the typical boundary conditions can be expressed as

Simply supported (S): $\sigma_x = u_z = 0$ (82)

Clamped (C): $u_x = u_z = 0$ (83)

Free (F): $\sigma_x = \tau_{zx} = 0$ (84)

at $x = 0$ or l.

An analytical solution for simply supported boundary conditions can be obtained by assuming

$$\begin{Bmatrix} \sigma_z \\ u_x \\ u_z \\ \tau_{zx} \end{Bmatrix} = \begin{Bmatrix} -c_{55}^0 \bar{\sigma}_z(\zeta)\sin(n\pi\xi) \\ h\bar{u}_x(\zeta)\cos(n\pi\xi) \\ h\bar{u}_z(\zeta)\sin(n\pi\xi) \\ c_{55}^0 \bar{\tau}_{xz}(\zeta)\cos(n\pi\xi) \end{Bmatrix} \exp(\mathrm{i}\omega t) \tag{85}$$

where ω is the circular frequency, $\zeta = z/h$ and $\xi = x/l$ are the dimensionless coordinates, n is an integer (half-wave number along the x-direction), and c_{55}^0 represents the elastic constant at $z = 0$. The succeeding analysis is almost the same as that described in Section 4.1 for rectangular plates, and is omitted here for brevity.

For a plate with other boundary conditions, say clamped or free boundary conditions, it is generally difficult to obtain an exact solution to Eq. (80). Hence, we will use the DQM to eliminate the partial derivatives with respect to x therein. In the DQM, the n-th order partial derivative of any continuous function $F(x)$ at a given point x_i is approximated as a linear sum of weighted function values at all of the discrete points in the domain of x:

$$\left.\frac{\partial^n F(x)}{\partial x^n}\right|_{x=x_i} = \sum_{j=1}^{M} g_{ij}^{(n)} F(x_j), \; (n=1, 2, \cdots, M-1; \; i=1, 2, \cdots, M) \tag{86}$$

where M is the number of discrete points, $g_{ij}^{(n)}$ are weight coefficients depending on x_i $(i=1, 2, \cdots, M)$ only, and the corresponding expressions are [69]

$$g_{ir}^{(1)} = \frac{\prod_{j=1, j\neq i}^{M} (x_i - x_j)}{(x_i - x_r) \prod_{j=1, j\neq r}^{M} (x_r - x_j)} \tag{87a}$$

$$g_{ir}^{(n)} = n\left[g_{ii}^{(n-1)} g_{ir}^{(1)} - \frac{g_{ir}^{(n-1)}}{x_i - x_r} \right], \; (n=2, 3, \cdots, M) \tag{87b}$$

where $i, r = 1, 2, \cdots, M$, but $i \neq r$, $g_{ii}^{(n)}$ are determined as

$$g_{ii}^{(n)} = -\sum_{r=1, r\neq i}^{M} g_{ir}^{(n)}, \; (i=1, 2, \cdots, M; \; n=1, 2, \cdots, M-1) \tag{87c}$$

By virtue of the rule in Eq. (86), we obtain the following discrete forms of Eqs. (80) and (81), with respect to the variable ξ,

$$\frac{\mathrm{d}Z_i}{\mathrm{d}\zeta} = -\frac{\rho}{\rho^0}\Omega^2 W_i - s\sum_{j=1}^{M} g_{ij}^{(1)} T_j$$
$$\frac{\mathrm{d}U_i}{\mathrm{d}\zeta} = -s\sum_{j=1}^{M} g_{ij}^{(1)} W_j + \frac{c_{55}^0}{c_{55}} T_i$$
$$\frac{\mathrm{d}W_i}{\mathrm{d}\zeta} = \frac{c_{55}^0}{c_{33}} Z_i - \frac{c_{13}}{c_{33}} s\sum_{j=1}^{M} g_{ij}^{(1)} U_j \tag{88}$$
$$\frac{\mathrm{d}T_i}{\mathrm{d}\zeta} = \frac{\alpha}{c_{55}^0} s^2 \sum_{j=1}^{M} g_{ij}^{(2)} U_j - \frac{\rho}{\rho^0}\Omega^2 U_i - \frac{c_{13}}{c_{33}} s\sum_{j=1}^{M} g_{ij}^{(1)} Z_j$$

and

$$X_i = \frac{c_{13}}{c_{33}} Z_i - \frac{\alpha}{c_{55}^0} s\sum_{j=1}^{M} g_{ij}^{(1)} U_j, \quad Y_i = \frac{c_{23}}{c_{33}} Z_i + \frac{c_{12}c_{33} - c_{13}c_{23}}{c_{33}c_{55}^0} s\sum_{j=1}^{M} g_{ij}^{(1)} U_j \tag{89}$$

where $i = 1, 2, \cdots, M$, $\Omega = h\omega\sqrt{\rho^0 / c_{55}^0}$, ρ^0 is the mass density at $z = 0$, $s = h/l$ is the thickness-to-span ratio, and

$$Z_i = \sigma_z(\zeta, \xi_i)/c_{55}^0,\ U_i = u_x(\zeta, \xi_i)/h,\ W_i = u_z(\zeta, \xi_i)/h$$
$$T_i = \tau_{zx}(\zeta, \xi_i)/c_{55}^0,\ X_i = \sigma_x(\zeta, \xi_i)/c_{55}^0,\ Y_i = \sigma_y(\zeta, \xi_i)/c_{55}^0 \tag{90}$$

in which ξ_i are the sampling points coordinates determined by

$$\xi_i = \frac{1 - \cos[(i-1)\pi/(M-1)]}{2} \quad (i = 1, 2, \cdots, M) \tag{91}$$

These points are known as the Chebyshev-Gauss-Lobatto points.

At this stage, we have derived a discrete form of the state equation, as shown in Eq. (88). However, for a practical problem, one should take account of the edge boundary conditions before solving it. To illustrate the idea, we consider, for instance, a cantilever laminate with clamped at $\xi = 0$ and free at $\xi = 1$:

$$U_1 = W_1 = X_M = T_M = 0 \tag{92}$$

From the third condition in Eq. (92) and the first equation in Eq. (89), we have

$$Z_M = \frac{c_{33}\alpha}{c_{13}c_{55}^0} s\sum_{j=1}^{M} g_{Mj}^{(1)} U_j \tag{93}$$

Then we can derive from Eq. (88) the final form of discrete state equation

$$\frac{\mathrm{d}Z_i}{\mathrm{d}\zeta} = -\frac{\rho}{\rho^0}\Omega^2 W_i - s\sum_{j=1}^{M-1} g_{ij}^{(1)} T_j,\ (i = 1, 2, \cdots, M-1)$$

$$\frac{\mathrm{d}U_i}{\mathrm{d}\zeta} = -s\sum_{j=2}^{M} g_{ij}^{(1)} W_j + \frac{c_{55}^0}{c_{55}} T_i,\ (i = 2, 3, \cdots, M)$$

$$\frac{\mathrm{d}W_i}{\mathrm{d}\zeta} = \frac{c_{55}^0}{c_{33}} Z_i - \frac{c_{13}}{c_{33}} s\sum_{j=2}^{M} g_{ij}^{(1)} U_j,\ (i = 2, 3, \cdots, M-1) \tag{94}$$

$$\frac{\mathrm{d}W_M}{\mathrm{d}\zeta} = \left(\frac{\alpha}{c_{13}} - \frac{c_{13}}{c_{33}}\right) s\sum_{j=2}^{M} g_{Mj}^{(1)} U_j,$$

$$\frac{\mathrm{d}T_i}{\mathrm{d}\zeta} = \frac{\alpha}{c_{55}^0} s^2 \sum_{j=2}^{M}\left[g_{ij}^{(2)} - g_{iM}^{(1)} g_{Mj}^{(1)}\right] U_j - \frac{\rho}{\rho^0}\Omega^2 U_i - \frac{c_{13}}{c_{33}} s\sum_{j=1}^{M-1} g_{ij}^{(1)} Z_j,$$
$$(i = 1, 2, \cdots, M-1)$$

which can be rewritten in the following matrix form

$$\frac{\mathrm{d}\bar{\mathbf{V}}_d}{\mathrm{d}\zeta} = \bar{\mathbf{A}}_d \bar{\mathbf{V}}_d \tag{95}$$

where $\bar{\mathbf{V}}_d = [Z_1, Z_2, \cdots, Z_{M-1}, U_2, U_3, \cdots, U_M, W_2, W_3, \cdots, W_M, T_1, T_2, \cdots, T_{M-1}]^{\mathrm{T}}$ is the discrete dimensionless state vector, $\bar{\mathbf{A}}_d$ is the coefficient matrix whose elements can be obtained from Eq. (94) easily. Similar to the analysis in Section 4.1, we can arrive at

$$\mathbf{V}_d(1) = \mathbf{T}_d \mathbf{V}_d(0) \tag{96}$$

where $\mathbf{T}_d$ is the global transfer matrix.

For free vibration problem, both the upper and lower surfaces are tractions free, thus the frequency equation can be obtained from Eq. (96) as follows

$$\begin{vmatrix} S_{1,M} & \cdots & \cdots & S_{1,3M-3} \\ \vdots & & & \vdots \\ S_{M-1,M} & \cdots & \cdots & S_{M-1,3M-3} \\ S_{3M-2,M} & \cdots & \cdots & S_{3M-2,3M-3} \\ \vdots & & & \vdots \\ S_{4M-4,M} & \cdots & \cdots & S_{4M-4,3M-3} \end{vmatrix} = 0 \tag{97}$$

where $S_{k,l}$ is the element of $\mathbf{T}_d$. It is seen that the scale of the secular equation is much larger than that of the analytical solution, which actually increases with the number of discrete points in the DQM. However, it still does not change with the number of layers that should be artificially divided using the layerwise approximation. Thus, the state-space-based semi-analytical method is very efficient for FGM structures [70,71].

The method has been extended to multi-directional functionally graded beams and plates by Lü et al. [72-74], to composite beam with interfacial slip by Xu and Wu [75], to cylindrical shells by Alibeigloo and Shakeri [76,77], to laminated piezoelectric plates in cylindrical bending by Zhou et al. [78], to FGM piezoelectric beams by Li and Shi [79], and to hybrid smart structures by Alibeigloo and Madoliat [80].

Conclusion

As shown in this chapter, efficient analytical solutions can be obtained for FGM plates and structures based on state-space formulations along with the layerwise approximation. Relevant formulas for anisotropic materials are presented. The extension to multi-field coupled media is illustrated for a multiferroic plate. A semi-analytical solution, which can deal with arbitrary boundary conditions, is also put forward. The methodology presented in the chapter should be very useful in analyzing statics, dynamics and buckling of structures made of various advanced materials.

Acknowledgments

Our work on functionally graded structures has been supported by the National Natural Science Foundation of China (Nos. 10432030, 10725210 and 10832009), and the National Basic Research Program of China (No. 2009CB623200).

References

Thomson, W. T. Matrix solution for the vibration of non-uniform beams. *J. Appl. Mech.* 1950, *Vol* 17, 337-339.

Thomson, W. T. Transmission of elastic waves through a stratified solid medium. *J. Appl. Phys.* 1950, *Vol* 21, 89-93.

Pestel, E. C., Leckie, F. A. *Matrix Methods in Elasto Mechanics*, McGraw-Hill: NY, 1963.

Kennett, B. L. N. *Seismic Wave Propagation in Stratified Media*, Cambridge University Press: Cambridge, 1983.

Das, Y. C., Setlur, A. V. Method of initial functions in two-dimensional elastodynamic problems. *J. Appl. Mech.* 1970, *Vol* 37, 137-140.

Bufler, H. Theory of elasticity of a multilayered medium. *J. Elast.* 1971, *Vol* 1, 125-143.

Bahar, L. Y. Transfer matrix approach to layered systems. *J. Engng. Mech.* 1972, *Vol* 98, 1159-1172.

Kameswara Rao, N. S. V., Das, Y. C. A mixed method in elasticity. *J. Appl. Mech.* 1977, *Vol* 44, 51-56.

Fan, J. R., Ye, J. Q. A series solution of the exact equation for thick orthotropic plates. *Int. J. Solids Struct.* 1990, *Vol* 26, 773-778.

Steele, C. R., Kim, Y. Y. Modified mixed variational principle and the state-vector equation for elastic bodies and shells of revolution. *J. Appl. Mech.* 1992, *Vol* 59, 587-595.

Tarn, J. Q. A state space formalism for anisotropic elasticity. Part I: Rectilinear anisotropy. *Int. J. Solids Struct.* 2002, *Vol* 39, 5143-5155.

Ye, J. Q. Laminated Composite Plates and Shells: 3D Modelling, Springer: London, 2003.

Ezzat, M. A. State space approach to solids and fluids. *Can. J. Phys.* 2008, *Vol* 86, 1241-1250.

Ding, H. J., Chen, W. Q., Xu, R. Q. On the bending, vibration and stability of laminated rectangular plates with transversely isotropic layers. *Appl. Math. Mech.* 2001, *Vol* 22, 17-24.

Ding, H. J., Chen, W. Q., Zhang, L. C. *Elasticity of Transversely Isotropic Materials*, Springer: Dordrecht, 2006.

Chen, W. Q., Ding, H. J. A state-space based stress analysis of multilayered spherical shell with spherical isotropy. *J. Appl. Mech.* 2001, *Vol* 68, 109-114.

Chen, W. Q., Ding, H. J. Free vibration of multi-layered spherically isotropic hollow spheres. *Int. J. Mech. Sci.* 2001, *Vol* 43, 667-680.

Ding, H. J., Chen W. Q., Xu, R. Q. New state space formulations for transversely isotropic piezoelasticity with application. *Mech. Res. Commun.* 2000, *Vol* 27, 319-326.

Chen, W. Q. Free vibration analysis of laminated piezoceramic hollow spheres. *J. Acoust. Soc. Am.* 2001, *Vol* 109, 41-50.

Chen, W. Q., Ding, H. J., Xu, R. Q. Three-dimensional static analysis of multi-layered piezoelectric hollow spheres via the state space method. *Int. J. Solids Struct.*, 2001, *Vol* 38, 4921-4936.

Ding, H. J., Chen, W. Q. *Three Dimensional Problems of Piezoelasticity*, Nova Science Publishers: NY, 2001.

Chen, W. Q., Lee, K. Y. Alternative state space formulations for magnetoelectric thermoelasticity with transverse isotropy and the application to bending analysis of nonhomogeneous plates. *Int. J. Solids Struct.* 2003, *Vol* 40, 5689-5705.

Zhong, W. X. *A New Systematic Methodology for Theory of Elasticity*, Dalian University of Technology Press: Dalian, China, 1995. (in Chinese)

Yao, W. A., Zhong, W. X., Lim, C. W. *Symplectic Elasticity*, World Scientific: Hackensack, NJ, 2009.

Tanigawa, Y. Some basic thermoelastic problems for nonhomogeneous structural materials. *Appl. Mech. Rev.* 1995, *Vol* 48, 287-300.

Chen, W. Q., Wang, X., Ding, H. J. Free vibration of a fluid-filled hollow sphere of a functionally graded material with spherical isotropy. *J. Acoust. Soc. Am.* 1999, *Vol* 106, 2588-2594.

Ye, G. R., Chen, W. Q., Cai, J. B. A uniformly heated functionally graded cylindrical shell with transverse isotropy. *Mech. Res. Commun.* 2001, *Vol* 28, 535-542.

Chen, W. Q., Ding, H. J., Liang, J. The exact elasto-electric field of a rotating piezoceramic spherical shell with a functionally graded property. *Int. J. Solids Struct.* 2001, *Vol* 38, 7015-7027.

Tarn, J. Q. Exact solutions for functionally graded anisotropic cylinders subjected to thermal and mechanical loads. *Int. J. Solids Struct.* 2001, *Vol* 38, 8189-8206.

Birman, V., Byld, L. W. Modeling and analysis of functionally graded materials and structures. *Appl. Mech. Rev.* 2007, *Vol* 60, 195-216.

Chen, W. Q., Ding, H. J. Bending of functionally graded piezoelectric rectangular plates. *Acta Mech. Solida Sinica* 2000, *Vol* 13, 312-319.

Chen, W. Q., Ye, G. R., Cai, J. B., Ding, H. J. Free vibrations of a functionally graded spherically isotropic hollow sphere. *Acta Mech. Sinica* 2001, *Vol* 33, 768-775. (in Chinese)

Chen, W. Q., Lu, Y., Ye, G. R., Cai, J. B. 3D electroelastic fields in a functionally graded piezoceramic hollow sphere under mechanical and electric loadings. *Arch. Appl. Mech.* 2002, *Vol* 72, 39-51.

Chen, W. Q., Ding, H. J. On free vibration of a functionally graded piezoelectric rectangular plate. *Acta Mech.* 2002, *Vol* 153, 207-216.

Chen, W. Q., Wang, L. Z., Lu, Y. Free vibrations of functionally graded piezoceramic hollow spheres with radial polarization. *J. Sound Vib.* 2002, *Vol* 251, 103-114.

Chen, W. Q., Bian, Z. G., Ding, H. J. Three-dimensional analysis of a thick FGM rectangular plate in thermal environment. *J. Zhejiang Univ. (SCI.)*, 2003, *Vol* 4, 1-7.

Chen, W. Q., Bian, Z. G., Lv, C. F., Ding, H. J. 3D free vibration analysis of a functionally graded piezoelectric hollow cylinder filled with compressible fluid. *Int. J. Solids Struct.* 2004, *Vol* 41, 947-964.

Chen, W. Q., Lee, K. Y., Ding, H. J. On free vibration of non-homogeneous transversely isotropic magneto-electro-elastic plates. *J. Sound Vib.* 2005, *Vol* 279, 237-251.

Chen, W. Q., Lv, C. F., Bian, Z. G. Elasticity solution for free vibration of laminated beams. *Compos. Struct.* 2003, *Vol* 62, 75-82.

Chen, W. Q., Lv, C. F., Bian, Z. G. Free vibration analysis of generally laminated beams via state-space-based differential quadrature. *Compos. Struct.* 2004, *Vol* 63, 417-425.

Chen, W. Q., Lee, K. Y. On free vibration of cross-ply laminates in cylindrical bending. *J. Sound Vib.* 2004, *Vol* 273, 667-676.

Chen, W. Q., Lü, C. F., Bian, Z. G. A mixed method for bending and free vibration of beams resting on a Pasternak elastic foundation. *Appl. Math. Modelling* 2004, *Vol* 28, 877-890.

Chen, W. Q., Lüe, C. F. 3D free vibration analysis of cross-ply laminated plates with one pair of opposite edges simply supported. *Compos. Struct.* 2005, *Vol* 69, 77-87.

Lü, C. F., Huang, Z. Y., Chen, W. Q. Semi-analytical solutions for free vibration of anisotropic laminated plates in cylindrical bending. *J. Sound Vib.* 2007, *Vol* 304, 987-995.

Friedland, B. Control System Design: An Introduction to State Space Methods, Dover: NY, 2005.

Moler, C., Loan, C. V. Nineteen dubious ways to compute the exponential of a matrix, twenty-five years later. *SIAM Rev.* 2003, *Vol* 45, 3-49.

Zhong, Z., Shang, E. T. Three-dimensional exact analysis of simply supported functionally gradient piezoelectric plate. *Int. J. Solids Struct.* 2003, *Vol* 40, 5335-5352.

Pagano, N. J. Exact solutions for composite laminates in cylindrical bending. *J. Compos. Mater.* 1969, *Vol* 3, 398-411.

Ding, H. J., Xu, R. Q., Chen, W. Q., Chi, Y. W. Free axisymmetric vibration of transversely isotropic laminated circular plates. *Acta Mech. Solida Sinica* 1998, *Vol* 11, 209-215.

Ding, H. J., Xu, R. Q. Free axisymmetric vibration of laminated transversely isotropic annular plates. *J. Sound Vib.* 2000, *Vol* 230, 1031-1044.

Fan, J. R., Zhang, J. Y. Exact solutions for thick laminated shells. *Sci. China (A)* 1992, *Vol* 35, 1343-1355.

Chen, W. Q., Bian, Z. G., Ding, H. J. Three-dimensional vibration analysis of fluid-filled orthotropic FGM cylindrical shells. *Int. J. Mech. Sci.* 2004, *Vol* 46, 159-171.

Chen, W. Q., and Lee, K. Y. State-space approach for statics and dynamics of angle-ply laminated cylindrical panels in cylindrical bending. *Int. J. Mech. Sci.* 2005, *Vol* 47, 374-387.

Lekhnitshii, S. G. *Theory of Elasticity of an Anisotropic Body*, Mir Publishers: Moscow, 1981.

Hu, H. C. On the general theory of elasticity for a spherically isotropic medium. *Acta Sci. Sinica* 1954, *Vol* 3, 247-260.

Shul'ga, N. A., Grigorenko, A. Y., Efimova, T. L. Free non-axisymmetric oscillations of a thick-walled, nonhomogeneous, transversely isotropic, hollow sphere. *Sov. Appl. Mech.* 1988, *Vol* 24, 439-444.

Fan, J. R., Zhang, J. Analytical solutions for thick doubly curved laminated shells. *J. Engng. Mech.* 1992, *Vol* 118, 1338-1356.

Li, J. Y. Magnetoelectroelastic multi-inclusion and inhomogeneity problems and their applications in composite materials. *Int. J. Engng. Sci.* 2000, *Vol* 38, 1993-2011.

Soh, A. K., Liu, J. X. On the constitutive equations of magnetoelectroelastic solids. *J. Intell. Mater. Sys. Struct.* 2005, *Vol* 16, 597-602.

Wang, J. G., Chen, L. F., Fang, S. S. State vector approach to analysis of multilayered magneto-electro-elastic plates. *Int. J. Solids Struct.* 2003, *Vol* 40, 1669-1680.

Tarn, J. Q. A state space formalism for piezothermoelasticity. *Int. J. Solids Struct.* 2002, *Vol* 39, 5173-5184.

Chen, W. Q., Zhou, Y. Y., Lü, C. F., Ding, H. J. Bending of multiferroic laminated rectangular plates with imperfect interlaminar bonding. *Euro. J. Mech. A-Solids* 2009, *Vol* 28, 720-727.

Fan, J. R., Ding, K. W. The exact solution for thick laminated cantilever cylindrical shell. *Appl. Math. Mech.* 1995, *Vol* 16, 143-159.

Chen, W. Q., Lv, C. F., Bian, Z. G. Elasticity solution for free vibration of laminated beams. *Compos. Struct.* 2003, *Vol* 62, 75-82.

Chen, W. Q., Lv, C. F., Bian, Z. G. Free vibration analysis of generally laminated beams via state-space-based differential quadrature. *Compos. Struct.* 2004, *Vol* 63, 417-425.

Chen, W. Q., Lee, K. Y. On free vibration of cross-ply laminates in cylindrical bending. *J. Sound Vib.* 2004, *Vol* 273, 667-676.

Chen, W. Q., Lüe, C. F. 3D free vibration analysis of cross-ply laminated plates with one pair of opposite edges simply supported. *Compos. Struct.* 2005, *Vol* 69, 77-87.

Bert, C. W., Malik, M. Differential quadrature method in computational mechanics: a review. *Appl. Mech. Rev.* 1996, *Vol* 49, 1-28.

Shu, C, Richards, B. E. Application of generalized differential quadrature to solve two-dimensional incompressible Naiver-Stocks equations. *Int. J. Num. Meth. Fluids* 1992, *Vol* 15, 791-798.

Lü, C. F., Chen, W. Q. Free vibration of orthotropic functionally graded beams with various end conditions. *Struct. Engng Mech.* 2005, *Vol* 20, 465-476.

Lü, C. F., Chen, W. Q., Zhong, Z. Two-dimensional thermoelasticity solution for functionally graded thick beams. *Science in China G* 2006, *Vol* 49, 451-460.

Lü, C. F., Chen, W. Q., Xu, R. Q., Lim, C. W. Semi-analytical elasticity solutions for bi-directional functionally graded beams. *Int. J. Solids Struct.* 2008, *Vol* 45, 258-275.

Lü, C. F., Chen, W. Q., Shao, J. W. Semi-analytical three-dimensional elasticity solutions for generally laminated composite plates. *Euro. J. Mech. A-Solids* 2008, *Vol* 27, 899-917.

Lü, C. F., Lim, C. W., Chen, W. Q. Semi-analytical analysis for multi-directional functionally graded plates: 3-D elasticity solutions. *Int. J. Num. Meth. Engng.* 2009, *Vol* 79, 25-44.

Xu, R. Q, Wu, Y. F. Free vibration and buckling of composite beams with interlayer slip by two-dimensional theory. *J. Sound Vib.* 2008, *Vol* 313, 875-890.

Alibeigloo, A., Shakeri, M. Elasticity solution for the free vibration analysis of laminated cylindrical panels using the differential quadrature method. *Compos. Struct.* 2007, *Vol* 81, 105-113.

Alibeigloo, A., Shakeri, M. Elasticity solution for static analysis of laminated cylindrical panel using differential quadrature method. *Engng. Struct.* 2009, *Vol* 31, 260-267.

Zhou, Y. Y., Chen, W. Q., Lu, C. F., Wang, J. Free vibration of cross-ply piezoelectric laminates in cylindrical bending with arbitrary edges. *Compos. Struct.* 2009, *Vol* 87, 93-100.

Li. Y., Shi, Z. F. Free vibration of a functionally graded piezoelectric beam via state-space based differential quadrature. *Compos. Struct.* 2009, *Vol* 87, 257-264.

Alibeigloo, A., Madoliat, R. Static analysis of cross-ply laminated plates with integrated surface piezoelectric layers using differential quadrature. *Compos. Struct.* 2009, *Vol* 88, 342-353.

About the Authors

Weiqiu Chen received his B.S. degree in engineering mechanics in 1990 and his Ph.D. degree in solid mechanics in 1996 from Zhejiang University, Hangzhou, China. From 1997 to 1999, he was a Postdoctoral Fellow at The University of Tokyo, Tokyo, Japan. Since 1996, he has been working in the Department of Civil Engineering, Zhejiang University, where he was promoted to be an Associate Professor and Full Professor in 1999 and 2000, respectively. He is now a full professor and vice director of the Institute of Applied Mechanics, Zhejiang University. His research interests include mechanics of piezoelectric materials and devices, vibration and waves in structures, and analytical methods in applied mechanics. He has coauthored 3 monographs, two in English and one in Chinese. He has published nearly 200 journal papers, among which about 150 are in international journals.

Haojiang Ding received his B.S. degree in applied mathematics and mechanics in 1957 from Peking University, Beijing, China. After his graduation, he has been working at Department of Engineering Mechanics, Zheiang University, Hangzhou, China until his retirement in 2006. He was promoted to be an Associate Professor and Full Professor in 1981 and 1985, respectively. His research interests include theory of piezoelasticity, anisotropic elasticity, mechanics of functionally graded materials and structures, analytical methods in mechanics, and numerical methods such as FEM and BEM. He has coauthored 3 monographs (two in English and one in Chinese) and several textbooks. He has published more than 300 journal papers, among which about 200 are in international journals.

In: Mechanics of Functionally Graded Materials and Structures ISBN 978-1-61668-920-9
Editors: Z. Zhong, L. Wu and W. Chen

Chapter 6

THE ANALYSIS OF SMART FGM MATERIALS AND STRUCTURES BY FINITE ELEMENT METHOD

Xu Rongqiao[1]
Department of Civil Engineering,
Zhejiang University,
Zijingang Campus, Hangzhou, China

ABSTRACT

This work surveys the application of the finite element method in functionally graded materials and structures firstly. Some graded isoparametric elements are then formulated based on the variational principle of piezoelectricity. This work also presents a graded harmonic element for axisymmetric piezoelectricity with non-symmetric loads. Finally some numerical results are shown to demonstrate the proposed elements.

1. A BRIEF REVIEW OF FEM IN FGMS

The finite element method (FEM) has been widely applied in many fields including functionally graded materials and structures due to its versatility. Since the FGMs were developed for thermal barrier originally, the FEM was used to analyze or optimize its mechanical behaviors under various thermal loadings. For examples, Williamson et al [1] formulated an elastic-plastic finite element to study residual stresses at graded ceramic-metal interfaces during cooling, in which the graded microstructure was treated as a series of perfectly bonded layers, each having slightly different properties. The same finite element was employed by Drake et al [2] to optimize the graded interfaces for residual stress reduction. Miller et al [3] also studied the residual stress during the fabrication of the functionally graded materials made of $NiAl/Al_2O_3$. Torres et al [4] showed the significant effect of the size of the graded interfaces on the thermal stress in the FEM simulation of composite material $NiAl/TiB_2$. In these works, traditionally homogeneous elements, in which

[1] e-mail: xurongqiao@zju.edu.cn.

the material properties were constant and did not varied with the coordinates within the element, were utilized to study the residual stresses due to thermal loading. Recently, Na and Kim [5] presented three-dimensional thermal buckling and post-buckling analyses of functionally graded material plates subjected to uniform or non-uniform temperature rise by using three-dimensional finite element method based on the Green-Lagrange nonlinear strain displacement relation. Cooley and Palazotto [6] analyzed the shell panels under thermal loading and compared performance of traditional homogeneous materials to FGMs using ABAQUS. Kordkheili and Naghdabadi [7] formulated a finite element to study the geometrically non-linear thermoelastic behavior of plates and shells made of functionally graded materials using the updated Lagrangian approach, in which the non-linear heat transfer equation was also solved for thermal distribution through the thickness by the Rayleigh-Ritz method. Zhao and Liu [8] used least square spectral element method based on discrete-ordinates equation to solve multidimensional radiative heat transfer problems in semitransparent graded index media. Akpobi and Edobor [9] presented a finite element-eigenvalue method to solve the finite element models of time dependent temperature field problems in functionally graded materials (FGMs). Lee et al [10] studied the thermal fatigue properties of $Mg/Mg\text{-}Al_{18}B_4O_{33}$ functionally graded material by the finite element method and exhibited a good agreement with the results of the thermal cycling test. Santos et al [11] developed a semi-analytical axisymmetric finite element model to conduct the thermoelastic analysis of functionally graded cylindrical shells subjected to transient thermal shock loading based on the three-dimensional linear elasticity theory.

With the development of the technology of the FGM coating, the FEM was also introduced into this field. Haque and Choy [12] investigated the effect of the coating on thermal residual stresses generated at the fiber/matrix interface due to differences in the thermal expansion coefficient between the various materials within the coating system. Stephens et al [13] analyzed initial yielding behavior due to the indentation and friction process between an elastic cylindrical surface and hard coating/functionally graded substrate system by finite element modeling. Limarga et al [14] studied residual stress in a plasma-sprayed functionally graded thermal barrier coating by finite element method. Hou et al [15] established an elastic finite element method numerical model for predicting thermal residual stresses in graded ceramic-ceramic materials to determine interface conditions favorable for achieving residual stress reductions. They also investigated the effects of different composition profiles on stress distributions established during the uniform cooling from an assumed sintering temperature. Zhang et al [16] analyzed the thermo-mechanical behaviors and the distribution of residual stresses in functionally graded ZrO_2/NiCoCrAlY coatings due to thermal spraying by thermo-mechanical finite element and showed that the level of residual stresses at the critical interface regions was obviously influenced by coating to substrate thickness ratio, the presence of interlayer and cooling rate.

Nowadays, the optimizations of the FGMs were paid more attention with the development of the technology of material fabricated and advancement of the hardware and software of computers. Zhong et al [17] presented an overall research program to develop a non-sticking, oxidation resistant, and wear resistant coating system for glass molding dies and forming tools, in which the FEM was integrated to predict an optimal coating architecture with minimized residual stress and optimized stress distribution. Chen and He [18] simulated the thermal residual stresses generated in MgO/Ni system functionally graded composites during post-sintering stage and proposed the compositional distribution optimization formulation. Zhou and Tang [19] proposed a new method to effectively lower the high temperature, smooth the thermal stress and reduce the thermal deformation of mirrors in high-power laser systems utilizing FGMs, in which FEM was used as analysis tools to obtain optimal design of FGM mirrors. Based on three-dimensional FEM, Lee et al [20] presented numerical analysis to reduce the number of FGM layers for joining Si_3N_4-Al_2O_3 using polytypoid interlayer by estimating the position of crack, and find the optimized number of layers and its thicknesses of FGM joint. Zhang et al [21] designed the SiC/C functionally graded materials for ablation resistance application. In the abovementioned works, the objective function was the stress distribution and the distribution of the material properties were chosen as design variables.

The FEM was also used to predict the effective material properties of FGMs according to their micro-structures. Pindera et al [22] concluded that the standard micromechanical analyses, based on the concept of the representative volume element, was not available to determine average or effective properties of FGMs when the temperature gradient was large with respect to the dimension of the reinforcement. Pindera and Dunn [23] evaluated the higher-order theory for functionally graded materials based on the finite element method. Grujicic and Zhang [24] implemented a Voronoi cell finite element method based on the commercial finite element package ABAQUS, which was used to determine the effective microstructure-dependent elastic properties of FGMs. The resulting effective elastic properties of Ni/MgO and Ni_3Al/TiC were in significantly better agreement with the experimental values than the values predicted by the commonly used Eshelby's equivalent inclusion method and the self consistent method. Cho and Ha [25] also numerically investigated thermomechanical properties and responses of dual-phase functionally graded materials by conventional averaging approaches and Voronoi cell element. More recently, Ajdari et al [26] investigated numerically the compressive uniaxial and biaxial behavior of functionally graded Voronoi structures, such as bio-inspired functionally graded materials for manufacturing implants or bone replacement.

The conception of graded finite element may refer back to Gu et al [27], who proposed a finite element based method for calculating stress intensity factors of functionally graded materials (FGMs). In their work, the material variation was achieved by using corresponding material properties at Gauss integration points (different Gauss points have different properties). Consequently, Santare and Lambros1 [28] firstly introduced the term "Graded Finite Element" and used it to analyze inhomogeneous materials. Furthermore, Kim and Paulino [29] presented the conception of the "Isoparametric Graded Finite Elements", in which the material properties within the element were also interpolated from these values at the nodes of the element through the same shape function of displacements. They also presented some numerical examples of isotropic and orthotropic materials. This kind of element was then used by Kim and Paulino [30] to analyze the stress intensity factors of FGMs. Moreover, this element was extended by Buttlar et al [31] to study the behavior of asphalt pavements, which exhibited vertically graded material properties due to temperature gradients and aging related stiffness gradients.

Some structural elements, such as beam, plate and shell elements, are also formulated for FGM structures to improve the efficiency and speed of the calculation. Chakraborty et al [32] developed a new beam element to study the thermoelastic behavior of functionally graded beam structures based on the first-order shear deformation theory and accounted for varying elastic and thermal properties along its thickness. Chakraborty and Gopalakrishnan [33] formulated a new spectral finite element with three degrees of freedom per node to analyze the wave propagation behavior in a functionally graded (FG) beam subjected to high frequency impulse loading, which can be either thermal or mechanical. Wu and Chen [34] presented a refined three-node triangular element, which satisfies the interelement C-1 weak-continuity conditions and based on the higher-order laminated beam theory proposed by themselves. Das et al [35] presented a new triangular finite element based on {3, 2}-order single-layer theory for modeling thick sandwich panels with or without a functionally graded core subjected to thermo-mechanical loading using a hybrid energy functional. Eason and

Caraballo [36] compared two finite element formulations based on first-order shear deformation theory undergoing thermo-mechanical loading. Naei et al [37] presented the buckling analysis of radially-loaded circular plate with variable thickness made of functionally graded material based on Love-Kichhoff hypothesis and Sander's non-linear strain-displacement relation for thin plates using FEM. Kim et al [38] extended four-node quasi-conforming shell element to the case of geometrically non-linear behavior of the FGM plates and shells. Oyekoya et al [39] derived two new Mindlin-type plate bending elements for the modeling of functionally graded plate subjected to various loading conditions such as tensile loading, in-plane bending and out-of-plane bending. Pradyumna and Bandyopadhyay [40] carried out the free vibration analysis of functionally graded curved panels using a C^0 finite element based on a higher-order formulation, in which a realistic parabolic distribution of transverse shear strains through the shell thickness was assumed to avoid the use of shear correction factor. Ramkumar and Ganesan [41] investigated the buckling and vibration behaviors of thin-walled box columns in a thermal environment made of functionally graded materials using FEM based on classical laminated plate theory. The effects of temperature dependent material properties on the critical buckling and the free vibration response were studied. Shaker et al [42] used both a first and second order reliability method with a nine-noded isoparametric Lagrangian element based on the third order shear deformation theory to investigate sensitivity of the fundamental frequency of FGM plates to material uncertainties, which may be occurred during manufacturing processes. MATLAB was applied to derive the derivatives of the stiffness and mass matrices symbolically for a considerable reduction in calculation time. Al-Ajmi et al [43] developed a finite element model for laminated beams with functionally graded viscoelastic layer as constrained damping and discussed the obtained natural frequencies and modal loss factors.

Recently, combined with cohesive zone model (CZM), FEM was usually employed to investigate the fracture or the crack growth in composite materials including FGMs. For instance, Jin et al [44] studied mode I crack growth in ceramic/metal (FGMs) using three-dimensional interface-cohesive elements based upon a new phenomenological cohesive fracture model. Comi and Mariani [45] simulated the fracture processes in quasi-brittle functionally graded materials (FGMs) using a finite element accounting for material gradation and following the propagation of cohesive cracks in the graded medium. The proposed methodology was checked by four points bending tests on notched beams, which featured a mechanical gradient parallel and perpendicular to the notch. Kirugulige and Tippur [46] used optical interferometry along with high-speed photography to measure mixed-mode dynamic crack growth behavior in a FGM beams with single edge cracks initially aligned in the direction of the compositional gradient and subjected to one-point eccentric impact loading. They also developed a new cohesive element based on a bilinear traction-separation law to simulate the crack paths and crack speeds and got qualitative agreement with the observed results.

Nowadays, the concept of FGM was introduced into smart materials and structures in which FEM also played an important role. He et al [47, 48] studied the active control of FGM plates and shells bonded with piezoelectric layers. Ng et al [49] also presented a flat-shell element to investigate the active control of functionally graded material shells in frequency domain through integrated piezoelectric sensor/actuator layer. Ray and Sachade [50] derived a finite element model for the static analysis of functionally graded plates integrated with a layer of piezoelectric fiber reinforced composite (PFRC) material, which acted as the distributed actuator of the FG plates. They declared that the activated PFRC layer was more effective in controlling the deformations of the FG plates when the layer is attached to the surface of the FG plate with minimum stiffness than when it is attached to the surface of the same with maximum stiffness. Bhangale and Ganesan [51] carried out the studies on the free vibration of functionally graded, anisotropic and linear magneto-electro- elastic plates by semi-analytical finite element method, in which a series solution was assumed in the plane of the plate and finite element procedure was adopted across the thickness of the plate. Panda and Ray [52] investigated the static behaviors of functionally graded plates integrated with patches of piezoelectric fiber reinforced composite based on the first order shear deformation theory and the Von Karman type geometric nonlinearity. They analyzed the effects of volume fraction index of the FG material of the substrate plates, the locations of the PFRC patches on the control authority of the patches, and the variation of piezoelectric fiber orientation angle.

With the application of FGMs as implants, their performance in the biological environment were paid more attentions. Wang et al [53] investigated the thermal- mechanical behavior of hydroxyapatite/titanium (HA/Ti) functionally graded dental implants with the three-dimensional finite element method, in which the effect of temperature variation due to daily oral activity was included. In addition, Shin [54] proposed a two-dimensional adaptive mesh generation algorithm based on the discretization by which continuous material variation inside an object is converted into step-wise variation. Kou and Tan [55] presented a systematic approach for integrated computer-aided design and finite element analysis of functionally-graded-material objects whose distributions were encoded into the proposed heterogeneous feature tree structure. Haghighi et al [56] presented a coupled differential quadrature and finite element method for 3-D transient heat transfer analysis of functionally graded thick plates. The in-plane spatial derivatives were discretized using the FEM and the resulting transient-variable-coefficient partial differential equations were discretized in the transverse direction and temporal domain employing the incremental and unconditionally stable DQM.

2. Generic Finite Element Formulation of Piezoelectricity

Since it is convenient for formulation of finite element in matrix notation, we render the governing equation of piezoelectricity as

$$\bar{\mathbf{L}}\bar{\boldsymbol{\sigma}} + \bar{\mathbf{f}} = \boldsymbol{\rho}\ddot{\bar{\mathbf{u}}} \tag{1}$$

$$\bar{\boldsymbol{\varepsilon}} = \bar{\mathbf{L}}^{\mathrm{T}}\bar{\mathbf{u}} \tag{2}$$

$$\bar{\boldsymbol{\sigma}} = \bar{\mathbf{C}}\bar{\boldsymbol{\varepsilon}} \tag{3}$$

For three-dimensional theory in Cartesian coordinates, we have

$$\bar{\boldsymbol{\sigma}} = \begin{bmatrix} \sigma_x & \sigma_y & \sigma_z & \tau_{yz} & \tau_{xz} & \tau_{xy} & D_x & D_y & D_z \end{bmatrix}^{\mathrm{T}} \tag{4}$$

$$\bar{\boldsymbol{\varepsilon}} = \begin{bmatrix} \varepsilon_x & \varepsilon_y & \varepsilon_z & \gamma_{yz} & \gamma_{xz} & \gamma_{xy} & -E_x & -E_y & -E_z \end{bmatrix}^{\mathrm{T}} \tag{5}$$

$$\bar{\mathbf{u}} = \begin{bmatrix} u & v & w & \phi \end{bmatrix}^{\mathrm{T}} \tag{6}$$

$$\bar{\mathbf{f}} = \begin{bmatrix} f_x & f_y & f_z & -f_e \end{bmatrix}^{\mathrm{T}} \tag{7}$$

$$\boldsymbol{\rho} = \mathrm{diag}(\rho, \rho, \rho, 0) \tag{8}$$

$$\overline{\mathbf{C}} = \begin{bmatrix} \mathbf{C} & \mathbf{e}^{\mathrm{T}} \\ \mathbf{e} & \boldsymbol{\kappa} \end{bmatrix} \tag{9}$$

$$\overline{\mathbf{L}} = \begin{bmatrix} \partial/\partial x & 0 & 0 & 0 & \partial/\partial z & \partial/\partial y & 0 & 0 & 0 \\ 0 & \partial/\partial y & 0 & \partial/\partial z & 0 & \partial/\partial x & 0 & 0 & 0 \\ 0 & 0 & \partial/\partial z & \partial/\partial y & \partial/\partial x & 0 & 0 & 0 & 0 \\ 0 & 0 & 0 & 0 & 0 & 0 & \partial/\partial x & \partial/\partial y & \partial/\partial z \end{bmatrix} \tag{10}$$

in which σ, τ, ε and γ denote the stresses and strains, D and E denote the electrical displacements and electrical fields, u, v, w and ϕ are displacements and electrical potential, f_x, f_y, f_z and f_e are volume forces and free charge density, ρ is mass density, and $\mathbf{C}$, $\mathbf{e}$ and $\boldsymbol{\kappa}$ are the elastic coefficients matrix, piezoelectric coefficient matrix and dielectric coefficients matrix. It is noticeable that the electric field is quasi-static and therefore the last diagonal element of the matrix $\boldsymbol{\rho}$ is zero. Since eqs.(1)-(3) are similar with the governing equation of elasticity, the symbols $\overline{\boldsymbol{\sigma}}$, $\overline{\boldsymbol{\varepsilon}}$, $\overline{\mathbf{u}}$ and $\overline{\mathbf{f}}$ are called generalized stresses, generalized strains and generalized displacements, generalized volume forces, respectively.

The corresponding boundary conditions can also be rewritten as

$$\mathbf{n}\overline{\boldsymbol{\sigma}} = \hat{\mathbf{T}} \text{ on } \Gamma_\sigma \tag{11}$$

$$\overline{\mathbf{u}} = \hat{\mathbf{u}} \text{ on } \Gamma_u \tag{12}$$

where Γ_σ and Γ_u denote the bounded surface of the piezoelectricity Ω on which the generalized surface traction $\hat{\mathbf{T}}$ and generalized displacements $\hat{\mathbf{u}}$ are specified, respectively. The generalized surface traction $\hat{\mathbf{T}}$ consists of the surface traction $\hat{T}_x$, $\hat{T}_y$, $\hat{T}_z$ and the surface charge density $\hat{T}_e$. The matrix $\mathbf{n}$ represents the outward normal of the surface Γ_σ and if n_x, n_y and n_z denote the three components of the outward normal it can be rendered as

$$\mathbf{n} = \begin{bmatrix} n_x & 0 & 0 & 0 & n_z & n_y & 0 & 0 & 0 \\ 0 & n_y & 0 & n_z & 0 & n_x & 0 & 0 & 0 \\ 0 & 0 & n_z & n_y & n_x & 0 & 0 & 0 & 0 \\ 0 & 0 & 0 & 0 & 0 & 0 & n_x & n_y & n_z \end{bmatrix} \tag{13}$$

The equivalent variational form of the governing equations (1)-(3), (11) and (12) are given by Tiersten [57], namely,

$$\delta\int_{t_0}^{t_1}\left[\frac{1}{2}\int_\Omega(\overline{\boldsymbol{\varepsilon}}^{\mathrm{T}}\overline{\mathbf{C}}\overline{\boldsymbol{\varepsilon}} - \dot{\overline{\mathbf{u}}}^{\mathrm{T}}\boldsymbol{\rho}\dot{\overline{\mathbf{u}}})\,\mathrm{d}\Omega - \int_\Omega \overline{\mathbf{f}}^{\mathrm{T}}\overline{\mathbf{u}}\,\mathrm{d}\Omega - \int_{\Gamma_\sigma}\hat{\mathbf{T}}^{\mathrm{T}}\overline{\mathbf{u}}\,\mathrm{d}\Gamma\right]\mathrm{d}t = 0 \tag{14}$$

The generalized displacements are interpolated through the node variables in finite element method as follows

$$\bar{\mathbf{u}} = \bar{\mathbf{N}}\bar{\boldsymbol{\delta}}^e \tag{15}$$

in which $\bar{\boldsymbol{\delta}}^e$ and $\bar{\mathbf{N}}$ are the generalized displacement vector of the nodes and the shape function matrix, respectively, which are both dependent on the details of the element. Using Eq.(2) the generalized strain is

$$\bar{\boldsymbol{\varepsilon}} = \bar{\mathbf{L}}^{\mathrm{T}}\bar{\mathbf{u}} = \bar{\mathbf{L}}^{\mathrm{T}}\bar{\mathbf{N}}\bar{\boldsymbol{\delta}}^e = \bar{\mathbf{B}}\bar{\boldsymbol{\delta}}^e \tag{16}$$

where the matrix $\bar{\mathbf{B}} = \bar{\mathbf{L}}^{\mathrm{T}}\bar{\mathbf{N}}$ establishes the relation between the generalized strain with generalized displacement of nodes and is called generalized strain matrix. Substitution of Eqs.(15) and (16) into Eq.(14) gives

$$\bar{\mathbf{M}}^e\ddot{\bar{\boldsymbol{\delta}}}^e + \bar{\mathbf{K}}^e\bar{\boldsymbol{\delta}}^e = \mathbf{f}^e \tag{17}$$

in which $\bar{\mathbf{M}}^e$, $\bar{\mathbf{K}}^e$ and $\mathbf{f}^e$ are the mass matrix, stiffness matrix and equivalent node force vector of elements, respectively, and determined by

$$\bar{\mathbf{M}}^e = \int_{\Omega^e} \bar{\mathbf{N}}^{\mathrm{T}}\boldsymbol{\rho}\bar{\mathbf{N}}\,\mathrm{d}\Omega,\ \bar{\mathbf{K}}^e = \int_{\Omega^e} \bar{\mathbf{B}}^{\mathrm{T}}\bar{\mathbf{C}}\bar{\mathbf{B}}\,\mathrm{d}\Omega \tag{18}$$

$$\mathbf{f}^e = \int_{\Omega^e} \bar{\mathbf{N}}^{\mathrm{T}}\bar{\mathbf{f}}\,\mathrm{d}\Omega + \int_{\Gamma_\sigma^e} \bar{\mathbf{N}}^{\mathrm{T}}\hat{\mathbf{T}}\,\mathrm{d}\Gamma \tag{19}$$

where Ω^e and Γ_σ^e denote the volume and partial surface subjected to generalized surface traction of the element.

3. Graded Hexahedral Isoparametric Element of Piezoelectricity

In conventional isoparametric element, we introduced a natural coordinates $\boldsymbol{\xi} = [\xi, \eta, \zeta]^{\mathrm{T}}$ and establish a map between it and the Cartesian coordinates $\mathbf{x} = [x, y, z]^{\mathrm{T}}$ through the transformation function $\mathbf{N}_i$ as

$$\mathbf{x} = \sum_i \mathbf{N}_i\boldsymbol{\xi}_i \tag{20}$$

in which $\boldsymbol{\xi}_i$ is the natural coordinates of the node i. An original hexahedron in Cartesian coordinates will becomes a cubic in the natural coordinates (Figure 1) if the transformation function $\mathbf{N}_i$ and $\boldsymbol{\xi}_i$ are

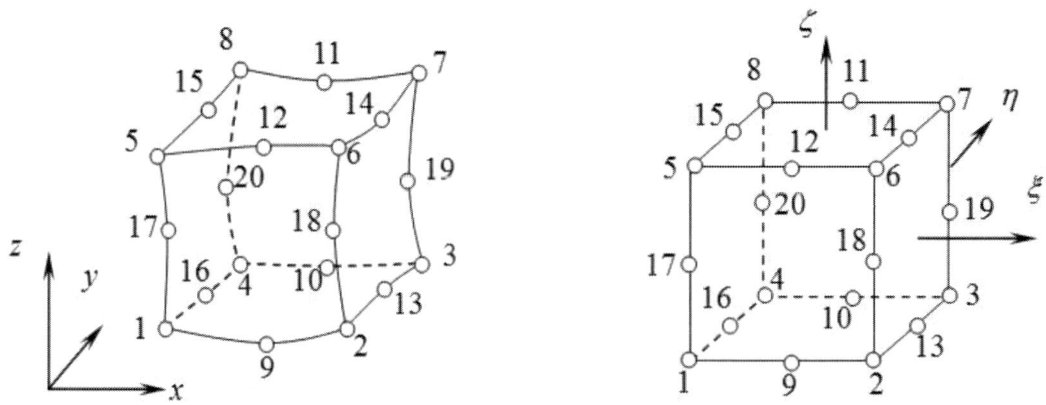

Figure 1. Graded Isoparametric Hexahedral element.

$$\mathbf{N}_i = N_i \mathbf{I}_3 \tag{21}$$

$$\begin{aligned} N_i &= (1-\xi^2)(1+\eta_i\eta)(1+\zeta_i\zeta)/4 \quad (i=9,10,11,12) \\ N_i &= (1-\eta^2)(1+\xi_i\xi)(1+\zeta_i\zeta)/4 \quad (i=13,14,15,16) \\ N_i &= (1-\zeta^2)(1+\xi_i\xi)(1+\eta_i\eta)/4 \quad (i=17,18,19,20) \end{aligned} \tag{22}$$

$$\begin{aligned} N_1 &= \bar{N}_1 - (N_9 + N_{16} + N_{17})/2, \quad N_2 = \bar{N}_2 - (N_9 + N_{13} + N_{18})/2 \\ N_3 &= \bar{N}_3 - (N_{10} + N_{13} + N_{19})/2, \quad N_4 = \bar{N}_4 - (N_{10} + N_{16} + N_{20})/2 \\ N_5 &= \bar{N}_5 - (N_{12} + N_{15} + N_{17})/2, \quad N_6 = \bar{N}_6 - (N_{12} + N_{14} + N_{18})/2 \\ N_7 &= \bar{N}_7 - (N_{11} + N_{14} + N_{19})/2, \quad N_8 = \bar{N}_8 - (N_{11} + N_{15} + N_{20})/2 \end{aligned} \tag{23}$$

$$\bar{N}_i = (1+\xi_i\xi)(1+\eta_i\eta)(1+\zeta_i\zeta)/8 \quad (i=1,2,\cdots,8) \tag{24}$$

$$\begin{aligned} \xi_{1,2,\cdots,8} &= -1, \quad 1, \quad 1, \quad -1, \quad -1, \quad 1, \quad 1, \quad -1 \\ \eta_{1,2,\cdots,8} &= -1, \quad -1, \quad 1, \quad 1, \quad -1, \quad -1, \quad 1, \quad 1 \\ \zeta_{1,2,\cdots,8} &= -1, \quad -1, \quad -1, \quad -1, \quad 1, \quad 1, \quad 1, \quad 1 \end{aligned} \tag{25}$$

$$\begin{aligned} \xi_{9,10,\cdots,20} &= 0, \quad 0, \quad 0, \quad 0, \quad 1, \quad 1, \quad -1, \quad -1, \quad -1, \quad 1, \quad 1, \quad -1 \\ \eta_{9,10,\cdots,20} &= -1, \quad 1, \quad 1, \quad -1, \quad 0, \quad 0, \quad 0, \quad 0, \quad -1, \quad -1, \quad 1, \quad 1 \\ \zeta_{9,10,\cdots,20} &= -1, \quad -1, \quad 1, \quad 1, \quad -1, \quad 1, \quad 1, \quad -1, \quad 0, \quad 0, \quad 0, \quad 0 \end{aligned} \tag{26}$$

in which $\mathbf{I}_3$ denote an identity matrix of 3×3. The shape function $\bar{\mathbf{N}}$ in Eq.(15) can be defined as

$$\bar{\mathbf{N}} = \left[\bar{\mathbf{N}}_1 \quad \bar{\mathbf{N}}_2 \quad \cdots \quad \bar{\mathbf{N}}_{20} \right] \tag{27}$$

$$\bar{\mathbf{N}}_i = N_i \mathbf{I}_4 \ (i=1,2,\cdots,20) \tag{28}$$

in which $\mathbf{I}_4$ denote an identity matrix of 4×4. Because the interpolation functions N_i of the coordinates transformation are same as those of displacements and electric potential, this element is called isoparametric element. The attractiveness of the elements in Figure1 lies in that the elements can have any number of nodes from 8 to 20 if the taken $N_i=0$ when the node i ($i\geq9$) is absent.

Since the shape function $\bar{\mathbf{N}}$ is explicitly expressed by natural coordinates (ξ,η,ζ) in Eqs.(22)-(24), the mass matrix and stiffness matrix in Eq.(18) can be rewritten as follows for convenience of integration,

$$\bar{\mathbf{M}}^e=\int_{-1}^{1}\int_{-1}^{1}\int_{-1}^{1}\bar{\mathbf{N}}^{\mathrm{T}}\boldsymbol{\rho}\bar{\mathbf{N}}|\mathbf{J}|\mathrm{d}\xi\,\mathrm{d}\eta\,\mathrm{d}\zeta\,,\ \bar{\mathbf{K}}^e=\int_{-1}^{1}\int_{-1}^{1}\int_{-1}^{1}\bar{\mathbf{B}}^{\mathrm{T}}\bar{\mathbf{C}}\bar{\mathbf{B}}|\mathbf{J}|\mathrm{d}\xi\,\mathrm{d}\eta\,\mathrm{d}\zeta \tag{29}$$

in which $\mathbf{J}$ is the Jacobi matrix of the transformation of two coordinates, i.e.,

$$\mathbf{J}=\frac{\partial\mathbf{x}}{\partial\boldsymbol{\xi}}=\begin{bmatrix}\partial x/\partial\xi & \partial y/\partial\xi & \partial z/\partial\xi\\ \partial x/\partial\eta & \partial y/\partial\eta & \partial z/\partial\eta\\ \partial x/\partial\zeta & \partial y/\partial\zeta & \partial z/\partial\zeta\end{bmatrix}=\sum_{i=1}^{20}\begin{bmatrix}N_{i,\xi}x_i & N_{i,\xi}y_i & N_{i,\xi}z_i\\ N_{i,\eta}x_i & N_{i,\eta}y_i & N_{i,\eta}z_i\\ N_{i,\zeta}x_i & N_{i,\zeta}y_i & N_{i,\zeta}z_i\end{bmatrix} \tag{30}$$

in which $N_{i,\xi}=\partial N_i/\partial\xi$ and (x_i,y_i,z_i) are the Cartesian coordinates of the node i. Moreover, the partial derivative N_i with respect to the Cartesian coordinates x, y and z in the strain matrix $\bar{\mathbf{B}}$ should also be transformed as

$$\frac{\partial N_i}{\partial\mathbf{x}}=\mathbf{J}^{-1}\frac{\partial N_i}{\partial\boldsymbol{\xi}} \tag{31}$$

If the material is graded, namely, the material properties $\boldsymbol{\rho}$ and $\bar{\mathbf{C}}$ are functions with respect to the coordinates. We assume that they are also interpolated by their values at the nodes through the shape function N_i, i.e.,

$$\boldsymbol{\rho}=\sum_{i=1}^{20}N_i\boldsymbol{\rho}_i\,,\ \bar{\mathbf{C}}=\sum_{i=1}^{20}N_i\bar{\mathbf{C}}_i \tag{32}$$

in which $\boldsymbol{\rho}_i$ and $\bar{\mathbf{C}}_i$ are the material properties at the node i. Then the integral in Eq.(29) can be carried out by Gaussian quadrature method. For example, the stiffness matrix can be given by

$$\mathbf{K}^e=\sum_{i=1}^{m}\sum_{j=1}^{n}\sum_{k=1}^{l}W_iW_jW_k(\bar{\mathbf{B}}^{\mathrm{T}}\bar{\mathbf{C}}\bar{\mathbf{B}}|\mathbf{J}|)_{ijk} \tag{33}$$

where m, n, l are the number the Gaussian quadrature points in ξ, η and ζ direction, W_i is corresponding weight of the Gaussian points, and the items in the parenthesis are evaluated at the corresponding Gaussian points. The mass matrix can computed in a similar manner.

4. Graded Isoparametric Quadrilateral Axisymmetric Element of Piezoelectricity

A quadrilateral axisymmetric element with curvature edges in cylindrical coordinates $\mathbf{r} = [r, z]^{\mathrm{T}}$ is shown in Figure 2(a) and its representation in the natural coordinates system $\boldsymbol{\xi} = [\xi, \eta]^{\mathrm{T}}$ will become square if the transformation of the two coordinate system is

$$r = \sum_{i=1}^{8} N_i(\xi,\eta) r_i \ , \ z = \sum_{i=1}^{8} N_i(\xi,\eta) z_i \tag{34}$$

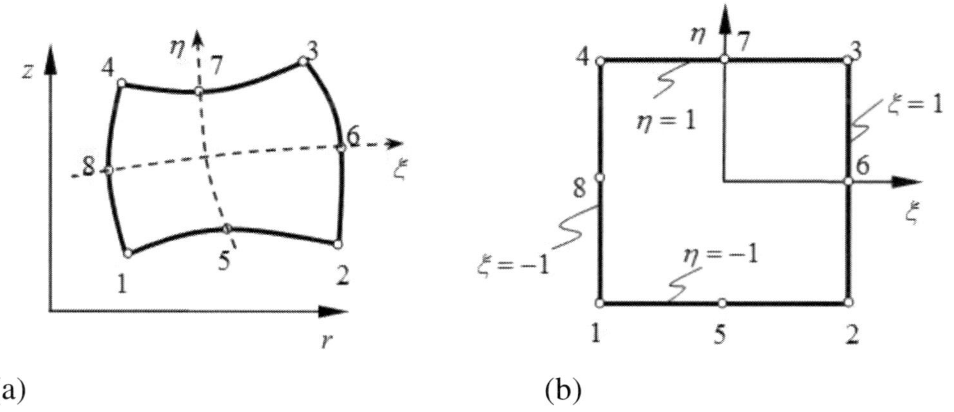

Figure 2. Graded isoparametric axisymmetric element.

in which (r_i, z_i) is the physical coordinates of the node i and the shape functions $N_i(\xi,\eta)$ are

$$\begin{aligned}
N_1 &= (1-\xi)(1-\eta)/4 - (N_8 + N_5)/2, \quad & N_5 &= (1-\xi^2)(1-\eta)/2 \\
N_2 &= (1+\xi)(1-\eta)/4 - (N_5 + N_6)/2, \quad & N_6 &= (1+\xi)(1-\eta^2)/2 \\
N_3 &= (1+\xi)(1+\eta)/4 - (N_6 + N_7)/2, \quad & N_7 &= (1-\xi^2)(1+\eta)/2 \\
N_4 &= (1-\xi)(1+\eta)/4 - (N_7 + N_8)/2, \quad & N_8 &= (1-\xi)(1-\eta^2)/2
\end{aligned} \tag{35}$$

The displacements and electrical potential inside the element are assumed that

$$\bar{\mathbf{u}} = \begin{bmatrix} u & w & \phi \end{bmatrix}^{\mathrm{T}} = \bar{\mathbf{N}} \boldsymbol{\delta}^e \tag{36}$$

where u, w and ϕ are displacement components in radial and axial direction and electric potential, respectively, and the shape function $\bar{\mathbf{N}}$ is

$$\bar{\mathbf{N}} = \left[\bar{\mathbf{N}}_1 \quad \bar{\mathbf{N}}_2 \quad \cdots \quad \bar{\mathbf{N}}_8\right],\ \bar{\mathbf{N}}_i = N_i \mathbf{I}_3 \ (i = 1, 2, \cdots, 8) \tag{37}$$

The elements in Figure2 can also have any number of nodes from 4 to 8 if the taken $N_i = 0$ when the node i ($i \geq 5$) is absent.

According to the strain-displacement relation and electrical displacement- electrical potential relation in the axisymmetric case, we have

$$\bar{\boldsymbol{\varepsilon}} = \left[\varepsilon_r \quad \varepsilon_\theta \quad \varepsilon_z \quad \gamma_{rz} \quad -E_r \quad -E_z\right]^{\mathrm{T}} = \bar{\mathbf{B}}\bar{\boldsymbol{\delta}}^e \tag{38}$$

where

$$\bar{\mathbf{B}} = \left[\bar{\mathbf{B}}_1 \quad \bar{\mathbf{B}}_2 \quad \cdots \quad \bar{\mathbf{B}}_8\right],\ \bar{\mathbf{B}}_i = \begin{bmatrix} N_{i,r} & 0 & 0 \\ N_i/r & 0 & 0 \\ 0 & N_{i,z} & 0 \\ N_{i,z} & N_{i,r} & 0 \\ 0 & 0 & N_{i,r} \\ 0 & 0 & N_{i,z} \end{bmatrix} \quad (i = 1, 2, \cdots 8) \tag{39}$$

in which the partial derivatives of N_i with respect to r and z are also given by

$$\frac{\partial N_i}{\partial \mathbf{r}} = \mathbf{J}^{-1} \frac{\partial N_i}{\partial \boldsymbol{\xi}} \tag{40}$$

in which $\mathbf{J}$ is the Jacobian matrix of the coordinate transformation, i.e.,

$$\mathbf{J} = \frac{\partial \mathbf{r}}{\partial \boldsymbol{\xi}} = \begin{bmatrix} \partial r/\partial \xi & \partial z/\partial \xi \\ \partial r/\partial \eta & \partial z/\partial \eta \end{bmatrix} = \sum_{i=1}^{8} \begin{bmatrix} N_{i,\xi}\, r_i & N_{i,\xi}\, z_i \\ N_{i,\eta}\, r_i & N_{i,\eta}\, z_i \end{bmatrix} \tag{41}$$

The stiffness matrix and mass matrix of elements are

$$\mathbf{K}^e = 2\pi \int_{-1}^{1}\int_{-1}^{1} \bar{\mathbf{B}}^{\mathrm{T}}\bar{\mathbf{C}}\bar{\mathbf{B}} r |\mathbf{J}| \mathrm{d}\xi \mathrm{d}\eta,\ \mathbf{M}^e = 2\pi \int_{-1}^{1}\int_{-1}^{1} \bar{\mathbf{N}}^{\mathrm{T}}\boldsymbol{\rho}\bar{\mathbf{N}} r |\mathbf{J}| \mathrm{d}\xi \mathrm{d}\eta \tag{42}$$

where $\boldsymbol{\rho} = \mathrm{diag}(\rho, \rho, 0)$ and the matrix $\bar{\mathbf{C}}$ is the constitutive matrix of linear piezoelectric body in axisymmetric case.

For non-homogeneous material, the entries of the matrix $\bar{\mathbf{C}}$ are the function of the coordinates (r, z). In a similar manner, we use the same interpolation functions N_i to define the constitutive matrix $\bar{\mathbf{C}}$ and mass density $\boldsymbol{\rho}$, namely

$$\bar{\mathbf{C}} = \sum_{i=1}^{8} N_i \bar{\mathbf{C}}_i \, , \; \boldsymbol{\rho} = \sum_{i=1}^{8} N_i \boldsymbol{\rho}_i \tag{43}$$

in which the matrix $\bar{\mathbf{C}}_i$ and $\boldsymbol{\rho}_i$ denotes the material properties at the node i.

The integral of Eq.(42) is also implemented using Gaussian quadrature method, i.e.,

$$\mathbf{K}^e = 2\pi \sum_{i=1}^{m} \sum_{j=1}^{n} W_i W_j (\bar{\mathbf{B}}^{\mathrm{T}} \bar{\mathbf{C}} \bar{\mathbf{B}} r |\mathbf{J}|)_{ij} \tag{44}$$

where m and n are the number the Gaussian quadrature points in ξ and η direction, W_i is corresponding weight of the Gaussian poi nts, and the items in the parenthesis are evaluated at the corresponding Gaussian points.

5. Graded Harmonic Elements of Piezoelectricity

If an axisymmetric piezoelectricity is subjected to non-symmetric loads, not only the radial and axial displacements will have to be considered but also a tangential component associated with the tangential angular direction θ. The abovementioned element is not available any more. However, the non-symmetric loads can be expanded in Fourier series as well as the displacements and electric potential and the original three-dimensional problem becomes a series of two-dimensional ones. The first application of this concept to the analysis of axisymmetric elasticity was made by Wilson [58] and here it is extended for graded piezoelectricity.

The non-symmetric traction $\hat{\mathbf{T}}$ is resolved into four components which are three traction components in radial (r), circumferential (θ) and axial (z) direction, and the surface charge density denoted by T_r, T_θ, T_z and T_e, respectively. Since the four components T_r, T_θ, T_z and T_e are also non-symmetric about the axis z, they can be expanded into Fourier series with respect to θ, namely

$$\begin{aligned}
T_r(r,\theta,z) &= T_{r0}(r,z) + \sum_{m=1}^{\infty} T_{rm}^c(r,z)\cos(m\theta) + \sum_{m=1}^{\infty} T_{rm}^s(r,z)\sin(m\theta) \\
T_\theta(r,\theta,z) &= T_{\theta 0}(r,z) + \sum_{m=1}^{\infty} T_{\theta m}^s(r,z)\sin(m\theta) + \sum_{m=1}^{\infty} T_{\theta m}^c(r,z)\cos(m\theta) \\
T_z(r,\theta,z) &= T_{z0}(r,z) + \sum_{m=1}^{\infty} T_{zm}^c(r,z)\cos(m\theta) + \sum_{m=1}^{\infty} T_{zm}^s(r,z)\sin(m\theta) \\
T_e(r,\theta,z) &= T_{e0}(r,z) + \sum_{m=1}^{\infty} T_{em}^c(r,z)\cos(m\theta) + \sum_{m=1}^{\infty} T_{em}^s(r,z)\sin(m\theta)
\end{aligned} \tag{45}$$

where the first items on the right hand side are independent on the coordinate θ, in which T_{r0}, T_{z0}, T_{e0} are axisymmetrical loads and $T_{\theta 0}$ is torsional load. The other items are dependent on the coordinate θ and therefore are non-symmetrical. The first series on the right hand side of Eq.(45) produce symmetric deformation and electric potential about the plane $\theta = 0$ while the second series on the right hand side of Eq.(45) correspond to antisymmetric deformation and potential. Then the displacement components u_r, u_θ, u_z and electric potential ϕ are also expanded into Fourier series as follows

$$\begin{aligned}
u_r(r,\theta,z) &= u_{r0}(r,z) + \sum_{m=1}^{\infty} u_{rm}^c(r,z)\cos(m\theta) + \sum_{m=1}^{\infty} u_{rm}^s(r,z)\sin(m\theta) \\
u_\theta(r,\theta,z) &= u_{\theta 0}(r,z) + \sum_{m=1}^{\infty} u_{\theta m}^s(r,z)\sin(m\theta) + \sum_{m=1}^{\infty} u_{\theta m}^c(r,z)\cos(m\theta) \\
u_z(r,\theta,z) &= u_{z0}(r,z) + \sum_{m=1}^{\infty} u_{zm}^c(r,z)\cos(m\theta) + \sum_{m=1}^{\infty} u_{zm}^s(r,z)\sin(m\theta) \\
\phi(r,\theta,z) &= \phi_0(r,z) + \sum_{m=1}^{\infty} \phi_m^c(r,z)\cos(m\theta) + \sum_{m=1}^{\infty} \phi_m^s(r,z)\sin(m\theta)
\end{aligned} \tag{46}$$

in which u_{r0}, u_{z0} and ϕ_0 are axisymmetrical components and $u_{\theta 0}$ is torsional components in circumferential direction. The other eight coefficients u_{rm}^c, u_{rm}^s, $u_{\theta m}^c$, $u_{\theta m}^s$, u_{zm}^c, u_{zm}^s, ϕ_m^c, ϕ_m^s are the function of r and z only.

By using of the orthogonality property of the trigonometric functions, the original three-dimensional problem is divided into a series of uncoupled two-dimensional problems in which the displacement and potential amplitude u_{rm}^c, u_{rm}^s, $u_{\theta m}^c$, $u_{\theta m}^s$, u_{zm}^c, u_{zm}^s, ϕ_m^c, ϕ_m^s are unknowns. For a typical wave number m, the symmetric loads are rewritten in matrix form

$$\hat{\mathbf{T}}_m = \begin{bmatrix} T_{rm}\cos(m\theta) & T_{\theta m}\sin(m\theta) & T_{zm}\cos(m\theta) & T_{em}\cos(m\theta) \end{bmatrix}^{\mathrm{T}} \tag{47}$$

The corresponding displacements are

$$\bar{\mathbf{u}}_m = \begin{bmatrix} u_{rm}\cos(m\theta) & u_{\theta m}\sin(m\theta) & u_{zm}\cos(m\theta) & \phi_m\cos(m\theta) \end{bmatrix}^{\mathrm{T}} \tag{48}$$

which can be interpolated by shape functions in an analogous method as axisymmetric element, namely

$$\bar{\mathbf{u}}_m = \bar{\mathbf{N}}^{(m)}\bar{\boldsymbol{\delta}}_m^e \tag{49}$$

where

$$\bar{\mathbf{N}}^{(m)} = \begin{bmatrix} \bar{\mathbf{N}}_1^{(m)} & \bar{\mathbf{N}}_2^{(m)} & \cdots & \bar{\mathbf{N}}_8^{(m)} \end{bmatrix} \tag{50}$$

$$\bar{\mathbf{N}}_i^{(m)} = \mathrm{diag}[N_i\cos(m\theta), N_i\sin(m\theta), N_i\cos(m\theta), N_i\cos(m\theta)] \tag{51}$$

The generalized strain of the element is

$$\bar{\boldsymbol{\varepsilon}}_m = \begin{bmatrix} \varepsilon_r & \varepsilon_\theta & \varepsilon_z & \gamma_{\theta z} & \gamma_{rz} & \gamma_{r\theta} & -E_r & -E_\theta & -E_z \end{bmatrix}^{\mathrm{T}} = \bar{\mathbf{B}}^{(m)} \bar{\boldsymbol{\delta}}_m^e \quad (52)$$

where

$$\bar{\mathbf{B}}^{(m)} = \begin{bmatrix} \bar{\mathbf{B}}_1^{(m)} & \bar{\mathbf{B}}_2^{(m)} & \cdots & \bar{\mathbf{B}}_8^{(m)} \end{bmatrix} \tag{53}$$

$$\bar{\mathbf{B}}_i^{(m)} = \begin{bmatrix} N_{i,r}\cos(m\theta) & 0 & 0 & 0 \\ N_i\cos(m\theta)/r & mN_i\cos(m\theta)/r & 0 & 0 \\ 0 & 0 & N_{i,z}\cos(m\theta) & 0 \\ 0 & N_{i,z}\sin(m\theta) & -mN_i\sin(m\theta)/r & 0 \\ N_{i,z}\cos(m\theta) & 0 & N_{i,r}\cos(m\theta) & 0 \\ -mN_i\sin(m\theta)/r & (N_{i,r}-N_i/r)\sin(m\theta) & 0 & 0 \\ 0 & 0 & 0 & N_{i,r}\cos(m\theta) \\ 0 & 0 & 0 & -mN_i\sin(m\theta)/r \\ 0 & 0 & 0 & N_{i,z}\cos(m\theta) \end{bmatrix}$$

$$(i = 1, 2, \cdots 8) \tag{54}$$

The stiffness matrix and mass matrix corresponding to wave number m are

$$\mathbf{K}_m^e = \int_0^{2\pi}\int_{-1}^{1}\int_{-1}^{1} [\bar{\mathbf{B}}^{(m)}]^{\mathrm{T}} \bar{\mathbf{C}} \bar{\mathbf{B}}^{(m)} r |\mathbf{J}| \, \mathrm{d}\xi \, \mathrm{d}\eta \, \mathrm{d}\theta \tag{55}$$

$$\mathbf{M}_m^e = \int_0^{2\pi}\int_{-1}^{1}\int_{-1}^{1} [\bar{\mathbf{N}}^{(m)}]^{\mathrm{T}} \boldsymbol{\rho} \bar{\mathbf{N}}^{(m)} r |\mathbf{J}| \, \mathrm{d}\xi \, \mathrm{d}\eta \, \mathrm{d}\theta \tag{56}$$

The equivalent node force of the tractions is

$$\mathbf{f}_m^e = \int_{\Gamma_\sigma^e} [\bar{\mathbf{N}}^{(m)}]^{\mathrm{T}} \hat{\mathbf{T}}_m \, \mathrm{d}\Gamma = \int_0^{2\pi}\int_L [\bar{\mathbf{N}}^{(m)}]^{\mathrm{T}} \hat{\mathbf{T}}_m r \, \mathrm{d}L \, \mathrm{d}\theta \tag{57}$$

In Eqs.(55)-(57), the integrands include trigonometric functions and the following identical equations will be utilized

$$\int_0^{2\pi} \sin^2(m\theta)\,\mathrm{d}\theta = \int_0^{2\pi} \cos^2(m\theta)\,\mathrm{d}\theta = \pi \tag{58}$$

to calculate the stiffness and mass matrix as well as the equivalent node force. The finial equation of symmetric force for wave number m is

$$\bar{\mathbf{M}}_m \ddot{\bar{\boldsymbol{\delta}}}_m + \bar{\mathbf{K}}_m \bar{\boldsymbol{\delta}}_m = \mathbf{f}_m \tag{59}$$

So far the original three-dimensional problem is converted to a series of uncoupled two-dimensional ones. The formulae for anti-symmetric forces can be readily obtained in a similar manner.

6. NUMERICAL EXAMPLES

Example 1. Graded Axisymmetric Element of Piezoelectricity

We consider a functionally graded piezoelectric plate of a radius 50*mm* and thickness 1*mm*. The material is composed of Ag and PTZ, whose properties are listed in Tab.1. The material properties of the FGM composed of PZT and Ag are determined by

$$c_{ij} = c_{ij}^{\mathrm{Ag}} V^{\mathrm{Ag}} + c_{ij}^{\mathrm{PZT}} (1 - V^{\mathrm{Ag}}),\ e_{ij} = e_{ij}^{\mathrm{PZT}} (1 - V^{\mathrm{Ag}}),\ \kappa_{ij} = \frac{1 + 2V^{\mathrm{Ag}}}{1 - V^{\mathrm{Ag}}} \kappa_{ij}^{\mathrm{PZT}} \tag{60}$$

in which V^{Ag} denotes the volume fraction of the phase Ag. At the top and bottom surfaces of the plate the volume fraction of the phase Ag is 15% and at the middle plane the volume of the phase Ag is 0%. The variation of the volume fraction of the phase Ag inside the plate is parabolic along the thickness of the plate. The top and bottom surfaces are subjected to a voltage 500V while the middle plane of the plate is connected to earth (0V). Figs. 3-5 shows the first three frequencies of the simply supported plate of the presented element as well as those by ANSYS. Figure 6 plots the maximum deflection of the plate. The mesh of the FEM model corresponding to the presented element is 2×50 while those of ANSYS (PLANE223) are 2×100, 4×100, 6×100, 8×100 and 10×100. It is clearly shown that the results of the ANSYS approach to the results by the presented element with the increase of the number of the elements in the thickness of the plate.

Table 1. The material properties of PZT-5 and Ag

PZT-5				Ag
$c_{11} = 120GPa$	$c_{33} = 111GPa$	$e_{15} = 12.3C/m^2$	$\varepsilon_{11} = 919.1\varepsilon_0$	$E = 72.4GPa$
$c_{12} = 75.2GPa$	$c_{44} = 21.1GPa$	$e_{31} = -5.35C/m^2$	$\varepsilon_{33} = 826.6\varepsilon_0$	$\mu = 0.37$
$c_{13} = 75.1GPa$	$\rho = 7500kg/m^3$	$e_{33} = 15.8C/m^2$	$\varepsilon_0 = 8.854\times10^{-12}\ F/m$	$\rho = 10500kg/m^3$

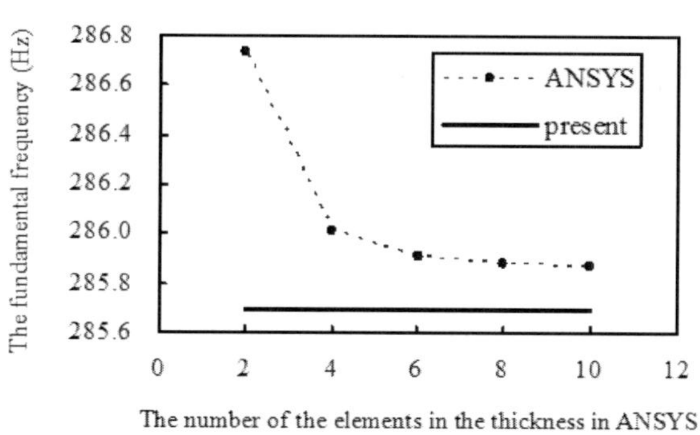

Figure 3. The fundamental frequency of the simply supported circular plate.

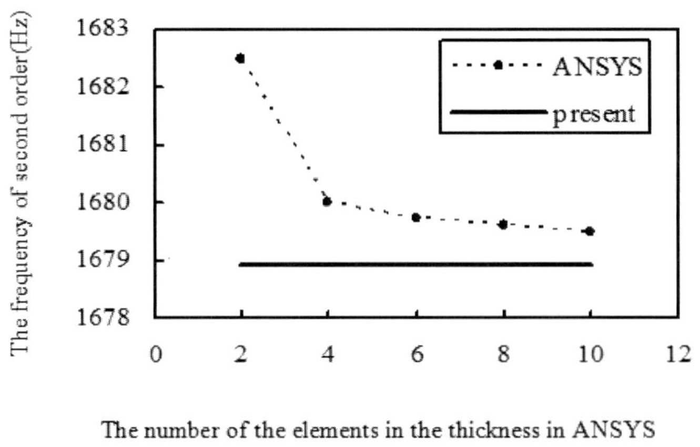

Figure 4. The second order frequency of the simply supported circular plate

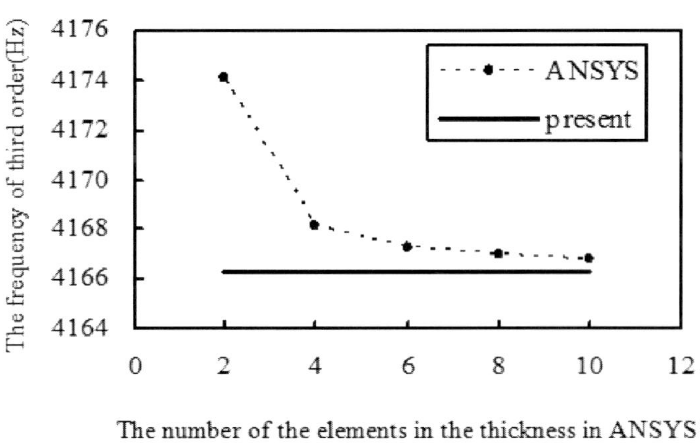

Figure 5. The third order frequency of the simply supported circular plate.

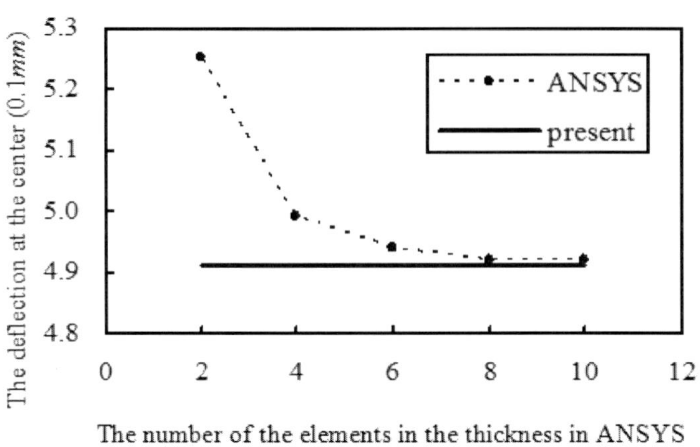

Figure 6. The deflection at the center of the simply supported circular plate.

Example 2. Graded Harmonic Element of Piezoelectricity

We first analyze the free vibration of the above-mentioned circular plate in Example 1 by the graded harmonic element. Figure 7 shows the first four mode shapes and corresponding frequencies of non-symmetric mode. The layout of the mesh of the present harmonic element is 2×50. The frequencies obtained by ANSYS using three dimensional element (SOLID226) are also presented in the parenthesis and the good agreement is observed.

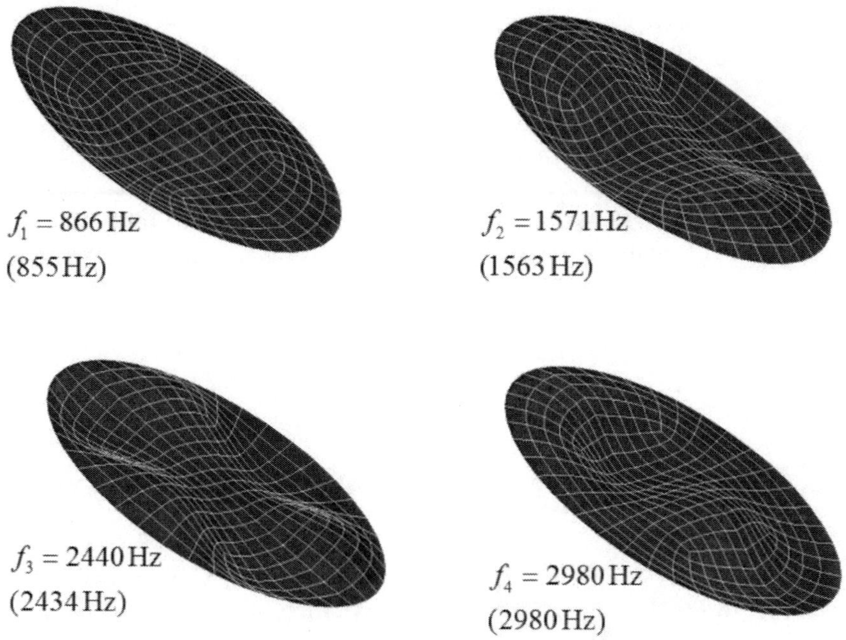

Figure 7. The first four mode shapes and frequencies of non-symmetric modes (the values in parenthesis are obtained by ANSYS).

CONCLUSION

This work proposed the formulae for some graded isoparametric elements based on the variational principle of piezoelectricity. They include three-dimensional element, axisymmetric element and harmonic element. Some numerical results are also presented to demonstrate these elements. The numerical examples show that the graded elements and harmonic element can give the accuracy results using relative coarse mesh compared to the model using homogeneous elements or three- dimensional elements.

ACKNOWLEDGMENTS

This work was supported by National Natural Science Foundation of China (Nos. 10872180 and 10725210).

References

Williamson, R. L., Rabin, B. H., Drake, J. T. Finite-element analysis of thermal residual-stresses at graded ceramic-metal interfaces. Part 1. Model description and geometrical effects. *Journal of Applied Physics* 1993, *Vol* 74(2), 1310-1320.

Drake, J. T., Williamson, R. L., Rabin, B. H. Finite-element analysis of thermal residual-stresses at graded ceramic-metal interfaces. Part 2. Interface optimization for residual-stress reduction. *Journal of Applied Physics* 1993, *Vol* 74(2), 1321-1326.

Miller, D. P., Lannutti, J. J., Noebe, R. D. Fabrication and properties of functionally graded $NiAl/Al_2O_3$ composites. *Journal of Material Research* 1993, *Vol* 8(8), 2004-2013.

Torres, R. D., Mustoe, G. G. W., Reimanis, L. E., Moore, J. J. Evaluation of residual stresses developed in a functionally graded material using the finite element technique. *Symposium on Processing and Fabrication of Advanced Materials IV, Cleveland, Ohio*, 1996, 431-438.

Na, K. S., Kim, J. H. Thermal postbuckling investigations of functionally graded plates using 3-D finite element method. *Finite Elements in Analysis and Design*, 2006, *Vol* 42 (8-9), 749-756.

Cooley, W. G., Palazotto, A. Finite element analysis of functionally graded shell panels under thermal loading. *Proceedings of the ASME Aerospace Division* 2005, 517-526.

Kordkheili, S. A. H., Naghdabadi, R. Geometrically non-linear thermoelastic analysis of functionally graded shells using finite element method. *International Journal for Numerical Methods in Engineering* 2007, *Vol* 72(8), 964-986.

Zhao J. M., Liu L. H. Solution of radiative heat transfer in graded index media by least square spectral element method. *International Journal of Heat and Mass Transfer* 2007, *Vol* 50(13-14), 2634-2642.

Akpobi, J. A., Edobor, C. O. Model development of the application of finite element-eigenvalue method to the determination of transient temperature field in functionally graded materials. *International Conference on Engineering Research and Development, Benin City, Nigeria*, 2007, *Vol* 18-19: 253-261.

Lee, W., Yang, J., Choi, K., Park Y., Park B., Park I., Park Y. A study on the thermal fatigue properties of $Mg/Mg\text{-}Al_{18}B_4O_{33}$ functionally graded material by thermal cycling test and finite element method. *Journal of the Korean Institute of Metals and Materials* 2008, *Vol* 46(8), 538-544.

Santos, H., Soares, C. M. M., Soares, C. A. M., Reddy, J. N. A semi-analytical finite element model for the analysis of cylindrical shells made of functionally graded materials under thermal shock. *Composite Structures* 2008, *Vol* 86(1-3), 10-21.

Haque, S., Choy, K. L. Finite element modelling of the effect of a functionally graded protective coating for SiC monofilaments on Ti-based composite behaviour. *Materials Science and Engineering A-Structural Materials Properties Microstructure and Processing*, 2000, *Vol* 291(1-2): 97-109.

Stephens, L. S., Liu, Y., Meletis, E. I. Finite element analysis of the initial yielding behavior of a hard coating/substrate system with functionally graded interface under indentation and friction. *Journal of Tribology* 2000, *Vol* 122(2), 381-387.

Limarga, A. M., Widjaja, S., Yip, T. H. Finite element analysis of residual stress in a plasma-sprayed functionally graded thermal barrier coating. *International Thermal Spray Conference (ITSC 2001)*, Singapore, 2001, 943-950.

Hou, L. K., Pan, W., Wang, R.G., Xie D. Finite element analysis of thermal residual stress in ceramic functional graded material: Interface optimization for residual reduction. *Materials Science Forum, The 7th International Symposium on Functionally Grade Materials (FGM2002)*, 2003, *Vol* 423-4, 677-680.

Zhang, X. C., Xu, B. S., Wang, H. D., Wu Y. X. Modeling of the residual stresses in plasma-spraying functionally graded ZrO2/NiCoCrAlY coatings using finite element method. *Materials and Design* 2006, *Vol* 27(4), 308-315.

Zhong, D., Mustoe, G. G. W., Moore, J. J., Thiel, S., Disam, J. Finite element analysis in design of a graded coating system for glass forming dies and tools. *Ceramic Transactions* 2001, *Vol* 114, 545-552.

Chen, F. Y., He, W. Q. Finite element design of MgO/Ni system functionally graded materials. *Journal of materials Processing Technology* 2007, *Vol* 182(1-3), 181-184.

Zhou, L. P., Tang, D. W. A functionally graded structural design of mirrors for reducing their thermal deformations in high-power laser systems by finite element method. *Optics and Laser Technology* 2007, *Vol* 39(5), 980-986.

Lee, J. C., Park, J. H., Ryu, S. H., Hong, H. J., Riu, D. H., Ahn, S. H., Lee C. S. Reduction of functionally graded material layers for Si_3N_4-Al_2O_3 system using three-dimensional finite element modeling. *Materials Transactions* 2008, *Vol* 49(4), 829-834.

Zhang, G. B., Guo, Q. G., Wang, K. J., Zhang, H., Song, Y., Shi, J. L., Liu, L. Finite element design of SiC/C functionally graded materials for ablation resistance application. *Materials Science and Engineering A-Structural Materials Properties Microstructure and Processing* 2008, *Vol* 488(1-2), 45-49.

Pindera, M. J., Aboudi, J., Arnold, S. M. Limitations of the uncoupled, RVE-based micromechanical approach in the analysis of functionally graded composites. *Mechanics of Materials* 1995, *Vol* 20, 77-94.

Pindera, M. J., Dunn, P. Evaluation of the higher-order theory for functionally graded materials via the finite-element method. *Composites Part B - Engineering* 1997, *Vol* 28(1-2), 109-119.

Grujicic, M., Zhang, Y. Determination of effective elastic properties of functionally graded materials using Voronoi cell finite element method. *Materials Science and Engineering A-Structural Materials Properties Microstructure and Processing* 1998, *Vol* 251(1-2), 64-76.

Cho, J. R., Ha, D. Y. Averaging and finite-element discretization approaches in the numerical analysis of functionally graded materials. *Materials Science and Engineering A-Structural Materials Properties Microstructure and Processing* 2001, *Vol* 302(2), 187-196.

Ajdari, A, Canavan, P, Nayeb-Hashemi, H, Warner G. Mechanical properties of functionally graded 2-D cellular structures: A finite element simulation. *Materials Science and Engineering A-Structural Materials Properties Microstructure and Processing* 2009, *Vol* 499(1-2), 434-439.

Gu, P., Dao, M., Asaro, R. J. A simplified method for calculating the crack tip field of functionally graded materials using the domain integral, *Journal of Applied Mechanics* 1999, *Vol* 66, 101–108.

Santare, M. H., Lambros1, J. Use of Graded Finite Elements to Model the Behavior of Nonhomogeneous Materials. *Journal of Applied Mechanics* 2000, *Vol* 67, 819-822.

Kim, J. H., Paulino, G. H. Isoparametric graded finite elements for nonhomogeneous isotropic and orthotropic materials. *Journal of Applied Mechanics* 2002, *Vol* 69, 502-514.

Kim, J. H., Paulino, G. H. Finite element evaluation of mixed mode stress intensity factors in functionally graded materials. *International Journal for Numerical Methods in Engineering* 2002, *Vol* 53(8), 1903-1935.

Buttlar, W. G., Paulino G. H., Song S. H. Application of graded finite elements for asphalt pavements. *Journal of Engineering Mechanics* 2006, *Vol* 132(3), 240-249.

Chakraborty, A., Gopalakrishnan, S., Reddy, J. N. A new beam finite element for the analysis of functionally graded materials. *International Journal of Mechanical Science* 2003, *Vol* 45(3), 519-539.

Chakraborty, A., Gopalakrishnan, S. A spectrally formulated finite element for wave propagation analysis in functionally graded beams. *International Journal of Solids and Structures* 2003, *Vol* 40(10), 2421-2448.

Wu, Z., Chen, W. J. A higher-order theory and refined three-node triangular element for functionally graded plates. *European Journal of Mechanics A-Solids* 2006, *Vol* 25(3), 447-463.

Das, M., Barut, A., Madenci, E., Ambur, D. R. A triangular plate element for thermo-elastic analysis of sandwich panels with a functionally graded core. *International Journal for Numerical Methods in Engineering* 2006, *Vol* 68(9), 940-966.

Eason, T. G., Caraballo, S. A. A comparison of finite element beam formulations for thermoelastic behavior of functionally graded structures. *Proceedings of the ASME Aerospace Division, ASME International Mechanical Engineering Congress and Exposition*, Orlando, 2005, 443-449.

Naei, M. H., Masoumi, A., Shamekhi, A. Buckling analysis of circular functionally graded material plate having variable thickness under uniform compression by finite-element method. *Proceedings of the Institution of Mechanical Engineers Part C - Journal of Mechanical Engineering Science* 2007, *Vol* 221(11), 1241-1247.

Kim, K. D., Lomboy, G. R., Han, S. C. Geometrically non-linear analysis of functionally graded material (FGM) plates and shells using a four-node quasi-conforming shell element. *Journal of Composite Materials* 2008, *Vol* 42(5), 485-511.

Oyekoya, O. O., Mba, D. U., El-Zafrany, A. M. Structural Integrity of Functionally Graded Composite Structure using Mindlin-type Element. *Computer Modeling in Engineering and Science* 2008, *Vol* 34(1), 55-85.

Pradyumna, S., Bandyopadhyay, J. N. Free vibration analysis of functionally graded curved panels using a higher-order finite element formulation. *Journal of Sound and Vibration* 2008, *Vol* 318(1-2), 176-192.

Ramkumar, K., Ganesan, N. Finite-element buckling and vibration analysis of functionally graded box columns in thermal environments. *Proceedings of the Institution of Mechanical Engineers Part L - Journal of Materials Design and Applications* 2008, *Vol* 222(L1), 53-64.

Shaker, A., Abdelrahman, W., Tawfik, M., Sadek, E. Stochastic Finite element analysis of the free vibration of functionally graded material plates. *Computational Mechanics* 2008, *Vol* 41(5), 707-714.

Al-Ajmi, M. A., Alhazza, K. A., Majeed, M. A. Graded finite element modeling of constrained layer damping treatments with functionally graded viscoelastic material. *Journal of Intelligent Material Systems and Structures* 2008, *Vol* 19(4), 469-474.

Jin, Z. H., Paulino, G. H., Dodds, R. H. Finite element investigation of quasi-static crack growth in functionally graded materials using a novel cohesive zone fracture model. *Journal of Applied Mechanics* 2002, *Vol* 69(3), 370-379.

Comi, C., Mariani, S. Extended finite element simulation of quasi-brittle fracture in functionally graded materials. *Computer Methods in Applied Mechanics and Engineering* 2007, *Vol* 196(41-44), 4013-4026.

Kirugulige, M., Tippur, H. V. Mixed-mode dynamic crack growth in a functionally graded particulate composite: Experimental measurements and finite element simulations. *Journal of Applied Mechanics* 2008, *Vol* 75(5), 051102.

He, X.Q., Ng, T. Y., Sivashanker, S., Liew, K. M., Active control of FGM plates with integrated piezoelectric sensors and actuators. *International Journal of Solids and Structures* 2001, *Vol* 38(9), 1641-1655.

He, X. Q., Liew, K. M., Ng, T. Y., Sivashanker, S. A FEM model for the active control ofcurved FGM shells using piezoelectric sensor/actuator layers. *International Journal for Numerical Methods in Engineering* 2002, *Vol* 54(6), 853-870.

Ng, T. Y., He, X. Q., Liew, K. M. Finite element modeling of active control of functionally graded shells in frequency domain via piezoelectric sensors and actuators. *Computational Mechanics* 2002, *Vol* 28 (1), 1-9.

Ray, M. C., Sachade, H. M. Finite element analysis of smart functionally graded plates. *International Journal of Solids and Structures* 2006, *Vol* 43(18-19), 5468-5484.

Bhangale, R. K., Ganesan, N. Free vibration of simply supported functionally graded and layered magneto-electro-elastic plates by finite element method. *Journal of Sound and Vibration* 2006, *Vol* 294(4-5), 1016-1038.

Panda, S., Ray, M. C. Nonlinear finite element analysis of functionally graded plates integrated with patches of piezoelectric fiber reinforced composite. *Finite Elements in Analysis and Design* 2008, *Vol* 44(8), 493-504.

Wang, F., Lee, H. P., Lu, C. Thermal-mechanical study of functionally graded dental implants with the finite element method. *Journal of Biomedical Materials Research Part A* 2007, *Vol* 80A(1), 146-158.

Shin, K. H. Adaptive mesh generation for Finite Element Analysis of Functionally Graded Materials. Proceedings of the ASME Computers and Information in Engineering Division, ASME International Mechanical Engineering Congress and Exposition, Orlando, 2005, 199-206.

Kou, X. Y., Tan, S. T. A systematic approach for integrated computer-aided design and finite element analysis of functionally-graded-material objects. *Materials and Design* 2007, *Vol* 28(10), 2549-2565.

Haghighi, M. R. G., Eghtesad, M., Malekzadeh, P. A coupled differential quadrature and finite element method for 3-D transient heat transfer analysis of functionally graded thick plates. *Numerical Heat Transfer Part B-Fundamentals* 2008, *Vol* 53(4), 358-373.

Tiersten H. F. *Linear Piezoelectric Plate Vibrations*, 1969, New York: Plenum.

Wilson E. L. Structural analysis of axisymmetric solids. *Journal of AIAA* 1965, *Vol* 3(12), 2269-2274.

About the Author

Rongqiao Xu received the Ph.D. degree from Zhejiang University, China, 1999. He is an Associate Professor of the Civil Engineering Department of Zhejiang University. He was a Teaching Staff of Nagaoka University of Technology in Japan and Research Fellow of City University of Hong Kong. His research interests include solid mechanics, smart structures, composite structures, structural health monitoring and numerical methods in structural analysis.

In: Mechanics of Functionally Graded Materials and Structures ISBN 978-1-61668-920-9
Editors: Z. Zhong, L. Wu and W. Chen

Chapter 7

TRANSIENT RESPONSES OF MULTI-FIELD COUPLED FUNCTIONALLY GRADED CYLINDRICAL AND SPHERICAL SHELLS

***Huiming Wang*[1*] *and Haojiang Ding*[2]**
[1]Department of Mechanics, Zhejiang University
Hangzhou, P.R. China
[2]Department of Civil Engineering, Zhejiang University
Hangzhou, P.R. China

ABSTRACT

Transient response is an important physical quantity in modern engineering analysis. In this chapter, the transient responses of multi-field coupled functionally graded cylindrical and spherical shells under radial deformation are investigated. The cylindrical and spherical shells are assumed to be graded along the radial direction as a power function. For functionally graded magneto-electro-elastic (MEE) cylindrical and spherical shells, the governing equations are finally transformed into two Volterra integral equations of the second kind with respect to two functions of time. The recursive formulae are constructed and the integral equations are solved effectively. The present methodology can also be employed to solve the transient responses of functionally graded cylindrical and spherical shells made of piezoelectric and purely elastic materials. For functionally graded piezoelectric cylindrical and spherical shells, the governing equations are transformed into a Volterra integral equation of the second kind with respect to a function of time. For functionally graded purely elastic cylindrical and spherical shells, no Volterra integral equation is involved and the obtained solution is a closed-form solution.

---

[1] e-mail: wanghuiming@zju.edu.cn.

1. INTRODUCTION

Elastodynamics is an important branch of solid mechanics. When the structure is subjected to the dynamic loads, all the physical quantities in the structure vary as the functions of time and are well known as transient responses. The investigation and application of the transient response can be widely found in various regions, such as dynamic analysis, structural health monitoring and vibration control etc. Many structure forms can be found in modern engineering structures, systems and devices etc. Among them, cylindrical and spherical shells are common forms. Subsequently, the topic of transient responses of cylindrical and spherical shells has attracted increasing attention and great achievements have been made.

The investigation for transient responses of homogeneous cylindrical and spherical shells has been carried out for more than half a century. For purely elastic materials, some classic works can be cited [1-8]. It is well known that the analysis for transient responses is time consuming. Lately, with the development of computer technique, a number of investigations involve the analysis of transient responses for cylindrical and spherical shells. The works on the transient responses of purely elastic materials under the mechanical and thermal excitations have been carried out by many researchers [9-23]. At the beginning of this century, the study for transient responses was penetrated into analyzing the transient responses of homogeneous cylindrical and spherical shells made of intelligent materials, for instance, piezoelectric, pyroelectric and magnetoelectroelastic materials. Now, the analysis of transient response for intelligent media has been an increasing interested research field and quite a number of literatures have been reported [24-33].

Functionally graded materials (FGMs) are known as a new class of materials, of which the material properties vary continuously in one or more spatial directions. Due to the distinct advantages in mechanical properties and functional characteristics, FGMs have received worldwide attention and a lot of theoretical and basic investigations have been carried out for cylindrical and spherical shells [34-49].

In this chapter, we focus our works on the analytical and semi-analytical solutions of the transient responses of multi-field coupled functionally graded cylindrical and spherical shells under the radial deformation. The solution is based on the three-dimensional elastic theory and no additional assumption is employed. In the analysis, the material properties are assumed to vary as a power law function of radial position.

2. BASIC EQUATIONS

The basic equations of the magneto-electro-elastic (MEE) media for thermoelastic dymamic problem can be expressed as [50,51]

$$\begin{aligned} &\sigma_{ij,j} = \rho u_{i,tt}, \\ &D_{i,i} = 0, \\ &B_{i,i} = 0. \end{aligned} \tag{1a-c}$$

$$\sigma_{ij} = c_{ijkl}\varepsilon_{kl} - e_{kij}E_k - q_{kij}H_k - \lambda_{ij}\Theta,$$
$$D_i = e_{ikl}\varepsilon_{kl} + \kappa_{ik}E_k + g_{ik}H_k + p_i\Theta, \qquad (2a\text{-}c)$$
$$B_i = q_{ikl}\varepsilon_{kl} + g_{ik}E_k + \mu_{ik}H_k + m_i\Theta,$$

$$\varepsilon_{ij} = \tfrac{1}{2}\left(u_{i,j} + u_{j,i}\right),$$
$$E_i = -\Phi_{,i}, \qquad (3a\text{-}c)$$
$$H_i = -\Psi_{,i},$$

where a comma in the subscript denotes the partial differential with respect to the following variable. u_i, Φ, Ψ and Θ denote the displacement, electric potential, magnetic potential and temperature difference, respectively. σ_{ij}, D_i and B_i are the components of stress, electric displacement and magnetic induction, respectively. c_{ijkl}, e_{kij}, q_{kij}, κ_{ij}, g_{ij} and μ_{ij} are elastic, piezoelectric, piezomagnetic, dielectric, electromagnetic and magnetic constants, respectively. λ_{ij}, p_i and m_i are thermal-stress, pyroelectric and pyromagnetic coefficients, respectively.

If we set all the coefficients related to magnetic field equal to zero and omit Eqs. (1c), (2c) and (3c), the governing equations for pyroelectric media under thermal environment are then obtained. If we further set all the coefficients related to electric field equal to zero and omit Eqs. (1b), (2b) and (3b), the governing equations of thermoelastic problem for purely elastic media are then derived. In the above mentioned cases, if we set the temperature difference $\Theta = 0$, the governing equations for the multi-field coupled elastodynamic problem are then constructed.

In this investigation, the dynamic problem for multi-field coupled cylindrical and spherical shells made of functionally graded materials will be considered. The inhomogeneity of the material is assumed to vary as a power function of radial position. In the following, we describe the problem for cylindrical shell in cylindrical coordinates (r,θ,z) and that for spherical shell in spherical coordinates (r,θ,φ). Then the material properties can be expressed as

$$c_{ijkl}(r) = \bar{c}_{ijkl}\left(r/b\right)^{2\beta}, \quad e_{kij}(r) = \bar{e}_{kij}\left(r/b\right)^{2\beta}, \quad q_{kij}(r) = \bar{q}_{kij}\left(r/b\right)^{2\beta},$$
$$\lambda_{ij}(r) = \bar{\lambda}_{ij}\left(r/b\right)^{2\beta}, \quad \kappa_{ik}(r) = \bar{\kappa}_{ik}\left(r/b\right)^{2\beta}, \quad g_{ik}(r) = \bar{g}_{ik}\left(r/b\right)^{2\beta},$$
$$\mu_{ik}(r) = \bar{\mu}_{ik}\left(r/b\right)^{2\beta}, \quad p_i(r) = \bar{p}_i\left(r/b\right)^{2\beta}, \quad m_i(r) = \bar{m}_i\left(r/b\right)^{2\beta}, \qquad (4)$$
$$\rho(r) = \bar{\rho}\left(r/b\right)^{2\beta}.$$

where all the quantities with a "–" are known constants. b and β are real numbers. For uncoupled thermal stress theory, thermal effect can be treated with the way similar as the mechanical load. So, in the following consideration, for the sake of simplicity, the temperature difference $\Theta = 0$ is adopted.

3. SOLUTION FOR MEE CYLINDRICAL SHELLS

In cylindrical coordinates (r, θ, z), considering the material properties in Eq.(4), the governing equations of radially polarized functionally graded MEE media (crystal class 2mm) can be expressed as [47]

$$\begin{aligned}
&\frac{\partial \sigma_{rr}}{\partial r} + \frac{\partial \sigma_{r\theta}}{r\partial \theta} + \frac{\partial \sigma_{zr}}{\partial z} + \frac{\sigma_{rr} - \sigma_{\theta\theta}}{r} = \bar{\rho}\left(\frac{r}{b}\right)^{2\beta} \frac{\partial^2 u_r}{\partial t^2}, \\
&\frac{\partial \sigma_{r\theta}}{\partial r} + \frac{\partial \sigma_{\theta\theta}}{r\partial \theta} + \frac{\partial \sigma_{\theta z}}{\partial z} + \frac{2\sigma_{r\theta}}{r} = \bar{\rho}\left(\frac{r}{b}\right)^{2\beta} \frac{\partial^2 u_\theta}{\partial t^2}, \\
&\frac{\partial \sigma_{zr}}{\partial r} + \frac{\partial \sigma_{\theta z}}{r\partial \theta} + \frac{\partial \sigma_{zz}}{\partial z} + \frac{\sigma_{zr}}{r} = \bar{\rho}\left(\frac{r}{b}\right)^{2\beta} \frac{\partial^2 u_z}{\partial t^2},
\end{aligned} \tag{5a}$$

$$\frac{1}{r}\frac{\partial}{\partial r}(rD_r) + \frac{1}{r}\frac{\partial D_\theta}{\partial \theta} + \frac{\partial D_z}{\partial z} = 0, \tag{5b}$$

$$\frac{1}{r}\frac{\partial}{\partial r}(rB_r) + \frac{1}{r}\frac{\partial B_\theta}{\partial \theta} + \frac{\partial B_z}{\partial z} = 0. \tag{5c}$$

$$\begin{aligned}
&\sigma_{\theta\theta} = (r/b)^{2\beta}\left(\bar{c}_{11}\varepsilon_{\theta\theta} + \bar{c}_{12}\varepsilon_{zz} + \bar{c}_{13}\varepsilon_{rr} - \bar{e}_{31}E_r - \bar{q}_{31}H_r\right), \\
&\sigma_{zz} = (r/b)^{2\beta}\left(\bar{c}_{12}\varepsilon_{\theta\theta} + \bar{c}_{22}\varepsilon_{zz} + \bar{c}_{23}\varepsilon_{rr} - \bar{e}_{32}E_r - \bar{q}_{32}H_r\right), \\
&\sigma_{rr} = (r/b)^{2\beta}\left(\bar{c}_{13}\varepsilon_{\theta\theta} + \bar{c}_{23}\varepsilon_{zz} + \bar{c}_{33}\varepsilon_{rr} - \bar{e}_{33}E_r - \bar{q}_{33}H_r\right), \\
&\sigma_{zr} = (r/b)^{2\beta}\left(2\bar{c}_{44}\varepsilon_{zr} - \bar{e}_{24}E_z - \bar{q}_{24}H_z\right), \\
&\sigma_{r\theta} = (r/b)^{2\beta}\left(2\bar{c}_{55}\varepsilon_{r\theta} - \bar{e}_{15}E_\theta - \bar{q}_{15}H_\theta\right), \\
&\sigma_{\theta z} = (r/b)^{2\beta}\left(2\bar{c}_{66}\varepsilon_{\theta z}\right),
\end{aligned} \tag{6a}$$

$$D_\theta = (r/b)^{2\beta}\left(2\bar{e}_{15}\varepsilon_{r\theta} + \bar{\kappa}_{11}E_\theta + \bar{g}_{11}H_\theta\right),$$
$$D_z = (r/b)^{2\beta}\left(2\bar{e}_{24}\varepsilon_{zr} + \bar{\kappa}_{22}E_z + \bar{g}_{22}H_z\right), \tag{6b}$$
$$D_r = (r/b)^{2\beta}\left(\bar{e}_{31}\varepsilon_{\theta\theta} + \bar{e}_{32}\varepsilon_{zz} + \bar{e}_{33}\varepsilon_{rr} + \bar{\kappa}_{33}E_r + \bar{g}_{33}H_r\right),$$

$$B_\theta = (r/b)^{2\beta}\left(2\bar{q}_{15}\varepsilon_{r\theta} + \bar{g}_{11}E_\theta + \bar{\mu}_{11}H_\theta\right),$$
$$B_z = (r/b)^{2\beta}\left(2\bar{q}_{24}\varepsilon_{zr} + \bar{g}_{22}E_z + \bar{\mu}_{22}H_z\right), \tag{6c}$$
$$B_r = (r/b)^{2\beta}\left(\bar{q}_{31}\varepsilon_{\theta\theta} + \bar{q}_{32}\varepsilon_{zz} + \bar{q}_{33}\varepsilon_{rr} + \bar{g}_{33}E_r + \bar{\mu}_{33}H_r\right),$$

$$\varepsilon_{rr} = \frac{\partial u_r}{\partial r}, \quad \varepsilon_{\theta\theta} = \frac{1}{r}\frac{\partial u_\theta}{\partial \theta} + \frac{u_r}{r}, \quad \varepsilon_{zz} = \frac{\partial u_z}{\partial z}, \quad \varepsilon_{zr} = \frac{1}{2}\left(\frac{\partial u_z}{\partial r} + \frac{\partial u_r}{\partial z}\right),$$
$$\varepsilon_{r\theta} = \frac{1}{2}\left(\frac{1}{r}\frac{\partial u_r}{\partial \theta} + \frac{\partial u_\theta}{\partial r} - \frac{u_\theta}{r}\right), \quad \varepsilon_{\theta z} = \frac{1}{2}\left(\frac{1}{r}\frac{\partial u_z}{\partial \theta} + \frac{\partial u_\theta}{\partial z}\right), \tag{7a}$$

$$E_r = -\frac{\partial \Phi}{\partial r}, \quad E_\theta = -\frac{1}{r}\frac{\partial \Phi}{\partial \theta}, \quad E_z = -\frac{\partial \Phi}{\partial z}, \tag{7b}$$

$$H_r = -\frac{\partial \Psi}{\partial r}, \quad H_\theta = -\frac{1}{r}\frac{\partial \Psi}{\partial \theta}, \quad H_z = -\frac{\partial \Psi}{\partial z}, \tag{7c}$$

For radial vibration (axisymmetric plane strain elastodynamic problem), we have

$$u_r = u_r(r,t), \quad u_\theta = u_z = 0, \quad \Phi = \Phi(r,t), \quad \Psi = \Psi(r,t). \tag{8}$$

Substituting Eq. (8) into Eqs. (7a)-(7c) gives

$$\varepsilon_{rr} = \frac{\partial u_r}{\partial r}, \quad \varepsilon_{\theta\theta} = \frac{u_r}{r}, \quad \varepsilon_{zz} = \varepsilon_{zr} = \varepsilon_{r\theta} = \varepsilon_{\theta z} = 0, \tag{9a}$$

$$E_r = -\frac{\partial \Phi}{\partial r}, \quad E_\theta = E_z = 0, \tag{9b}$$

$$H_r = -\frac{\partial \Psi}{\partial r}, \quad H_\theta = H_z = 0, \tag{9c}$$

Correspondingly, Eqs. (6a)-(6c) can be rewritten as

$$\sigma_{\theta\theta} = \left(\frac{r}{b}\right)^{2\beta} \left(\bar{c}_{11}\frac{u_r}{r} + \bar{c}_{13}\frac{\partial u_r}{\partial r} + \bar{e}_{31}\frac{\partial \Phi}{\partial r} + \bar{q}_{31}\frac{\partial \Psi}{\partial r}\right),$$

$$\sigma_{zz} = \left(\frac{r}{b}\right)^{2\beta} \left(\bar{c}_{12}\frac{u_r}{r} + \bar{c}_{23}\frac{\partial u_r}{\partial r} + \bar{e}_{32}\frac{\partial \Phi}{\partial r} + \bar{q}_{32}\frac{\partial \Psi}{\partial r}\right), \tag{10a}$$

$$\sigma_{rr} = \left(\frac{r}{b}\right)^{2\beta} \left(\bar{c}_{13}\frac{u_r}{r} + \bar{c}_{33}\frac{\partial u_r}{\partial r} + \bar{e}_{33}\frac{\partial \Phi}{\partial r} + \bar{q}_{33}\frac{\partial \Psi}{\partial r}\right),$$

$$\sigma_{zr} = \sigma_{r\theta} = \sigma_{\theta z} = 0,$$

$$D_r = \left(\frac{r}{b}\right)^{2\beta} \left(\bar{e}_{31}\frac{u_r}{r} + \bar{e}_{33}\frac{\partial u_r}{\partial r} - \bar{\kappa}_{33}\frac{\partial \Phi}{\partial r} - \bar{g}_{33}\frac{\partial \Psi}{\partial r}\right), \quad D_\theta = D_z = 0, \tag{10b}$$

$$B_r = \left(\frac{r}{b}\right)^{2\beta} \left(\bar{q}_{31}\frac{u_r}{r} + \bar{q}_{33}\frac{\partial u_r}{\partial r} - \bar{g}_{33}\frac{\partial \Phi}{\partial r} - \bar{\mu}_{33}\frac{\partial \Psi}{\partial r}\right), \quad B_\theta = B_z = 0, \tag{10c}$$

With the aid of Eqs. (10a)-(10c), Eqs. (5a)-(5c) can be simplified as

$$\frac{\partial \sigma_{rr}}{\partial r} + \frac{\sigma_{rr} - \sigma_{\theta\theta}}{r} = \bar{\rho}\left(\frac{r}{b}\right)^{2\beta} \frac{\partial^2 u_r}{\partial t^2}, \tag{11a}$$

$$\frac{1}{r}\frac{\partial}{\partial r}(rD_r) = 0, \tag{11b}$$

$$\frac{1}{r}\frac{\partial}{\partial r}(rB_r) = 0. \tag{11c}$$

For the sake of convenience, the following non-dimensional quantities and variables are introduced as

$$\sigma_i = \frac{\sigma_{ii}}{\bar{c}_{33}}, \quad D = \frac{D_r}{e_0}, \quad B = \frac{B_r}{q_0}, \quad u = \frac{u_r}{b}, \quad \phi = \frac{\Phi}{\Phi_0}, \quad \psi = \frac{\Psi}{\Psi_0},$$

$$\xi = \frac{r}{b}, \quad \tau = \frac{c_v}{b}t, \quad \tilde{c}_{ij} = \frac{\bar{c}_{ij}}{\bar{c}_{33}}, \quad \tilde{e}_i = \frac{\bar{e}_{3i}}{e_0}, \quad \tilde{q}_i = \frac{\bar{q}_{3i}}{q_0}, \quad \tilde{g}_3 = \frac{\bar{g}_{33}}{g_0}, \tag{12}$$

$$s = a/b, \quad e_0 = \sqrt{\bar{c}_{33}\bar{\kappa}_{33}}, \quad q_0 = \sqrt{\bar{c}_{33}\bar{\mu}_{33}}, \quad g_0 = \sqrt{\bar{\kappa}_{33}\bar{\mu}_{33}},$$

$$\Phi_0 = b\sqrt{\bar{c}_{33}/\bar{\kappa}_{33}}, \quad \Psi_0 = b\sqrt{\bar{c}_{33}/\bar{\mu}_{33}}, \quad c_v = \sqrt{\bar{c}_{33}/\bar{\rho}},$$

In Eq. (12) and there after, *a* denotes the inner radius and *b* denotes the outer radius of the cylindrical or spherical shells, respectively. The non-dimensional governing equations are then derived as [45]

$$\sigma_\theta = \xi^{2\beta}\left(\tilde{c}_{11}\frac{u}{\xi}+\tilde{c}_{13}\frac{\partial u}{\partial \xi}+\tilde{e}_1\frac{\partial \phi}{\partial \xi}+\tilde{q}_1\frac{\partial \psi}{\partial \xi}\right),$$

$$\sigma_z = \xi^{2\beta}\left(\tilde{c}_{12}\frac{u}{\xi}+\tilde{c}_{23}\frac{\partial u}{\partial \xi}+\tilde{e}_2\frac{\partial \phi}{\partial \xi}+\tilde{q}_2\frac{\partial \psi}{\partial \xi}\right), \tag{13a}$$

$$\sigma_r = \xi^{2\beta}\left(\tilde{c}_{13}\frac{u}{\xi}+\frac{\partial u}{\partial \xi}+\tilde{e}_3\frac{\partial \phi}{\partial \xi}+\tilde{q}_3\frac{\partial \psi}{\partial \xi}\right),$$

$$D = \xi^{2\beta}\left(\tilde{e}_1\frac{u}{\xi}+\tilde{e}_3\frac{\partial u}{\partial \xi}-\frac{\partial \phi}{\partial \xi}-\tilde{g}_3\frac{\partial \psi}{\partial \xi}\right), \tag{13b}$$

$$B = \xi^{2\beta}\left(\tilde{q}_1\frac{u}{\xi}+\tilde{q}_3\frac{\partial u}{\partial \xi}-\tilde{g}_3\frac{\partial \phi}{\partial \xi}-\frac{\partial \psi}{\partial \xi}\right), \tag{13c}$$

$$\frac{\partial \sigma_r}{\partial \xi}+\frac{\sigma_r-\sigma_\theta}{\xi} = \xi^{2\beta}\frac{\partial^2 u}{\partial \tau^2}, \tag{14}$$

$$\frac{1}{\xi}\frac{\partial}{\partial \xi}(\xi D) = 0, \tag{15}$$

$$\frac{1}{\xi}\frac{\partial}{\partial \xi}(\xi B) = 0. \tag{16}$$

The boundary conditions are prescribed as

$$\sigma_r(s,\tau) = p_a(\tau), \quad \sigma_r(1,\tau) = p_b(\tau), \tag{17a}$$

$$\phi(s,\tau) = \phi_a(\tau), \quad \psi(s,\tau) = \psi_a(\tau), \tag{17b}$$

$$\phi(1,\tau) = \phi_b(\tau), \quad \psi(1,\tau) = \psi_b(\tau), \tag{17c}$$

where $p_a(\tau)$ and $p_b(\tau)$ are non-dimensional pressures acting on the internal and external surfaces of the cylindrical shell, respectively. $\phi_a(\tau)$, $\psi_a(\tau)$ and $\phi_b(\tau)$, $\psi_b(\tau)$ are non-dimensional electric potential and magnetic potential imposed on the internal and external cylindrical surfaces, respectively.

The initial conditions are ($\tau = 0$)

$$u(\xi,0) = u_0(\xi), \; \dot{u}(\xi,0) = v_0(\xi), \tag{18}$$

where $u_0(\xi)$ and $v_0(\xi)$ are given functions of radial position. Here, $u_0(\xi)$ and $v_0(\xi)$ represent the initial displacement and velocity fields, respectively.

The solutions of Eqs. (15) and (16) are

$$D(\xi,\tau)=\frac{1}{\xi}\eta(\tau),\quad B(\xi,\tau)=\frac{1}{\xi}\chi(\tau), \tag{19}$$

where $\eta(\tau)$ and $\chi(\tau)$ are undetermined functions of time. With the aid of Eq. (19), Eqs. (13b) and (13c) can be rewritten as

$$\begin{aligned}\frac{\partial\phi}{\partial\xi}&=A_{11}\frac{u}{\xi}+A_{12}\frac{\partial u}{\partial\xi}-A_3\frac{1}{\xi^{2\beta+1}}\eta(\tau)+A_4\frac{1}{\xi^{2\beta+1}}\chi(\tau),\\ \frac{\partial\psi}{\partial\xi}&=A_{21}\frac{u}{\xi}+A_{22}\frac{\partial u}{\partial\xi}+A_4\frac{1}{\xi^{2\beta+1}}\eta(\tau)-A_3\frac{1}{\xi^{2\beta+1}}\chi(\tau).\end{aligned} \tag{20}$$

where

$$\begin{aligned}&A_{11}=(\tilde{e}_1-\tilde{q}_1\tilde{g}_3)/X,\quad A_{12}=(\tilde{e}_3-\tilde{q}_3\tilde{g}_3)/X,\quad A_3=1/X,\quad A_4=\tilde{g}_3/X,\\ &A_{21}=(\tilde{q}_1-\tilde{e}_1\tilde{g}_3)/X,\quad A_{22}=(\tilde{q}_3-\tilde{e}_3\tilde{g}_3)/X,\quad X=1-\tilde{g}_3\tilde{g}_3.\end{aligned} \tag{21}$$

The substitution of Eq. (20) into Eq. (13a) gives

$$\begin{aligned}\sigma_\theta&=\xi^{2\beta}\left(\hat{c}_{11}\frac{u}{\xi}+\hat{c}_{13}\frac{\partial u}{\partial\xi}\right)+\hat{e}_1\frac{\eta(\tau)}{\xi}+\hat{q}_1\frac{\chi(\tau)}{\xi},\\ \sigma_z&=\xi^{2\beta}\left(\hat{c}_{12}\frac{u}{\xi}+\hat{c}_{23}\frac{\partial u}{\partial\xi}\right)+\hat{e}_2\frac{\eta(\tau)}{\xi}+\hat{q}_2\frac{\chi(\tau)}{\xi},\\ \sigma_r&=\xi^{2\beta}\left(\hat{c}_{13}\frac{u}{\xi}+\hat{c}_{33}\frac{\partial u}{\partial\xi}\right)+\hat{e}_3\frac{\eta(\tau)}{\xi}+\hat{q}_3\frac{\chi(\tau)}{\xi},\end{aligned} \tag{22}$$

where

$$\begin{aligned}&\hat{c}_{11}=\tilde{c}_{11}+\tilde{e}_1A_{11}+\tilde{q}_1A_{21},\quad \hat{c}_{12}=\tilde{c}_{12}+\tilde{e}_2A_{11}+\tilde{q}_2A_{21},\\ &\hat{c}_{13}=\tilde{c}_{13}+\tilde{e}_3A_{11}+\tilde{q}_3A_{21},\quad \hat{c}_{23}=\tilde{c}_{23}+\tilde{e}_2A_{12}+\tilde{q}_2A_{22},\\ &\hat{c}_{33}=\tilde{c}_{33}+\tilde{e}_3A_{12}+\tilde{q}_3A_{22},\\ &\hat{e}_1=\tilde{q}_1A_4-\tilde{e}_1A_3,\quad \hat{e}_2=\tilde{q}_2A_4-\tilde{e}_2A_3,\quad \hat{e}_3=\tilde{q}_3A_4-\tilde{e}_3A_3,\\ &\hat{q}_1=\tilde{e}_1A_4-\tilde{q}_1A_3,\quad \hat{q}_2=\tilde{e}_2A_4-\tilde{q}_2A_3,\quad \hat{q}_3=\tilde{e}_3A_4-\tilde{q}_3A_3,\end{aligned} \tag{23}$$

Substituting σ_r and σ_θ in Eq. (22) into Eq. (14) derives

$$\frac{\partial^2 u}{\partial \xi^2}+(2\beta+1)\frac{1}{\xi}\frac{\partial u}{\partial \xi}-\frac{\mu_1^2}{\xi^2}u=\frac{1}{c_L^2}\frac{\partial^2 u}{\partial \tau^2}-\frac{\xi^{-2(\beta+1)}}{\hat{c}_{33}}\left[\hat{e}_1\eta(\tau)+\hat{q}_1\chi(\tau)\right], \tag{24}$$

where

$$\mu_1=\sqrt{(\hat{c}_{11}-2\beta\hat{c}_{13})/\hat{c}_{33}}, \quad c_L=\sqrt{\hat{c}_{33}}. \tag{25}$$

Eq. (17a) can be rewritten as

$$\begin{aligned}\xi=s:&\quad \hat{c}_{13}\frac{u}{\xi}+\hat{c}_{33}\frac{\partial u}{\partial \xi}=s^{-2\beta}\left[p_a(\tau)-\hat{e}_3\frac{\eta(\tau)}{s}-\hat{q}_3\frac{\chi(\tau)}{s}\right],\\ \xi=1:&\quad \hat{c}_{13}\frac{u}{\xi}+\hat{c}_{33}\frac{\partial u}{\partial \xi}=p_b(\tau)-\hat{e}_3\eta(\tau)-\hat{q}_3\chi(\tau),\end{aligned} \tag{26}$$

To solve the problem, a new dependent variable $w(\xi,\tau)$ is introduced as

$$u(\xi,\tau)=\xi^{-\beta}w(\xi,\tau). \tag{27}$$

Then Eqs. (24), (26) and (18) become

$$\frac{\partial^2 w}{\partial \xi^2}+\frac{1}{\xi}\frac{\partial w}{\partial \xi}-\frac{\mu^2}{\xi^2}w=\frac{1}{c_L^2}\frac{\partial^2 w}{\partial \tau^2}+X_1(\xi)\eta(\tau)+X_2(\xi)\chi(\tau), \tag{28}$$

$$\left.\frac{\partial w}{\partial \xi}+h\frac{w}{\xi}\right|_{\xi=s}=p_1(\tau);\quad \left.\frac{\partial w}{\partial \xi}+h\frac{w}{\xi}\right|_{\xi=1}=p_2(\tau). \tag{29}$$

$$w(\xi,0)=u_1(\xi),\ \dot{w}(\xi,0)=v_1(\xi), \tag{30}$$

where

$$\begin{aligned}&h=\frac{\hat{c}_{13}}{\hat{c}_{33}}-\beta,\quad \mu=\sqrt{\mu_1^2+\beta^2},\quad u_1(\xi)=\xi^{\beta}u_0(\xi),\quad v_1(\xi)=\xi^{\beta}v_0(\xi),\\ &p_1(\tau)=\frac{s^{-\beta}}{\hat{c}_{33}}[p_a(\tau)-\frac{\hat{e}_3}{s}\eta(\tau)-\frac{\hat{q}_3}{s}\chi(\tau)],\quad X_1(\xi)=\frac{\hat{e}_1}{\hat{c}_{33}}\xi^{-(\beta+2)},\\ &p_2(\tau)=\frac{1}{\hat{c}_{33}}[p_b(\tau)-\hat{e}_3\eta(\tau)-\hat{q}_3\chi(\tau)],\quad X_2(\xi)=\frac{\hat{q}_1}{\hat{c}_{33}}\xi^{-(\beta+2)}.\end{aligned} \tag{31}$$

In solving the dynamic problem, a key step is the homogenization of the boundary conditions. For this purpose, we assume

$$w(\xi,\tau)=w_1(\xi,\tau)+w_2(\xi,\tau), \tag{32}$$

where $w_2(\xi,\tau)$ is the part which satisfies the inhomogeneous boundary conditions only. By inspecting Eq. (29), we learn that $w_2(\xi,\tau)$ can be assumed as

$$w_2(\xi,\tau)=\alpha_1(\xi-1)^m p_1(\tau)+\alpha_2(\xi-s)^m p_2(\tau), \tag{33}$$

where m is an arbitrary natural number. α_1 and α_2 can be determined by substituting Eq. (33) into Eq. (29).

$$\alpha_1=[m(s-1)^{m-1}+h(s-1)^m/s]^{-1},\quad \alpha_2=[m(1-s)^{m-1}+h(1-s)^m]^{-1}. \tag{34}$$

Generally, we take $m=2$ in Eq. (26) if both the denominators of α_1 and α_2 are nonzero; otherwise, $m=3$ or 4 etc. can be adopted. With the aid of the expressions of $p_1(\tau)$ and $p_2(\tau)$ in Eq. (31), Eq. (33) can be rewritten as

$$w_2(\xi,\tau)=f_0(\xi,\tau)+f_1(\xi)\eta(\tau)+f_2(\xi)\chi(\tau), \tag{35}$$

where

$$\begin{aligned}
&f_0(\xi,\tau)=\vartheta_1(\xi)p_a(\tau)+\vartheta_2(\xi)p_b(\tau),\\
&f_1(\xi)=-\hat{e}_3[\vartheta_1(\xi)/s+\vartheta_2(\xi)],\quad f_2(\xi)=-\hat{q}_3[\vartheta_1(\xi)/s+\vartheta_2(\xi)],\\
&\vartheta_1(\xi)=s^{-\beta}\alpha_1(\xi-1)^m/\hat{c}_{33},\quad \vartheta_2(\xi)=\alpha_2(\xi-s)^m/\hat{c}_{33},
\end{aligned} \tag{36}$$

The substitution of Eq. (32) into Eqs. (28)-(30) gives

$$\frac{\partial^2 w_1(\xi,\tau)}{\partial\xi^2}+\frac{1}{\xi}\frac{\partial w_1(\xi,\tau)}{\partial\xi}-\frac{\mu^2}{\xi^2}w_1(\xi,\tau)=\frac{1}{c_L^2}\frac{\partial^2 w_1(\xi,\tau)}{\partial\tau^2}+g(\xi,\tau), \tag{37}$$

$$\frac{\partial w_1(\xi,\tau)}{\partial\xi}+h\frac{w_1(\xi,\tau)}{\xi}=0\ \ (\xi=s,\ 1), \tag{38}$$

$$\begin{aligned}
&w_1(\xi,0)=u_2(\xi)-f_1(\xi)\eta(0)-f_2(\xi)\chi(0),\\
&\dot{w}_1(\xi,0)=v_2(\xi)-f_1(\xi)\dot{\eta}(0)-f_2(\xi)\dot{\chi}(0),
\end{aligned} \tag{39}$$

where

$$\begin{aligned}
&g(\xi,\tau)=g_0(\xi,\tau)+g_1(\xi)\eta(\tau)+g_2(\xi)\chi(\tau)+[f_1(\xi)\ddot{\eta}(\tau)+f_2(\xi)\ddot{\chi}(\tau)]/c_L^2,\\
&u_2(\xi)=u_1(\xi)-f_0(\xi,0),\quad v_2(\xi)=v_1(\xi)-\dot{f}_0(\xi,0),\\
&g_0(\xi,\tau)=(\Re+\frac{\partial^2}{c_L^2\partial\tau^2})f_0(\xi,\tau),\quad \Re=\frac{\mu^2}{\xi^2}-\frac{1}{\xi}\frac{\mathrm{d}}{\mathrm{d}\xi}-\frac{\mathrm{d}^2}{\mathrm{d}\xi^2},\\
&g_1(\xi)=\Re f_1(\xi)+X_1(\xi),\quad g_2(\xi)=\Re f_2(\xi)+X_2(\xi).
\end{aligned} \tag{40}$$

By means of the separation of variables technique, the solution of Eq. (37) can be assumed as [45]

$$w_1(\xi,\tau)=\sum_{i=1}^{\infty}R_i(\xi)\Omega_i(\tau), \tag{41}$$

where $\Omega_i(\tau)$ are unknown functions of τ, and $R_i(\xi)$ is known function named as eigenfunction

$$R_i(\xi)=J_\mu(k_i\xi)Y(\mu,k_i,s)-Y_\mu(k_i\xi)J(\mu,k_i,s), \tag{42}$$

where

$$J(\mu,k_i,\xi)=\frac{\mathrm{d}J_\mu(k_i\xi)}{\mathrm{d}\xi}+h\frac{J_\mu(k_i\xi)}{\xi},\quad Y(\mu,k_i,\xi)=\frac{\mathrm{d}Y_\mu(k_i\xi)}{\mathrm{d}\xi}+h\frac{Y_\mu(k_i\xi)}{\xi}, \tag{43}$$

and k_i, arranged in an ascending order, are a series of positive roots of the following eigenequation:

$$J(\mu,k_i,s)Y(\mu,k_i,1)-J(\mu,k_i,1)Y(\mu,k_i,s)=0, \tag{44}$$

in which $J_\mu(k_i\xi)$ and $Y_\mu(k_i\xi)$ are Bessel functions of the first and second kind, and of order μ, respectively. It is noted here that $R_i(\xi)$ has the following orthogonal property

$$\int_s^1 \xi R_i(\xi)R_j(\xi)\mathrm{d}\xi = N_i\delta_{ij}, \tag{45}$$

where δ_{ij} is the Kronecker delta, and

$$N_i=\frac{1}{2k_i^2}\left\{\left[\frac{\mathrm{d}R_i(1)}{\mathrm{d}\xi}\right]^2-s^2\left[\frac{\mathrm{d}R_i(s)}{\mathrm{d}\xi}\right]^2+k_i^2[R_i^2(1)-s^2R_i^2(s)]-\mu^2[R_i^2(1)-R_i^2(s)]\right\}. \tag{46}$$

By means of Eq. (41) and (45), Eq. (37) can be simplified as

$$\frac{\mathrm{d}^2\Omega_i(\tau)}{\mathrm{d}\tau^2}+\omega_i^2\Omega_i(\tau)=q_i(\tau), \tag{47}$$

where

$$q_i(\tau) = h_{0i}(\tau) + h_{1i}\eta(\tau) + h_{2i}\chi(\tau) + h_{3i}\ddot{\eta}(\tau) + h_{4i}\ddot{\chi}(\tau),$$
$$\omega_i = k_i c_L, \quad h_{0i}(\tau) = -\frac{c_L^2}{N_i}\int_s^1 \xi g_0(\xi,\tau)R_i(\xi)\mathrm{d}\xi,$$
$$h_{1i} = -\frac{c_L^2}{N_i}\int_s^1 \xi g_1(\xi)R_i(\xi)\mathrm{d}\xi, \quad h_{2i} = -\frac{c_L^2}{N_i}\int_s^1 \xi g_2(\xi)R_i(\xi)\mathrm{d}\xi,$$
$$h_{3i} = -\frac{1}{N_i}\int_s^1 \xi f_1(\xi)R_i(\xi)\mathrm{d}\xi, \quad h_{4i} = -\frac{1}{N_i}\int_s^1 \xi f_2(\xi)R_i(\xi)\mathrm{d}\xi, \tag{48}$$

The solution of Eq. (47) is

$$\Omega_i(\tau) = \Omega_i(0)\cos\omega_i\tau + \frac{\dot{\Omega}_i(0)}{\omega_i}\sin\omega_i\tau + \frac{1}{\omega_i}\int_0^\tau q_i(p)\sin\omega_i(\tau-p)\mathrm{d}p, \tag{49}$$

By employing the integration-by-parts formula, Eq.(49) can be rewritten as

$$\Omega_i(\tau) = \Omega_{1i}(\tau) + h_{3i}\eta(\tau) + h_{4i}\chi(\tau) + I_{1i}\int_0^\tau \eta(p)\sin\omega_i(\tau-p)\mathrm{d}p$$
$$+ I_{2i}\int_0^\tau \chi(p)\sin\omega_i(\tau-p)\mathrm{d}p, \tag{50}$$

where

$$\Omega_{1i}(\tau) = \left[\Omega_i(0) - h_{3i}\eta(0) - h_{4i}\chi(0)\right]\cos\omega_i\tau + \frac{1}{\omega_i}\left[\dot{\Omega}_i(0) - h_{3i}\dot{\eta}(0) - h_{4i}\dot{\chi}(0)\right]\sin\omega_i\tau$$
$$+\frac{1}{\omega_i}\int_0^\tau h_{0i}(p)\sin\omega_i(\tau-p)\mathrm{d}p. \tag{51}$$

and

$$I_{1i} = \frac{h_{1i}}{\omega_i} - h_{3i}\omega_i, \quad I_{2i} = \frac{h_{2i}}{\omega_i} - h_{4i}\omega_i. \tag{52}$$

Utilizing Eqs.(41) and (45), Eq. (39) can be reduced as.

$$\Omega_i(0) = h_{3i}\eta(0) + h_{4i}\chi(0) + h_{5i}, \quad \dot{\Omega}_i(0) = h_{3i}\dot{\eta}(0) + h_{4i}\dot{\chi}(0) + h_{6i}, \tag{53}$$

where

$$h_{5i} = \frac{1}{N_i}\int_s^1 \xi u_2(\xi)R_i(\xi)\mathrm{d}\xi, \quad h_{6i} = \frac{1}{N_i}\int_s^1 \xi v_2(\xi)R_i(\xi)\mathrm{d}\xi. \tag{54}$$

With the aid of Eq. (53), Eq. (51) is then totally determined as

$$\Omega_{1i}(\tau)=h_{5i}\cos\omega_i\tau+\frac{h_{6i}}{\omega_i}\sin\omega_i\tau+\frac{1}{\omega_i}\int_0^\tau h_{0i}(p)\sin\omega_i(\tau-p)\mathrm{d}p. \tag{55}$$

Substituting Eq. (35) and (41) into Eq. (32) and then into Eq. (27) derives

$$u(\xi,\tau)=\xi^{-\beta}[\sum_{i=1}^{\infty}R_i(\xi)\Omega_i(\tau)+f_0(\xi,\tau)+f_1(\xi)\eta(\tau)+f_2(\xi)\chi(\tau)]. \tag{56}$$

Integrating Eq. (20) at the space interval $[s,\xi]$ and utilizing Eq. (17b) gives

$$\begin{aligned}\phi(\xi,\tau)&=\phi_a(\tau)+\phi_0(\xi,\tau)+\phi_1(\xi)\eta(\tau)+\phi_2(\xi)\chi(\tau)+\sum_{i=1}^{\infty}\phi_{3i}(\xi)\Omega_i(\tau),\\ \psi(\xi,\tau)&=\psi_a(\tau)+\psi_0(\xi,\tau)+\psi_1(\xi)\eta(\tau)+\psi_2(\xi)\chi(\tau)+\sum_{i=1}^{\infty}\psi_{3i}(\xi)\Omega_i(\tau),\end{aligned} \tag{57}$$

where

$$\begin{aligned}&\phi_0(\xi,\tau)=A_{11}H_{01}(\xi,\tau)+A_{12}H_{02}(\xi,\tau),\quad \psi_0(\xi,\tau)=A_{21}H_{01}(\xi,\tau)+A_{22}H_{02}(\xi,\tau),\\ &\phi_1(\xi)=A_{11}H_{11}(\xi)+A_{12}H_{12}(\xi)-A_3H(\xi),\quad \psi_1(\xi)=A_{21}H_{11}(\xi)+A_{22}H_{12}(\xi)+A_4H(\xi),\\ &\phi_2(\xi)=A_{11}H_{21}(\xi)+A_{12}H_{22}(\xi)+A_4H(\xi),\quad \psi_2(\xi)=A_{21}H_{21}(\xi)+A_{22}H_{22}(\xi)-A_3H(\xi),\\ &\phi_{3i}(\xi)=A_{11}H_{31i}(\xi)+A_{12}H_{32i}(\xi),\quad \psi_{3i}(\xi)=A_{21}H_{31i}(\xi)+A_{22}H_{32i}(\xi),\\ &H_{01}(\xi,\tau)=\int_s^\xi \xi^{-(\beta+1)}f_0(\xi,\tau)\mathrm{d}\xi,\quad H_{11}(\xi,\tau)=\int_s^\xi \xi^{-(\beta+1)}f_1(\xi)\mathrm{d}\xi,\\ &H_{21}(\xi,\tau)=\int_s^\xi \xi^{-(\beta+1)}f_2(\xi)\mathrm{d}\xi,\quad H_{31i}(\xi,\tau)=\int_s^\xi \xi^{-(\beta+1)}R_i(\xi)\mathrm{d}\xi,\\ &H_{02}(\xi)=\xi^{-\beta}f_0(\xi,\tau)-s^{-\beta}f_0(s,\tau),\quad H_{12}(\xi)=\xi^{-\beta}f_1(\xi)-s^{-\beta}f_1(s),\\ &H_{22}(\xi)=\xi^{-\beta}f_2(\xi)-s^{-\beta}f_2(s),\quad H_{32i}(\xi)=\xi^{-\beta}R_i(\xi)-s^{-\beta}R_i(s),\\ &H(\xi)=\begin{cases}-\left(\xi^{-2\beta}-s^{-2\beta}\right)/(2\beta) & (\beta\neq 0),\\ -(\ln\xi-\ln s) & (\beta=0).\end{cases}\end{aligned} \tag{58}$$

Setting $\xi=1$ in Eq. (57) and utilizing Eq. (17c), we obtain

$$\begin{aligned}\phi_b(\tau)&=\phi_a(\tau)+\phi_0(1,\tau)+\phi_1(1)\eta(\tau)+\phi_2(1)\chi(\tau)+\sum_{i=1}^{\infty}\phi_{3i}(1)\Omega_i(\tau),\\ \psi_b(\tau)&=\psi_a(\tau)+\psi_0(1,\tau)+\psi_1(1)\eta(\tau)+\psi_2(1)\chi(\tau)+\sum_{i=1}^{\infty}\psi_{3i}(1)\Omega_i(\tau),\end{aligned} \tag{59}$$

Substituting Eq. (50) into Eq. (59), we have

$$\begin{aligned}&Z_{11}\eta(\tau)+Z_{12}\chi(\tau)+\sum_{i=1}^{\infty}\int_0^\tau[Z_{13i}\eta(p)+Z_{14i}\chi(p)]\sin\omega_i(\tau-p)\mathrm{d}p=F_1(\tau),\\ &Z_{21}\eta(\tau)+Z_{22}\chi(\tau)+\sum_{i=1}^{\infty}\int_0^\tau[Z_{23i}\eta(p)+Z_{24i}\chi(p)]\sin\omega_i(\tau-p)\mathrm{d}p=F_2(\tau).\end{aligned} \tag{60}$$

where

$$\begin{aligned}
&Z_{11}=\phi_1(1)+\sum_{i=1}^{\infty}\phi_{3i}(1)h_{3i},\quad Z_{12}=\phi_2(1)+\sum_{i=1}^{\infty}\phi_{3i}(1)h_{4i},\\
&Z_{21}=\psi_1(1)+\sum_{i=1}^{\infty}\psi_{3i}(1)h_{3i},\quad Z_{22}=\psi_2(1)+\sum_{i=1}^{\infty}\psi_{3i}(1)h_{4i},\\
&Z_{13i}=\phi_{3i}(1)I_{1i},\quad Z_{14i}=\phi_{3i}(1)I_{2i},\quad Z_{23i}=\psi_{3i}(1)I_{1i},\quad Z_{24i}=\psi_{3i}(1)I_{2i},\\
&F_1(\tau)=\phi_b(\tau)-\phi_a(\tau)-\phi_0(1,\tau)-\sum_{i=1}^{\infty}\phi_{3i}(1)\Omega_{1i}(\tau),\\
&F_2(\tau)=\psi_b(\tau)-\psi_a(\tau)-\psi_0(1,\tau)-\sum_{i=1}^{\infty}\psi_{3i}(1)\Omega_{1i}(\tau).
\end{aligned}\tag{61}$$

Eq. (60) is a system of second kind Volterra integral equations, which can be solved efficiently by newly developed numerical method [45]. Then the mechanical, electric and magnetic fields for functionally graded magneto-electro-elastic hollow cylinder can be determined completely. For the sake of convenience to the reader, the numerical method for solving Eq. (60) is included in Appendix A.

4. Solution for MEE Spherical Shells

In spherical coordinates (r,θ,φ), considering the material properties in Eq.(4), the governing equations of the radially polarized functionally graded MEE spherically isotropic media can be expressed as [52,53]

$$\begin{aligned}
&\frac{\partial\sigma_{rr}}{\partial r}+\frac{\partial\sigma_{r\theta}}{r\partial\theta}+\frac{1}{r\sin\theta}\frac{\partial\sigma_{\varphi r}}{\partial\varphi}+\frac{1}{r}\left(2\sigma_{rr}-\sigma_{\theta\theta}-\sigma_{\varphi\varphi}+\sigma_{r\theta}\cot\theta\right)=\bar{\rho}\left(\frac{r}{b}\right)^{2\beta}\frac{\partial^2 u_r}{\partial t^2},\\
&\frac{\partial\sigma_{r\theta}}{\partial r}+\frac{\partial\sigma_{\theta\theta}}{r\partial\theta}+\frac{1}{r\sin\theta}\frac{\partial\sigma_{\theta\varphi}}{\partial\varphi}+\frac{1}{r}\left[\left(\sigma_{\theta\theta}-\sigma_{\varphi\varphi}\right)\cot\theta+3\sigma_{r\theta}\right]=\bar{\rho}\left(\frac{r}{b}\right)^{2\beta}\frac{\partial^2 u_\theta}{\partial t^2},\\
&\frac{\partial\sigma_{\varphi r}}{\partial r}+\frac{\partial\sigma_{\theta\varphi}}{r\partial\theta}+\frac{1}{r\sin\theta}\frac{\partial\sigma_{\varphi\varphi}}{\partial\varphi}+\frac{1}{r}\left(2\sigma_{\theta\varphi}\cot\theta+3\sigma_{\varphi r}\right)=\bar{\rho}\left(\frac{r}{b}\right)^{2\beta}\frac{\partial^2 u_\varphi}{\partial t^2},
\end{aligned}\tag{62a}$$

$$\frac{1}{r^2}\frac{\partial}{\partial r}(r^2 D_r)+\frac{1}{r\sin\theta}\frac{\partial\left(\sin\theta D_\theta\right)}{\partial\theta}+\frac{1}{r\sin\theta}\frac{\partial D_\varphi}{\partial\varphi}=0,\tag{62b}$$

$$\frac{1}{r^2}\frac{\partial}{\partial r}(r^2 B_r)+\frac{1}{r\sin\theta}\frac{\partial\left(\sin\theta B_\theta\right)}{\partial\theta}+\frac{1}{r\sin\theta}\frac{\partial B_\varphi}{\partial\varphi}=0,\tag{62c}$$

$$\begin{aligned}
&\sigma_{\theta\theta}=\left(r/b\right)^{2\beta}\left(\bar{c}_{11}\varepsilon_{\theta\theta}+\bar{c}_{12}\varepsilon_{\varphi\varphi}+\bar{c}_{13}\varepsilon_{rr}-\bar{e}_{31}E_r-\bar{q}_{31}H_r\right),\\
&\sigma_{\varphi\varphi}=\left(r/b\right)^{2\beta}\left(\bar{c}_{12}\varepsilon_{\theta\theta}+\bar{c}_{11}\varepsilon_{\varphi\varphi}+\bar{c}_{13}\varepsilon_{rr}-\bar{e}_{31}E_r-\bar{q}_{31}H_r\right),\\
&\sigma_{rr}=\left(r/b\right)^{2\beta}\left(\bar{c}_{13}\varepsilon_{\theta\theta}+\bar{c}_{13}\varepsilon_{\varphi\varphi}+\bar{c}_{33}\varepsilon_{rr}-\bar{e}_{33}E_r-\bar{q}_{33}H_r\right),\\
&\sigma_{\varphi r}=\left(r/b\right)^{2\beta}\left(2\bar{c}_{44}\varepsilon_{zr}-\bar{e}_{15}E_\varphi-\bar{q}_{15}H_\varphi\right),\\
&\sigma_{r\theta}=\left(r/b\right)^{2\beta}\left(2\bar{c}_{44}\varepsilon_{r\theta}-\bar{e}_{15}E_\theta-\bar{q}_{15}H_\theta\right),\\
&\sigma_{\theta\varphi}=\left(r/b\right)^{2\beta}\left(2\bar{c}_{66}\varepsilon_{\theta\varphi}\right),
\end{aligned}\tag{63a}$$

$$D_\theta = (r/b)^{2\beta}\left(2\bar{e}_{15}\varepsilon_{r\theta} + \bar{\kappa}_{11}E_\theta + \bar{g}_{11}H_\theta\right),$$
$$D_\varphi = (r/b)^{2\beta}\left(2\bar{e}_{15}\varepsilon_{\varphi r} + \bar{\kappa}_{11}E_\varphi + \bar{g}_{11}H_\varphi\right), \tag{63b}$$
$$D_r = (r/b)^{2\beta}\left(\bar{e}_{31}\varepsilon_{\theta\theta} + \bar{e}_{31}\varepsilon_{\varphi\varphi} + \bar{e}_{33}\varepsilon_{rr} + \bar{\kappa}_{33}E_r + \bar{g}_{33}H_r\right),$$

$$B_\theta = (r/b)^{2\beta}\left(2\bar{q}_{15}\varepsilon_{r\theta} + \bar{g}_{11}E_\theta + \bar{\mu}_{11}H_\theta\right),$$
$$B_\varphi = (r/b)^{2\beta}\left(2\bar{q}_{15}\varepsilon_{zr} + \bar{g}_{11}E_\varphi + \bar{\mu}_{11}H_\varphi\right), \tag{63c}$$
$$B_r = (r/b)^{2\beta}\left(\bar{q}_{31}\varepsilon_{\theta\theta} + \bar{q}_{31}\varepsilon_{\varphi\varphi} + \bar{q}_{33}\varepsilon_{rr} + \bar{g}_{33}E_r + \bar{\mu}_{33}H_r\right),$$

$$\varepsilon_{rr} = \frac{\partial u_r}{\partial r}, \quad \varepsilon_{\theta\theta} = \frac{1}{r}\frac{\partial u_\theta}{\partial\theta} + \frac{u_r}{r}, \quad \varepsilon_{\varphi\varphi} = \frac{1}{r\sin\theta}\frac{\partial u_\varphi}{\partial\varphi} + \frac{u_r}{r} + \frac{u_\theta}{r}\cot\theta,$$
$$\varepsilon_{\theta\varphi} = \frac{1}{2}\left(\frac{1}{r}\frac{\partial u_\varphi}{\partial\theta} - \frac{u_\varphi}{r}\cot\theta + \frac{1}{r\sin\theta}\frac{\partial u_\theta}{\partial\varphi}\right), \tag{64a}$$
$$\varepsilon_{\varphi r} = \frac{1}{2}\left(\frac{1}{r\sin\theta}\frac{\partial u_r}{\partial\varphi} + \frac{\partial u_\varphi}{\partial r} - \frac{u_\varphi}{r}\right), \quad \varepsilon_{r\theta} = \frac{1}{2}\left(\frac{1}{r}\frac{\partial u_r}{\partial\theta} + \frac{\partial u_\theta}{\partial r} - \frac{u_\theta}{r}\right),$$

$$E_r = -\frac{\partial\Phi}{\partial r}, \quad E_\theta = -\frac{1}{r}\frac{\partial\Phi}{\partial\theta}, \quad E_\varphi = -\frac{1}{r\sin\theta}\frac{\partial\Phi}{\partial\varphi}, \tag{64b}$$

$$H_r = -\frac{\partial\Psi}{\partial r}, \quad H_\theta = -\frac{1}{r}\frac{\partial\Psi}{\partial\theta}, \quad H_\varphi = -\frac{1}{r\sin\theta}\frac{\partial\Psi}{\partial\varphi}. \tag{64c}$$

For radial vibration (spherically symmetric elastodynamic problem), we have

$$u_r = u_r(r,t), \quad u_\theta = u_\varphi = 0, \quad \Phi = \Phi(r,t), \quad \Psi = \Psi(r,t). \tag{65}$$

Substituting Eq. (65) into Eqs. (64a)-(64c) gives

$$\varepsilon_{rr} = \frac{\partial u_r}{\partial r}, \quad \varepsilon_{\theta\theta} = \varepsilon_{\varphi\varphi} = \frac{u_r}{r}, \quad \varepsilon_{\varphi r} = \varepsilon_{r\theta} = \varepsilon_{\theta\varphi} = 0, \tag{66a}$$

$$E_r = -\frac{\partial\Phi}{\partial r}, \quad E_\theta = E_\varphi = 0, \tag{66b}$$

$$H_r = -\frac{\partial\Psi}{\partial r}, \quad H_\theta = H_\varphi = 0, \tag{66c}$$

Then Eqs. (63a)-(63c) are reduced as

$$\sigma_{\theta\theta}=\sigma_{\varphi\varphi}=\left(\frac{r}{b}\right)^{2\beta}\left[\left(\bar{c}_{11}+\bar{c}_{12}\right)\frac{u_r}{r}+\bar{c}_{13}\frac{\partial u_r}{\partial r}+\bar{e}_{31}\frac{\partial\Phi}{\partial r}+\bar{q}_{31}\frac{\partial\Psi}{\partial r}\right],$$
$$\sigma_{rr}=\left(\frac{r}{b}\right)^{2\beta}\left(2\bar{c}_{13}\frac{u_r}{r}+\bar{c}_{33}\frac{\partial u_r}{\partial r}+\bar{e}_{33}\frac{\partial\Phi}{\partial r}+\bar{q}_{33}\frac{\partial\Psi}{\partial r}\right),\tag{67a}$$
$$\sigma_{\varphi r}=\sigma_{r\theta}=\sigma_{\theta\varphi}=0,$$

$$D_r=\left(\frac{r}{b}\right)^{2\beta}\left(2\bar{e}_{31}\frac{u_r}{r}+\bar{e}_{33}\frac{\partial u_r}{\partial r}-\bar{\kappa}_{33}\frac{\partial\Phi}{\partial r}-\bar{g}_{33}\frac{\partial\Psi}{\partial r}\right),\quad D_\theta=D_\varphi=0,\tag{67b}$$

$$B_r=\left(\frac{r}{b}\right)^{2\beta}\left(2\bar{q}_{31}\frac{u_r}{r}+\bar{q}_{33}\frac{\partial u_r}{\partial r}-\bar{g}_{33}\frac{\partial\Phi}{\partial r}-\bar{\mu}_{33}\frac{\partial\Psi}{\partial r}\right),\quad B_\theta=B_\varphi=0,\tag{67c}$$

Correspondingly, Eqs. (62a)-(62c) can be simplified as

$$\frac{\partial\sigma_{rr}}{\partial r}+\frac{2\left(\sigma_{rr}-\sigma_{\theta\theta}\right)}{r}=\bar{\rho}\left(\frac{r}{b}\right)^{2\beta}\frac{\partial^2 u_r}{\partial t^2},\tag{68a}$$

$$\frac{1}{r^2}\frac{\partial}{\partial r}(r^2 D_r)=0,\tag{68b}$$

$$\frac{1}{r^2}\frac{\partial}{\partial r}(r^2 B_r)=0.\tag{68c}$$

Utilizing the non-dimensional quantities and variables introduced in Eq. (12), the governing equations for functionally graded MEE spherical shell under radial deformation are then constructed as [46]

$$\sigma_\theta=\sigma_\varphi=\xi^{2\beta}\left[\left(\tilde{c}_{11}+\tilde{c}_{12}\right)\frac{u}{\xi}+\tilde{c}_{13}\frac{\partial u}{\partial\xi}+\tilde{e}_1\frac{\partial\phi}{\partial\xi}+\tilde{q}_1\frac{\partial\psi}{\partial\xi}\right],$$
$$\sigma_r=\xi^{2\beta}\left(2\tilde{c}_{13}\frac{u}{\xi}+\frac{\partial u}{\partial\xi}+\tilde{e}_3\frac{\partial\phi}{\partial\xi}+\tilde{q}_3\frac{\partial\psi}{\partial\xi}\right),\tag{69a}$$

$$D=\xi^{2\beta}\left(2\tilde{e}_1\frac{u}{\xi}+\tilde{e}_3\frac{\partial u}{\partial\xi}-\frac{\partial\phi}{\partial\xi}-\tilde{g}_3\frac{\partial\psi}{\partial\xi}\right),\tag{69b}$$

$$B=\xi^{2\beta}\left(2\tilde{q}_1\frac{u}{\xi}+\tilde{q}_3\frac{\partial u}{\partial\xi}-\tilde{g}_3\frac{\partial\phi}{\partial\xi}-\frac{\partial\psi}{\partial\xi}\right),\tag{69c}$$

$$\frac{\partial\sigma_r}{\partial\xi}+\frac{2\left(\sigma_r-\sigma_\theta\right)}{\xi}=\xi^{2\beta}\frac{\partial^2 u}{\partial\tau^2},\tag{70a}$$

$$\frac{1}{\xi^2}\frac{\partial}{\partial \xi}(\xi^2 D) = 0\,, \tag{70b}$$

$$\frac{1}{\xi^2}\frac{\partial}{\partial \xi}(\xi^2 B) = 0\,. \tag{70c}$$

The boundary conditions are considered here as

$$\sigma_r(s,\tau) = p_a(\tau), \quad \sigma_r(1,\tau) = p_b(\tau), \tag{71a}$$

$$\phi(s,\tau) = \phi_a(\tau), \quad \psi(s,\tau) = \psi_a(\tau), \tag{71b}$$

$$\phi(1,\tau) = \phi_b(\tau), \quad \psi(1,\tau) = \psi_b(\tau), \tag{71c}$$

where $p_a(\tau)$ and $p_b(\tau)$ are non-dimensional pressures acting on the internal and external surfaces of the spherical shell, respectively. $\phi_a(\tau)$, $\psi_a(\tau)$ and $\phi_b(\tau)$, $\psi_b(\tau)$ are non-dimensional electric potential and magnetic potential imposed on the internal and external spherical surfaces, respectively.

The initial conditions are completed as ($\tau = 0$)

$$u(\xi,0) = u_0(\xi)\,,\ \dot{u}(\xi,0) = v_0(\xi)\,, \tag{72}$$

where the initial displacement field $u_0(\xi)$ and the initial velocity field $v_0(\xi)$ are known functions of radial position.

The solutions of Eqs. (70b) and (70c) are

$$D(\xi,\tau) = \frac{1}{\xi^2}\eta(\tau), \quad B(\xi,\tau) = \frac{1}{\xi^2}\chi(\tau)\,, \tag{73}$$

Same as in Eq. (19), $\eta(\tau)$ and $\chi(\tau)$ in Eq. (73) are undetermined functions of time. With the aid of Eq. (73), Eqs. (67b) and (67c) can be rewritten as

$$\begin{aligned} \frac{\partial \phi}{\partial \xi} &= 2A_{11}\frac{u}{\xi} + A_{12}\frac{\partial u}{\partial \xi} - A_3\frac{1}{\xi^{2\beta+2}}\eta(\tau) + A_4\frac{1}{\xi^{2\beta+2}}\chi(\tau), \\ \frac{\partial \psi}{\partial \xi} &= 2A_{21}\frac{u}{\xi} + A_{22}\frac{\partial u}{\partial \xi} + A_4\frac{1}{\xi^{2\beta+2}}\eta(\tau) - A_3\frac{1}{\xi^{2\beta+2}}\chi(\tau). \end{aligned} \tag{74}$$

where

$$\begin{aligned} &A_{11} = (\tilde{e}_1 - \tilde{q}_1\tilde{g}_3)/X, \quad A_{12} = (\tilde{e}_3 - \tilde{q}_3\tilde{g}_3)/X, \quad A_3 = 1/X\,, \quad A_4 = \tilde{g}_3/X\,, \\ &A_{21} = (\tilde{q}_1 - \tilde{e}_1\tilde{g}_3)/X, \quad A_{22} = (\tilde{q}_3 - \tilde{e}_3\tilde{g}_3)/X, \quad X = 1 - \tilde{g}_3\tilde{g}_3. \end{aligned} \tag{75}$$

The substitution of Eq. (74) into Eq. (69a) derives

$$\begin{aligned}\sigma_\theta = \sigma_\varphi = \xi^{2\beta}\left[\left(\hat{c}_{11}+\hat{c}_{12}\right)\frac{u}{\xi}+\hat{c}_{13}\frac{\partial u}{\partial \xi}\right]+\hat{e}_1\frac{\eta(\tau)}{\xi^2}+\hat{q}_1\frac{\chi(\tau)}{\xi^2},\\ \sigma_r = \xi^{2\beta}\left(2\hat{c}_{13}\frac{u}{\xi}+\hat{c}_{33}\frac{\partial u}{\partial \xi}\right)+\hat{e}_3\frac{\eta(\tau)}{\xi^2}+\hat{q}_3\frac{\chi(\tau)}{\xi^2},\end{aligned} \tag{76}$$

where

$$\begin{aligned}&\hat{c}_{11}=\tilde{c}_{11}+\tilde{e}_1A_{11}+\tilde{q}_1A_{21},\quad \hat{c}_{12}=\tilde{c}_{12}+\tilde{e}_1A_{11}+\tilde{q}_1A_{21},\\ &\hat{c}_{13}=\tilde{c}_{13}+\tilde{e}_3A_{11}+\tilde{q}_3A_{21},\quad \hat{c}_{33}=\tilde{c}_{33}+\tilde{e}_3A_{12}+\tilde{q}_3A_{22},\\ &\hat{e}_1=\tilde{q}_1A_4-\tilde{e}_1A_3,\quad \hat{e}_3=\tilde{q}_3A_4-\tilde{e}_3A_3,\\ &\hat{q}_1=\tilde{e}_1A_4-\tilde{q}_1A_3,\quad \hat{q}_3=\tilde{e}_3A_4-\tilde{q}_3A_3,\end{aligned} \tag{77}$$

Substituting σ_r and σ_θ in Eq. (76) into Eq. (70a) derives

$$\frac{\partial^2 u}{\partial \xi^2}+(2\beta+2)\frac{1}{\xi}\frac{\partial u}{\partial \xi}-\frac{\mu_1^2}{\xi^2}u=\frac{1}{c_L^2}\frac{\partial^2 u}{\partial \tau^2}-\frac{\xi^{-(2\beta+3)}}{\hat{c}_{33}}\left[2\hat{e}_1\eta(\tau)+2\hat{q}_1\chi(\tau)\right], \tag{78}$$

where

$$\mu_1=\sqrt{2\frac{\hat{c}_{11}+\hat{c}_{12}-(2\beta+1)\hat{c}_{13}}{\hat{c}_{33}}},\quad c_L=\sqrt{\hat{c}_{33}}. \tag{79}$$

Eq. (71a) can be rewritten as

$$\begin{aligned}&\xi=s:\quad 2\hat{c}_{13}\frac{u}{\xi}+\hat{c}_{33}\frac{\partial u}{\partial \xi}=s^{-2\beta}\left[p_a(\tau)-\hat{e}_3\frac{\eta(\tau)}{s^2}-\hat{q}_3\frac{\chi(\tau)}{s^2}\right],\\ &\xi=1:\quad 2\hat{c}_{13}\frac{u}{\xi}+\hat{c}_{33}\frac{\partial u}{\partial \xi}=p_b(\tau)-\hat{e}_3\eta(\tau)-\hat{q}_3\chi(\tau),\end{aligned} \tag{80}$$

In this technique, a new dependent variable $w(\xi,\tau)$ is introduced as

$$u(\xi,\tau)=\xi^{-(\beta+\frac{1}{2})}w(\xi,\tau). \tag{81}$$

With this treatment, Eqs. (78), (80) and (72) become

$$\frac{\partial^2 w}{\partial \xi^2}+\frac{1}{\xi}\frac{\partial w}{\partial \xi}-\frac{\mu^2}{\xi^2}w=\frac{1}{c_L^2}\frac{\partial^2 w}{\partial \tau^2}+X_1(\xi)\eta(\tau)+X_2(\xi)\chi(\tau), \tag{82}$$

$$\left.\frac{\partial w}{\partial \xi}+h\frac{w}{\xi}\right|_{\xi=s}=p_1(\tau);\quad \left.\frac{\partial w}{\partial \xi}+h\frac{w}{\xi}\right|_{\xi=1}=p_2(\tau). \tag{83}$$

$$w(\xi,0)=u_1(\xi),\ \dot{w}(\xi,0)=v_1(\xi), \tag{84}$$

where

$$\begin{aligned}
&h=2\frac{\hat{c}_{13}}{\hat{c}_{33}}-\left(\beta+\frac{1}{2}\right),\quad \mu=\sqrt{\mu_1^2+\left(\beta+\tfrac{1}{2}\right)^2},\\
&X_1(\xi)=\frac{2\hat{e}_1}{\hat{c}_{33}}\xi^{-(\beta+\frac{5}{2})},\quad X_2(\xi)=\frac{2\hat{q}_1}{\hat{c}_{33}}\xi^{-(\beta+\frac{5}{2})},\\
&p_1(\tau)=\frac{s^{-\beta+\frac{1}{2}}}{\hat{c}_{33}}[p_a(\tau)-\frac{\hat{e}_3}{s^2}\eta(\tau)-\frac{\hat{q}_3}{s^2}\chi(\tau)],\\
&p_2(\tau)=\frac{1}{\hat{c}_{33}}[p_b(\tau)-\hat{e}_3\eta(\tau)-\hat{q}_3\chi(\tau)],\\
&u_1(\xi)=\xi^{\beta+\frac{1}{2}}u_0(\xi),\quad v_1(\xi)=\xi^{\beta+\frac{1}{2}}v_0(\xi).
\end{aligned} \tag{85}$$

In order to homogenize the boundary conditions, here we also set

$$w(\xi,\tau)=w_1(\xi,\tau)+w_2(\xi,\tau), \tag{86}$$

where $w_2(\xi,\tau)$ is the part which satisfies the inhomogeneous boundary conditions only. By inspecting Eq. (83), we learn that $w_2(\xi,\tau)$ can be assumed as

$$w_2(\xi,\tau)=\alpha_1(\xi-1)^m p_1(\tau)+\alpha_2(\xi-s)^m p_2(\tau), \tag{87}$$

The treatment for m and the expressions of α_1 and α_2 are same as those for functionally graded cylindrical shells and finally we get

$$w_2(\xi,\tau)=f_0(\xi,\tau)+f_1(\xi)\eta(\tau)+f_2(\xi)\chi(\tau), \tag{88}$$

where

$$\begin{aligned}
&f_0(\xi,\tau)=\vartheta_1(\xi)p_a(\tau)+\vartheta_2(\xi)p_b(\tau),\\
&f_1(\xi)=-\hat{e}_3[\vartheta_1(\xi)/s^2+\vartheta_2(\xi)],\quad f_2(\xi)=-\hat{q}_3[\vartheta_1(\xi)/s^2+\vartheta_2(\xi)],\\
&\vartheta_1(\xi)=s^{-\beta+\frac{1}{2}}\alpha_1(\xi-1)^m/\hat{c}_{33},\quad \vartheta_2(\xi)=\alpha_2(\xi-s)^m/\hat{c}_{33},
\end{aligned}\tag{89}$$

Substituting Eq. (88) into Eqs. (82)-(84) gives

$$\frac{\partial^2 w_1(\xi,\tau)}{\partial\xi^2}+\frac{1}{\xi}\frac{\partial w_1(\xi,\tau)}{\partial\xi}-\frac{\mu^2}{\xi^2}w_1(\xi,\tau)=\frac{1}{c_L^2}\frac{\partial^2 w_1(\xi,\tau)}{\partial\tau^2}+g(\xi,\tau),\tag{90}$$

$$\frac{\partial w_1(\xi,\tau)}{\partial\xi}+h\frac{w_1(\xi,\tau)}{\xi}=0\ \ (\xi=s,1),\tag{91}$$

$$\begin{aligned}
&w_1(\xi,0)=u_2(\xi)-f_1(\xi)\eta(0)-f_2(\xi)\chi(0),\\
&\dot{w}_1(\xi,0)=v_2(\xi)-f_1(\xi)\dot{\eta}(0)-f_2(\xi)\dot{\chi}(0),
\end{aligned}\tag{92}$$

where

$$\begin{aligned}
&g(\xi,\tau)=g_0(\xi,\tau)+g_1(\xi)\eta(\tau)+g_2(\xi)\chi(\tau)+[f_1(\xi)\ddot{\eta}(\tau)+f_2(\xi)\ddot{\chi}(\tau)]/c_L^2,\\
&u_2(\xi)=u_1(\xi)-f_0(\xi,0),\quad v_2(\xi)=v_1(\xi)-\dot{f}_0(\xi,0),\\
&g_0(\xi,\tau)=(\Re+\frac{\partial^2}{c_L^2\partial\tau^2})f_0(\xi,\tau),\quad \Re=\frac{\mu^2}{\xi^2}-\frac{1}{\xi}\frac{\mathrm{d}}{\mathrm{d}\xi}-\frac{\mathrm{d}^2}{\mathrm{d}\xi^2},\\
&g_1(\xi)=\Re f_1(\xi)+X_1(\xi),\quad g_2(\xi)=\Re f_2(\xi)+X_2(\xi).
\end{aligned}\tag{93}$$

Following the same way in Section 3, the solution of $w_1(\xi,\tau)$ can be obtained by the separation of variables method [46]. Finally, the solution of $u(\xi,\tau)$ is presented as

$$u(\xi,\tau)=\xi^{-(\beta+\frac{1}{2})}[\sum_{i=1}^{\infty}R_i(\xi)\Omega_i(\tau)+f_0(\xi,\tau)+f_1(\xi)\eta(\tau)+f_2(\xi)\chi(\tau)].\tag{94}$$

where the expressions of $R_i(\xi)$ and $\Omega_i(\tau)$ are similar with those for functionally graded MEE cylindrical shells. While in the applications, the constants, such as μ, h, $h_{ji}(j=1,2,\cdots,6)$, I_{1i} and I_{2i}, should be replaced by those for functionally graded MEE spherical shells.

Integrating Eq. (74) at the space interval $[s,\xi]$ and utilizing Eq. (71b) gives

$$\begin{aligned}
&\phi(\xi,\tau)=\phi_a(\tau)+\phi_0(\xi,\tau)+\phi_1(\xi)\eta(\tau)+\phi_2(\xi)\chi(\tau)+\sum_{i=1}^{\infty}\phi_{3i}(\xi)\Omega_i(\tau),\\
&\psi(\xi,\tau)=\psi_a(\tau)+\psi_0(\xi,\tau)+\psi_1(\xi)\eta(\tau)+\psi_2(\xi)\chi(\tau)+\sum_{i=1}^{\infty}\psi_{3i}(\xi)\Omega_i(\tau),
\end{aligned}\tag{95}$$

where

$$\begin{aligned}
&\phi_0(\xi,\tau)=2A_{11}H_{01}(\xi,\tau)+A_{12}H_{02}(\xi,\tau),\quad \psi_0(\xi,\tau)=2A_{21}H_{01}(\xi,\tau)+A_{22}H_{02}(\xi,\tau),\\
&\phi_1(\xi)=2A_{11}H_{11}(\xi)+A_{12}H_{12}(\xi)-A_3H(\xi),\quad \psi_1(\xi)=2A_{21}H_{11}(\xi)+A_{22}H_{12}(\xi)+A_4H(\xi),\\
&\phi_2(\xi)=2A_{11}H_{21}(\xi)+A_{12}H_{22}(\xi)+A_4H(\xi),\quad \psi_2(\xi)=2A_{21}H_{21}(\xi)+A_{22}H_{22}(\xi)-A_3H(\xi),\\
&\phi_{3i}(\xi)=2A_{11}H_{31i}(\xi)+A_{12}H_{32i}(\xi),\quad \psi_{3i}(\xi)=2A_{21}H_{31i}(\xi)+A_{22}H_{32i}(\xi),\\
&H_{01}(\xi,\tau)=\int_s^{\xi}\xi^{-(\beta+\frac{3}{2})}f_0(\xi,\tau)\mathrm{d}\xi,\quad H_{11}(\xi,\tau)=\int_s^{\xi}\xi^{-(\beta+\frac{3}{2})}f_1(\xi)\mathrm{d}\xi,\\
&H_{21}(\xi,\tau)=\int_s^{\xi}\xi^{-(\beta+\frac{3}{2})}f_2(\xi)\mathrm{d}\xi,\quad H_{31i}(\xi,\tau)=\int_s^{\xi}\xi^{-(\beta+\frac{3}{2})}R_i(\xi)\mathrm{d}\xi,\\
&H_{02}(\xi)=\xi^{-(\beta+\frac{1}{2})}f_0(\xi,\tau)-s^{-(\beta+\frac{1}{2})}f_0(s,\tau),\quad H_{12}(\xi)=\xi^{-(\beta+\frac{1}{2})}f_1(\xi)-s^{-(\beta+\frac{1}{2})}f_1(s),\\
&H_{22}(\xi)=\xi^{-(\beta+\frac{1}{2})}f_2(\xi)-s^{-(\beta+\frac{1}{2})}f_2(s),\quad H_{32i}(\xi)=\xi^{-(\beta+\frac{1}{2})}R_i(\xi)-s^{-(\beta+\frac{1}{2})}R_i(s),\\
&H(\xi)=\begin{cases}-\left[\xi^{-(2\beta+1)}-s^{-(2\beta+1)}\right]/(2\beta+1) & (\beta\neq-1/2),\\ -(\ln\xi-\ln s) & (\beta=-1/2).\end{cases}
\end{aligned} \tag{96}$$

Setting $\xi=1$ in Eq. (95) and utilizing Eq. (71c), we obtain the same governing equations for $\eta(\tau)$ and $\chi(\tau)$ as shown in Eq. (60).

5. Some Special Applications of the Present Solution

The present solutions are developed for functionally graded magneto-electro-elastic cylindrical and spherical shells. They can be directly degenerated into those for some practical cases as:

If we set $\bar{q}_{kij}=0$, $\bar{g}_{ij}=0$ and $\bar{\mu}_{ij}=0$ in Eq. (4) and omit the equations related to magnetic fields, such as Eqs. (1c), (2c) and (3c), the present solutions then become that for functionally graded radially polarized piezoelectric cylindrical and spherical shells. For this case, instead of Eq. (60), only one Volterra integral equation for the unknown function $\eta(\tau)$ is involved.

If we further set $\bar{e}_{kij}=0$ and $\bar{\kappa}_{ij}=0$ in Eq. (4) and omit the equations related to electric fields, such as Eqs. (1b), (2b) and (3b), the present solution becomes that for functionally graded radially polarized purely elastic cylindrical and spherical shells. Especially, for this case, the Volterra integral equation is avoided and the obtained solutions are known as closed solutions.

If we set the inhomogeneity parameter β=0, the present solution becomes that for homogeneous radially polarized MEE cylindrical and spherical shells under radial vibration.

If the temperature change Θ is considered, the present solutions can also be modified into those for functionally graded magneto-thermo-electro-elastic cylindrical and spherical shells.

6. Numerical Results

Several numerical results will be performed in this subsection. It is noted here that if the inhomogeneity parameter is set as $\beta=0$, the functionally graded materials then become traditional homogeneous materials. As an illustrative example, the material constants used here for homogeneous magneto-electro-elastic material are taken as those listed in Table 1 in Ref. [54] and see Table 1 in this subsection. It should also be mentioned that such material constants are the prescient values by the micromechanical analysis for $CoFe_2O_4$ and $BaTiO_3$ composites [51].

Table 1 material constants

Material property	MEE
$\bar{c}_{11}$ (Pa)	218×10^9
$\bar{c}_{12}$ (Pa)	120×10^9
$\bar{c}_{13}$ (Pa)	120×10^9
$\bar{c}_{33}$ (Pa)	215×10^9
$\bar{e}_{31}$ (C/m^2)	-2.5
$\bar{e}_{32}$ (C/m^2)	-2.5
$\bar{e}_{33}$ (C/m^2)	7.5
$\bar{q}_{31}$ (N/(Am))	265
$\bar{q}_{32}$ (N/(Am))	265
$\bar{q}_{33}$ (N/(Am))	345
$\bar{\kappa}_{33}$ (C^2/(Nm^2))	5.8×10^{-9}
$\bar{g}_{33}$ (Ns/(VC))	2.82×10^{-9}
$\bar{\mu}_{33}$ (Ns^2/C^2)	95×10^{-6}

In all the following numerical analysis, we suppose the cylindrical and spherical shells are initially at rest. That is $u_0(\xi)=0$ and $v_0(\xi)=0$.

6.1. Functionally Graded Magneto-Electro-Elastic Cylindrical Shells

6.1.1. The Effect of Material Inhomogeneity

The dynamic behaviors of transient response in the functionally graded MEE cylindrical shells for different values of material inhomogeneity β are illustrated graphically in this subsection. In this case, the inner-to-outer radius ratio of the cylindrical shell is set as $s=0.2$. Also, we assume both the electric and magnetic potentials at the internal and external surfaces equal to zero. The MEE cylindrical shell is stress free at the external surface and is subjected to dynamic pressures at the internal surface. The boundary conditions are then expressed as

$$
\begin{aligned}
&p_a(\tau) = e^{-5\tau} - 1, \quad p_b(\tau) = 0; \\
&\phi_a(\tau) = 0, \quad \phi_b(\tau) = 0, \\
&\psi_a(\tau) = 0, \quad \psi_b(\tau) = 0.
\end{aligned}
\tag{97}
$$

where $e^{(\cdot)}$ denotes exponential function.

Figure 1 depicts the transient responses of radial stress σ_r at the middle surface (ξ=0.6) in functionally graded MEE cylindrical shells for β= -1, 0 and 1, respectively. It is noted here that the case for β=0 is just that for homogeneous MEE cylindrical shell. The curves display that the amplitude of compressive radial stresses σ_r at the radial point ξ=0.6 increases notably with the increase of β. The responses of hoop stress σ_θ at the internal surface (ξ=0.2) is shown in Figure 2. From the presented curves, we learn that the tensile hoop stress at the internal surface decreases rapidly with the increase of β. That is, for internally pressurized cylindrical shells, it is beneficial to use a functionally graded material with negative inhomogeneity parameter β.

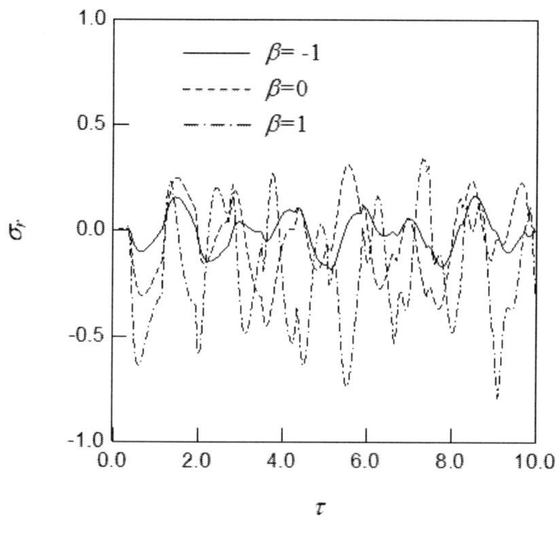

Figure 1. Transient responses of radial stress σ_r in MEE cylindrical shells at ξ=0.6 for different values of β.

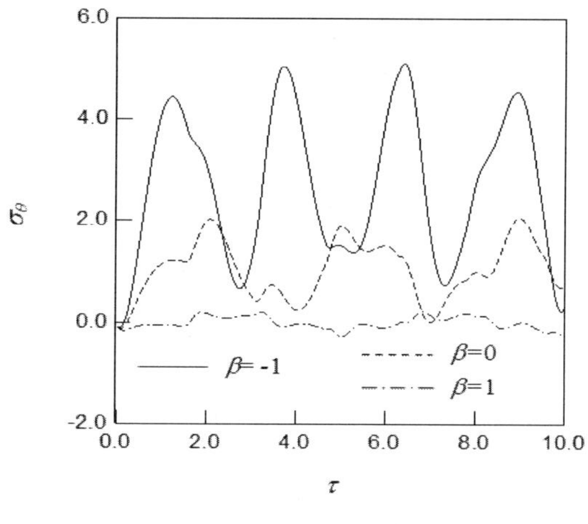

Figure 2. Transient responses of hoop stress σ_θ in MEE cylindrical shells at ξ=0.2 for Different values of β.

Figures 3 and 4 present the distributions of electric potential ϕ and magnetic potential ψ in MEE cylindrical shells at two very beginning times τ=0.2 and 0.5, respectively. We clearly found that both the amplitude and distribution form of ϕ and ψ are sensitive to the material inhomogeneity parameter β.

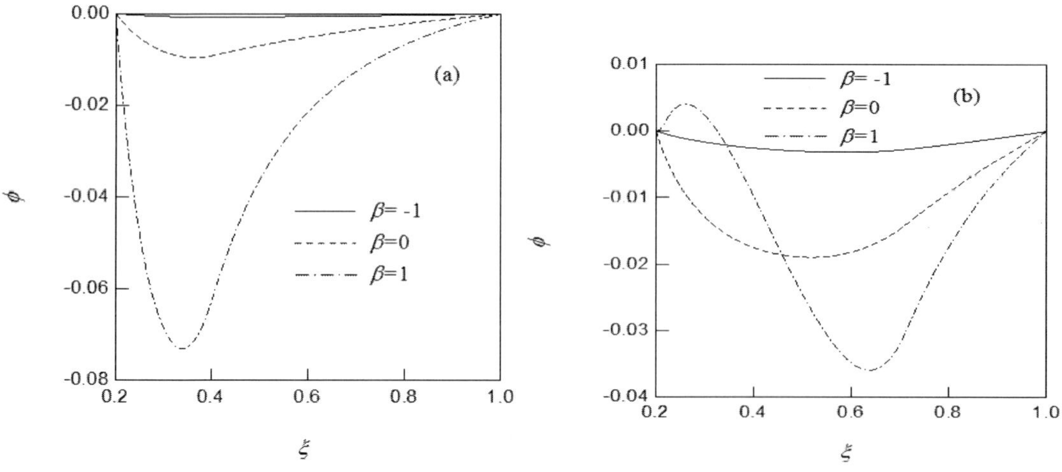

Figure 3. Distributions of electric potential ϕ in MEE cylindrical shells for different values of β (a) τ=0.2; (b) τ=0.5.

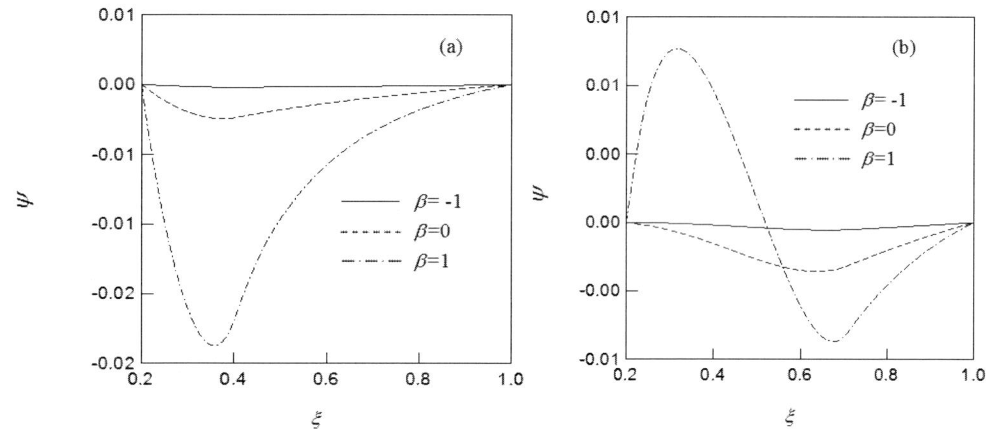

Figure 4. Distributions of magnetic potential ψ in MEE cylindrical shells for different values of β (a) τ=0.2; (b) τ=0.5.

6.1.2. The Effect of the Dynamic Mechanical Load

The dynamic behaviors of transient response in the functionally graded MEE cylindrical shells subjected to two different dynamic mechanical loads are considered in this example. In this calculation, the material inhomogeneity and the inner-to-outer radius ratio are adopted as $\beta = 1.0$ and $s = 0.2$, respectively. The electric and magnetic boundary conditions are prescribed as

$$\begin{aligned} &\phi_a(\tau) = 0, \quad \phi_b(\tau) = 0, \\ &\psi_a(\tau) = 0, \quad \psi_b(\tau) = 0. \end{aligned} \tag{98a}$$

Two different dynamic load forms are considered here as

Load A: $p_a(\tau) = e^{-2\tau} - 1, \quad p_b(\tau) = 0;$ (98b)

Load B: $p_a(\tau) = e^{-10\tau} - 1, \quad p_b(\tau) = 0.$ (98c)

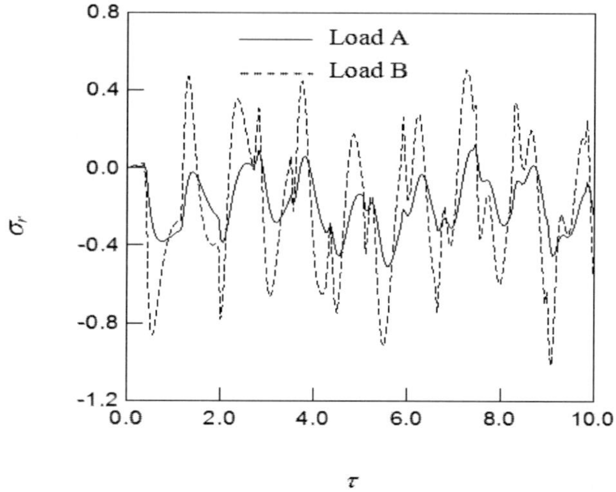

Figure 5. Transient responses of radial stress σ_r at ξ=0.6 in MEE cylindrical shells for different dynamic mechanical loads.

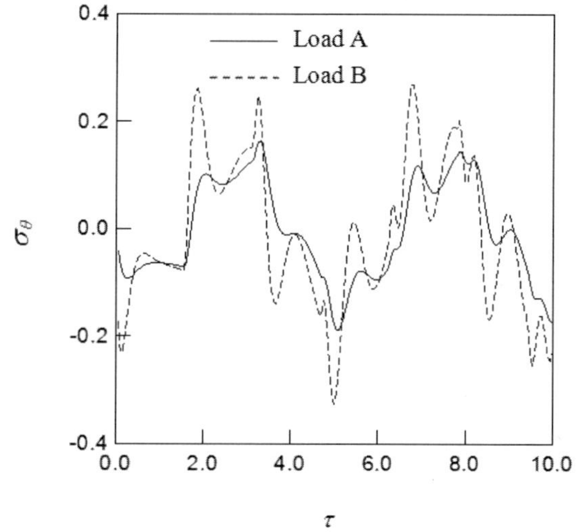

Figure 6. Transient responses of hoop stress σ_θ at ξ=0.2 in MEE cylindrical shells for different dynamic mechanical loads.

Figure 5 shows the transient responses of radial stress σ_r at the middle surface ξ=0.6 in MEE cylindrical shells subjected two different dynamic mechanical loads at the internal surface. Figure 6 shows the dynamic responses of hoop stress σ_θ at the internal surface ξ=0.2. It is observed that the dynamic responses for both radial and hoop stresses for Load B appear more sharply than those for Load A. Figures 7 and 8 illustrate the distributions of electric potential ϕ and magnetic potential ψ in MEE cylindrical shells at the time τ=0.2. We find that both the electric potential ϕ and magnetic potential ψ are negative for Loads A and B. We also notice that the absolute values of ϕ and ψ for Load B are larger than those for Load A.

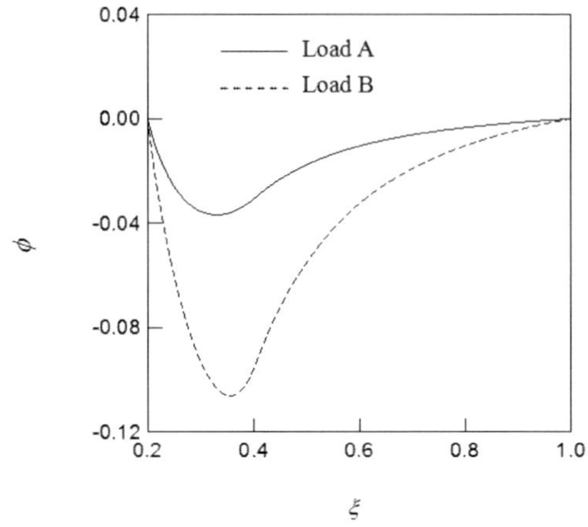

Figure 7. Distributions of electric potential ϕ in MEE cylindrical shells at the time τ=0.2.

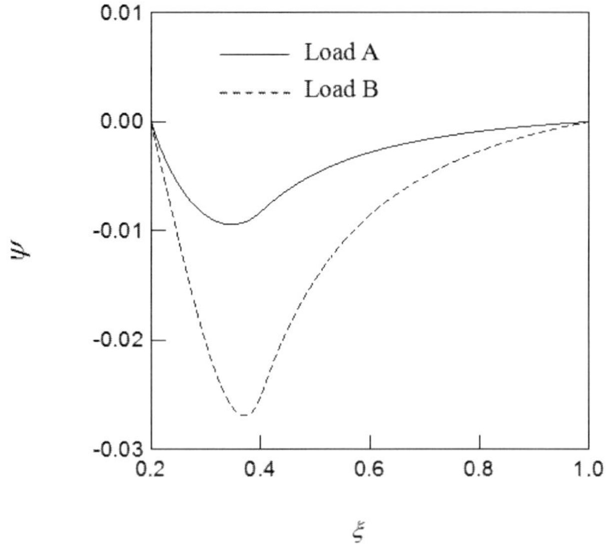

Figure 8. Distributions of magnetic potential ψ in MEEcylindrical shells at the time τ=0.2.

6.2. Functionally Graded Magneto-Electro-Elastic Spherical Shells

6.2.1. The Effect of Inner-to-Outer Radius Ratio

The dynamic responses of functionally graded magneto-electro-elastic spherical shells with different inner-to-outer radius ratios are analyzed in this example. The material inhomogeneity parameter is employed as β=0.5. We suppose the spherical shells are traction free at both the internal and external surfaces and are subjected to the electric potential excitation at the external surface. The boundary conditions are

$$\begin{aligned} &p_a(\tau)=0, \quad p_b(\tau)=0; \\ &\phi_a(\tau)=0, \quad \phi_b(\tau)=\sin(10\tau), \\ &\psi_a(\tau)=0, \quad \psi_b(\tau)=0. \end{aligned} \tag{99}$$

For electrically excited MEE spherical shells, Figures 9-12 show that both the transient responses of the mechanical and electrical fields are sensitive to the parameter of inner-to-outer radius ratio.

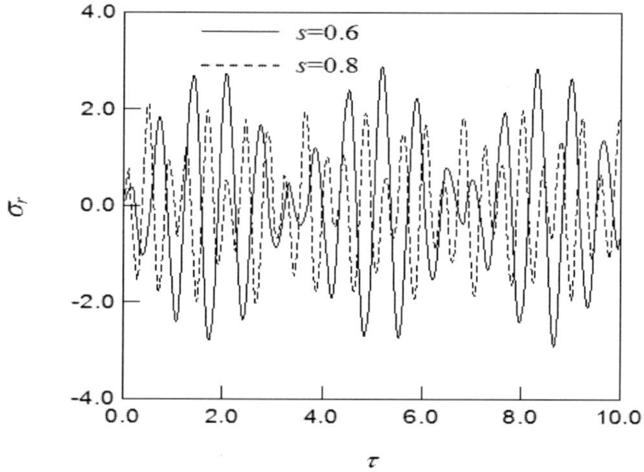

Figure 9. Transient responses of radial stress σ_r at ξ=(1+s)/2 in MEE spherical shells for different inner-to-outer radius ratios.

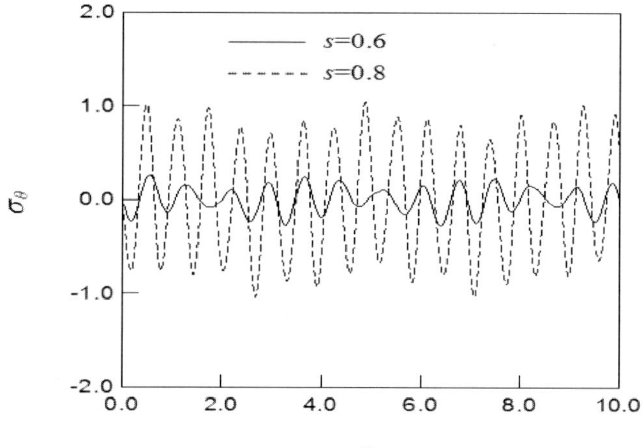

Figure 10. Transient responses of hoop stress σ_θ at ξ=1 in MEE spherical shells for different inner-to-outer radius ratios.

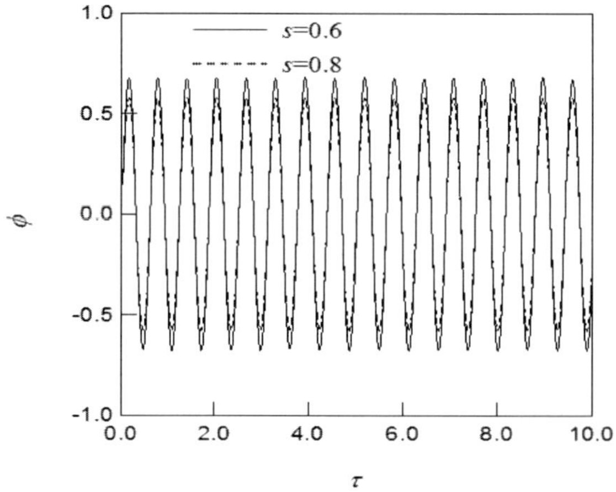

Figure 11. Transient responses of electric potential ϕ at ξ=(1+s)/2 in MEE spherical shells for different inner-to-outer radius ratios.

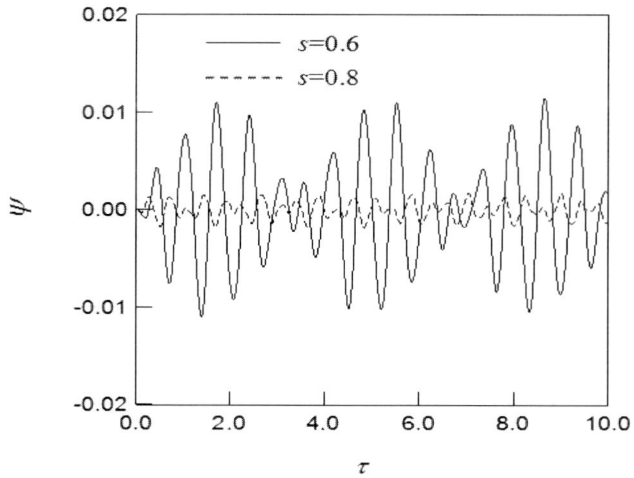

Figure 12. Transient responses of magnetic potential ψ at ξ=(1+s)/2 in MEE spherical shells for different inner-to-outer radius ratios.

6.2.2. The Effect of the Dynamic Electric Load

The effect of the dynamic electric load on the transient responses of functionally graded magneto-electro-elastic spherical shells will further be examined. The computation parameters are taken as β=1 and s=0.5. The mechanical and magnetic boundary conditions are prescribed as

$$\begin{aligned} &p_a(\tau)=0, \quad p_b(\tau)=0, \\ &\psi_a(\tau)=0, \quad \psi_b(\tau)=0. \end{aligned} \tag{100a}$$

Two different electrical loads varying as sine form are considered.

Load C: $\phi_a(\tau) = 0, \quad \phi_b(\tau) = \sin(2\tau);$ (100b)

Load D: $\phi_a(\tau) = 0, \quad \phi_b(\tau) = \sin(5\tau);$ (100c)

Figures 13-17 show the transient responses of the mechanical and electrical fields in MEE spherical shells for two electrical excitations varying as sine form. It is observed that the amplitudes of the responses of radial stress σ_r, hoop stress σ_θ, as well as magnetic potential ψ are sensitive to the angular frequency of the electrical excitation. Interestingly, there is little difference between the amplitudes of the responses of electric potential ϕ for the two electrical excitations.

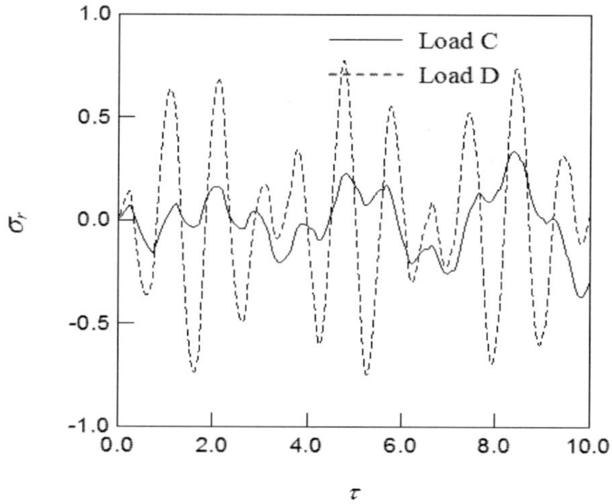

Figure 13. Transient responses of radial stress σ_r at ξ=0.75 in MEE spherical shells for different electrical excitations.

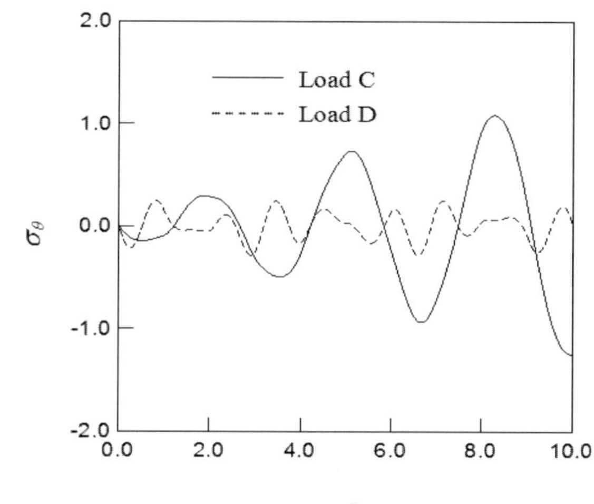

Figure 14. Transient responses of hoop stress σ_θ at ξ=0.5 in MEE spherical shells for different electrical excitations.

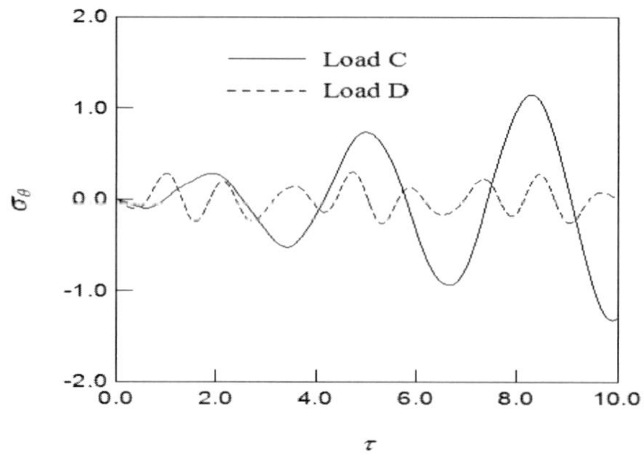

Figure 15. Transient responses of hoop stress σ_θ at ξ=1 in MEE spherical shells for different electrical excitations.

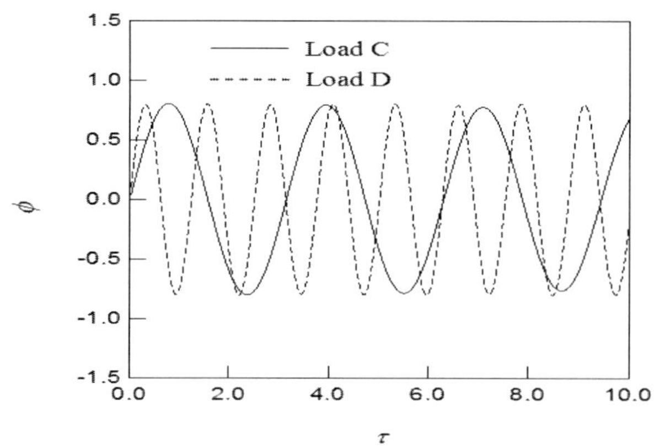

Figure 16. Transient responses of electric potential ϕ at ξ=0.75 in MEE spherical shells for different electrical excitations.

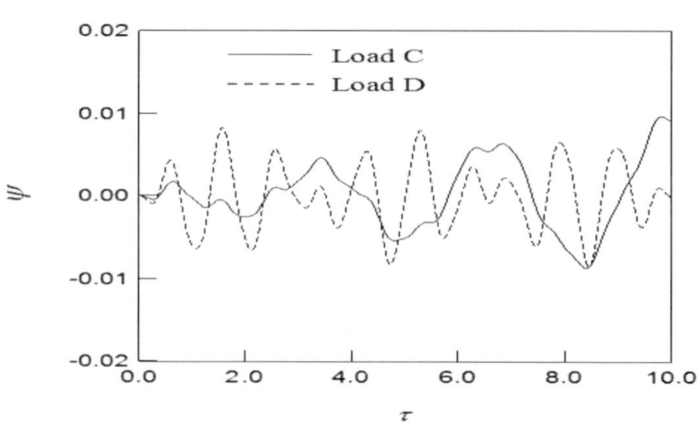

Figure 17. Transient responses of magnetic potential ψ at ξ=0.75 in MEE spherical shells for different electrical excitations.

CONCLUSION

The analytical solution is developed for the radial vibration of the multi-field coupled functionally graded cylindrical and spherical shells excited by the dynamic mechanical, electrical and magnetic loads. The parameter investigations are illustrated in the given examples. The potential applications of the present solution can be found in the exact analysis for the dynamic behavior of the multi-field coupled functionally graded structures and devices.

ACKNOWLEDGMENTS

The work was supported by the National Natural Science Foundation of China (Nos. 10872179 and 10725210), the Zhejiang Provincial Natural Science Foundation of China (No. Y7080298) and Zijin and XinXing Plan of Zhejiang University.

APPENDIX A.

We first divide the total time interval $[0, \tau_n]$ into n equal subintervals with discrete time points $\tau_0 = 0,\ \tau_1,\ \tau_2,\ \cdots\ \tau_n$. At the each interval $[\tau_{j-1}, \tau_j]$, a cubic Hermite polynomial is introduced to approximate the unknown functions $\eta(\tau)$ and $\chi(\tau)$ as

$$\begin{aligned}\eta(\tau) &= H_{0j}(\tau)\eta(\tau_{j-1}) + H_{1j}(\tau)\eta(\tau_j) + H_{2j}(\tau)\dot{\eta}(\tau_{j-1}) + H_{3j}(\tau)\dot{\eta}(\tau_j)\\ \chi(\tau) &= H_{0j}(\tau)\chi(\tau_{j-1}) + H_{1j}(\tau)\chi(\tau_j) + H_{2j}(\tau)\dot{\chi}(\tau_{j-1}) + H_{3j}(\tau)\dot{\chi}(\tau_j)\\ &\qquad (\tau_{j-1} \le \tau \le \tau_j;\quad j = 1, 2\cdots n).\end{aligned} \tag{A1}$$

where

$$\begin{aligned}&H_{0j}(\tau) = (1 + 2\frac{\tau - \tau_{j-1}}{\tau_j - \tau_{j-1}})(\frac{\tau - \tau_j}{\tau_j - \tau_{j-1}})^2, \quad H_{1j}(\tau) = (1 + 2\frac{\tau_j - \tau}{\tau_j - \tau_{j-1}})(\frac{\tau - \tau_{j-1}}{\tau_j - \tau_{j-1}})^2,\\ &H_{2j}(\tau) = (\tau - \tau_{j-1})(\frac{\tau - \tau_j}{\tau_j - \tau_{j-1}})^2, \quad H_{3j}(\tau) = (\tau - \tau_j)(\frac{\tau - \tau_{j-1}}{\tau_j - \tau_{j-1}})^2 \quad (j = 1, 2\cdots n).\end{aligned} \tag{A2}$$

From Eq. (60), we have

$$\begin{aligned}&Z_{11}\dot{\eta}(\tau) + Z_{12}\dot{\chi}(\tau) + \sum_{i=1}^{\infty}\int_0^{\tau}[Z_{13i}\eta(p) + Z_{14i}\chi(p)]\omega_i \cos\omega_i(\tau - p)\mathrm{d}p = \dot{F}_1(\tau),\\ &Z_{21}\dot{\eta}(\tau) + Z_{22}\dot{\chi}(\tau) + \sum_{i=1}^{\infty}\int_0^{\tau}[Z_{23i}\eta(p) + Z_{24i}\eta(p)]\omega_i \cos\omega_i(\tau - p)\mathrm{d}p = \dot{F}_2(\tau).\end{aligned} \tag{A3}$$

Setting $\tau = \tau_j$ in Eqs (60) and (A3) and utilizing Eq. (A1) derives

$$
\begin{aligned}
F_1(\tau_j) &= Z_{11}\eta(\tau_j) + \sum_{i=1}^{\infty} Z_{13i} \sum_{k=1}^{j} [L_{0ijk}\eta(\tau_{k-1}) + L_{1ijk}\eta(\tau_k) + L_{2ijk}\dot{\eta}(\tau_{k-1}) + L_{3ijk}\dot{\eta}(\tau_k)] \\
&\quad + Z_{12}\chi(\tau_j) + \sum_{i=1}^{\infty} Z_{14i} \sum_{k=1}^{j} [L_{0ijk}\chi(\tau_{k-1}) + L_{1ijk}\chi(\tau_k) + L_{2ijk}\dot{\chi}(\tau_{k-1}) + L_{3ijk}\dot{\chi}(\tau_k)], \\
F_2(\tau_j) &= Z_{21}\eta(\tau_j) + \sum_{i=1}^{\infty} Z_{23i} \sum_{k=1}^{j} [L_{0ijk}\eta(\tau_{k-1}) + L_{1ijk}\eta(\tau_k) + L_{2ijk}\dot{\eta}(\tau_{k-1}) + L_{3ijk}\dot{\eta}(\tau_k)] \\
&\quad + Z_{22}\chi(\tau_j) + \sum_{i=1}^{\infty} Z_{24i} \sum_{k=1}^{j} [L_{0ijk}\chi(\tau_{k-1}) + L_{1ijk}\chi(\tau_k) + L_{2ijk}\dot{\chi}(\tau_{k-1}) + L_{3ijk}\dot{\chi}(\tau_k)],
\end{aligned}
\tag{A4}
$$

$$
\begin{aligned}
\dot{F}_1(\tau_j) &= Z_{11}\dot{\eta}(\tau_j) + \sum_{i=1}^{\infty} Z_{13i} \sum_{k=1}^{j} [K_{0ijk}\eta(\tau_{k-1}) + K_{1ijk}\eta(\tau_k) + K_{2ijk}\dot{\eta}(\tau_{k-1}) + K_{3ijk}\dot{\eta}(\tau_k)] \\
&\quad + Z_{12}\dot{\chi}(\tau_j) + \sum_{i=1}^{\infty} Z_{14i} \sum_{k=1}^{j} [K_{0ijk}\chi(\tau_{k-1}) + K_{1ijk}\chi(\tau_k) + K_{2ijk}\dot{\chi}(\tau_{k-1}) + K_{3ijk}\dot{\chi}(\tau_k)], \\
\dot{F}_2(\tau_j) &= Z_{21}\dot{\eta}(\tau_j) + \sum_{i=1}^{\infty} Z_{23i} \sum_{k=1}^{j} [K_{0ijk}\eta(\tau_{k-1}) + K_{1ijk}\eta(\tau_k) + K_{2ijk}\dot{\eta}(\tau_{k-1}) + K_{3ijk}\dot{\eta}(\tau_k)] \\
&\quad + Z_{22}\dot{\chi}(\tau_j) + \sum_{i=1}^{\infty} Z_{24i} \sum_{k=1}^{j} [K_{0ijk}\chi(\tau_{k-1}) + K_{1ijk}\chi(\tau_k) + K_{2ijk}\dot{\chi}(\tau_{k-1}) + K_{3ijk}\dot{\chi}(\tau_k)],
\end{aligned}
\tag{A5}
$$

where

$$
L_{lijk} = \int_{\tau_{k-1}}^{\tau_k} H_{lk}(p)\sin\omega_i(\tau_j - p)\mathrm{d}p, \quad K_{lijk} = \int_{\tau_{k-1}}^{\tau_k} \omega_i H_{lk}(p)\cos\omega_i(\tau_j - p)\mathrm{d}p, \tag{A6}
$$
$$
(l = 0,1,2,3; \quad k = 1,2\cdots j; \quad j = 1,2\cdots n).
$$

Then $\eta(\tau_j)$, $\dot{\eta}(\tau_j)$, $\chi(\tau_j)$ and $\dot{\chi}(\tau_j)$ $(j = 0,1,2\cdots n)$ can be obtained step by step by means of Eqs. (A4) and (A5).

References

Huth, J. H. and Cole, J. D. Elastic-stress waves produced by pressure loads on a spherical shell. *ASME J. Appl. Mech.*, 1955, *Vol* 22, 473-478.

Baker, W. E., Hu, W. C. L. and Jackson, T. R. Elastic response of thin spherical shells to axisymmetric blast loading. *ASME J. Appl. Mech.*, 1966, *Vol* 33, 800-806.

Cinelli, G. Dynamic vibrations and stresses in elastic cylinders and spheres. *ASME J. Appl. Mech.*, 1966, *Vol* 33, 825-830.

Chou, P. C. and Koenig, H. A. A unified approach to cylindrical and spherical elastic waves by method of characteristics. *ASME J. Appl. Mech.*, 1966, *Vol* 33, 159-167.

Rose, J. L., Chou, S. C. and Chou, P. C. Vibration analysis of thick-walled spheres and cylinders. *J. Acoust. Soc. Am.*, 1973, *Vol* 53, 771-776.

Pao, Y. H. and Ceranoglu, A. N. Determination of transient responses of a thick-walled spherical shell by the ray theory. *ASME J. Appl. Mech.*, 1978, *Vol* 45, 114-122.

Tusi, T., and Kraus, H. J. Thermal stress-wave propagation in hollow elastic spheres. *Acoust. Soc. Am.*, 1965, *Vol* 37, 730-737.

Ho, C. H. Stress focusing effect in a uniformly heated cylindrical rod. *ASME J. Appl. Mech.*, 1976, *Vol* 43, 464-468.

Wang, X. and Gong, Y. N. A theoretical solution for axially symmetric problem in elastodynamics. *Acta Mech. Sin.*, 1991, *Vol* 7, 275-282.

Wang, X. An elastodynamic solution for an anisotropic hollow sphere. *Int. J. Solids Struct.*, 1994, *Vol* 31, 903-911.

Wang, X. Thermal shock in a hollow cylinder caused by rapid arbitrary heating. *J. Sound Vib.*, 1995, *Vol* 183, 899-906.

Wang, X., Zhang, K., Zhang, W. and Chen, J. B. Theoretical solution and finite element solution for an orthotropic thick cylindrical shell under impact load. *J. Sound Vib.*, 2000, *Vol* 236, 129-140.

Wang, X., Zhang, W. and Chan, J. B. Dynamic thermal stress in a transversely isotropic hollow sphere. *J. Thermal Stresses*, 2002, *Vol* 24, 335-346.

Wang, X., Wang, C., Lu, G. and Zhou, B. M. Thermal stress-focusing in a transversely isotropic sphere and an isotropic sphere. *J. Thermal Stresses*, 2002, *Vol* 25, 31-44.

Hata, T. Thermal shock in a hollow sphere caused by rapid uniform heating. *ASME J. Appl. Mech.*, 1991, *Vol* 58, 64-69.

Hata, T. Stress-focusing effect in a uniformly heated solid sphere. *ASME J. Appl. Mech.*, 1991, *Vol* 58, 58-63.

Hata, T. Stress-focusing effect in a uniformly heated transversely isotropic sphere. *Int. J. Solids Struct.*, 1993, *Vol* 30, 1419-1428.

Cho, H. and Kardomateas, G. A. Thermal shock stresses due to heat convection at a bounding surface in a thick orthotropic cylindrical shell. *Int. J. Solids Struct.*, 2001, *Vol* 38, 2769-2788.

Ding, H. J., Wang, H. M. and Chen, W. Q. A theoretical solution of cylindrically isotropic cylindrical tube for axisymmetric plane strain dynamic thermoelastic problem. *Acta Mech. Solida Sin.*, 2001, *Vol* 14, 357-363.

Ding, H. J., Wang, H. M. and Chen, W. Q. A theoretical solution of cylindrical shells for axisymmetric plane strain elastodynamic problems. *Appl. Math. Mech.*, 2002, *Vol* 23, 138-145.

Ding, H. J., Wang, H. M. and Chen, W. Q. On stress-focusing effect in a uniformly heated solid sphere. *ASME J. Appl. Mech.*, 2003, *Vol* 70, 304-309.

El-Raheb, M. Transient response in a finite hollow cylinder from time-delayed prescribed motion at the boundary. *J. Sound Vib.*, 2005, *Vol* 285, 491-500.

Savruk, M. P., Onyshko, L. I. and Senyuk, M. M. A plane dynamic axisymmetric problem for a hollow cylinder. *Mater. Sci.*, 2008, *Vol* 44, 1-9.

Ding, H. J., Wang, H. M. and Hou, P. F. The transient responses of piezoelectric hollow cylinders for axisymmetric plane strain problems. *Int. J. Solids Struct.*, 2003, *Vol* 40, 105-123.

Ding, H. J., Wang, H. M. and Ling, D. S. Analytical solution of a pyroelectric hollow cylinder for piezothermoelastic axisymmetric dynamic problems. *J. Thermal Stresses*, 2003, *Vol* 26, 261-276.

Ding, H. J., Wang, H. M. and Chen, W. Q. Transient responses in a piezoelectric spherically isotropic hollow sphere for symmetric problems. *ASME J. Appl. Mech.*, 2003, *Vol* 70, 436-445.

Hou, P. F. and Leung, A. Y. T. The transient responses of magneto-electro-elastic hollow cylinders. *Smart Mater. Struct.*, 2004, *Vol* 13, 762-776.

Dai, H. L. and Wang, X. Dynamic responses of piezoelectric hollow cylinders in a axial magnetic field. *Int. J. Solids Struct.*, 2004, *Vol* 41, 5231-5246.

Wang, H. M. and Ding, H. J. Transient responses of a magneto-electro-elastic hollow sphere for fully coupled spherically symmetric problem. *Eur. J. Mech./A Solids*, 2006, *Vol* 25, 965-980.

Dai, H. L., Fu, Y. M. and Liu, T. X. Electromagnetoelastic dynamic response of transversely isotropic piezoelectric hollow spheres in a uniform magnetic field. *ASME J. Appl. Mech.*, 2007, *Vol* 74, 65-73.

Higuchi, M., Kawamura, R. and Tanigawa, Y. Magneto-thermo-elastic stresses induced by a transient magnetic field in a conducting solid circular cylinder. *Int. J. Solids Struct.*, 2007, *Vol* 44, 5316-5335.

Higuchi, M., Kawamura, R. and Tanigawa, Y. Dynamic and quasi-static behaviors of magneto-thermo-elastic stresses in a conducting solid circular cylinder subjected to an arbitrary variation of magnetic field. *J. Thermal Stresses*, 2007, *Vol* 30, 395-416.

Higuchi, M., Kawamura, R. and Tanigawa, Y. Dynamic and quasi-static behaviors of magneto-thermo-elastic stresses in a conducting hollow circular cylinder subjected to an arbitrary variation of magnetic field. *Int. J. Mech. Sci.*, 2008, *Vol* 50, 365-379.

Ding, H. J., Wang, H. M. and Chen, W. Q. Analytical thermo-elastodynamic solutions for a nonhomogeneous transversely isotropic hollow sphere. *Arch. Appl. Mech.*, 2002, *Vol* 72, 545-553.

Ding, H. J., Wang, H. M. and Chen, W. Q. A solution of a non-homogeneous orthotropic cylindrical shell for axisymmetric plane strain dynamic thermoelastic problems. *J. Sound Vib.*, 2003, *Vol* 263, 815-829.

El-Raheb, M. Transient waves in an inhomogeneous hollow infinite cylinder. *Int. J. Solids Struct.*, 2005, *Vol* 42, 5356-5376.

El-Raheb, M. Transient response of a hollow cylinder with radial and axial material inhomogeneity. *J. Acoust. Soc. Am.*, 2006, *Vol* 119, 2738-2749.

Hosseini, S. M., Akhlaghi, M. and Shakeri, M. Dynamic response and radial wave propagation velocity in thick hollow cylinder made of functionally graded materials. *Eng. Comput.*, 2007, *Vol* 24, 288-303.

Sarma K.V. Torsional wave motion of a finite inhomogeneous piezoelectric cylindrical Shell. *Int. J. Engng. Sci.,* 1980, *Vol* 18, 449-454.

Hou, P. F., Wang, H. M. and Ding, H. J. Analytical solution for the axisymmetric plane strain electroelastic dynamics of a special non-homogeneous piezoelectric hollow cylinder. *Int. J. Eng. Sci.*, 2003, *Vol* 41, 1849-1868.

Ding, H. J., Wang, H. M. and Chen, W. Q. Analytical solution for the electroelastic dynamics of a nonhomogeneous spherically isotropic piezoelectric hollow sphere. *Arch. Appl. Mech.*, 2003, *Vol* 73, 49-62.

Ding, H. J., Wang, H. M. and Chen, W. Q. Dynamic response of a pyroelectric hollow sphere under radial deformation. *Eur. J. Mech. - A/Solids*, 2003, *Vol* 22, 617-631.

Ding, H. J., Wang, H. M. and Chen, W. Q. Dynamic responses of a functionally graded pyroelectric hollow sphere for spherically symmetric problems. *Int. J. Mech. Sci.*, 2003, *Vol* 45, 1029-1051.

Ding, H. J., Wang, H. M. and Chen, W. Q. Analytical solution of a special non-homogeneous pyroelectric hollow cylinder for piezothermoelastic axisymmetric plane strain dynamic problems. *Appl. Math. Comput.*, 2004, *Vol* 151, 423-441.

Wang, H. M. and Ding, H. J. Transient responses of a special non-homogeneous magneto-electro-elastic hollow cylinder for fully coupled axisymmetric plane strain problem. *Acta Mech.*, 2006, *Vol* 184, 137-157.

Wang, H. M. and Ding, H. J. Spherically symmetric transient responses of functionally graded magneto-electro-elastic hollow sphere. *Struct. Eng. Mech.*, 2006, *Vol* 23, 525-542.

Hou, P. F., Ding, H. J. and Leung, A. Y. T. The transient responses of a special nom-homogeneous magneto-electro-elastic hollow cylinder for axisymmetric plane strain problem. *J. Sound Vib.*, 2006, *Vol* 291, 19-47.

Dai, H. L. and Fu, Y. M. Magnetothermoelastic interactions in hollow structures of functionally graded material subjected to mechanical loads. *Int. J. Press. Ves. Pip.*, 2007, *Vol* 84, 132-138.

Dai, H. L., Fu, Y. M. and Yang, J. H. Electromagnetoelastic behaviors of functionally graded piezoelectric solid cylinder and sphere. *Acta Mech. Sin.*, 2007, *Vol* 23, 55-63.

Li, J. Y. Magnetoelectroelastic multi-inclusion and inhomogeneity problems and their applications in composite materials. *Int. J. Eng. Sci.*, 2000, *Vol* 38, 1993-2011.

Aboudi, J. Micromechanical analysis of fully coupled electro-magneto- thermo-elastic multiphase composites. *Smart Mater. Struct.*, 2001, *Vol* 10, 867-877.

Ding, H. J. and Chen, W. Q. *Three dimensional problems of piezoelasticity*. New York, Nova Science Publishers, Inc. 2001.

Wang, X. and Zhong, Z. The general solution of spherically isotropic megnetoelectroelastic media and its applications. *Eur. J. Mech. A/Solids*, 2003, *Vol* 22, 953-969.

Buchanan, G. R. Free vibration of an infinite magneto-electro-elastic cylinder. *J. Sound Vib.*, 2003, *Vol* 268, 413-426.

ABOUT THE AUTHORS

Huiming Wang received the B.S. and M.S. degrees in engineering mechanics from Dalian University of Technology, Dalian, China in 1994 and 1997 respectively and the Ph.D. degree in solid mechanics from Zhejiang University, Hangzhou, China in 2003. He is an associate professor of the Engineering Mechanics Department, Zhejiang University. His research focuses on elastodynamics of piezoelectric structures and systems.

In: Mechanics of Functionally Graded Materials and Structures ISBN 978-1-61668-920-9
Editors: Z. Zhong, L. Wu and W. Chen

Chapter 8

FRACTURE BEHAVIOR OF FUNCTIONALLY GRADED PIEZOELECTRIC/PIEZOMAGNETIC MATERIALS

***Zhen-Gong Zhou*[1] *and Lin-Zhi Wu*[1]**
Center for Composite Materials, Harbin Institute of Technology, Harbin, China

ABSTRACT

In this chapter, the dynamic behavior of an interface crack and the basic solution of a Mode-I limited-permeable crack in functionally graded piezoelectric/piezomagnetic materials were investigated. At the same time, the electric permittivity and the magnetic permeability of air in crack are considered in the present paper. By using the Fourier transform technique, these problems can be solved with the help of a pair of dual integral equations in which the unknown variable is the displacement jump across the crack surfaces. These equations were solved by using the Schmidt method. The relations among the electric filed, the magnetic flux field and the stress field near the crack tips can be obtained. From the solutions of the present paper, it is found that the singular stresses, the singular electric displacements and the singular magnetic flux at crack tips in functionally graded piezoelectric/piezomagnetic materials carry the same forms as those in homogeneous piezoelectric/piezomagnetic materials but the magnitudes of the intensity factors are dependent on the gradient of functionally graded piezoelectric/piezomagnetic materials. It can be also obtained that the effects of the electric and magnetic boundary conditions on the electric displacement and magnetic flux fields near crack tips are large from the solution of the present paper.

1. INTRODUCTION

The piezoelectric/piezomagnetic composites possess piezoelectric, piezomagnetic and magneto-electric effects. In some cases, the coupling effect of piezoelectric/piezomagnetic

[1] e-mail: wlz@hit.edu.cn.

composites can be a hundred times larger than that in a single-phase magnetoelectric material, thereby making the composite sensitive to elastic, electric and magnetic fields. Consequently, they are extensively used as electric packaging, sensors and actuators, e.g., magnetic field probes, acoustic/ultrasonic devices, hydrophones, and transducers with the responsibility of electro-magneto-mechanical energy conversion (Wu and Huang, 2000). The study of fracture mechanics of piezoelectric/piezomagnetic materials plays an important role in the design of piezoelectric/piezomagnetic devices and piezoelectric/piezomagnetic composite structures, for instance as dampers in controlling structural vibration, as sensors and actuators in adhesively bonded joints, and as sensors in non-destructive testing. In these structural applications, fracture of piezoelectric/piezomagnetic ceramics is a critical failure mode for the structures to lose their structural capacity and functionality. Understanding of fracture of piezoelectric/piezomagnetic ceramics under various loading conditions is vital for further advancement of the smart structures technology. In this field, fracture parameters, such as stress intensity factors, electric displacement intensity factors, magnetic flux intensity factors and energy release rate were studied by many researchers (Sih and Song, 2003; Song and Sih, 2003; Wang and Mai, 2003; Gao, et al., 2003a, b, c, d; Spyropoulos, et al., 2003) under different electric boundary conditions on the crack surfaces. The development of piezoelectric/piezomagnetic composites has its roots in the early work of Van Suchtelen (1972) who proposed that the combination of piezoelectric/piezomagnetic phases may exhibit a new material property—the magnetoelectric coupling effect. Since then, many researchers have studied magnetoelectric coupling effect in $BaTiO_3$-$CoFe_2O_4$ composites (Wu and Huang, 2000; Sih and Song, 2003; Song and Sih, 2003; Wang and Mai, 2003; Gao, et al., 2003a, b, c, d; Spyropoulos, et al., 2003; Liu, et al., 2001; Wang and Mai, 2004; Harshe and Dougherty, 1993; Avellaneda and Harshe, 1994; Nan, 1994; Benveniste, 1995; Huang and Kuo, 1997; Li, 2000). For instances, Liu et al. (2001) studied the generalized 2D problem of an infinite magnetoelectroelastic plane with an elliptical hole using Green's functions; Chung and Ting (1995) obtained the two-dimensional Green functions for a magnetoelectroelastic anisotropic medium with an elliptical cavity or rigid inclusion; Pan (2002) derived the three-dimensional Green functions in anisotropic magnetoelectroelastic bimaterials; Gao et al. (2003b, 2003c) and Wang and Mai (2004) also studied the fracture problem of piezoelectric/piezomagnetic composites by Stroh formalism; Chen et al. (2004) obtained the exact three-dimensional expressions for the full-space magneto-electro-thermo-elastic field for a penny-shaped crack subject to a uniform load on the crack surfaces with six harmonic functions; Wang and Shen (2002) obtained the general solution of three-dimensional problems in magnetoelectroelastic media through five potential functions.

The development of piezoelectric/piezomagnetic composites has its roots in the early work of Van Suchtelen (1972) who proposed that the combination of piezoelectric/piezomagnetic phases may exhibit a new material property—the magnetoelectric coupling effect. Since then, only a few researchers studied magnetoelectric coupling effect in $BaTiO_3$-$CoFe_2O_4$ composites such as Wu and Huang (2000), Liu et al. (2001), Chung and Ting (1995), Pan (2002), Gao et al. (2003b and 2003c), Wang and Mai (2004), Chen et al. (2004), Wang and Shen (2002), Harshe (1993), Avellaneda and Harshe (1994), Nan (1994), Benveniste (1995), Huang and Kuo (1997), and Li (2000). Multiple-crack interaction problems in magnetoelectroelastic solids were also investigated by the 'pseudo-traction-electric displacement' method (Tian and Gabbert, 2004). Recently, Zhou and Wang (2004), Zhou et al. (2004, 2005a, b) investigated the static fracture behaviors of a single crack or two cracks in piezoelectric/piezomagnetic materials by Schmidt method (Morse and Feshbach, 1958).

Although many experts, such as Gao et al. (1997), Hao and Shen (1994), Sosa (1992), Suo et al. (1992), Zhang and Tong (1996), Zhang et al. (1997), Zhong et al. (1997) and

McMeeking (1989) have studied the fracture problem of piezoelectric materials, there are still arguments about the electric boundary conditions along the crack surfaces. Some authors such as Parton (1976) and Mikhailov and Parton (1990) argued that since the thickness of the crack is very small, the electric potential and the electric displacement should be continuous across the crack surfaces. This is the so-called permeable crack model. Others assumed that since air occupies the crack gap and the permittivity of the air inside the crack is far less than those of piezoelectric materials (Pak, 1990), the electric potential and the electric displacement are not continuous across the crack surface, as shown in Deeg's paper (1980). This is the so-called impermeable crack model. It was worth noting that different electric boundary conditions on the crack surfaces led to very different results (Soh et al. 2000). Strictly speaking, even if the permittivity of the air inside the crack is quite small, the flux of an electric field through the crack gap should not be zero, so it is more reasonable to assume the electric boundary condition along the crack surfaces take the following form (Hao and Shen, 1994; Hao 2001; Dascalu and Homentcovschi, 2002):

$$D_z^+ = D_z^-,\ D_z^+\left(w^+ - w^-\right) = \varepsilon_0\left(\phi^+ - \phi^-\right) \tag{1-1}$$

in which D_z, ϕ, ε_0 and $\left(w^+ - w^-\right)$ are the electric displacement component along z-axis, the electric potential, the permittivity of the air inside the crack, and the opening displacement component of crack surfaces, respectively. This electric boundary condition was firstly proposed by Hao and Shen (1994) as the limited-permeable crack model, which will be reduced to permeable boundary conditions when $w^+ - w^- = 0$ and to impermeable ones when $\varepsilon_0 = 0$.

On the other hand, the development of functionally graded materials has demonstrated that they have the potential to reduce the stress concentration and increase of fracture toughness. Consequently, some applications of functionally graded piezoelectric materials have been made by Takagi, et al. (2003) and Jin (2003). The fracture problems of functionally graded piezoelectric materials have been considered by Chen, et al., (2003); Jin and Zhong (2002); Soon (2003); Li and Weng (2002) and Ueda (2006). Li and Weng (2002) first considered the static anti-plane problem of a finite crack in a functionally graded piezoelectric material strip. The problems of electriomechanical impact of a mode-I impermeable crack in a functionally graded piezoelectric medium and a functionally graded piezoelectric strip were studied by Chen, et al., (2003) and Ueda (2006). The concept of functionally graded materials was firstly extended to the piezoelectric/piezomagnetic materials to improve the reliability of piezoelectric/piezomagnetic materials and structures by Zhou and Wang (2004) and Zhou, et al. (2005b) until recent years. However, to our knowledge, the results of fracture problems of piezoelectric/piezomagnetic composite materials are not richer than ones of piezoelectric materials and the electric permittivity and magnetic permeability of the air inside the crack in piezoelectric/piezomagnetic materials have not been considered before. Other, the magneto-electro-elastic behavior of magnetoelectroelastic composite materials with two collinear mode-I permeable cracks has not been studied by the non-local theory in the open literature. Here, we will report these works.

This part summarizes the recent development of fracture problems of functionally graded piezoelectric/piezomagnetic materials and structure. The review covers dynamic response of crack to harmonic stress waves in functionally graded piezoelectric and piezoelectric/piezomagnetic materials.

2. An Interface Crack in Functionally Graded Piezoelectric/Piezomagnetic Materials

Here, the concept of functionally graded materials was extended to the piezoelectric/piezomagnetic materials by Zhou and Wang (2008). The dynamic magneto-electro-elastic behavior of a permeable interface crack between two dissimilar functionally graded piezoelectric/piezomagnetic material half infinite planes subjected to the harmonic anti-plane shear stress waves was investigated using the Schmidt method (Morse and Feshbach, 1958). To make the analysis tractable, it was assumed that the material properties vary exponentially with coordinate vertical to the crack. Fourier transform technique is applied and a mixed boundary value problem is reduced to a pair of dual integral equations. To solve the dual integral equations, the jump of the displacements across the crack surface was expanded in a series of Jacobi polynomials. Numerical solutions (Zhou and Wang, 2008) were obtained for the dynamic stress, the electric displacement and the magnetic flux intensity factors for permeable crack surface conditions. The paper (Zhou and Wang, 2008) was rewritten here.

2.1. Formulation of the Problem

Consider an interface crack of length $2l$ between two dissimilar functionally graded piezoelectric/piezomagnetic material half infinite planes as shown in Figure II-1. It wais assumed that the propagation direction of the harmonic elastic anti-plane shear stress wave is vertical to the crack in functionally graded piezoelectric/piezomagnetic materials. Let ω be the circular frequency of the incident wave. $w_0^{(i)}(x,y,t)$, $\phi_0^{(i)}(x,y,t)$ and $\psi_0^{(i)}(x,y,t)\,(i=1,2)$ are the mechanical displacement, the electric potential and the magnetic potential, respectively. $\tau_{zk0}^{(i)}(x,y,t)$, $D_{k0}^{(i)}(x,y,t)$ and $B_{k0}^{(i)}(x,y,t)$ ($k=x,y$, $i=1,2$) are the anti-plane shear stress field, in-plane electric displacement field and in-plane magnetic flux, respectively. Also note that all quantities with superscript i corresponds to the upper half plane 1 (for $i=1$) and the lower half plane 2 (for $i=2$) as shown in Figure2-1, respectively. Because of the incident waves is the harmonic anti-plane shear stress wave, all field quantities of $w_0^{(i)}(x,y,t)$, $\phi_0^{(i)}(x,y,t)$, $\psi_0^{(i)}(x,y,t)$, $\tau_{zk0}^{(i)}(x,y,t)$, $D_{k0}^{(i)}(x,y,t)$ and $B_{k0}^{(i)}(x,y,t)$ can be assumed to be of the forms as follows

$$\begin{aligned}&[w_0^{(i)}(x,y,t),\ \phi_0^{(i)}(x,y,t),\ \psi_0^{(i)}(x,y,t),\tau_{zk0}^{(i)}(x,y,t),D_{k0}^{(i)}(x,y,t),\\&B_{k0}^{(i)}(x,y,t)]=[w^{(i)}(x,y),\ \phi^{(i)}(x,y),\ \psi^{(i)}(x,y),\tau_{zk}^{(i)}(x,y),D_k^{(i)}(x,y),\\&B_k^{(i)}(x,y)]e^{-i\omega t}\end{aligned}\tag{2-1}$$

In what follows, the time dependence of $e^{-i\omega t}$ will be suppressed but understood. The functionally graded piezoelectric/piezomagnetic materials boundary-value problem for the harmonic anti-plane shear waves is considerably simplified if we consider only the out-of-plane displacement, the in-plane electric fields and the in-plane magnetic fields. As discussed in Soh's (Soh, et al., 2000) and Srivastava's (Srivastava, et al., 1983) works, since no opening displacement exists for the present anti-plane problem, the crack surfaces can be assumed to be in perfect contact. Accordingly, permeable condition will be enforced in the present study, i.e., the electric potential, the magnetic potential, the normal electric displacement and the magnetic flux are assumed to be continuous across the crack surfaces. Therefore the boundary conditions of the present problem are (In this paper, we just consider the perturbation field.)

$$\begin{cases} \tau_{yz}^{(1)}(x,0) = \tau_{yz}^{(2)}(x,0) = -\tau_0, |x| \le l \\ \tau_{yz}^{(1)}(x,0) = \tau_{yz}^{(2)}(x,0), w^{(1)}(x,0) = w^{(2)}(x,0), |x| > l \end{cases} \tag{2-2}$$

$$\begin{cases} \phi^{(1)}(x,0) = \phi^{(2)}(x,0), D_y^{(1)}(x,0) = D_y^{(2)}(x,0) \\ \psi^{(1)}(x,0) = \psi^{(2)}(x,0), B_y^{(1)}(x,0) = B_y^{(2)}(x,0) \end{cases}, |x| \le \infty \tag{2-3}$$

$$\begin{cases} w^{(1)}(x,y) = w^{(2)}(x,y) = 0 \\ \phi^{(1)}(x,y) = \phi^{(2)}(x,y) = 0 \quad \text{for } (x^2 + y^2)^{1/2} \to \infty \\ \psi^{(1)}(x,y) = \psi^{(2)}(x,y) = 0 \end{cases} \tag{2-4}$$

where $-\tau_0$ is a magnitude of the incident wave.

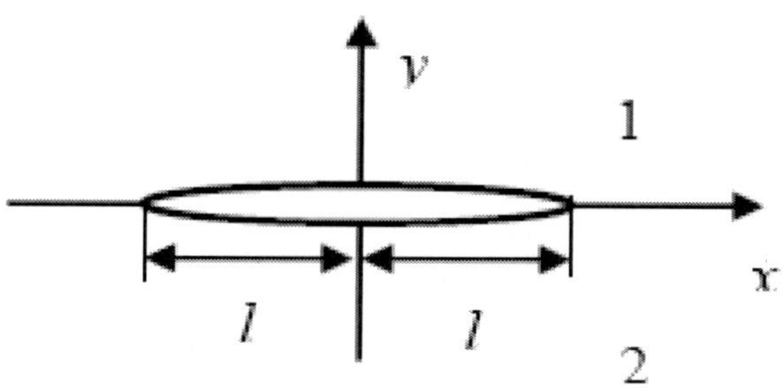

Figure2-1. An interface crack in functionally graded piezoelectric/piezomagnetic materials

Crack problems in the non-homogeneous piezoelectric/piezomagnetic materials do not appear to be analytically tractable for arbitrary variations of material properties. Usually, one tries to generate the forms of non- homogeneities for which the problem becomes tractable. Similar to the treatment of the crack problem for isotropic non-homogeneous materials in Refs. (Delale and Erdogan, 1988), we assume the material properties are described by:

$$\begin{cases} c_{44}^{(1)} = c_{440}^{(1)} e^{\beta^{(1)} y}, e_{15}^{(1)} = e_{150}^{(1)} e^{\beta^{(1)} y}, \varepsilon_{11}^{(1)} = \varepsilon_{110}^{(1)} e^{\beta^{(1)} y} \\ q_{15}^{(1)} = q_{150}^{(1)} e^{\beta^{(1)} y}, d_{11}^{(1)} = d_{110}^{(1)} e^{\beta^{(1)} y}, \mu_{11}^{(1)} = \mu_{110}^{(1)} e^{\beta^{(1)} y}, \rho^{(1)}(y) = \rho_0^{(1)} e^{\beta^{(1)} y} \\ c_{44}^{(2)} = c_{440}^{(2)} e^{\beta^{(2)} y}, e_{15}^{(2)} = e_{150}^{(2)} e^{\beta^{(2)} y}, \varepsilon_{11}^{(2)} = \varepsilon_{110}^{(2)} e^{\beta^{(2)} y} \\ q_{15}^{(2)} = q_{150}^{(2)} e^{\beta^{(2)} y}, d_{11}^{(2)} = d_{110}^{(2)} e^{\beta^{(2)} y}, \mu_{11}^{(2)} = \mu_{110}^{(2)} e^{\beta^{(2)} y}, \rho^{(2)}(y) = \rho_0^{(2)} e^{\beta^{(2)} y} \end{cases} \tag{2-5}$$

where $c_{440}^{(i)}$, $e_{150}^{(i)}$, $\varepsilon_{110}^{(i)}$, $q_{150}^{(i)}$, $d_{150}^{(i)}$, $\mu_{110}^{(i)}$, $\rho_0^{(i)}$ and $\beta^{(i)}$ are the shear modulus, the piezoelectric coefficient, the dielectric parameter, the piezomagnetic coefficient, the magnetioelectric coefficient, the magnetic permeability, the mass density and the functionally graded parameter of two dissimilar functionally graded piezoelectric/piezomagnetic material half planes, respectively.

The constitutive equations for the mode III crack can be expressed as

$$\tau_{zk}^{(i)} = c_{44}^{(i)} w^{(i)}{}_{,k} + e_{15}^{(i)} \phi^{(i)}{}_{,k} + q_{15}^{(i)} \psi^{(i)}{}_{,k}, \ (k = x, y, \ i = 1, 2) \tag{2-6}$$

$$D_k^{(i)} = e_{15}^{(i)} w^{(i)}{}_{,k} - \varepsilon_{11}^{(i)} \phi^{(i)}{}_{,k} - d_{11}^{(i)} \psi^{(i)}{}_{,k}, \ (k = x, y, \ i = 1, 2) \tag{2-7}$$

$$B_k^{(i)} = q_{15}^{(i)} w^{(i)}{}_{,k} - d_{11}^{(i)} \phi^{(i)}{}_{,k} - \mu_{11}^{(i)} \psi^{(i)}{}_{,k}, \ (k = x, y, \ i = 1, 2) \tag{2-8}$$

The anti-plane governing equations are

$$c_{440}^{(i)}(\nabla^2 w^{(i)} + \beta^{(i)} \frac{\partial w^{(i)}}{\partial y}) + e_{150}^{(i)}(\nabla^2 \phi^{(i)} + \beta^{(i)} \frac{\partial \phi^{(i)}}{\partial y}) + q_{150}^{(i)}(\nabla^2 \psi^{(i)} + \beta^{(i)} \frac{\partial \psi^{(i)}}{\partial y}) = -\rho_0^{(i)} \omega^2 w^{(i)} \tag{2-9}$$

$$e_{150}^{(i)}(\nabla^2 w^{(i)} + \beta^{(i)} \frac{\partial w^{(i)}}{\partial y}) - \varepsilon_{110}^{(i)}(\nabla^2 \phi^{(i)} + \beta^{(i)} \frac{\partial \phi^{(i)}}{\partial y}) - d_{110}^{(i)}(\nabla^2 \psi^{(i)} + \beta^{(i)} \frac{\partial \psi^{(i)}}{\partial y}) = 0 \tag{2-10}$$

$$q_{150}^{(i)}(\nabla^2 w^{(i)} + \beta^{(i)} \frac{\partial w^{(i)}}{\partial y}) - d_{110}^{(i)}(\nabla^2 \phi^{(i)} + \beta^{(i)} \frac{\partial \phi^{(i)}}{\partial y}) - \mu_{110}^{(i)}(\nabla^2 \psi^{(i)} + \beta^{(i)} \frac{\partial \psi^{(i)}}{\partial y}) = 0 \tag{2-11}$$

where $-\rho_0^{(i)} \omega^2 w^{(i)}(x, y) e^{-i\omega t} = \rho_0^{(i)} \dfrac{\partial^2 w_0^{(i)}(x, y, t)}{\partial t^2} = \rho_0^{(i)} \dfrac{\partial^2 (w^{(i)}(x, y) e^{-i\omega t})}{\partial t^2}$ and $\nabla^2 = \partial^2 / \partial x^2 + \partial^2 / \partial y^2$ is the two dimensional Laplace operator.

2.2. Solutions

Because of the assumed symmetry in geometry and loading, it was sufficient to consider the problem for $0 \le x < \infty$, $-\infty \le y < \infty$ only. The solutions of above governing equations (2-9)-(2-11) was solved by using the Fourier integral transform technique can be expressed as follows:

$$\begin{cases} w^{(1)}(x, y) = \dfrac{2}{\pi} \displaystyle\int_0^\infty A_1(s) e^{-\gamma_1^{(1)} y} \cos(sx) ds \\ \phi^{(1)}(x, y) = a_0^{(1)} w^{(1)}(x, y) + \dfrac{2}{\pi} \displaystyle\int_0^\infty B_1(s) e^{-\gamma_2^{(1)} y} \cos(sx) ds, \ (y \ge 0) \\ \psi^{(1)}(x, y) = a_1^{(1)} w^{(1)}(x, y) + \dfrac{2}{\pi} \displaystyle\int_0^\infty C_1(s) e^{-\gamma_2^{(1)} y} \cos(sx) ds \end{cases} \tag{2-12}$$

$$\begin{cases} w^{(2)}(x,y) = \dfrac{2}{\pi}\displaystyle\int_0^\infty A_2(s)e^{\gamma_1^{(2)}y}\cos(sx)ds \\ \phi^{(2)}(x,y) = a_0^{(2)}w^{(2)}(x,y) + \dfrac{2}{\pi}\displaystyle\int_0^\infty B_2(s)e^{\gamma_2^{(2)}y}\cos(sx)ds \ , (y \le 0) \\ \psi^{(2)}(x,y) = a_1^{(2)}w^{(2)}(x,y) + \dfrac{2}{\pi}\displaystyle\int_0^\infty C_2(s)e^{\gamma_2^{(2)}y}\cos(sx)ds \end{cases} \tag{2-13}$$

where $A_1(s)$, $B_1(s)$, $C_1(s)$, $A_2(s)$, $B_2(s)$ and $C_2(s)$ are unknown functions, where

$$\gamma_1^{(1)} = \frac{\beta^{(1)} + \sqrt{\beta^{(1)2} + 4[s^2 - \omega^2 / c_1^2]}}{2}, \ \gamma_2^{(1)} = \frac{\beta^{(1)} + \sqrt{\beta^{(1)2} + 4s^2}}{2},$$

$$c_1 = \sqrt{\mu_0^{(1)} / \rho_0^{(1)}}, \ \mu_0^{(1)} = c_{440}^{(1)} + a_0^{(1)}e_{150}^{(1)} + a_1^{(1)}q_{150}^{(1)}, \ a_0^{(1)} = \frac{\mu_{110}^{(1)}e_{150}^{(1)} - d_{110}^{(1)}q_{150}^{(1)}}{\varepsilon_{110}^{(1)}\mu_{110}^{(1)} - d_{110}^{(1)2}},$$

$$a_1^{(1)} = \frac{q_{150}^{(1)}\varepsilon_{110}^{(1)} - d_{11}e_{150}^{(1)}}{\varepsilon_{110}^{(1)}\mu_{110}^{(1)} - d_{110}^{(1)2}}, \ \gamma_1^{(2)} = \frac{\beta^{(2)} + \sqrt{\beta^{(2)2} + 4[s^2 - \omega^2 / c_2^2]}}{2},$$

$$\gamma_2^{(2)} = \frac{\beta^{(2)} + \sqrt{\beta^{(2)2} + 4s^2}}{2}, \ c_2 = \sqrt{\mu_0^{(2)} / \rho_0^{(2)}},$$

$$\mu_0^{(2)} = c_{440}^{(2)} + a_0^{(2)}e_{150}^{(2)} + a_1^{(2)}q_{150}^{(2)}, \ a_0^{(2)} = \frac{\mu_{110}^{(2)}e_{150}^{(2)} - d_{110}^{(2)}q_{150}^{(1)}}{\varepsilon_{110}^{(2)}\mu_{110}^{(2)} - d_{110}^{(2)2}}, \ a_1^{(2)} = \frac{q_{150}^{(2)}\varepsilon_{110}^{(2)} - d_{110}^{(2)}e_{150}^{(2)}}{\varepsilon_{110}^{(2)}\mu_{110}^{(2)} - d_{110}^{(2)2}}.$$

From Eqs. (2-6)-(2-8), we have

$$\tau_{yz}^{(1)}(x,y) = -\frac{2e^{\beta^{(1)}y}}{\pi}\int_0^\infty \{\mu_0^{(1)}\gamma_1^{(1)}A_1(s)e^{-\gamma_1^{(1)}y}$$

$$+\gamma_2^{(1)}[e_{150}^{(1)}B_1(s) + q_{150}^{(1)}C_1(s))]e^{-\gamma_2^{(1)}y}\}\cos(sx)ds \tag{2-14}$$

$$D_y^{(1)}(x,y) = \frac{2e^{\beta^{(1)}y}}{\pi}\int_0^\infty \gamma_2^{(1)}[\varepsilon_{110}^{(1)}B_1(s) + d_{110}^{(1)}C_1(s)]e^{-\gamma_2^{(1)}y}\cos(sx)ds \tag{2-15}$$

$$B_y^{(1)}(x,y) = \frac{2e^{\beta^{(1)}y}}{\pi}\int_0^\infty \gamma_2^{(1)}[d_{110}^{(1)}B_1(s) + \mu_{110}^{(1)}C_1(s)]e^{-\gamma_2^{(1)}y}\cos(sx)ds \tag{2-16}$$

$$\tau_{yz}^{(2)}(x,y)=\frac{2e^{\beta^{(2)}y}}{\pi}\int_0^\infty\{\mu_0^{(2)}\gamma_1^{(2)}A_2(s)e^{\gamma_1^{(2)}y}$$
$$+\gamma_2^{(2)}[e_{150}^{(2)}B_2(s)+q_{150}^{(2)}C_2(s))]e^{-\gamma_2^{(2)}y}\}\cos(sx)ds \tag{2-17}$$

$$D_y^{(2)}(x,y)=-\frac{2e^{\beta^{(2)}y}}{\pi}\int_0^\infty\gamma_2^{(2)}[\varepsilon_{110}^{(2)}B_2(s)+d_{110}^{(2)}C_2(s)]e^{\gamma_2^{(2)}y}\cos(sx)ds \tag{2-18}$$

$$B_y^{(2)}(x,y)=-\frac{2e^{\beta^{(2)}y}}{\pi}\int_0^\infty\gamma_2^{(2)}[d_{110}^{(2)}B_2(s)+\mu_{110}^{(2)}C_2(s)]e^{-\gamma_2^{(2)}y}\cos(sx)ds \tag{2-19}$$

To solve the problem, the jump of the displacements across the crack surfaces is defined as follows:

$$f(x)=w^{(1)}(x,0)-w^{(2)}(x,0) \tag{2-20}$$

Substituting Eqs. (2-12)-(2-13) into Eq. (2-20), and applying the Fourier transform and the boundary conditions Eq. (2-3), we have

$$\bar{f}(s)=A_1(s)-A_2(s) \tag{2-21}$$

$$a_0^{(1)}A_1(s)-a_0^{(2)}A_2(s)+B_1(s)-B_2(s)=0 \tag{2-22}$$

$$a_1^{(1)}A_1(s)-a_1^{(2)}A_2(s)+C_1(s)-C_2(s)=0 \tag{2-23}$$

Substituting Eqs. (2-14)-(2-19) into the boundary conditions Eqs. (2-2)-(2-4), it can be obtained

$$\mu_0^{(1)}\gamma_1^{(1)}A_1(s)+\gamma_2^{(1)}[e_{150}^{(1)}B_1(s)+q_{150}^{(1)}C_1(s))]$$
$$+\mu_0^{(2)}\gamma_1^{(2)}A_2(s)+\gamma_2^{(2)}[e_{150}^{(2)}B_2(s)+q_{150}^{(2)}C_2(s))]=0 \tag{2-24}$$

$$\gamma_2^{(1)}[\varepsilon_{110}^{(1)}B_1(s)+d_{110}^{(1)}C_1(s)]+\gamma_2^{(2)}[\varepsilon_{110}^{(2)}B_2(s)+d_{110}^{(2)}C_2(s)]=0 \tag{2-25}$$

$$\gamma_2^{(1)}[d_{110}^{(1)}B_1(s)+\mu_{110}^{(1)}C_1(s)]+\gamma_2^{(2)}[d_{110}^{(2)}B_2(s)+\mu_{110}^{(2)}C_2(s)]=0 \tag{2-26}$$

An over bar indicates the Fourier transform throughout the paper. By solving six equations (2-21)-(2-26) with six unknown functions and substituting the solutions into Eqs. (2-14)-(2-16) and applying the boundary conditions Eqs. (2-2)-(2-3), it can be obtained:

$$\frac{2}{\pi}\int_0^\infty\bar{f}(s)\cos(sx)ds=0,\ x>l \tag{2-27}$$

$$\frac{2}{\pi}\int_0^\infty g_1(s)\bar{f}(s)\cos(sx)ds = -\tau_0 \,,\; 0 \le x \le l \tag{2-28}$$

where $g_1(s)$ is a known function (see Appendix-II). $\lim_{s\to\infty} g_1(s)/s = \beta_1$. β_1 is a constant which depends on the properties of the materials (see Appendix-II). However, β_1 is independent of the functionally graded parameters $\beta^{(1)}$ and $\beta^{(2)}$. When the properties of the upper and the lower half planes are continuous along the crack line, $\beta_1 = -c_{440}^{(1)}/2$. To determine the unknown function $\bar{f}(s)$, the above a pair of dual integral equations (2-27)-(2-28) must be solved.

2.3. Solution of the Dual Integral Equations

The Schmidt method (Morse and Feshbach, 1958) was used to solve the dual integral equations. The jump of the displacements across the crack surfaces was represented by the following series:

$$f(x) = \sum_{n=1}^\infty b_n P_{2n-2}^{(\frac{1}{2},\frac{1}{2})}(\frac{x}{l})(1-\frac{x^2}{l^2})^{\frac{1}{2}} \,,\; \text{for}\; 0 \le x \le l \tag{2-29}$$

$$f(x) = w^{(1)}(x,0) - w^{(2)}(x,0) = 0 \,,\; \text{for}\; x > l \tag{2-30}$$

where b_n are unknown coefficients to be determined and $P_n^{(1/2,1/2)}(x)$ is a Jacobi polynomial (Abramowitz and Stegun, 1968). The Fourier transform of Eqs.(2-29)-(2-30) is (Erdelyi, 1954):

$$\bar{f}(s) = \sum_{n=1}^\infty b_n G_n \frac{1}{s} J_{2n-1}(sl) \,,\; G_n = 2\sqrt{\pi}(-1)^{n-1}\frac{\Gamma(2n-\frac{1}{2})}{(2n-2)!} \tag{2-31}$$

where $\Gamma(x)$ and $J_n(x)$ are the Gamma and Bessel functions, respectively.

Substituting Eq.(2-31) into Eqs. (2-27)-(2-28), equation (2-27) has been automatically satisfied. After integration with respect to x over $[0,x]$, Eq. (2-28) was reduced to following form:

$$\frac{2}{\pi}\sum_{n=1}^\infty b_n G_n \int_0^\infty \frac{1}{s^2} g_1(s) J_{2n-1}(sl)\sin(sx)ds = -\tau_0 x \,,\; 0 \le x \le l \tag{2-32}$$

From the relationship (Abramowitz and Stegun, 1968)

$$\int_0^\infty \frac{1}{s} J_n(sa)\sin(bs)ds = \begin{cases} \dfrac{\sin[n\sin^{-1}(b/a)]}{n}, a > b \\ \dfrac{a^n \sin(n\pi/2)}{n[b+\sqrt{b^2-a^2}]^n}, b > a \end{cases} \tag{2-33}$$

the semi-infinite integral in Eq. (2-32) can be modified as:

$$\int_0^\infty \frac{1}{s}[\beta_1 + (\frac{g_1(s)}{s} - \beta_1)]J_{2n-1}(sl)\sin(sx)ds$$

$$= \frac{\beta_1}{2n-1}\sin[(2n-1)\sin^{-1}(\frac{x}{l})] + \int_0^\infty \frac{1}{s}\frac{g_1(s)-\beta_1 s}{s} J_{2n-1}(sl)\sin(sx)ds \tag{2-34}$$

It can be seen that the integrands in the right end of Eq. (2-34) tend rapidly to zero. Thus the semi-infinite integral in Eq. (2-34) can be numerical evaluated easily. Equation (2-32) can now be solved for the coefficients b_n by the Schmidt method (Morse and Feshbach, 1958) . For brevity, (2-32) can be rewritten as

$$\sum_{n=1}^\infty b_n E_n(x) = U(x), 0 \le x \le l \tag{2-35}$$

where $E_n(x)$ and $U(x)$ are known functions and the coefficients b_n are to be determined. A set of functions $P_n(x)$ which satisfy the orthogonality condition

$$\int_0^l P_m(x)P_n(x)dx = N_n\delta_{mn}\ ,\ N_n = \int_0^l P_n^2(x)dx \tag{2-36}$$

can be constructed from the function $E_n(x)$, such that

$$P_n(x) = \sum_{i=1}^n \frac{M_{in}}{M_{nn}} E_i(x) \tag{2-37}$$

where M_{ij} is the cofactor of the element d_{ij} of D_n, which is defined as

$$D_n = \begin{bmatrix} d_{11}, d_{12}, d_{13}, \ldots, d_{1n} \\ d_{21}, d_{22}, d_{23}, \ldots, d_{2n} \\ d_{31}, d_{32}, d_{33}, \ldots, d_{3n} \\ \cdots\cdots\cdots\cdots \\ \cdots\cdots\cdots\cdots \\ \cdots\cdots\cdots\cdots \\ d_{n1}, d_{n2}, d_{n3}, \ldots, d_{nn} \end{bmatrix},\ d_{ij} = \int_0^l E_i(x)E_j(x)dx \tag{2-38}$$

Using Eqs.(2-35)-(2-38), we obtain

$$b_n = \sum_{j=n}^{\infty} q_j \frac{M_{nj}}{M_{jj}} \text{ with } q_j = \frac{1}{N_j} \int_0^l U(x) P_j(x) dx \tag{2-39}$$

2.4. Intensity Factors

The coefficients b_n are known, so that the entire perturbation stress field, the perturbation electric displacement field and the magnetic flux can be obtained. However, in fracture mechanics, it is of importance to determine the perturbation stress $\tau_{yz}^{(1)}$, the perturbation electric displacement $D_y^{(1)}$ and the magnetic flux $B_y^{(1)}$ in the vicinity of the crack's tips. In the case of the present study, $\tau_{yz}^{(1)}$, $D_y^{(1)}$ and $B_y^{(1)}$ along the crack line can be expressed, respectively, as follows:

$$\tau_{yz}^{(1)}(x,0) = \frac{2}{\pi} \sum_{n=1}^{\infty} b_n G_n \int_0^{\infty} \frac{1}{s} g_1(s) J_{2n-1}(sl) \cos(xs) ds$$
$$= \frac{2\beta_1}{\pi} \sum_{n=1}^{\infty} b_n G_n \int_0^{\infty} J_{2n-1}(sl) \cos(xs) ds + \frac{2}{\pi} \sum_{n=1}^{\infty} b_n G_n \int_0^{\infty} [\frac{1}{s} g_1(s) - \beta_1] J_{2n-1}(sl) \cos(xs) ds \tag{2-40}$$

$$D_y^{(1)}(x,0) = \frac{2}{\pi} \sum_{n=1}^{\infty} b_n G_n \int_0^{\infty} \frac{g_2(s)}{s} J_{2n-1}(sl) \cos(xs) ds$$
$$= \frac{2\beta_2}{\pi} \sum_{n=1}^{\infty} b_n G_n \int_0^{\infty} J_{2n-1}(sl) \cos(xs) ds + \frac{2}{\pi} \sum_{n=1}^{\infty} b_n G_n \int_0^{\infty} [\frac{1}{s} g_2(s) - \beta_2] J_{2n-1}(sl) \cos(xs) ds \tag{2-41}$$

$$B_y^{(1)}(x,0) = \frac{2}{\pi} \sum_{n=1}^{\infty} b_n G_n \int_0^{\infty} \frac{g_3(s)}{s} J_{2n-1}(sl) \cos(xs) ds$$
$$= \frac{2\beta_3}{\pi} \sum_{n=1}^{\infty} b_n G_n \int_0^{\infty} J_{2n-1}(sl) \cos(xs) ds + \frac{2}{\pi} \sum_{n=1}^{\infty} b_n G_n \int_0^{\infty} [\frac{1}{s} g_3(s) - \beta_3] J_{2n-1}(sl) \cos(xs) ds \tag{2-42}$$

where $g_2(s)$ and $g_3(s)$ are known functions (see Appendix II). $\lim_{s\to\infty} g_2(s)/s = \beta_2$. $\lim_{s\to\infty} g_3(s)/s = \beta_3$. Where β_2 and β_3 are two constants which depend on the properties of the materials (see Appendix II). When the properties of the upper and the lower half planes are continuous along the crack line, $\beta_2 = -e_{150}^{(1)}/2$ and $\beta_3 = -q_{150}^{(1)}/2$.

From the relationship (Abramowitz and Stegun, 1968)

$$\int_0^\infty J_n(sa)\cos(bs)ds = \begin{cases} \dfrac{\cos[n\sin^{-1}(b/a)]}{\sqrt{a^2-b^2}}, a>b \\ -\dfrac{a^n \sin(n\pi/2)}{\sqrt{b^2-a^2}[b+\sqrt{b^2-a^2}]^n}, b>a \end{cases} \tag{2-43}$$

the singular parts of the stress field, the electric displacement and the magnetic flux near the crack tips in Eqs. (2-40)-(2-42) can be expressed respectively as follows ($x > l$):

$$\tau = -\frac{2\beta_1}{\pi}\sum_{n=1}^{\infty} b_n G_n H_n(x) \tag{2-44}$$

$$D = -\frac{2\beta_2}{\pi}\sum_{n=1}^{\infty} b_n G_n H_n(x) \tag{2-45}$$

$$B = -\frac{2\beta_3}{\pi}\sum_{n=1}^{\infty} b_n G_n H_n(x) \tag{2-46}$$

where $H_n(x) = \dfrac{(-1)^{n-1} l^{2n-1}}{\sqrt{x^2-l^2}[x+\sqrt{x^2-l^2}]^{2n-1}}$

We obtain the stress intensity factor K as

$$K = \lim_{x\to l^+}\sqrt{2(x-l)}\cdot\tau = -\frac{4\beta_1}{\sqrt{\pi l}}\sum_{n=1}^{\infty} b_n \frac{\Gamma(2n-\frac{1}{2})}{(2n-2)!} \tag{2-47}$$

We obtain the electric displacement intensity factor K^D as

$$K^D = \lim_{x\to l^+}\sqrt{2(x-l)}\cdot D = -\frac{4\beta_2}{\sqrt{\pi l}}\sum_{n=1}^{\infty} b_n \frac{\Gamma(2n-\frac{1}{2})}{(2n-2)!} = \frac{\beta_2}{\beta_1}K \tag{2-48}$$

We obtain the magnetic flux intensity factor K^B as

$$K^B = \lim_{x\to l^+}\sqrt{2(x-l)}\cdot B = -\frac{4\beta_3}{\sqrt{\pi l}}\sum_{n=1}^{\infty} b_n \frac{\Gamma(2n-\frac{1}{2})}{(2n-2)!} = \frac{\beta_3}{\beta_1}K \tag{2-49}$$

2.5. Numerical Calculations and Discussion

From the literatures (Zhou and Wang, 2004; Zhou et al., 2004; 2005a and 2005b), it can be seen that the Schmidt method performs satisfactorily if the first ten terms of the infinite series (2-35) are retained. At $-l \le x \le l$, $y = 0$, it can be obtained that $\tau_{yz}^{(1)} / \tau_0$ is very close to negative unity. Hence, the solution of present paper can also be proved to satisfactory the boundary conditions Eq. (2-1). In all computations, according to Refs. (Sih and Song, 2003; Song and Sih, 2003), the constants of materials-I are assumed to be that

$$c_{440}^{(1)} = 44.0(GPa),\ e_{150}^{(1)} = 5.8(C/m^2),$$
$$\varepsilon_{110}^{(1)} = 5.64 \times 10^{-9}(C^2/Nm^2),\ q_{150}^{(1)} = 275.0(N/Am),$$
$$d_{110}^{(1)} = 0.005 \times 10^{-9}(Ns/VC),\ \mu_{110}^{(1)} = -297.0 \times 10^{-6}(Ns^2/C^2),\ \rho_0^{(1)} = 1500kg/m^3$$

and the constants of materials-II are assumed to be that $c_{440}^{(2)} = 34.0(GPa)$,

$$e_{150}^{(2)} = 4.8(C/m^2),\ \varepsilon_{110}^{(2)} = 4.64 \times 10^{-9}(C^2/Nm^2),$$
$$q_{150}^{(2)} = 195.0(N/Am),\ d_{110}^{(2)} = 0.004 \times 10^{-9}(Ns/VC),$$
$$\mu_{110}^{(2)} = -201.0 \times 10^{-6}(Ns^2/C^2),\ \rho_0^{(2)} = 1000kg/m^3.$$

The normalized non-homogeneity constants $\beta^{(i)} l\ (i = 1,2)$ are varied between –2 and 2, which covers most of the practical cases. The results of the present paper (Zhou and Wang, 2008) were shown in Figs. (2-2)-(2-12). From the results, the following observations are very significant:

From the results, it can be shown that the singular stress, electric displacements and the magnetic flux in functionally graded piezoelectric/piezomagnetic materials carry the same forms as those in the homogeneous piezoelectric/piezomagnetic materials or in the homogeneous piezoelectric materials but the magnitudes of the intensity factors depend significantly upon the gradient of the functionally graded piezoelectric/piezomagnetic materials properties as discussed in Refs (Li and Weng, 2002; Zhou and Wang, 2004).

The electro-magneto-elastic coupling effects can be obtained as shown in Eqs. (2-47)-(2-49). For the electric displacement and the magnetic flux intensity factors, they have the same changing tendency as the stress intensity factor as shown in Figs. (2-2)-(2-4). However, the amplitude values of the electric displacement filed, the magnetic flux field and the stress field are different. The amplitude values of the electric displacement and the magnetic flux fields are very small as shown in Figs. 2-3 and 2-4. The results of the electric displacement and the magnetic flux intensity factors can be directly obtained form the results of the stress intensity factors through Eqs. (2-47)-(2-49). This means that an applied mechanical load alone can produce the electric displacement and magnetic flux singularities. The results of the electric displacement and the magnetic flux intensity factors of the other cases have been omitted in the present paper.

The stress, the electric displacement and the magnetic flux intensity factors in the functionally graded materials tend to increase with the frequency reaching a peak and then to decrease in magnitude as shown in Figs. (2-2)-(2-6). When the material properties of the upper half plane and the lower half plane alone the crack line are continuous, it can be got the same conclusion as shown in Figs. (2-5)-(2-6). It can be also obtained that this conclusion is the same as the dynamic anti-plane shear fracture problem in the isotropic homogeneous materials. From the results, it can be concluded that the stress, the electric displacement and the magnetic fields near the crack tips can be deduced by adjusting the frequency of incident waves in engineering practices.

The stress intensity factors tend to decrease with increase in the functionally graded parameters $\beta^{(i)}l$ $(i = 1, 2)$ as shown in Figs.(2-7)-(2-12). When the material properties of the upper half plane and the lower half plane alone the crack line are continuous, it can be got the same conclusion as shown in Figs. (2-10)-(2-12). This means that, by adjusting the functionally graded parameters, the dynamic stress fields near the crack tips can be reduced.

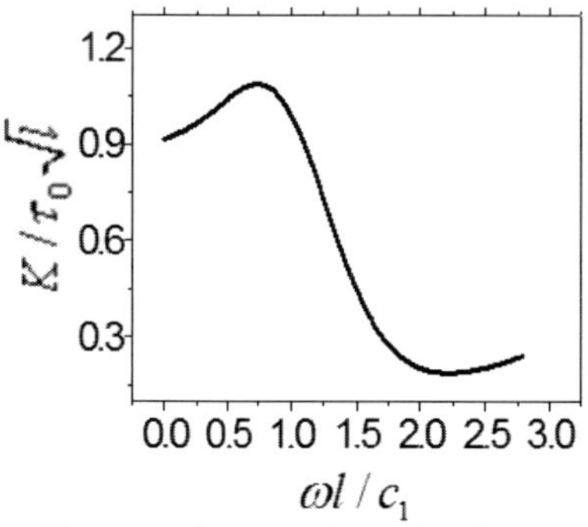

Fig. 2-2. The stress intensity factor versus $\omega l / c_1$ for $l = 1.0$, $\beta^{(1)}l = 0.2$ and $\beta^{(2)}l = 0.3$ (Material-I/Material-II).

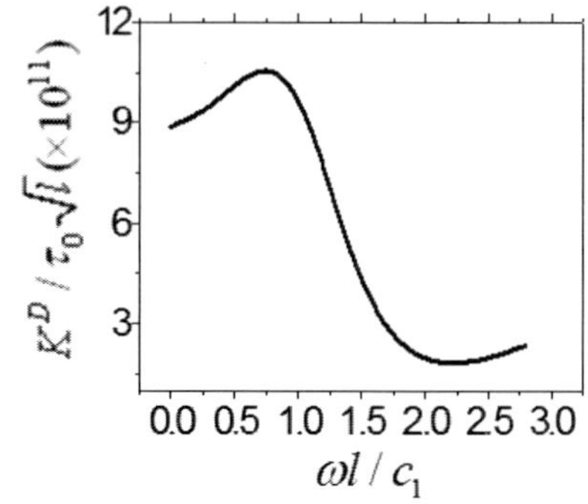

Fig. 2-3. The electric displacement intensity factor versus $\omega l / c_1$ for $l = 1.0$, $\beta^{(1)}l = 0.2$ and $\beta^{(2)}l = 0.3$ (Material-I/Material-II).

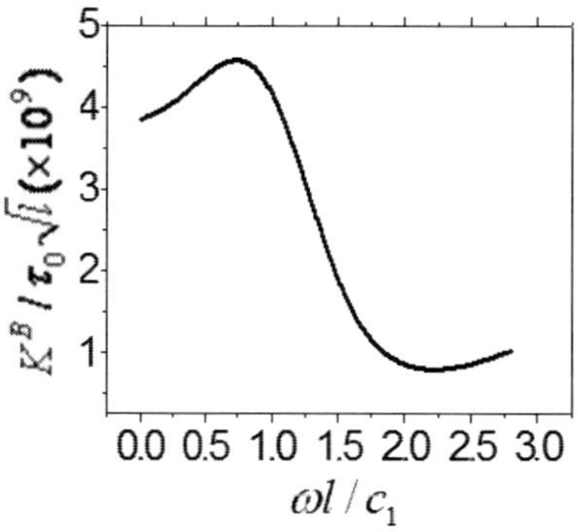

Fig. 2-4. The magnetic flux intensity factor versus $\omega l / c_1$ for $l = 1.0$ $\beta^{(1)}l = 0.2$ and $\beta^{(2)}l = 0.3$ (Material-I/Material-II).

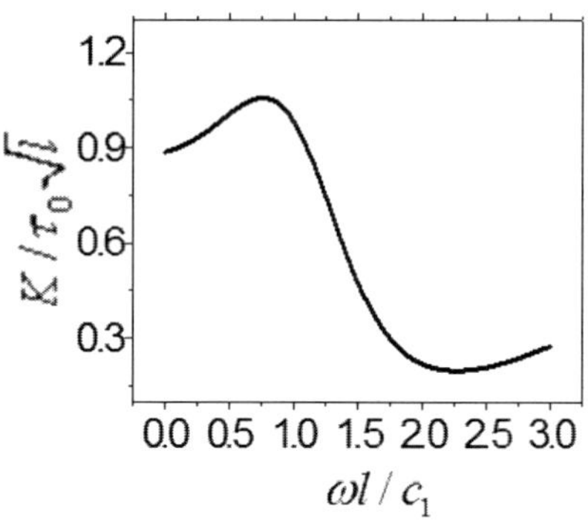

Fig. 2-5. The stress intensity factor versus $\omega l / c_1$ for $l = 1.0$ $\beta^{(1)}l = 0.4$ and $\beta^{(2)}l = 0.3$ (Material-I/Material-I).

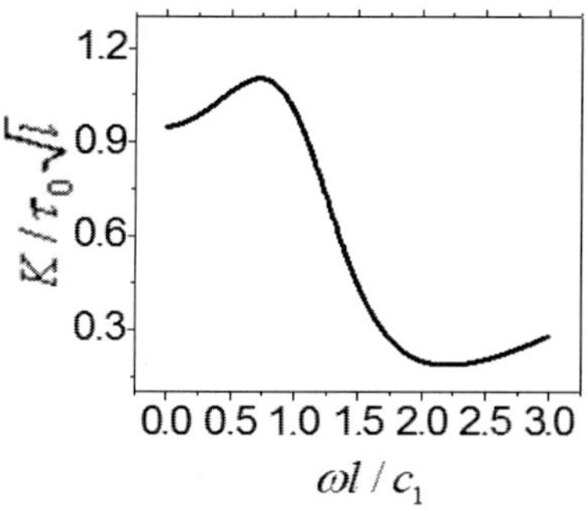

Fig. 2-6. The stress intensity factor versus $\omega l/c_1$ for $l=1.0$ $\beta^{(1)}l=0.4$ and $\beta^{(2)}l=0.0$ (Material-I/Material-I).

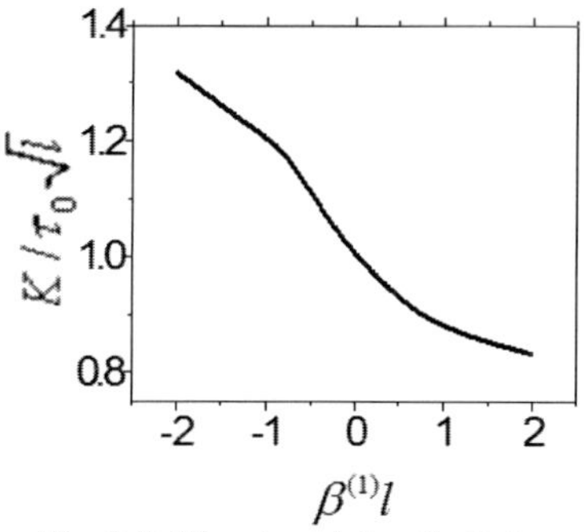

Fig. 2-7. The stress intensity factor versus $\beta^{(1)}l$ for $l=1.0$ $\omega l/c_1=0.3$ and $\beta^{(2)}l=0.3$ (Material-I/Material-II).

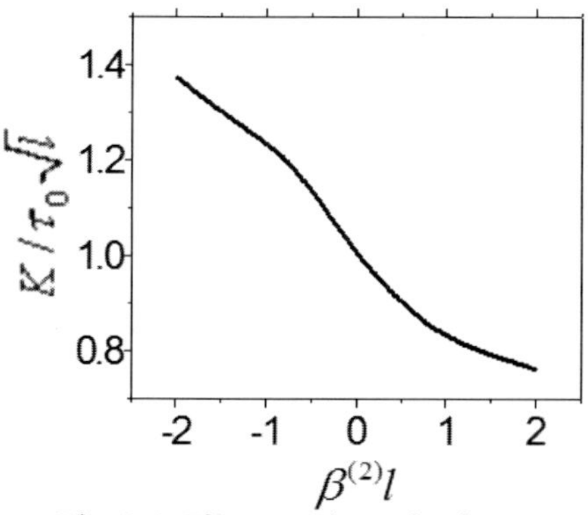

Fig. 2-8. The stress intensity factor versus $\beta^{(2)}l$ for $l=1.0$, $\omega l/c_1=0.3$ and $\beta^{(1)}l=0.4$ (Material-I/Material-II).

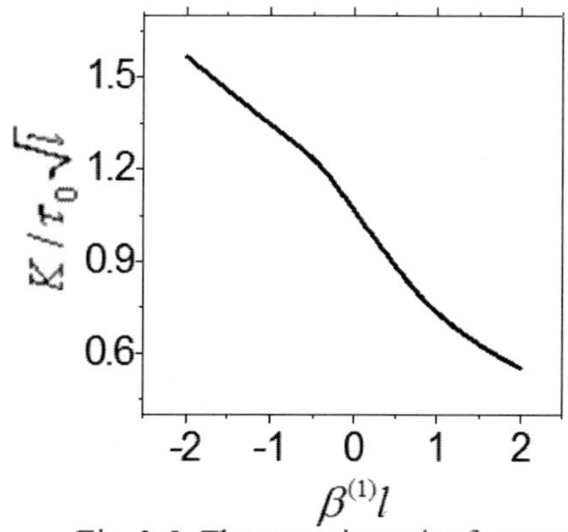

Fig. 2-9. The stress intensity factor versus $\beta^{(1)}l$ for $l=1.0$, $\omega l/c_1=0.3$ and $\beta^{(2)}l=\beta^{(1)}l$ (Material-I/Material-II).

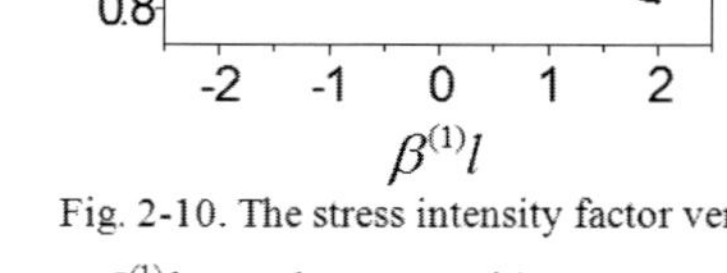

Fig. 2-10. The stress intensity factor versus $\beta^{(1)}l$ for $l=1.0$, $\omega l/c_1=0.3$ and $\beta^{(2)}l=0.3$ (Material-I/Material-I).

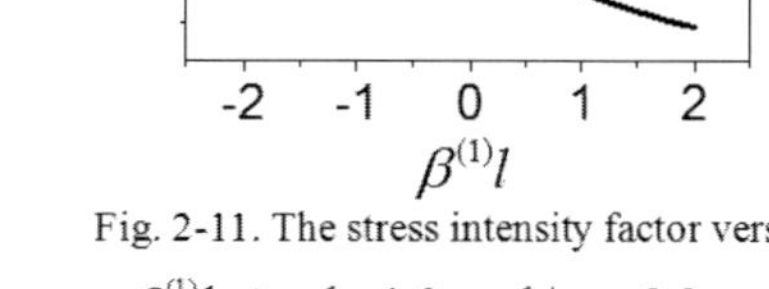

Fig. 2-11. The stress intensity factor versus $\beta^{(1)}l$ for $l=1.0$, $\omega l/c_1=0.3$ and $\beta^{(2)}l=0.0$ (Material-I/Material-I).

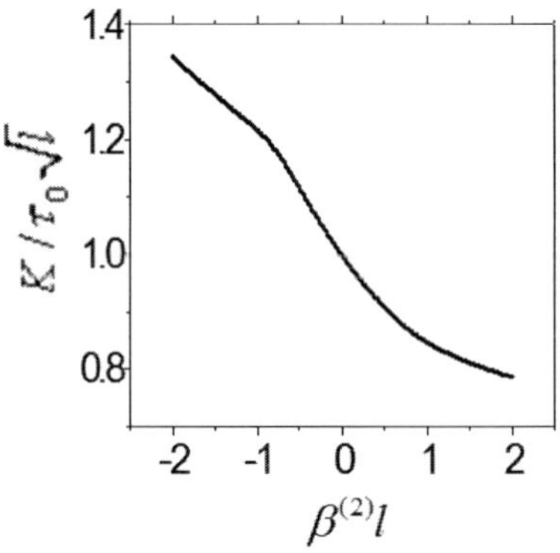

Fig. 2-12. The stress intensity factor versus $\beta^{(2)}l$ for $l=1.0$, $\omega l/c_1=0.3$ and $\beta^{(1)}l=0.4$ (Material-I/Material-I).

(v) The solution of the present paper can revert to the one of the problem which the material properties of the upper half plane and the lower half plane alone the crack line are continuous as shown in Figs. (2-5)-(2-6) and (2-10)-(2-12).

APPENDIX II

The functions of $g_1(s)$, $g_2(s)$ and $g_3(s)$ can be obtained by the operation of the follow matrixes.

$$[X_1]=\begin{bmatrix}1 & 0 & 0\\ a_0^{(1)} & 1 & 0\\ a_1^{(1)} & 0 & 1\end{bmatrix},\ [X_2]=\begin{bmatrix}-1 & 0 & 0\\ -a_0^{(2)} & -1 & 0\\ -a_1^{(2)} & 0 & -1\end{bmatrix},$$

$$[X_3]=\begin{bmatrix}\mu_0^{(1)}\gamma_1^{(1)} & \gamma_2^{(1)}e_{150}^{(1)} & \gamma_2^{(1)}q_{150}^{(1)}\\ 0 & \gamma_2^{(1)}\varepsilon_{110}^{(1)} & \gamma_2^{(1)}d_{110}^{(1)}\\ 0 & \gamma_2^{(1)}d_{110}^{(1)} & \gamma_2^{(1)}\mu_{110}^{(1)}\end{bmatrix},$$

$$[X_4]=\begin{bmatrix}\mu_0^{(2)}\gamma_1^{(2)} & \gamma_2^{(2)}e_{150}^{(2)} & \gamma_2^{(2)}q_{150}^{(2)}\\ 0 & \gamma_2^{(2)}\varepsilon_{110}^{(2)} & \gamma_2^{(2)}d_{110}^{(2)}\\ 0 & \gamma_2^{(2)}d_{110}^{(2)} & \gamma_2^{(2)}\mu_{110}^{(2)}\end{bmatrix},\ [X_5]=[X_1]-[X_2][X_4]^{-1}[X_3],$$

$$[X_6]=\begin{bmatrix}-\mu_0^{(1)}\gamma_1^{(1)} & -\gamma_2^{(1)}e_{150}^{(1)} & -\gamma_2^{(1)}q_{150}^{(1)}\\ 0 & \gamma_2^{(1)}\varepsilon_{110}^{(1)} & \gamma_2^{(1)}d_{110}^{(1)}\\ 0 & \gamma_2^{(1)}d_{110}^{(1)} & \gamma_2^{(1)}\mu_{110}^{(1)}\end{bmatrix},$$

$$[X_7]=\begin{bmatrix}x_{11}(s) & x_{12}(s) & x_{13}(s)\\ x_{21}(s) & x_{22}(s) & x_{23}(s)\\ x_{31}(s) & x_{32}(s) & x_{33}(s)\end{bmatrix}=[X_6][X_5]^{-1}.$$

$g_1(s) = x_{11}(s)$, $g_2(s) = x_{21}(s)$, $g_3(s) = x_{31}(s)$.

The constants of β_1, β_2 and β_3 can be obtained by the operation of the follow matrixes.

$$[Y_3] = \begin{bmatrix} \mu_0^{(1)} & e_{150}^{(1)} & q_{150}^{(1)} \\ 0 & \varepsilon_{110}^{(1)} & d_{110}^{(1)} \\ 0 & d_{110}^{(1)} & \mu_{110}^{(1)} \end{bmatrix}, \ [Y_4] = \begin{bmatrix} \mu_0^{(2)} & e_{150}^{(2)} & q_{150}^{(2)} \\ 0 & \varepsilon_{110}^{(2)} & d_{110}^{(2)} \\ 0 & d_{110}^{(2)} & \mu_{110}^{(2)} \end{bmatrix},$$

$$[Y_5] = [X_1] - [X_2][Y_4]^{-1}[Y_3], \ [Y_6] = \begin{bmatrix} -\mu_0^{(1)} & -e_{150}^{(1)} & -q_{150}^{(1)} \\ 0 & \varepsilon_{110}^{(1)} & d_{110}^{(1)} \\ 0 & d_{110}^{(1)} & \mu_{110}^{(1)} \end{bmatrix},$$

$$[Y_7] = \begin{bmatrix} y_{11} & y_{12} & y_{13} \\ y_{21} & y_{22} & y_{23} \\ y_{31} & y_{32} & y_{33} \end{bmatrix} = [Y_6][Y_5]^{-1}, \ \beta_1 = y_{11}, \ \beta_2 = y_{21}, \ \beta_3 = y_{31}.$$

3. Basic Solution of a Mode-I Limited-Permeable Crack

In this part, the limited-permeable crack model in piezoelectric materials (Hao and Shen, 1994; Hao 2001; Dascalu and Homentcovschi, 2002) as shown in equation (1-1) and the concept of functionally graded materials were extended to deal with the electric and magnetic boundary conditions along the crack surfaces for the Mode-I fracture problem of functionally graded piezoelectric/piezomagnetic materials (Zhou and Chen, 2008). The Schmidt method (Morse and Feshbach, 1958) was employed to investigate the behavior of a Mode-I crack in functionally graded piezoelectric/piezomagnetic materials subjected to a stress loading by the generalized Almansi's theorem. The Fourier transform was applied to reduce the mixed boundary-value problem to two pairs of dual integral equations about the jumps of displacements across the crack surfaces, which were solved by a direct expansion as a series of Jacobi polynomials. The solution (Zhou and Chen, 2008) shows that the electric and magnetic boundary conditions along the crack surfaces greatly affect the results of the electric displacement and magnetic flux fields near crack tips. The paper (Zhou and Chen, 2008) was rewritten here.

3.1. The Mode- I Crack

It is assumed that there is a Mode-I Griffith crack of length $2l$ along x-axis in a functionally graded piezoelectric/piezomagnetic material as shown in Figure 3-1. As discussed in the literature (Hao and Shen, 1994; Hao 2001; Dascalu and Homentcovschi, 2002), the electric permittivity and the magnetic permeability of the air inside the crack will be considered in the present study. It is assumed that a distributed normal stress loading $\sigma_{zz}(x,0) = -\tau_0(x)$. $\tau_0(x)$ is directly applied on the upper and lower crack surfaces, which is equivalent to investigating the perturbation fields for a remotely loaded cracked-body through the standard superposition technique in fracture mechanics. Therefore the boundary conditions along the crack surfaces can be written as follows:

$$\begin{cases} \sigma_{xz}^{(1)}(x,0) = \sigma_{xz}^{(2)}(x,0) = 0, \quad \sigma_{zz}^{(1)}(x,0) = \sigma_{zz}^{(2)}(x,0) = -\tau_0(x) \\ D_z(x,0)[w^{(1)}(x,0) - w^{(2)}(x,0)] = \varepsilon_0[\phi^{(1)}(x,0) - \phi^{(2)}(x,0)] \\ B_z(x,0)[w^{(1)}(x,0) - w^{(2)}(x,0)] = \mu_0[\psi^{(1)}(x,0) - \psi^{(2)}(x,0)] \\ D_z^{(1)}(x,0) = D_z^{(2)}(x,0), \quad B_z^{(1)}(x,0) = B_z^{(2)}(x,0) \end{cases}, \ |x| \le l \quad (3\text{-}1)$$

$$\begin{cases} u^{(1)}(x,0) = u^{(2)}(x,0), w^{(1)}(x,0) = w^{(2)}(x,0) \\ \sigma_{zz}^{(1)}(x,0) = \sigma_{zz}^{(2)}(x,0), \sigma_{xz}^{(1)}(x,0) = \sigma_{xz}^{(2)}(x,0) \\ \phi^{(1)}(x,0) = \phi^{(2)}(x,0) \\ \psi^{(1)}(x,0) = \psi^{(2)}(x,0) \\ D_z^{(1)}(x,0) = D_z^{(2)}(x,0) \\ B_z^{(1)}(x,0) = B_z^{(2)}(x,0) \end{cases}, \ |x| > l \quad (3\text{-}2)$$

where $\sigma_{ik}^{(j)}(x,z)$, $D_k^{(j)}(x,z)$ and $B_k^{(j)}(x,z)$ ($i = x,z$, $k = x,z$, $j = 1,2$) are the plane stresses, in-plane electric displacements and in-plane magnetic fluxes; $u^{(j)}(x,z)$ and $w^{(j)}(x,z)$ are the mechanical displacement in the x- and z-directions; $\phi^{(j)}(x,z)$ and $\psi^{(j)}(x,z)$ are the electric potential and the magnetic potential, respectively. It should be noted that all quantities with superscript j corresponds to the upper half plane 1 (for $j=1$) and the lower half plane 2 (for $j=2$) as shown in Figure 3-1. In equations (3-1)-(3-2), $D_z(x,0)$ and $B_z(x,0)$ are the electric displacement and the magnetic flux inside the crack, respectively, which depend on the external loading; ε_0 and μ_0 are the electric permittivity and the magnetic permeability of the air inside the crack, respectively. The electric and the magnetic boundary conditions expressed by equations (3-1)-(3-2) will be reduced to permeable boundary condition when $w^+ - w^- = 0$ and to impermeable one when $\varepsilon_0 = 0$ and $\mu_0 = 0$.

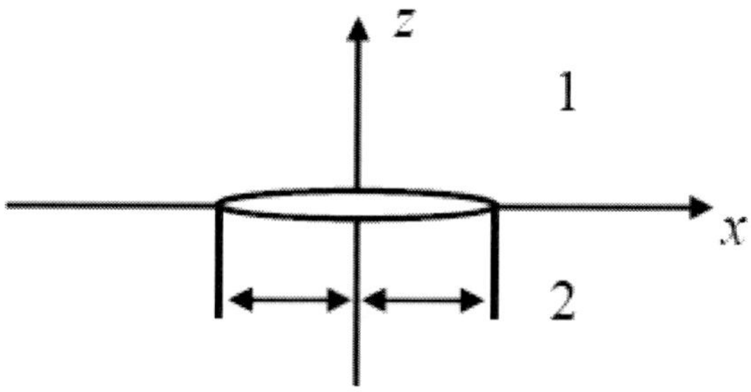

Figure 3.1. The coordinate system for a crack in functionally graded piezoelectric/piezomagnetic materials.

3.2. Basic equations of functionally graded piezoelectric/piezomagnetic materials

For the plane problem of linear, homogeneous, transversely isotropic functionally graded piezoelectric/piezomagnetic materials with vanishing body force, the basic equations are as follows (Song and Sih, 2003; Wang and Mai, 2003; Gao et al., 2003a; Gao et al., 2003d; Spyropoulos et al., 2003; Liu et al., 2001):

$$\begin{cases}\dfrac{\partial\sigma_{xx}^{(j)}(x,z)}{\partial x}+\dfrac{\partial\sigma_{xz}^{(j)}(x,z)}{\partial z}=0\\ \dfrac{\partial\sigma_{xz}^{(j)}(x,z)}{\partial x}+\dfrac{\partial\sigma_{zz}^{(j)}(x,z)}{\partial z}=0\\ \dfrac{\partial D_{x}^{(j)}(x,z)}{\partial x}+\dfrac{\partial D_{z}^{(j)}(x,z)}{\partial z}=0\\ \dfrac{\partial B_{x}^{(j)}(x,z)}{\partial x}+\dfrac{\partial B_{z}^{(j)}(x,z)}{\partial z}=0\end{cases}\tag{3-3}$$

$$\begin{cases}\sigma_{xx}^{(j)}=c_{11}^{*}\dfrac{\partial u^{(j)}(x,z)}{\partial x}+c_{13}^{*}\dfrac{\partial w^{(j)}(x,z)}{\partial z}+e_{31}^{*}\dfrac{\partial\varphi^{(j)}(x,z)}{\partial z}-f_{31}^{*}\dfrac{\partial\psi^{(j)}(x,z)}{\partial z}\\ \sigma_{zz}^{(j)}=c_{13}^{*}\dfrac{\partial u^{(j)}(x,z)}{\partial x}+c_{33}^{*}\dfrac{\partial w^{(j)}(x,z)}{\partial z}+e_{33}^{*}\dfrac{\partial\varphi^{(j)}(x,z)}{\partial z}-f_{33}^{*}\dfrac{\partial\psi^{(j)}(x,z)}{\partial z}\\ \sigma_{xz}^{(j)}=c_{44}^{*}(\dfrac{\partial u^{(j)}(x,z)}{\partial z}+\dfrac{\partial w^{(j)}(x,z)}{\partial x})+e_{15}^{*}\dfrac{\partial\varphi^{(j)}(x,z)}{\partial x}-f_{15}^{*}\dfrac{\partial\psi^{(j)}(x,z)}{\partial x}\\ D_{x}^{(j)}=e_{15}^{*}(\dfrac{\partial u^{(j)}(x,z)}{\partial z}+\dfrac{\partial w^{(j)}(x,z)}{\partial x})-\varepsilon_{11}^{*}\dfrac{\partial\varphi^{(j)}(x,z)}{\partial x}-g_{11}^{*}\dfrac{\partial\psi^{(j)}(x,z)}{\partial x}\\ D_{z}^{(j)}=e_{31}^{*}\dfrac{\partial u^{(j)}(x,z)}{\partial x}+e_{33}^{*}\dfrac{\partial w^{(j)}(x,z)}{\partial z}-\varepsilon_{33}^{*}\dfrac{\partial\varphi^{(j)}(x,z)}{\partial z}-g_{33}^{*}\dfrac{\partial\psi^{(j)}(x,z)}{\partial z}\\ B_{x}^{(j)}=f_{15}^{*}(\dfrac{\partial u^{(j)}(x,z)}{\partial z}+\dfrac{\partial w^{(j)}(x,z)}{\partial x})+g_{11}^{*}\dfrac{\partial\varphi^{(j)}(x,z)}{\partial x}-\mu_{11}^{*}\dfrac{\partial\psi^{(j)}(x,z)}{\partial x}\\ B_{z}^{(j)}=f_{31}^{*}\dfrac{\partial u^{(j)}(x,z)}{\partial x}+f_{33}^{*}\dfrac{\partial w^{(j)}(x,z)}{\partial z}+g_{33}^{*}\dfrac{\partial\varphi^{(j)}(x,z)}{\partial z}-\mu_{33}^{*}\dfrac{\partial\psi^{(j)}(x,z)}{\partial z}\end{cases}\tag{3-4}$$

where $c_{11}^*, c_{13}^*, c_{33}^*$ and c_{44}^* are the elastic stiffness constants; ε_{11}^* and ε_{33}^* are the dielectric constants; e_{15}^*, e_{31}^* and e_{33}^* are the piezoelectric constants; f_{15}^*, f_{31}^* and f_{33}^* are the piezomagnetic constants; g_{11}^* and g_{33}^* are the electromagnetic constants; and μ_{11}^* and μ_{33}^* are the magnetic permeabilities.

Crack problems in functionally graded materials do not appear to be analytically tractable for arbitrary variations of material properties. Usually, one tries to generate the forms of functionally graded materials for which the problem becomes tractable. Similar to the treatment of the crack problem for isotropic functionally graded materials (Delale and Erdogan, 1988; Fildis and Yahsi, 1996; Ozturk and Erdogan, 1997), we assume that the material properties are described by:

$$\begin{aligned}&(c_{11}^*, c_{13}^*, c_{33}^*, c_{44}^*, e_{15}^*, e_{31}^*, e_{33}^*, \varepsilon_{11}^*, \varepsilon_{33}^*, f_{15}^*, f_{31}^*, f_{33}^*, g_{11}^*, g_{33}^*)\\&=(c_{11}, c_{13}, c_{33}, c_{44}, e_{15}, e_{31}, e_{33}, \varepsilon_{11}, \varepsilon_{33}, f_{15}, f_{31}, f_{33}, g_{11}, g_{33}, \mu_{11}, \mu_{33})e^{\gamma x}\end{aligned} \tag{3-5}$$

where γ is a constant which measures the variation rate of material properties in functionally graded materials. When $\gamma = 0$, the material properties won't change which will reduce to the homogenous magnetoelectroelastic material case with a closed form solution reported recently (Zhou et al., 2007). The expressions of Eq. (3-5) are purely assumed for making the problem tractable without loss of generality.

Substituting Eqs. (3-4) into Eqs. (3-3), and using Eq. (3-5), the governing equations are obtained as follows:

$$\begin{aligned}&(c_{11}\frac{\partial^2}{\partial x^2}+c_{44}\frac{\partial^2}{\partial z^2})u^{(j)}(x,z)+(c_{13}+c_{44})\frac{\partial^2}{\partial x\partial z}w^{(j)}(x,z)\\&+(e_{31}+e_{15})\frac{\partial^2}{\partial x\partial z}\varphi^{(j)}(x,z)+(-f_{31}-f_{15})\frac{\partial^2}{\partial x\partial z}\psi^{(j)}(x,z)\\&+\gamma[c_{11}\frac{\partial u^{(j)}(x,z)}{\partial x}+c_{13}\frac{\partial w^{(j)}(x,z)}{\partial z}+e_{31}\frac{\partial\varphi^{(j)}(x,z)}{\partial z}-f_{31}\frac{\partial\psi^{(j)}(x,z)}{\partial z}]=0\end{aligned} \tag{3-6}$$

$$\begin{aligned}&(c_{13}+c_{44})\frac{\partial^2}{\partial x\partial z}u^{(j)}(x,z)+(c_{44}\frac{\partial^2}{\partial x^2}+c_{33}\frac{\partial^2}{\partial z^2})w^{(j)}(x,z)\\&+(e_{15}\frac{\partial^2}{\partial x^2}+e_{33}\frac{\partial^2}{\partial z^2})\varphi^{(j)}(x,z)+(-f_{15}\frac{\partial^2}{\partial x^2}-f_{33}\frac{\partial^2}{\partial z^2})\psi^{(j)}(x,z)\\&+\gamma[c_{44}(\frac{\partial u^{(j)}(x,z)}{\partial z}+\frac{\partial w^{(j)}(x,z)}{\partial x})+e_{15}\frac{\partial\varphi^{(j)}(x,z)}{\partial x}-f_{15}\frac{\partial\psi^{(j)}(x,z)}{\partial x}]=0\end{aligned} \tag{3-7}$$

$$\begin{aligned}&(e_{15}+e_{31})\frac{\partial^2}{\partial x\partial z}u^{(j)}(x,z)+(e_{15}\frac{\partial^2}{\partial x^2}+e_{33}\frac{\partial^2}{\partial z^2})w^{(j)}(x,z)\\&+(-\varepsilon_{11}\frac{\partial^2}{\partial x^2}-\varepsilon_{33}\frac{\partial^2}{\partial z^2})\varphi^{(j)}(x,z)-(g_{11}\frac{\partial^2}{\partial x^2}+g_{33}\frac{\partial^2}{\partial z^2})\psi^{(j)}(x,z)\\&+\gamma[e_{15}(\frac{\partial u^{(j)}(x,z)}{\partial z}+\frac{\partial w^{(j)}(x,z)}{\partial x})-\varepsilon_{11}\frac{\partial\varphi^{(j)}(x,z)}{\partial x}-g_{11}\frac{\partial\psi^{(j)}(x,z)}{\partial x}]=0\end{aligned} \tag{3-8}$$

$$(f_{15}+f_{31})\frac{\partial^2}{\partial x\partial z}u^{(j)}(x,z)+(f_{15}\frac{\partial^2}{\partial x^2}+f_{33}\frac{\partial^2}{\partial z^2})w^{(j)}(x,z)$$
$$+(g_{11}\frac{\partial^2}{\partial x^2}+g_{33}\frac{\partial^2}{\partial z^2})\varphi^{(j)}(x,z)-(\mu_{11}\frac{\partial^2}{\partial x^2}+\mu_{33}\frac{\partial^2}{\partial z^2})\psi^{(j)}(x,z)$$
$$+\gamma[f_{15}(\frac{\partial u^{(j)}(x,z)}{\partial z}+\frac{\partial w^{(j)}(x,z)}{\partial x})+g_{11}\frac{\partial\varphi^{(j)}(x,z)}{\partial x}-\mu_{11}\frac{\partial\psi^{(j)}(x,z)}{\partial x}]=0 \quad (3\text{-}9)$$

3.3. Solution Procedures

Equations (3-6)-(3-9) can be solved using the method proposed by Yang (2001). First, rewrite Eqs (3-6)-(3-9) as follows:

$$[MD]\begin{Bmatrix} u^{(j)}(x,z) \\ w^{(j)}(x,z) \\ \phi^{(j)}(x,z) \\ \psi^{(j)}(x,z) \end{Bmatrix}=0 \quad (3\text{-}10)$$

where the operator matrix $[MD]$ is

$$[MD]=\begin{bmatrix} c_{11}\frac{\partial^2}{\partial x^2}+c_{44}\frac{\partial^2}{\partial z^2}+\gamma c_{11}\frac{\partial}{\partial x} & (c_{13}+c_{44})\frac{\partial^2}{\partial x\partial z}+\gamma c_{13}\frac{\partial}{\partial z} & (e_{31}+e_{15})\frac{\partial^2}{\partial x\partial z}+\gamma e_{31}\frac{\partial}{\partial z} & (-f_{31}-f_{15})\frac{\partial^2}{\partial x\partial z}-\gamma f_{31}\frac{\partial}{\partial z} \\ (c_{13}+c_{44})\frac{\partial^2}{\partial x\partial z}+\gamma c_{44}\frac{\partial}{\partial z} & c_{44}\frac{\partial^2}{\partial x^2}+c_{33}\frac{\partial^2}{\partial z^2}+\gamma c_{44}\frac{\partial}{\partial x} & e_{15}\frac{\partial^2}{\partial x^2}+e_{33}\frac{\partial^2}{\partial z^2}+\gamma e_{15}\frac{\partial}{\partial x} & -f_{15}\frac{\partial^2}{\partial x^2}-f_{33}\frac{\partial^2}{\partial z^2}-\gamma f_{15}\frac{\partial}{\partial x} \\ (e_{15}+e_{31})\frac{\partial^2}{\partial x\partial z}+\gamma e_{15}\frac{\partial}{\partial z} & e_{15}\frac{\partial^2}{\partial x^2}+e_{33}\frac{\partial^2}{\partial z^2}+\gamma e_{15}\frac{\partial}{\partial x} & -\varepsilon_{11}\frac{\partial^2}{\partial x^2}-\varepsilon_{33}\frac{\partial^2}{\partial z^2}-\gamma\varepsilon_{11}\frac{\partial}{\partial x} & -g_{11}\frac{\partial^2}{\partial x^2}-g_{33}\frac{\partial^2}{\partial z^2}-\gamma g_{11}\frac{\partial}{\partial x} \\ (f_{15}+f_{31})\frac{\partial^2}{\partial x\partial z}+\gamma f_{15}\frac{\partial}{\partial z} & f_{15}\frac{\partial^2}{\partial x^2}+f_{33}\frac{\partial^2}{\partial z^2}+\gamma f_{15}\frac{\partial}{\partial x} & g_{11}\frac{\partial^2}{\partial x^2}+g_{33}\frac{\partial^2}{\partial z^2}+\gamma g_{11}\frac{\partial}{\partial x} & -\mu_{11}\frac{\partial^2}{\partial x^2}-\mu_{33}\frac{\partial^2}{\partial z^2}-\gamma\mu_{11}\frac{\partial}{\partial x} \end{bmatrix}$$

The determinant of matrix $[MD]$ is

$$\det[MD]$$

$$=d_1\frac{\partial^8}{\partial z^8}+d_2\frac{\partial^8}{\partial x^2\partial z^6}+d_3\frac{\partial^7}{\partial x\partial z^6}+d_4\frac{\partial^6}{\partial z^6}+d_5\frac{\partial^8}{\partial x^4\partial z^4}+d_6\frac{\partial^7}{\partial x^3\partial z^4}+d_7\frac{\partial^6}{\partial x^2\partial z^4}+d_8\frac{\partial^5}{\partial x\partial z^4}$$
$$+d_9\frac{\partial^8}{\partial x^6\partial z^2}+d_{10}\frac{\partial^7}{\partial x^5\partial z^2}+d_{11}\frac{\partial^6}{\partial x^4\partial z^2}+d_{12}\frac{\partial^5}{\partial x^3\partial z^2}+d_{13}\frac{\partial^4}{\partial x^2\partial z^2}$$
$$+d_{14}\frac{\partial^8}{\partial x^8}+d_{15}\frac{\partial^7}{\partial x^7}+d_{16}\frac{\partial^6}{\partial x^6}+d_{17}\frac{\partial^5}{\partial x^5}+d_{18}\frac{\partial^4}{\partial x^4}$$

where $d_i (i=1,2,3,...,21)$ are given in the appendix.

Based on the cofactors Δ_{ik} of matrix $\left[MD\right]$ $(i,k=1,2,3,4)$, using the method developed in the literature (Chen et al., 2004; Yang, 2001; Ding et al., 1996), the general solution of Eq. (3-10) is

$$[u^{(j)}(x,z),w^{(j)}(x,z),\phi^{(j)}(x,z),\psi^{(j)}(x,z)]^T=(\Delta_{i1},\Delta_{i2},\Delta_{i3},\Delta_{i4})^T F^{(j)}(x,z),$$
$$(i=1,2,3,4,\ j=1,2) \quad (3\text{-}11)$$

with $F^{(j)}(x,z)$ satisfying the following equation

$$\det\left[MD\right]F^{(j)}(x,z)=0,\ (j=1,2) \quad (3\text{-}12)$$

In the following analysis, we use only ($\Delta_{21},\Delta_{22},\Delta_{23},\Delta_{24}$) for the present problem, which can be expressed as follows:

$$\Delta_{21}=\alpha_1^{(1)}\frac{\partial^3}{\partial x^2\partial z}+\alpha_2^{(1)}\frac{\partial^4}{\partial x^3\partial z}+\alpha_3^{(1)}\frac{\partial^5}{\partial x^4\partial z}+\alpha_4^{(1)}\frac{\partial^6}{\partial x^5\partial z}$$
$$+\alpha_5^{(1)}\frac{\partial^4}{\partial x\partial z^3}+\alpha_6^{(1)}\frac{\partial^5}{\partial x^2\partial z^3}+\alpha_7^{(1)}\frac{\partial^6}{\partial x^3\partial z^3}+\alpha_8^{(1)}\frac{\partial^5}{\partial z^5}+\alpha_9^{(1)}\frac{\partial^6}{\partial x\partial z^5} \quad (3\text{-}13)$$

$$\Delta_{22}=\alpha_1^{(2)}\frac{\partial^3}{\partial x^3}+\alpha_2^{(2)}\frac{\partial^4}{\partial x^4}+\alpha_3^{(2)}\frac{\partial^5}{\partial x^5}+\alpha_4^{(2)}\frac{\partial^6}{\partial x^6}+\alpha_5^{(2)}\frac{\partial^3}{\partial x\partial z^2}+\alpha_6^{(2)}\frac{\partial^4}{\partial x^2\partial z^2}$$
$$+\alpha_7^{(2)}\frac{\partial^5}{\partial x^3\partial z^2}+\alpha_8^{(2)}\frac{\partial^6}{\partial x^4\partial z^2}+\alpha_9^{(2)}\frac{\partial^4}{\partial z^4}+\alpha_{10}^{(2)}\frac{\partial^5}{\partial x\partial z^4}+\alpha_{11}^{(2)}\frac{\partial^6}{\partial x^2\partial z^4}+\alpha_{12}^{(2)}\frac{\partial^6}{\partial z^6} \quad (3\text{-}14)$$

$$\Delta_{23}=\alpha_1^{(3)}\frac{\partial^3}{\partial x^3}+\alpha_2^{(3)}\frac{\partial^4}{\partial x^4}+\alpha_3^{(3)}\frac{\partial^5}{\partial x^5}+\alpha_4^{(3)}\frac{\partial^6}{\partial x^6}+\alpha_5^{(3)}\frac{\partial^3}{\partial x\partial z^2}+\alpha_6^{(3)}\frac{\partial^4}{\partial x^2\partial z^2}+\alpha_7^{(3)}\frac{\partial^5}{\partial x^3\partial z^2}$$
$$+\alpha_8^{(3)}\frac{\partial^6}{\partial x^4\partial z^2}+\alpha_9^{(3)}\frac{\partial^4}{\partial z^4}+\alpha_{10}^{(3)}\frac{\partial^5}{\partial x\partial z^4}+\alpha_{11}^{(3)}\frac{\partial^6}{\partial x^2\partial z^4}+\alpha_{12}^{(3)}\frac{\partial^6}{\partial z^6} \quad (3\text{-}15)$$

$$\Delta_{24}=\alpha_1^{(4)}\frac{\partial^3}{\partial x^3}+\alpha_2^{(4)}\frac{\partial^4}{\partial x^4}+\alpha_3^{(4)}\frac{\partial^5}{\partial x^5}+\alpha_4^{(4)}\frac{\partial^6}{\partial x^6}+\alpha_5^{(4)}\frac{\partial^3}{\partial x\partial z^2}+\alpha_6^{(4)}\frac{\partial^4}{\partial x^2\partial z^2}$$
$$+\alpha_7^{(4)}\frac{\partial^5}{\partial x^3\partial z^2}+\alpha_8^{(4)}\frac{\partial^6}{\partial x^4\partial z^2}+\alpha_9^{(4)}\frac{\partial^4}{\partial z^4}+\alpha_{10}^{(4)}\frac{\partial^5}{\partial x\partial z^4}+\alpha_{11}^{(4)}\frac{\partial^6}{\partial x^2\partial z^4}+\alpha_{12}^{(4)}\frac{\partial^6}{\partial z^6} \quad (3\text{-}16)$$

where $\alpha_k^{(i)}$ ($i=1,2,3,4;k=1,2,3,...,12$) are given in the appendix.

Performing Fourier transform with respect x, $F^{(j)}(x,z)$ can be expressed as follows:

$$F^{(j)}(x,z)=\frac{1}{2\pi}\int_{-\infty}^{\infty}f^{(j)}(s,z)e^{isx}ds \quad (3\text{-}17)$$

Substitution of Eq. (3-17) into Eq. (3-12) yields

$$a\frac{\partial^8 f^{(j)}(s,z)}{\partial z^8}-b\frac{\partial^6 f^{(j)}(s,z)}{\partial z^6}+c\frac{\partial^4 f^{(j)}(s,z)}{\partial z^4}-d\frac{\partial^2 f^{(j)}(s,z)}{\partial z^2}+ef^{(j)}(s,z)=0 \tag{3-18}$$

where

$$a=d_1,\ b=d_2s^2-id_3s-d_4,\ c=d_5s^4-id_6s^3-d_7s^2+id_8s,$$
$$d=d_9s^6-id_{10}s^5-d_{11}s^4+id_{12}s^3+d_{13}s^2,$$
$$e=d_{14}s^8-id_{15}s^7-d_{16}s^6+id_{17}s^5+d_{18}s^4.$$

Eq. (3-18) is a homogeneous equation, and the solution of $f^{(j)}(s,z)$ is a function of $\exp(-\lambda z)$, in which λ is the root of the following algebraic equation:

$$a\lambda^8-b\lambda^6+c\lambda^4-d\lambda^2+e=0 \tag{3-19}$$

which is determined by

$$\lambda_1^2=\frac{b}{4a}-\frac{1}{2}\sqrt{R_5+R_6}-\frac{1}{2}\sqrt{2R_5-R_6-\frac{\frac{b^3}{a^3}-\frac{4bc}{a^2}+\frac{8d}{a}}{4\sqrt{R_5+R_6}}} \tag{3-20}$$

$$\lambda_2^2=\frac{b}{4a}-\frac{1}{2}\sqrt{R_5+R_6}+\frac{1}{2}\sqrt{2R_5-R_6-\frac{\frac{b^3}{a^3}-\frac{4bc}{a^2}+\frac{8d}{a}}{4\sqrt{R_5+R_6}}} \tag{3-21}$$

$$\lambda_3^2=\frac{b}{4a}+\frac{1}{2}\sqrt{R_5+R_6}-\frac{1}{2}\sqrt{2R_5-R_6+\frac{\frac{b^3}{a^3}-\frac{4bc}{a^2}+\frac{8d}{a}}{4\sqrt{R_5+R_6}}} \tag{3-22}$$

$$\lambda_4^2=\frac{b}{4a}+\frac{1}{2}\sqrt{R_5+R_6}+\frac{1}{2}\sqrt{2R_5-R_6+\frac{\frac{b^3}{a^3}-\frac{4bc}{a^2}+\frac{8d}{a}}{4\sqrt{R_5+R_6}}} \tag{3-23}$$

where

$$R_1=2c^3-9bcd+27ad^2+27b^2e-72ace,\ R_2=c^2-3bd+12ae,$$
$$R_3=\sqrt{-4(R_2)^3+(R_1)^2},\ R_4=\frac{(R_1+R_3)^{\frac{1}{3}}}{2^{\frac{1}{3}}},\ R_5=\frac{b^2}{4a^2}-\frac{2c}{3a},\ R_6=\frac{R_2}{3aR_4}+\frac{R_4}{3a}.$$

Depending on the properties of λ^2, function $f^{(j)}(s,z)$ has five different general solutions (only for $j=1$ and $z\geq 0$)(Other cases can be obtained using a similar method, but they are omitted for brevity) as follows:

If $\lambda_1^2\neq\lambda_2^2\neq\lambda_3^2\neq\lambda_4^2>0$, then

$$f^{(1)}(s,z)=A_1(s)e^{-\lambda_1 z}+A_2(s)e^{-\lambda_2 z}+A_3(s)e^{-\lambda_3 z}+A_4(s)e^{-\lambda_4 z} \tag{3-24}$$

If $\lambda_1^2 \neq \lambda_2^2 \neq \lambda_3^2 = \lambda_4^2 > 0$, then

$$f^{(1)}(s,z)=A_1(s)e^{-\lambda_1 z}+A_2(s)e^{-\lambda_2 z}+A_3(s)e^{-\lambda_3 z}+A_4(s)sze^{-\lambda_3 z} \tag{3-25}$$

If $\lambda_1^2 \neq \lambda_2^2 = \lambda_3^2 = \lambda_4^2 > 0$, then

$$f^{(1)}(s,z)=A_1(s)e^{-\lambda_1 z}+A_2(s)e^{-\lambda_2 z}+A_3(s)sze^{-\lambda_2 z}+A_4(s)s^2z^2e^{-\lambda_2 z} \tag{3-26}$$

If $\lambda_1^2 = \lambda_2^2 = \lambda_3^2 = \lambda_4^2 > 0$, then

$$f^{(1)}(s,z)=A_1(s)e^{-\lambda_2 z}+A_2(s)ze^{-\lambda_2 z}+A_3(s)z^2e^{-\lambda_2 z}+A_4(s)z^3e^{-\lambda_2 z} \tag{3-27}$$

If $\lambda_1^2 > 0$, $\lambda_2^2 > 0$, $\lambda_1^2 \neq \lambda_2^2$ and $\lambda_3^2, \lambda_4^2 < 0$ or λ_3^2 and λ_4^2 being a pair of conjugate complex roots, and therefore λ_3 and λ_4 are a pair of conjugate complexes $-\delta \pm i\omega$, the solution of function $f^{(1)}(s,z)$ is

$$f^{(1)}(s,z)=A_1(s)e^{-\lambda_1 z}+A_2(s)e^{-\lambda_2 z}+A_3(s)e^{-\delta z}\cos(\omega z)+A_4(s)e^{-\delta z}\sin(\omega z) \tag{3-28}$$

where δ and $\omega > 0$ and $A_i(s)$ (i=1, 2, 3, 4) is a function of s to be determined by the boundary conditions.

Substituting the solution of auxiliary function $f^{(j)}(s,z)$ into Eqs. (3-4), (3-11), (3-17) and Eqs. (3-24)-(3-28), the mechanical and electric displacement, stresses, electric potential fields, magnetic flux and magnetic potential fields are calculated using Mathematica. For the case of $\lambda_1^2 \neq \lambda_2^2 \neq \lambda_3^2 \neq \lambda_4^2 > 0$, the displacements, stresses, electric displacements, electric potentials, magnetic fluxes and magnetic potentials are given as follows (Other cases can be obtained using a similar method, but they are omitted in the present paper for brevity.):

$$\begin{cases} u^{(1)}(x,z)=\dfrac{1}{2\pi}\sum_{i=1}^{4}\int_{-\infty}^{\infty}\beta_i^{(1)}(s)A_i(s)e^{-\lambda_i z}e^{isx}ds \\ w^{(1)}(x,z)=\dfrac{1}{2\pi}\sum_{i=1}^{4}\int_{-\infty}^{\infty}\beta_i^{(2)}(s)A_i(s)e^{-\lambda_i z}e^{isx}ds \\ \phi^{(1)}(x,z)=\dfrac{1}{2\pi}\sum_{i=1}^{4}\int_{-\infty}^{\infty}\beta_i^{(3)}(s)A_i(s)e^{-\lambda_i z}e^{isx}ds \\ \psi^{(1)}(x,z)=\dfrac{1}{2\pi}\sum_{i=1}^{4}\int_{-\infty}^{\infty}\beta_i^{(4)}(s)A_i(s)e^{-\lambda_i z}e^{isx}ds \end{cases}, \quad z \geq 0 \tag{3-29}$$

$$\begin{cases} u^{(2)}(x,z) = -\dfrac{1}{2\pi}\sum_{i=1}^{4}\int_{-\infty}^{\infty}\beta_i^{(1)}(s)B_i(s)e^{\lambda_i z}e^{isx}ds \\ w^{(2)}(x,z) = \dfrac{1}{2\pi}\sum_{i=1}^{4}\int_{-\infty}^{\infty}\beta_i^{(2)}(s)B_i(s)e^{\lambda_i z}e^{isx}ds \\ \phi^{(2)}(x,z) = \dfrac{1}{2\pi}\sum_{i=1}^{4}\int_{-\infty}^{\infty}\beta_i^{(3)}(s)B_i(s)e^{\lambda_i z}e^{isx}ds \\ \psi^{(2)}(x,z) = \dfrac{1}{2\pi}\sum_{i=1}^{4}\int_{-\infty}^{\infty}\beta_i^{(4)}(s)B_i(s)e^{\lambda_i z}e^{isx}ds \end{cases}, \quad z \le 0 \tag{3-30}$$

where $\beta_i^{(1)}(s) = \alpha_1^{(1)}s^2\lambda_i + \alpha_2^{(1)}is^3\lambda_i - \alpha_3^{(1)}s^4\lambda_i - \alpha_4^{(1)}is^5\lambda_i - \alpha_5^{(1)}is\lambda_i^3 + \alpha_6^{(1)}s^2\lambda_i^3$

$+\alpha_7^{(1)}is^3\lambda_i^3 - \alpha_8^{(1)}\lambda_i^5 - \alpha_9^{(1)}is\lambda_i^5$

$\beta_i^{(2)}(s) = -\alpha_1^{(2)}is^3 + \alpha_2^{(2)}s^4 + \alpha_3^{(2)}is^5 - \alpha_4^{(2)}s^6 + \alpha_5^{(2)}is\lambda_i^2 - \alpha_6^{(2)}s^2\lambda_i^2 - \alpha_7^{(2)}is^3\lambda_i^2$

$+\alpha_8^{(2)}s^4\lambda_i^2 + \alpha_9^{(2)}\lambda_i^4 + \alpha_{10}^{(2)}is\lambda_i^4 - \alpha_{11}^{(2)}s^2\lambda_i^4 + \alpha_{12}^{(2)}\lambda_i^6$

$\beta_i^{(3)}(s) = -\alpha_1^{(3)}is^3 + \alpha_2^{(3)}s^4 + \alpha_3^{(3)}is^5 - \alpha_4^{(3)}s^6 + \alpha_5^{(3)}is\lambda_i^2 - \alpha_6^{(3)}s^2\lambda_i^2 - \alpha_7^{(3)}is^3\lambda_i^2 + \alpha_8^{(3)}s^4\lambda_i^2 + \alpha_9^{(3)}\lambda_i^4$

$+\alpha_{10}^{(3)}is\lambda_i^4 - \alpha_{11}^{(3)}s^2\lambda_i^4 + \alpha_{12}^{(3)}\lambda_i^6$

$\beta_i^{(4)}(s) = -\alpha_1^{(4)}is^3 + \alpha_2^{(4)}s^4 + \alpha_3^{(4)}is^5 - \alpha_4^{(4)}s^6 + \alpha_5^{(4)}is\lambda_i^2 - \alpha_6^{(4)}s^2\lambda_i^2 - \alpha_7^{(4)}is^3\lambda_i^2$

$+\alpha_8^{(4)}s^4\lambda_i^2 + \alpha_9^{(4)}\lambda_i^4 + \alpha_{10}^{(4)}is\lambda_i^4 - \alpha_{11}^{(4)}s^2\lambda_i^4 + \alpha_{12}^{(4)}\lambda_i^6$

$$\begin{cases} \sigma_{zz}^{(1)}(x,z) = \dfrac{e^{\gamma x}}{2\pi}\sum_{i=1}^{4}\int_{-\infty}^{\infty}\chi_i^{(1)}(s)A_i(s)se^{-\lambda_i z}e^{isx}ds \\ \sigma_{xz}^{(1)}(x,z) = \dfrac{e^{\gamma x}}{2\pi}\sum_{i=1}^{4}\int_{-\infty}^{\infty}\chi_i^{(2)}(s)A_i(s)se^{-\lambda_i z}e^{isx}ds \\ D_z^{(1)}(x,z) = \dfrac{e^{\gamma x}}{2\pi}\sum_{i=1}^{4}\int_{-\infty}^{\infty}\chi_i^{(3)}(s)A_i(s)se^{-\lambda_i z}e^{isx}ds \\ B_z^{(1)}(x,z) = \dfrac{e^{\gamma x}}{2\pi}\sum_{i=1}^{4}\int_{-\infty}^{\infty}\chi_i^{(4)}(s)A_i(s)se^{-\lambda_i z}e^{isx}ds \\ D_x^{(1)}(x,z) = \dfrac{e^{\gamma x}}{2\pi}\sum_{i=1}^{4}\int_{-\infty}^{\infty}\chi_i^{(5)}(s)A_i(s)se^{-\lambda_i z}e^{isx}ds \\ B_x^{(1)}(x,z) = \dfrac{e^{\gamma x}}{2\pi}\sum_{i=1}^{4}\int_{-\infty}^{\infty}\chi_i^{(6)}(s)A_i(s)se^{-\lambda_i sz}e^{isx}ds \end{cases}, \quad z \ge 0 \tag{3-31}$$

$$\begin{cases} \sigma_{zz}^{(2)}(x,z) = -\dfrac{e^{\gamma x}}{2\pi}\sum_{i=1}^{4}\int_{-\infty}^{\infty}\chi_i^{(1)}(s)B_i(s)se^{\lambda_i z}e^{isx}ds \\ \sigma_{xz}^{(2)}(x,z) = \dfrac{e^{\gamma x}}{2\pi}\sum_{i=1}^{4}\int_{-\infty}^{\infty}\chi_i^{(2)}(s)B_i(s)se^{\lambda_i z}e^{isx}ds \\ D_z^{(2)}(x,z) = -\dfrac{e^{\gamma x}}{2\pi}\sum_{i=1}^{4}\int_{-\infty}^{\infty}\chi_i^{(3)}(s)B_i(s)se^{\lambda_i z}e^{isx}ds \\ B_z^{(2)}(x,z) = -\dfrac{e^{\gamma x}}{2\pi}\sum_{i=1}^{4}\int_{-\infty}^{\infty}\chi_i^{(4)}(s)B_i(s)se^{\lambda_i z}e^{isx}ds \end{cases}, \quad z \le 0 \tag{3-32}$$

where $\chi_i^{(1)}(s) = c_{13}is\beta_i^{(1)}(s) - c_{33}\lambda_i\,\beta_i^{(2)}(s) - e_{33}\lambda_i\,\beta_i^{(3)}(s) + f_{33}\lambda_i\,\beta_i^{(4)}(s)$,

$\chi_i^{(2)}(s) = -c_{44}\lambda_i\,\beta_i^{(1)}(s) + c_{44}is\beta_i^{(2)}(s) + e_{15}is\beta_i^{(3)}(s) - f_{15}is\beta_i^{(4)}(s)$,

$\chi_i^{(3)}(s) = e_{31}is\beta_i^{(1)}(s) - e_{33}\lambda_i\,\beta_i^{(2)}(s) + \varepsilon_{33}\lambda_i\,\beta_i^{(3)}(s) + g_{33}\lambda_i\,\beta_i^{(4)}(s)$,

$\chi_i^{(4)}(s) = f_{31}is\beta_i^{(1)}(s) - f_{33}\lambda_i\,\beta_i^{(2)}(s) - g_{33}\lambda_i\,\beta_i^{(3)}(s) + \mu_{33}\lambda_i\,\beta_i^{(4)}(s)$,

$\chi_i^{(5)}(s) = -e_{15}\lambda_i\,\beta_i^{(1)}(s) + e_{15}is\beta_i^{(2)}(s) - \varepsilon_{11}is\beta_i^{(3)}(s) - g_{11}is\beta_i^{(4)}(s)$,

$\chi_i^{(6)}(s) = -f_{15}\lambda_i\,\beta_i^{(1)}(s) + f_{15}is\beta_i^{(2)}(s) + g_{11}is\beta_i^{(3)}(s) - \mu_{11}is\beta_i^{(4)}(s)$.

Introduce the jumps of displacements, electric potential and magnetic potential across the crack surfaces as follows:

$$\begin{cases} f_1(x) = u^{(1)}(x,0) - u^{(2)}(x,0) \\ f_2(x) = w^{(1)}(x,0) - w^{(2)}(x,0) \\ f_3(x) = \phi^{(1)}(x,0) - \phi^{(2)}(x,0) \\ f_4(x) = \psi^{(1)}(x,0) - \psi^{(2)}(x,0) \end{cases} \tag{3-33}$$

we can prove that $f_1(x)$ is an odd function, $f_2(x)$, $f_3(x)$ and $f_4(x)$ are three even functions.

By substituting Eqs. (3-29) and (3-30) into Eq. (3-33), and the resultant equations into Eqs. (3-31) and (3-32), through Fourier transform with respect to x and using boundary conditions Eqs. (3-1)-(3-2), the following equations can be obtained

$$[X_1]\begin{bmatrix} A_1(s) \\ A_2(s) \\ A_3(s) \\ A_4(s) \end{bmatrix} + [X_1]\begin{bmatrix} B_1(s) \\ B_2(s) \\ B_3(s) \\ B_4(s) \end{bmatrix} = \begin{bmatrix} \bar{f}_1(s) \\ 0 \\ 0 \\ 0 \end{bmatrix} \tag{3-34}$$

$$[X_2]\begin{bmatrix} A_1(s) \\ A_2(s) \\ A_3(s) \\ A_4(s) \end{bmatrix} - [X_2]\begin{bmatrix} B_1(s) \\ B_2(s) \\ B_3(s) \\ B_4(s) \end{bmatrix} = \begin{bmatrix} \bar{f}_2(s) \\ \bar{f}_3(s) \\ \bar{f}_4(s) \\ 0 \end{bmatrix} = \begin{bmatrix} \bar{f}_2(s) \\ \dfrac{D_0\bar{f}_2(s)}{\varepsilon_0} \\ \dfrac{B_0\bar{f}_2(s)}{\mu_0} \\ 0 \end{bmatrix} \tag{3-35}$$

where matrices $[X_i]$ ($i=1,2$) are given in the appendix-III.

To reveal the effects of the electric permittivity and the magnetic permeability of the air inside the crack on the intensity factors, we assumed that $D_0 = D_z(x,0)$ and $B_0 = B_z(x,0)$ ($|x| \le l$) inside the crack were used as two different constants to make the problem tractable as discussed in the literature (Hao and Shen, 1994; Hao 2001; Dascalu and Homentcovschi, 2002). Here an over bar indicates Fourier transform.

By solving the eight equations of Eq. (3-34) and Eq. (3-35) with eight unknown functions and substituting the solutions into Eq. (3-31) and applying the boundary conditions Eqs. (3-1)-(3-2), it can be obtained(The solving processes can be shown in the appendix-III.):

$$\begin{cases} \sigma_{zz}^{(1)}(x,0) = \dfrac{e^{\gamma x}}{2\pi}\displaystyle\int_{-\infty}^{\infty} \beta_1(s)\bar{f}_2(s)e^{isx}ds = -\tau_0(x) \\ \sigma_{xz}^{(1)}(x,0) = \dfrac{e^{\gamma x}}{2\pi}\displaystyle\int_{-\infty}^{\infty} \beta_2(s)\bar{f}_1(s)e^{isx}ds = 0 \end{cases}, \ -l \le x \le l \tag{3-36}$$

$$\int_{-\infty}^{\infty} \bar{f}_1(s)e^{isx}ds = 0, \ \int_{-\infty}^{\infty} \bar{f}_2(s)e^{isx}ds = 0, \ |x| > l \tag{3-37}$$

where $\beta_j(s)(j=1,2)$ are known functions which are dependent on material properties as shown in the appendix. Moreover, $\lim_{s\to+\infty} \beta_1(s)/s = \beta_1^{(0)} = -\lim_{s\to-\infty} \beta_1(s)/s$, $\lim_{s\to+\infty} \beta_2(s)/s = \beta_2^{(0)} = -\lim_{s\to-\infty} \beta_2(s)/s$ and $\beta_j^{(0)}(j=1,2)$ are known constants which depend on material properties; D_0/ε_0 and B_0/μ_0 were given in Zhou et al. (2007) which are independent of the functionally graded parameter γ. Here, we just give these constants ($\beta_j^{(0)}(j=1,2)$) for $\lambda_1^2 \ne \lambda_2^2 \ne \lambda_3^2 \ne \lambda_4^2 > 0$. Other cases can be obtained using a similar method, but they are omitted for brevity. The above two pairs of dual integral equations (3-36) and (3-37) must be solved to determine the unknown functions $\bar{f}_1(s)$ and $\bar{f}_2(s)$.

3.4. Solution of the Dual Integral Equations

The Schmidt method (Morse and Feshbach, 1958) is used to solve the dual integral equations (3-36)-(3-37). The jumps of displacement across the crack surfaces are directly expanded as the following series:

$$f_1(x)=\begin{cases}\sum_{n=0}^{\infty}a_n P_n^{(\frac{1}{2},\frac{1}{2})}(\frac{x}{l})(1-\frac{x^2}{l^2})^{\frac{1}{2}}, & -l\le x\le l\\ 0, & |x|>l\end{cases} \tag{3-38}$$

$$f_2(x)=\begin{cases}\sum_{n=0}^{\infty}b_n P_n^{(\frac{1}{2},\frac{1}{2})}(\frac{x}{l})(1-\frac{x^2}{l^2})^{\frac{1}{2}}, & -l\le x\le l\\ 0, & |x|>l\end{cases} \tag{3-39}$$

where a_n and b_n are unknown coefficients, $P_n^{(1/2,1/2)}(x)$ is a Jacobi polynomial (Gradshteyn and Ryzhik, 1980). It is worth noting that Jacobi polynomials $P_n^{(1/2,1/2)}(x)$ is equivalent to the Chebyshev polynomials $U_n(x)$, i.e., $U_n(x)=\dfrac{(n+1)!\sqrt{\pi}}{2\Gamma(n+\frac{3}{2})}p_n^{(\frac{1}{2},\frac{1}{2})}(x)$ (Abramowitz and Stegun, 1968). The Fourier transforms of Eqs. (3-38)-(3-39) are as follows(Erdelyi, 1954):

$$\bar{f}_1(s)=\sum_{n=0}^{\infty}a_n G_n\frac{1}{s}J_{n+1}(sl)\,,\ G_n=2\sqrt{\pi}(-1)^n i^n\frac{\Gamma(n+1+\frac{1}{2})}{n!} \tag{3-40}$$

$$\bar{f}_2(s)=\sum_{n=0}^{\infty}b_n G_n\frac{1}{s}J_{n+1}(sl)\,, \tag{3-41}$$

where $\Gamma(x)$ and $J_n(x)$ are the Gamma and Bessel functions of order n, respectively.

Substituting Eqs. (3-40) and (3-41) into Eqs. (3-36) and (3-37), it can be shown that Eq. (3-37) is automatically satisfied. After integration with respect to x in $[-l,x]$, Eq. (3-36) was reduced to the following forms:

$$\frac{1}{2\pi}\sum_{n=0}^{\infty}\int_{-\infty}^{\infty}\frac{1}{s(is+\gamma)}\beta_1(s)b_nG_nJ_{n+1}(sl)[e^{isx+\gamma x}-e^{-isl-\gamma l}]ds=-\int_{-l}^{x}\tau_0(t)dt\,,-l\le x\le l \tag{3-42}$$

$$\sum_{n=0}^{\infty}\int_{-\infty}^{\infty}\frac{1}{s(is+\gamma)}\beta_2(s)a_n G_n J_{n+1}(sl)[e^{isx+\gamma x}-e^{-isl-\gamma l}]ds=0,\ -l\le x\le l \tag{3-43}$$

The semi-infinite integral in Eqs. (3-42) and (3-43) can be modified as follows:

$$\int_{-\infty}^{\infty}\frac{\beta_j(s)}{s(is+\gamma)}J_{n+1}(sl)[e^{isx+\gamma x}-e^{-isl-\gamma l}]ds=\begin{cases}\dfrac{-2i\beta_j^{(0)}}{n+1}\{e^{\gamma x}\cos[(n+1)\sin^{-1}(\frac{x}{l})]-(-1)^{\frac{n+1}{2}}e^{-\gamma l}\},n=1,3,5,7,...\\ \dfrac{2\beta_j^{(0)}}{n+1}\{e^{\gamma x}\sin[(n+1)\sin^{-1}(\frac{x}{l})]+(-1)^{\frac{n}{2}}e^{-\gamma l}\},n=0,2,4,6,...\end{cases}$$

$$+\int_0^{\infty}\frac{1}{s}[\frac{\beta_j(s)}{(is+\gamma)}+i\beta_j^{(0)}]J_{n+1}(sl)[e^{isx+\gamma x}-e^{-isl-\gamma l}]ds$$

$$+\int_{-\infty}^{0}\frac{1}{s}[\frac{\beta_j(s)}{(is+\gamma)}-i\beta_j^{(0)}]J_{n+1}(sl)[e^{isx+\gamma x}-e^{-isl-\gamma l}]ds,(j=1,2) \tag{3-44}$$

It can be derived that $a_n=0\,(n=0,1,2,3,....)$, i.e. $f_1(x)=0$.

The integrands of the semi-infinite integrals in the right hand side of Eq. (3-44) tend rapidly to zero for $s\to\infty$. So the semi-infinite integrals in the right hand side of Eq. (3-44) can be evaluated numerically. Equation (3-42) can now be solved for coefficients b_n using the Schmidt method (Morse and Feshbach, 1958, Itou, 1978). It can be seen in references (Morse and Feshbach, 1958, Itou, 1978). Here, it was omitted for brevity.

3.5. Intensity Factors

Once we determine coefficients a_n and b_n, we can obtain the stress, electric displacement and magnetic flux fields. However, from the viewpoint of fracture mechanics, it is important to determine stresses $\sigma_{zz}^{(1)}$, $\sigma_{xz}^{(1)}$, electric displacements $D_x^{(1)}$, $D_z^{(1)}$ and magnetic fluxes $B_x^{(1)}$, $B_z^{(1)}$ in the vicinity of crack tips, respectively. In the present study, $\sigma_{zz}^{(1)}$, $\sigma_{xz}^{(1)}$, $D_x^{(1)}$, $D_z^{(1)}$, $B_x^{(1)}$ and $B_z^{(1)}$ along the crack line can be expressed, respectively, as follows:

$$\begin{cases} \sigma_{zz}^{(1)}(x,0) = \dfrac{e^{\gamma x}}{2\pi}\sum_{n=0}^{\infty}\int_{-\infty}^{\infty}\dfrac{1}{s}\beta_1(s)b_nG_nJ_{n+1}(sl)e^{isx}ds \\ \sigma_{xz}^{(1)}(x,0) = \dfrac{e^{\gamma x}}{2\pi}\sum_{n=0}^{\infty}\int_{-\infty}^{\infty}\dfrac{1}{s}\beta_2(s)a_nG_nJ_{n+1}(sl)e^{isx}ds = 0 \\ D_z^{(1)}(x,0) = \dfrac{e^{\gamma x}}{2\pi}\sum_{n=0}^{\infty}\int_{-\infty}^{\infty}\dfrac{1}{s}\beta_3(s)b_nG_nJ_{n+1}(sl)e^{isx}ds \\ D_x^{(1)}(x,0) = \dfrac{e^{\gamma x}}{2\pi}\sum_{n=0}^{\infty}\int_{-\infty}^{\infty}\dfrac{1}{s}\beta_4(s)b_nG_nJ_{n+1}(sl)e^{isx}ds \\ B_z^{(1)}(x,0) = \dfrac{e^{\gamma x}}{2\pi}\sum_{n=0}^{\infty}\int_{-\infty}^{\infty}\dfrac{1}{s}\beta_5(s)b_nG_nJ_{n+1}(sl)e^{isx}ds \\ B_x^{(1)}(x,0) = \dfrac{e^{\gamma x}}{2\pi}\sum_{n=0}^{\infty}\int_{-\infty}^{\infty}\dfrac{1}{s}\beta_6(s)b_nG_nJ_{n+1}(sl)e^{isx}ds \end{cases} \tag{3-45}$$

where $\beta_j(s)(j=3,4,5,6)$ are known functions which are dependent on material properties as shown in the appendix-III. Moreover, $\lim_{s\to+\infty}\beta_j(s)/s = \beta_j^{(0)} = -\lim_{s\to-\infty}\beta_j(s)/s, (j=3,4,5,6)$. $\beta_j^{(0)}(j=3,4,5,6)$ are known constants which depend on material properties, D_0/ε_0 and B_0/μ_0 as shown in the iterature (Zhou et al., 2007). They do not depend on the functionally graded parameter γ.

The singular parts of the stress, electric displacements and magnetic flux fields near the right tip of the crack in Eq. (3-45) can be expressed for $x > l$, respectively, as follows:

$$\begin{cases} \sigma_{zz1}^{(1)} = \dfrac{\beta_1^{(0)}e^{\gamma x}}{2\pi}\sum_{n=0}^{\infty}b_nG_n[\int_0^{\infty}J_{n+1}(sl)e^{isx}ds - \int_{-\infty}^{0}J_{n+1}(sl)e^{isx}ds] = \dfrac{\beta_1^{(0)}e^{\gamma x}}{2\pi}\sum_{n=0}^{\infty}b_nG_nQ_n(x) \\ D_{z1}^{(1)} = \dfrac{\beta_3^{(0)}e^{\gamma x}}{2\pi}\sum_{n=0}^{\infty}b_nG_n[\int_0^{\infty}J_{n+1}(sl)e^{isx}ds - \int_{-\infty}^{0}J_{n+1}(sl)e^{isx}ds] = \dfrac{\beta_3^{(0)}e^{\gamma x}}{2\pi}\sum_{n=0}^{\infty}b_nG_nQ_n(x) \\ D_{x1}^{(1)} = \dfrac{\beta_4^{(0)}e^{\gamma x}}{2\pi}\sum_{n=0}^{\infty}b_nG_n[\int_0^{\infty}J_{n+1}(sl)e^{isx}ds - \int_{-\infty}^{0}J_{n+1}(sl)e^{isx}ds] = \dfrac{\beta_4^{(0)}e^{\gamma x}}{2\pi}\sum_{n=0}^{\infty}b_nG_nQ_n(x) \\ B_{z1}^{(1)} = \dfrac{\beta_5^{(0)}e^{\gamma x}}{2\pi}\sum_{n=0}^{\infty}b_nG_n[\int_0^{\infty}J_{n+1}(sl)e^{isx}ds - \int_{-\infty}^{0}J_{n+1}(sl)e^{isx}ds] = \dfrac{\beta_5^{(0)}e^{\gamma x}}{2\pi}\sum_{n=0}^{\infty}b_nG_nQ_n(x) \\ B_{x1}^{(1)} = \dfrac{\beta_6^{(0)}e^{\gamma x}}{2\pi}\sum_{n=0}^{\infty}b_nG_n[\int_0^{\infty}J_{n+1}(sl)e^{isx}ds - \int_{-\infty}^{0}J_{n+1}(sl)e^{isx}ds] = \dfrac{\beta_6^{(0)}e^{\gamma x}}{2\pi}\sum_{n=0}^{\infty}b_nG_nQ_n(x) \end{cases} \tag{3-46}$$

where $Q_n(x)=\begin{cases}\dfrac{-2(-1)^{\frac{n}{2}}l^{n+1}}{\sqrt{x^2-l^2}[x+\sqrt{x^2-l^2}]^{n+1}}, n=0,2,4,6,...\\ \dfrac{2i(-1)^{\frac{n+1}{2}}l^{n+1}}{\sqrt{x^2-l^2}[x+\sqrt{x^2-l^2}]^{n+1}}, n=1,3,5,7,...\end{cases}$

The singular parts of the stress, electric displacements and magnetic flux fields near the right tip of the crack in Eq. (3-45) can be expressed for $x<-l$, respectively, as follows:

$$\begin{cases}\sigma_{zz1}^{(1)}=\dfrac{\beta_1^{(0)}e^{\gamma x}}{2\pi}\sum_{n=0}^{\infty}b_nG_n[\int_0^{\infty}J_{n+1}(sl)e^{isx}ds-\int_{-\infty}^{0}J_{n+1}(sl)e^{isx}ds]=\dfrac{\beta_1^{(0)}e^{\gamma x}}{2\pi}\sum_{n=0}^{\infty}b_nG_nQ_n^*(x)\\ D_{z1}^{(1)}=\dfrac{\beta_3^{(0)}e^{\gamma x}}{2\pi}\sum_{n=0}^{\infty}b_nG_n[\int_0^{\infty}J_{n+1}(sl)e^{isx}ds-\int_{-\infty}^{0}J_{n+1}(sl)e^{isx}ds]=\dfrac{\beta_3^{(0)}e^{\gamma x}}{2\pi}\sum_{n=0}^{\infty}b_nG_nQ_n^*(x)\\ D_{x1}^{(1)}=\dfrac{\beta_4^{(0)}e^{\gamma x}}{2\pi}\sum_{n=0}^{\infty}b_nG_n[\int_0^{\infty}J_{n+1}(sl)e^{isx}ds-\int_{-\infty}^{0}J_{n+1}(sl)e^{isx}ds]=\dfrac{\beta_4^{(0)}e^{\gamma x}}{2\pi}\sum_{n=0}^{\infty}b_nG_nQ_n^*(x)\\ B_{z1}^{(1)}=\dfrac{\beta_5^{(0)}e^{\gamma x}}{2\pi}\sum_{n=0}^{\infty}b_nG_n[\int_0^{\infty}J_{n+1}(sl)e^{isx}ds-\int_{-\infty}^{0}J_{n+1}(sl)e^{isx}ds]=\dfrac{\beta_5^{(0)}e^{\gamma x}}{2\pi}\sum_{n=0}^{\infty}b_nG_nQ_n^*(x)\\ B_{x1}^{(1)}=\dfrac{\beta_6^{(0)}e^{\gamma x}}{2\pi}\sum_{n=0}^{\infty}b_nG_n[\int_0^{\infty}J_{n+1}(sl)e^{isx}ds-\int_{-\infty}^{0}J_{n+1}(sl)e^{isx}ds]=\dfrac{\beta_6^{(0)}e^{\gamma x}}{2\pi}\sum_{n=0}^{\infty}b_nG_nQ_n^*(x)\end{cases}\tag{3-47}$$

where $Q_n^*(x)=\begin{cases}\dfrac{-2(-1)^{\frac{n}{2}}l^{n+1}}{\sqrt{x^2-l^2}[|x|+\sqrt{x^2-l^2}]^{n+1}}, n=0,2,4,6,...\\ \dfrac{-2i(-1)^{\frac{n+1}{2}}l^{n+1}}{\sqrt{x^2-l^2}[|x|+\sqrt{x^2-l^2}]^{n+1}}, n=1,3,5,7,...\end{cases}$

From Eq. (3-46), the stress intensity factors, $K_{\mathrm{I}R}$, $K_{\mathrm{II}R}$, electric displacement intensity factors $K_{\mathrm{I}R}^{D}, K_{\mathrm{II}R}^{D}$ and magnetic flux intensity factors $K_{\mathrm{I}R}^{B}, K_{\mathrm{II}R}^{B}$ at the right tip of the crack can be expressed, respectively, as follows:

$$\begin{cases} K_{\mathrm{I}R} = \lim_{x \to l^+} \sqrt{2(x-l)} \cdot \sigma_{zz}^{(1)}(x,0) = -\dfrac{2\beta_1^{(0)} e^{\gamma l}}{\sqrt{\pi l}} \sum_{n=0}^{\infty} b_n \dfrac{\Gamma(n+1+\frac{1}{2})}{n!} \\ K_{\mathrm{II}R} = \lim_{x \to l^+} \sqrt{2(x-l)} \cdot \sigma_{xz}^{(1)}(x,0) = 0 \\ K_{\mathrm{I}R}^{D} = \lim_{x \to l^+} \sqrt{2(x-l)} \cdot D_z^{(1)}(x,0) = \dfrac{\beta_3^{(0)}}{\beta_1^{(0)}} K_{\mathrm{I}R} \\ K_{\mathrm{II}R}^{D} = \lim_{x \to l^+} \sqrt{2(x-l)} \cdot D_x^{(1)}(x,0) = \dfrac{\beta_4^{(0)}}{\beta_1^{(0)}} K_{\mathrm{I}R} \\ K_{\mathrm{I}R}^{B} = \lim_{x \to l^+} \sqrt{2(x-l)} \cdot B_z^{(1)}(x,0) = \dfrac{\beta_5^{(0)}}{\beta_1^{(0)}} K_{\mathrm{I}R} \\ K_{\mathrm{II}R}^{B} = \lim_{x \to l^+} \sqrt{2(x-l)} \cdot B_x^{(1)}(x,0) = \dfrac{\beta_6^{(0)}}{\beta_1^{(0)}} K_{\mathrm{I}R} \end{cases} \tag{3-48}$$

From Eq. (3-47), the stress intensity factors $K_{\mathrm{I}L}, K_{\mathrm{II}L}$, electric displacement intensity factors $K_{\mathrm{I}L}^{D}, K_{\mathrm{II}L}^{D}$ and magnetic flux intensity factors $K_{\mathrm{I}L}^{B}, K_{\mathrm{II}L}^{B}$ at the left tip of the crack can be expressed, respectively, as follows:

$$\begin{cases} K_{\mathrm{I}L} = \lim_{x \to -l^-} \sqrt{2(x+l)} \cdot \sigma_{zz}^{(1)}(x,0) = \dfrac{2\beta_1^{(0)} e^{-\gamma l}}{\sqrt{\pi l}} \sum_{n=0}^{\infty} (-1)^{n+1} b_n \dfrac{\Gamma(n+1+\frac{1}{2})}{n!} \\ K_{\mathrm{II}L} = \lim_{x \to -l^-} \sqrt{2(x+l)} \cdot \sigma_{xz}^{(1)}(x,0) = 0 \\ K_{\mathrm{I}L}^{D} = \lim_{x \to -l^-} \sqrt{2(x+l)} \cdot D_z^{(1)}(x,0) = \dfrac{\beta_3^{(0)}}{\beta_1^{(0)}} K_{\mathrm{I}L} \\ K_{\mathrm{II}L}^{D} = \lim_{x \to -l^-} \sqrt{2(x+l)} \cdot D_x^{(1)}(x,0) = \dfrac{\beta_4^{(0)}}{\beta_1^{(0)}} K_{\mathrm{I}L} \\ K_{\mathrm{I}L}^{B} = \lim_{x \to -l^-} \sqrt{2(x+l)} \cdot B_z^{(1)}(x,0) = \dfrac{\beta_5^{(0)}}{\beta_1^{(0)}} K_{\mathrm{I}L} \\ K_{\mathrm{II}L}^{B} = \lim_{x \to -l^-} \sqrt{2(x+l)} \cdot B_x^{(1)}(x,0) = \dfrac{\beta_6^{(0)}}{\beta_1^{(0)}} K_{\mathrm{I}L} \end{cases} \tag{3-49}$$

3.6. Discussion and Conclusions

It can be seen from previous works (Zhou and Wang, 2004; Zhou et al., 2004; 2005a and 2005b) that the Schmidt method possesses sufficient accuracy if the first ten terms of the infinite series in Eqs. (3-42) and (3-45) are retained. Crack surface loading $-\tau_0(x)$ will simply be assumed to be a polynomial of the following form (The properties of materials are non-symmetric about z -axis for $\gamma \neq 0$, so the stress loading on the crack surfaces should not be a uniform tensile stress.):

$$-\tau_0(x) = -p_0 - p_1(\frac{x}{l}) - p_2(\frac{x}{l})^2 - p_3(\frac{x}{l})^3 \tag{3-50}$$

Since the problem is linear, the results can be obtained through superposition in any suitable manner. Here the results are obtained by taking only one of the four input parameters p_0, p_1, p_2 and p_3 as nonzero each time in the calculation. The magnetoelectroelastic material properties are listed in Table I (Song and Sih, 2003; Tian and Gabbert, 2004). Although the values of D_0 and B_0 depend on the external loading, we assume in the calculation that D_0 and B_0 are two constants, which allows us to focus on the effect of the electric permittivity ε_0 and the magnetic permeability μ_0 of the air inside the crack on the stress, electric displacement and magnetic flux fields near crack tips. Since ε_0 and μ_0 are two variables, D_0 / ε_0 and B_0 / μ_0 can be used as two variables in the calculation.

The calculated stress, electric displacement and magnetic flux intensity factors at crack tips are plotted in Figs. 3-2 to 3-32, respectively. We discuss the results and draw our conclusions as follows:

In the present paper, the basic solution of a Mode-I crack in functionally graded piezoelectric/piezomagnetic materials is obtained by the generalized Almansi's theorem. This method is feasible for general cases, as discussed in Eqs. (3-24)-(3-28), and thus the obtained solution is valid in general cases. In contrast, the Eshelby-Stroh's method which is adopted in Gao et al. (2003b, 2003c) is valid only for the cases of non-degenerate materials. As discussed in literature (Hao and Shen, 1994; Hao, 2001; Dascalu and Homentcovschi, 2004), the electric permittivity and the magnetic permeability of the air inside the crack are considered in the present paper to mimic the real electromagnetic boundary conditions along the crack surfaces. When $\gamma = 0$, the solution can be reverted to a closed form one for homogeneous piezoelectric/piezomagnetic materials as shown in the literature (Zhou et al., 2007).

The current paper introduces the jumps of displacements across the crack surfaces as unknown variables in constructing the dual integral equations, which is quite different from the previous work (Wu and Huang, 2000; Sih and Song, 2003; Wang and Mai, 2003; Gao et al., 2003a; Gao et al., 2003d; Liu et al., 2001; Pan, 2002; Hao and Shen, 1994; Hao 2001; Dascalu and Homentcovschi, 2002), in which the unknown variables of dual integral equations are dislocation density functions. Furthermore, to solve the dual integral equations, the jumps of displacements across the crack surfaces are directly expanded in a series of Jacobi polynomials. This is the major difference between the current work and the available work in the literature.

The solution shows that the singular stress, the singular electric displacement and the singular magnetic flux in functionally graded piezoelectric/ piezomagnetic materials carry the same forms as those in homogeneous piezoelectric/piezomagnetic materials, except that the magnitudes of the stress, electric displacement and magnetic flux intensity factors depend significantly upon the gradient of functionally graded piezoelectric/piezomagnetic materials as reported by Li and Weng (2002.) for the fracture problem in functionally graded piezoelectric materials. The stress, electric displacement and magnetic flux intensity factors depend on the crack length and material properties for the Mode-I crack in functionally graded piezoelectric/piezomagnetic materials as shown in Eqs. (3-36) and (3-45). The electro-magneto-elastic coupling effects on the magnetoelectroelastic intensity factors are also obtained, as shown in Eqs. (3-48) and (3-49). The electric displacement and magnetic flux intensity factors shown in Figs. 3-2, 3-6 and 3-7 display the same varying tendencies as the stress intensity factors with variation in γl as shown in Eqs. (3-48) and (3-49). In addition, they are very small in magnitude as shown in Figs. 3-6 and 3-7. Therefore, some results of the electric displacement and magnetic flux intensity factors are omitted in the present paper for brevity. Actually, the results of electric displacement and magnetic flux intensity factors can be directly obtained from the results of stress intensity factors through Eqs. (3-48) and (3-49).

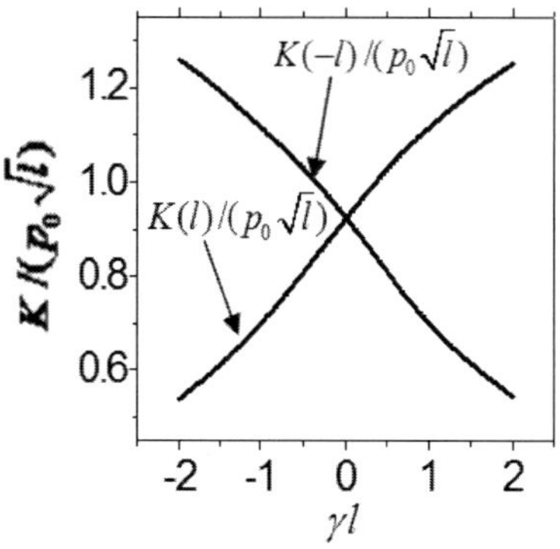

Fig. 3-2. Stress intensity factor versus γl for $l = 1.0$, $D_0/\varepsilon_0 = 5\times10^8$ and $B_0/\mu_0 = 5\times10^7$ ($\tau_0(x) = p_0$).

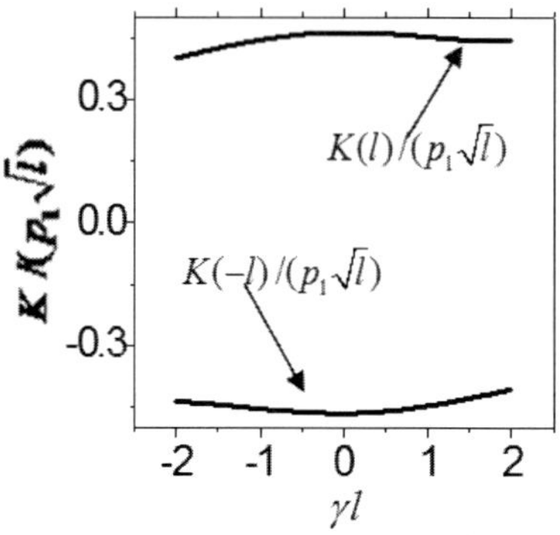

Fig. 3-3. Stress intensity factor versus γl for $D_0/\varepsilon_0 = 5\times10^8$, $l = 1.0$ and $B_0/\mu_0 = 5\times10^7$ ($\tau_0(x) = p_1(x/l)$).

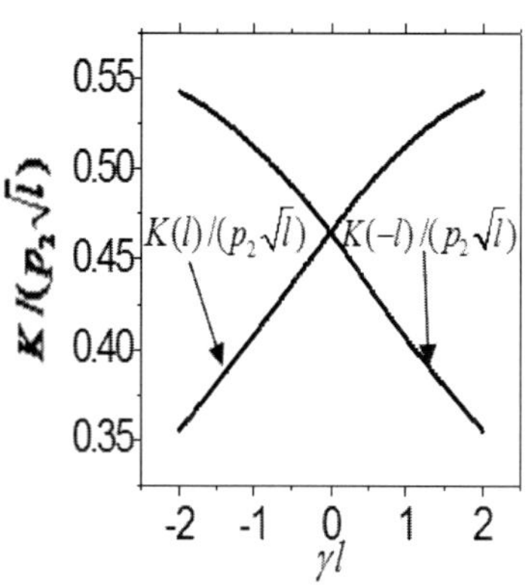

Fig. 3-4. Stress intensity factor versus γl for $D_0/\varepsilon_0 = 5\times10^8$, $l = 1.0$ and $B_0/\mu_0 = 5\times10^7$ ($\tau_0(x) = p_2(x/l)^2$).

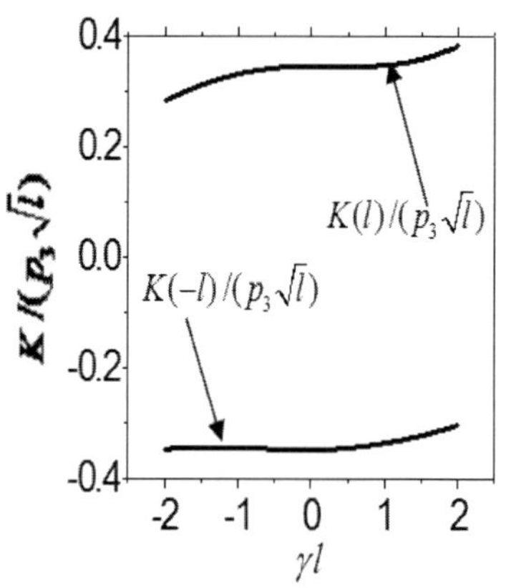

Fig. 3-5. Stress intensity factor versus γl for $D_0/\varepsilon_0 = 5\times10^8$, $l = 1.0$ and $B_0/\mu_0 = 5\times10^7$ ($\tau_0(x) = p_3(x/l)^3$).

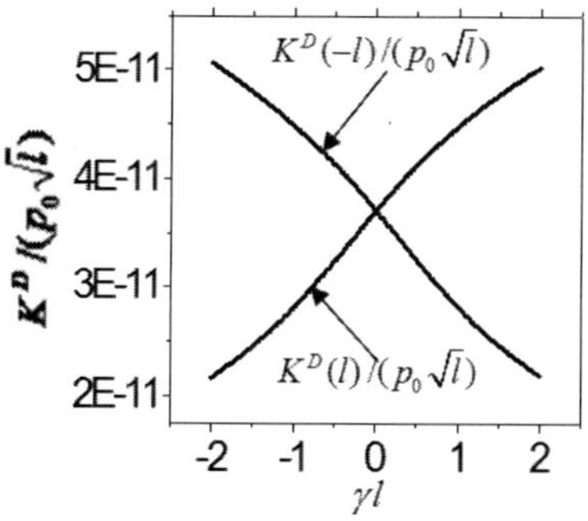

Fig. 3-6. Electric displacement intensity factor versus γl for $l=1.0$, $D_0/\varepsilon_0=5\times10^8$ and $B_0/\mu_0=5\times10^7$ ($\tau_0(x)=p_0$).

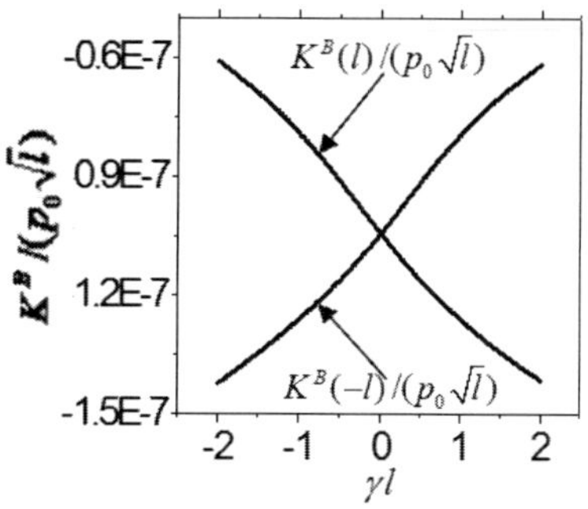

Fig. 3-7. Magnetic flux intensity factor versus γl for $l=1.0$, $D_0/\varepsilon_0=5\times10^8$ and $B_0/\mu_0=5\times10^7$ ($\tau_0(x)=p_0$).

For symmetric loading case, the results of stress, electric displacement and magnetic flux intensity factors at crack tips are symmetric about the line, $\gamma l=0$, as shown in Figs. 2, 4, 6 and 7. It can be also observed that $K(l)/(p_0\sqrt{l})=K(-l)/(p_0\sqrt{l})$, $K^D(l)/(p_0\sqrt{l})=K^B(-l)/(p_0\sqrt{l})$ and $K^B(l)/(p_0\sqrt{l})=K^B(-l)/(p_0\sqrt{l})$ for $\gamma l=0$ are shown in Figs. 3-2, 3-4, 3-6 and 3-7. It is the same as the closed form solution reported in Zhou et al.(2007). However, for anti-symmetric loading, the stress intensity factors are anti-symmetric about the point $K=0$ and $\gamma l=0$ as shown in Figs. 3-3 and 3-5. So for the symmetric loading, we only discuss the properties of stress, electric displacement and magnetic flux intensity factors at the right tip of the crack here and after.

For symmetric loading, the stress and electric displacement intensity factors at the right tip of the crack increase with the increase in the gradient parameter γl as shown in Figs. 3-2, 3-4 and 3-6. However, the stress and electric displacement intensity factors at the left tip of the crack decrease with the increase in the functionally graded parameter γl. It can be also obtained that the stress and electric displacement intensity factors at the right tip of the crack are larger than the corresponding ones at the left tip of the crack for $\gamma l>0$; while they are less than the corresponding ones at the left tip of the crack for $\gamma l<0$, as shown in Figs. 3-2, 3-4 and 3-6. This implies that the stress intensity on the stiffer side of the medium is always greater than that on the less stiff side. However, the magnitude of magnetic flux intensity factor at the right tip of the crack decreases with the increase in the gradient parameter, γl, as shown in Figure 3-7. This means that, by adjusting the gradient parameter for functionally graded piezoelectric/piezomagnetic materials, the stress, electric displacement and the magnetic flux fields near crack tips can be reduced.

For anti-symmetric loading, the stress intensity factor at the right tip of the crack increases until reaching the peak value at $\gamma l=0$, and then decreases with the increase in the gradient parameter, γl, as shown in Figs. 3-3 and 3-5. In this case, the stress intensity factor at the left tip of the crack is negative because the loading $\tau_0(x)<0$ for $x<0$.

As shown in Figure 3-8, it can be obtained that the stress intensity factor at the right tip of the crack increases with the increase in the crack length. However, the stress intensity factor at the left tip of the crack increases with the increase in the crack length. This phenomenon may be caused by the effect of functionally graded material parameter as discussed by Shbeeb and Binienda (1999). It should be further studied in future.

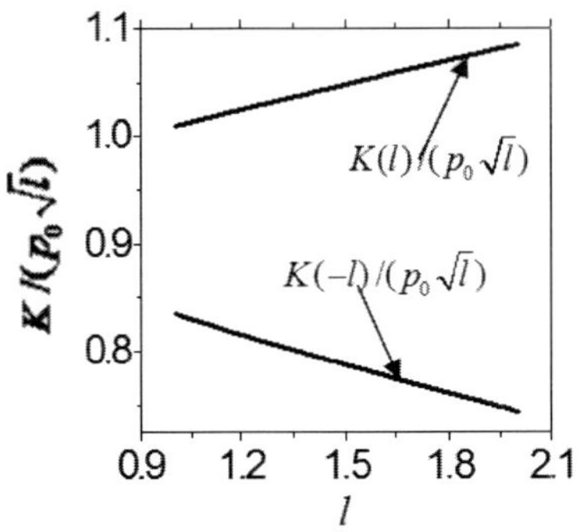

Fig. 3-8. Stress intensity factor versus l for $\gamma = 0.4$, $D_0/\varepsilon_0 = 5\times10^8$ and $B_0/\mu_0 = 5\times10^7$ ($\tau_0(x) = p_0$).

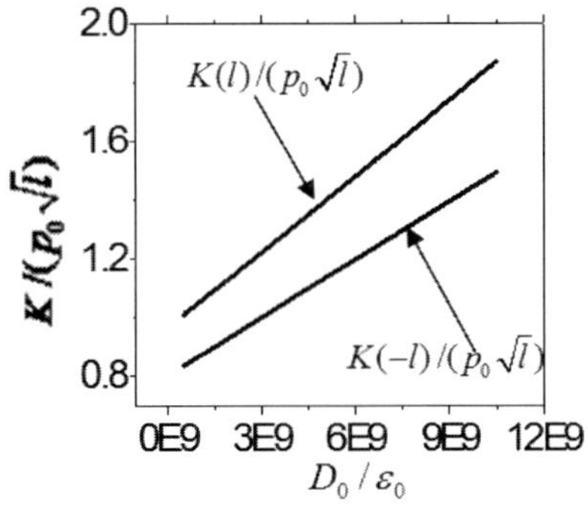

Fig. 3-9. Stress intensity factor versus D_0/ε_0 for $l = 1.0$, $\gamma l = 0.4$ and $B_0/\mu_0 = 5\times10^7$ ($\tau_0(x) = p_0$).

For symmetric loading, the stress intensity factors at both crack tips increase with the increase in D_0/ε_0 as shown in Figs. 3-9 and 3-11. This reveals that the stress intensity factors at both crack tips increase with the decrease in the electric permittivity ε_0 of the air inside the crack. For the anti-symmetric loading, the magnitudes of stress intensity factors at crack tips increase with the increase in D_0/ε_0 as shown in Figs. 3-10 and 3-12. This reveals, however, that the magnitudes of stress intensity factors at both crack tips increase with the decrease in the electric permittivity ε_0 of the air inside the crack. Therefore, we conclude that the effects of the electric permittivity ε_0 of air inside the crack on the near tip stress fields cannot be ignored.

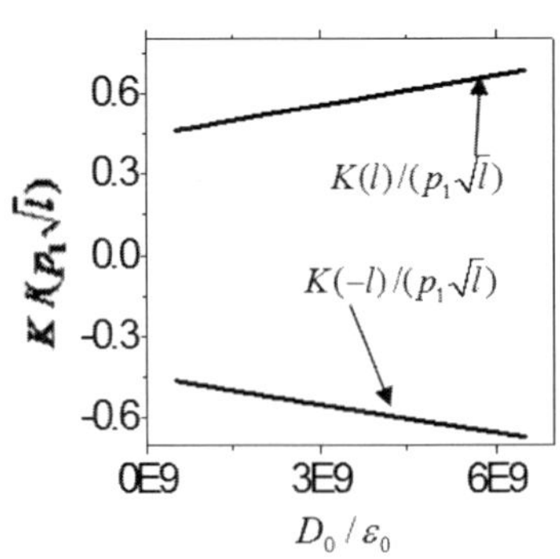

Fig. 3-10. Stress intensity factor versus D_0/ε_0 for $l = 1.0$, $\gamma l = 0.4$ and $B_0/\mu_0 = 5\times10^7$ ($\tau_0(x) = p_1(x/l)$).

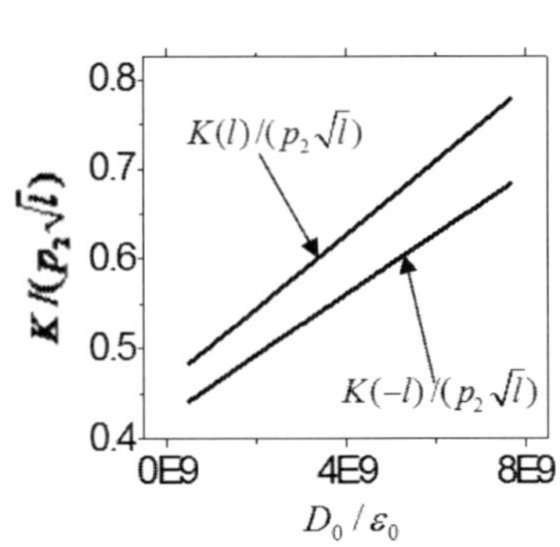

Fig. 3-11. Stress intensity factor versus D_0/ε_0 for $l = 1.0$, $\gamma l = 0.4$ and $B_0/\mu_0 = 5\times10^7$ ($\tau_0(x) = p_2(x/l)^2$).

For symmetric or anti-symmetric loading, the magnitudes of electric displacement intensity factors at both crack tips increase linearly with the increase in D_0/ε_0 as shown in Figs. (3-13)-(3-16). For the anti-symmetric loading, it can be also obtained that the electric displacement intensity factors at both crack tips are symmetric about the line $K^D = 0$, as shown in Figs. 3-14 and 3-16.

For symmetric or anti-symmetric loading as shown in Figs. (3-17)-(3-20), the magnitudes of magnetic flux intensity factors at the right crack tip increase linearly with the increase of D_0/ε_0. However, the magnitudes of magnetic flux intensity factors at the left crack tip decrease linearly with the increase of D_0/ε_0. It can be also obtained that the effects of the electric permittivity ε_0 of the air in the crack on electric displacement and magnetic flux fields near crack tips can not be ignored from solution of the present paper.

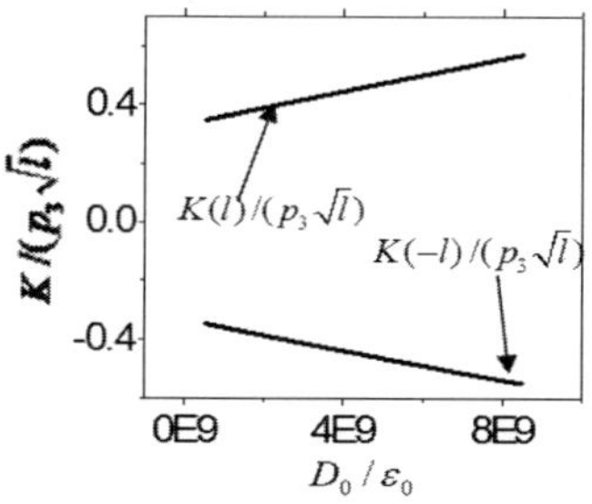

Fig. 3-12. Stress intensity factor versus D_0/ε_0 for $l=1.0$, $\gamma l = 0.4$ and $B_0/\mu_0 = 5\times10^7$ ($\tau_0(x) = p_3(x/l)^3$).

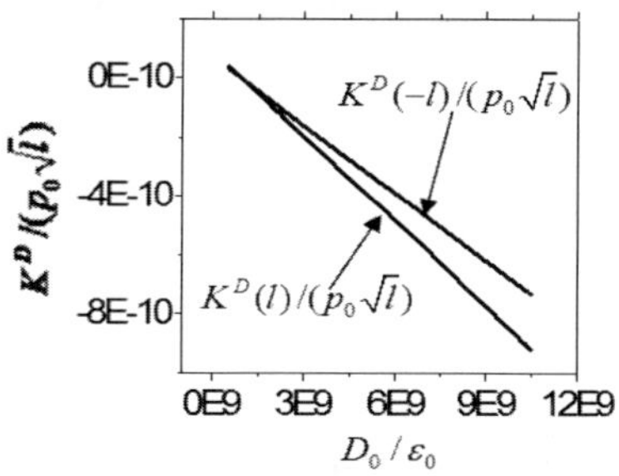

Fig. 3-13. Electric displacement intensity factor versus D_0/ε_0 for $l=1.0$, $\gamma l = 0.4$ and $B_0/\mu_0 = 5\times10^7$ ($\tau_0(x) = p_0$).

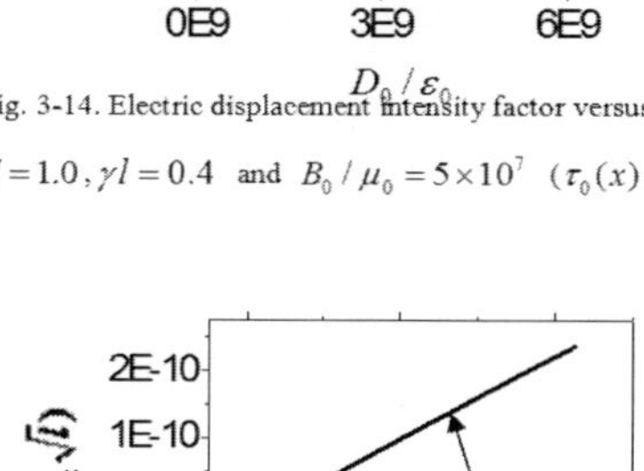

Fig. 3-14. Electric displacement intensity factor versus D_0/ε_0 for $l=1.0$, $\gamma l = 0.4$ and $B_0/\mu_0 = 5\times10^7$ ($\tau_0(x) = p_1(x/l)$).

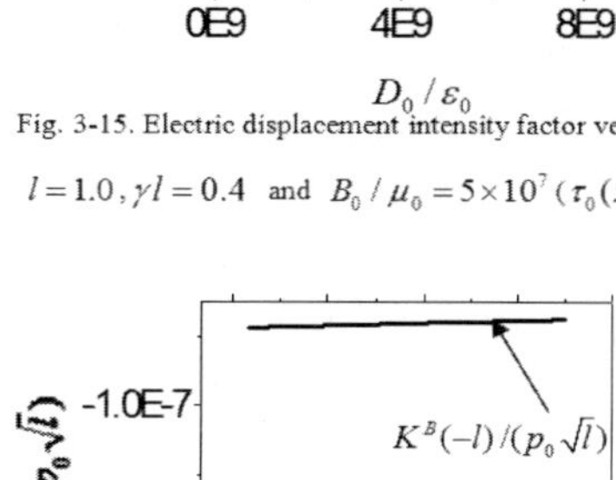

Fig. 3-15. Electric displacement intensity factor versus D_0/ε_0 for $l=1.0$, $\gamma l = 0.4$ and $B_0/\mu_0 = 5\times10^7$ ($\tau_0(x) = p_2(x/l)^2$).

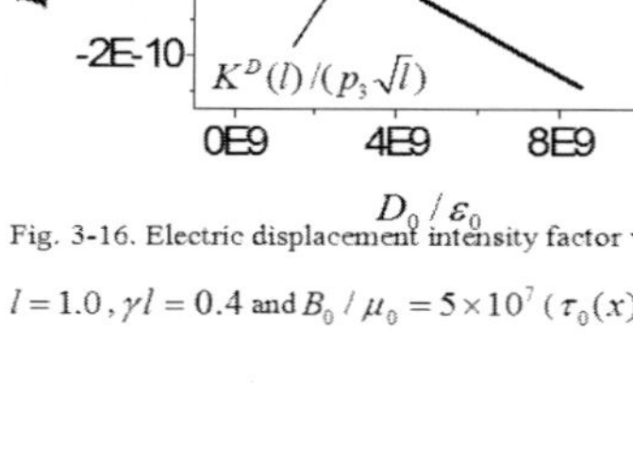

Fig. 3-16. Electric displacement intensity factor versus γl for $l=1.0$, $\gamma l = 0.4$ and $B_0/\mu_0 = 5\times10^7$ ($\tau_0(x) = p_3(x/l)^3$).

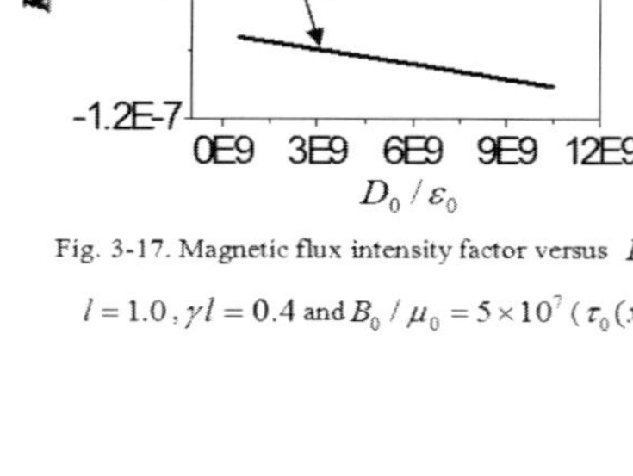

Fig. 3-17. Magnetic flux intensity factor versus D_0/ε_0 for $l=1.0$, $\gamma l = 0.4$ and $B_0/\mu_0 = 5\times10^7$ ($\tau_0(x) = p_0$).

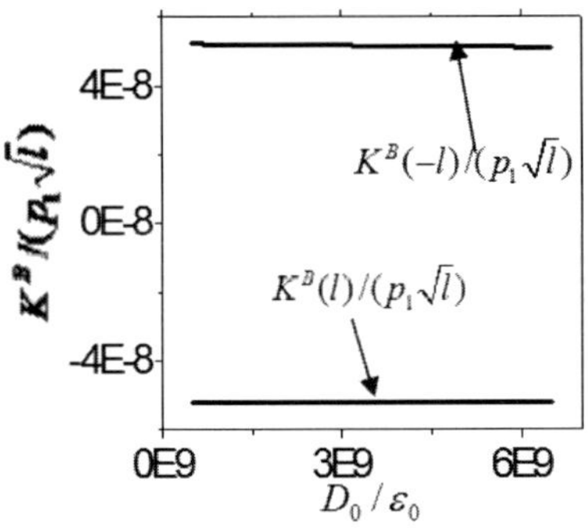

Fig. 3-18. Magnetic flux intensity factor versus D_0/ε_0 for $l=1.0$, $\gamma l=0.4$ and $B_0/\mu_0=5\times10^7$ ($\tau_0(x)=p_1(x/l)$).

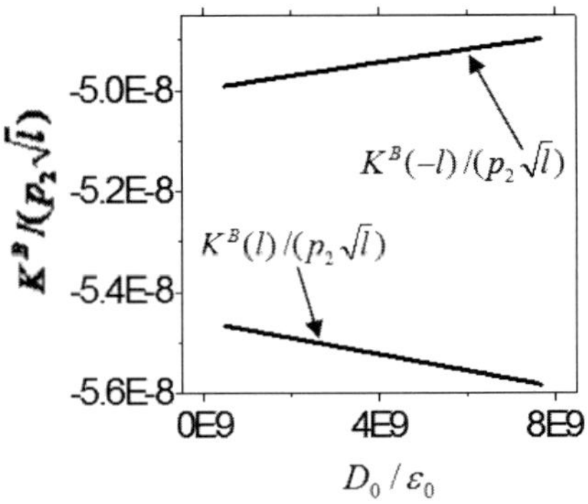

Fig. 3-19. Magnetic flux intensity factor versus D_0/ε_0 for $l=1.0$, $\gamma l=0.4$ and $B_0/\mu_0=5\times10^7$ ($\tau_0(x)=p_2(x/l)^2$).

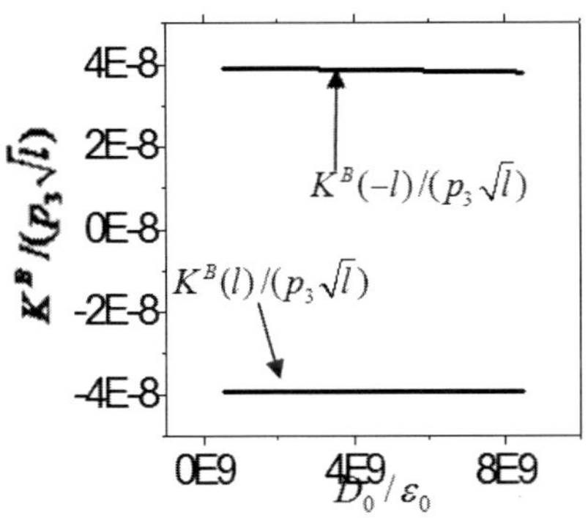

Fig. 3-20. Magnetic flux intensity factor versus D_0/ε_0 for $l=1.0$, $\gamma l=0.4$ and $B_0/\mu_0=5\times10^7$ ($\tau_0(x)=p_3(x/l)^3$).

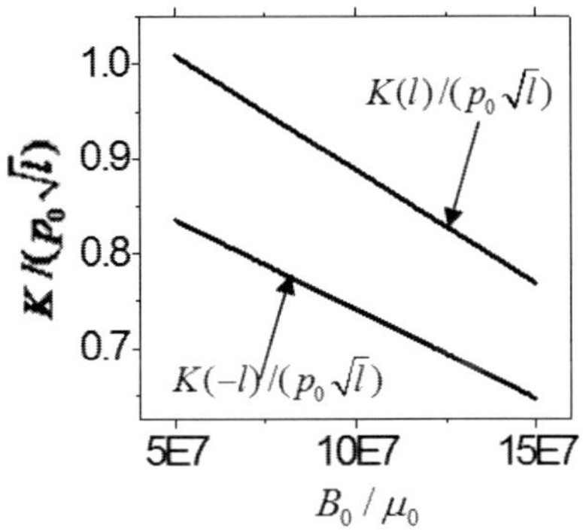

Fig. 3-21. Stress intensity factor versus B_0/μ_0 for $l=1.0$, $\gamma l=0.4$ and $D_0/\varepsilon_0=5\times10^8$ ($\tau_0(x)=p_0$).

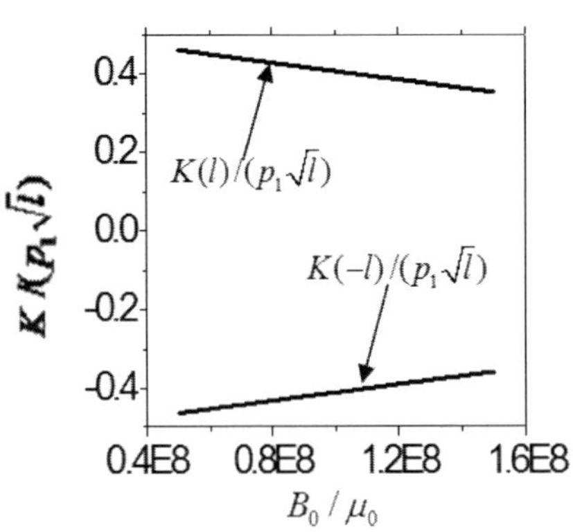

Fig. 3-22. Stress intensity factor versus B_0/μ_0 for $l=1.0$, $\gamma l=0.4$ and $D_0/\varepsilon_0=5\times10^8$ ($\tau_0(x)=p_1(x/l)$).

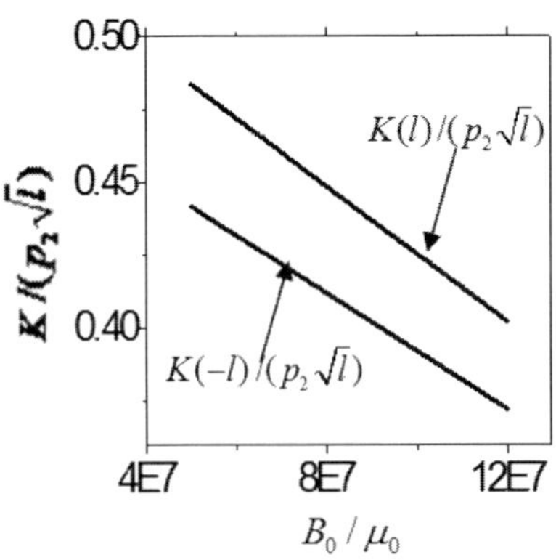

Fig. 3-23. Stress intensity factor versus B_0/μ_0 for $l=1.0$, $\gamma l=0.4$ and $D_0/\varepsilon_0=5\times10^8$ ($\tau_0(x)=p_2(x/l)^2$).

For symmetric or anti-symmetric loading as shown in Figs. (3-21)-(3-24), the magnitudes of stress intensity factors at both crack tips decrease linearly with the increase in B_0/μ_0. This reveals that the stress intensity factors at the crack tips increase with the increase in the magnetic permeability, μ_0, of the air inside the crack. We also conclude that the effects of the magnetic permeability, μ_0, of the air inside the crack on the near tip stress fields cannot be ignored.

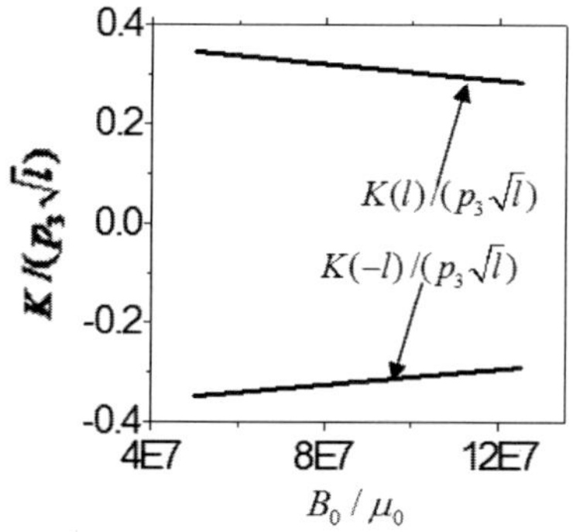

Fig. 3-24. Stress intensity factor versus B_0/μ_0 for $l = 1.0$, $\gamma l = 0.4$ and $D_0/\varepsilon_0 = 5\times10^8$ ($\tau_0(x) = p_3(x/l)^3$).

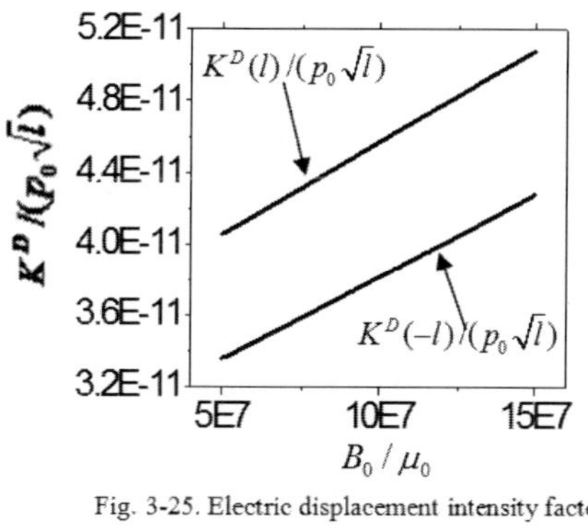

Fig. 3-25. Electric displacement intensity factor versus B_0/μ_0 for $l = 1.0, \gamma l = 0.4$ and $D_0/\varepsilon_0 = 5\times10^8$ ($\tau_0(x) = p_0$).

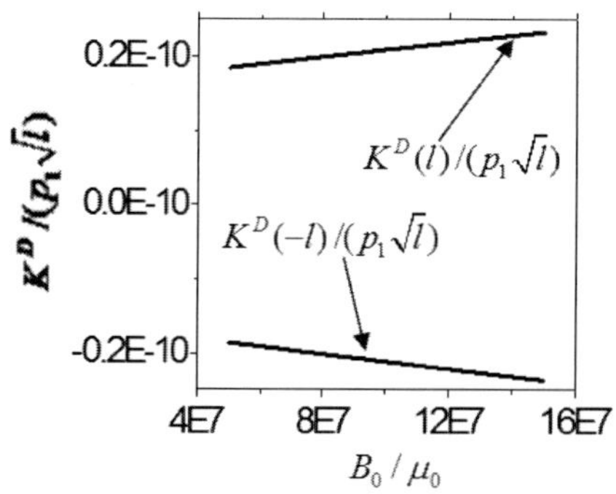

Fig. 3-26. Electric displacement intensity factor versus B_0/μ_0 for $l = 1.0, \gamma l = 0.4$ and $D_0/\varepsilon_0 = 5\times10^8$ ($\tau_0(x) = p_1(x/l)$).

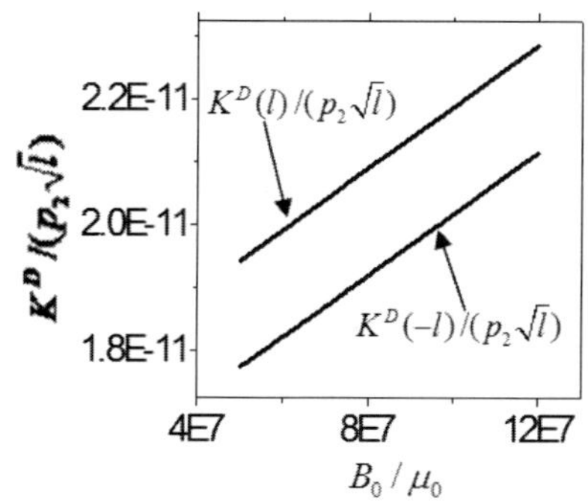

Fig. 3-27. Electric displacement intensity factor versus B_0/μ_0 for $l = 1.0, \gamma l = 0.4$ and $D_0/\varepsilon_0 = 5\times10^8$ ($\tau_0(x) = p_2(x/l)^2$).

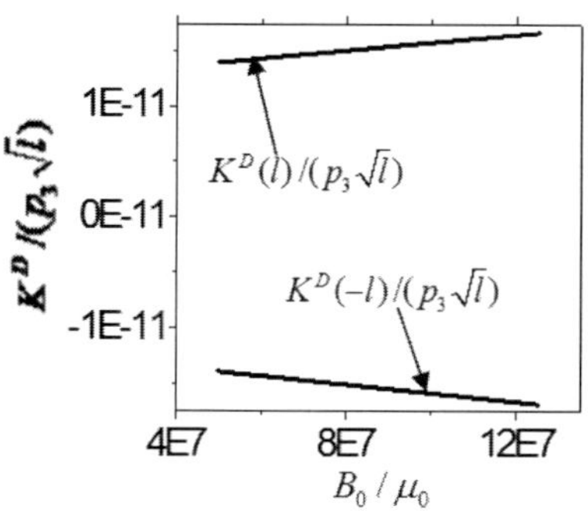

Fig. 3-28. Electric displacement intensity factor versus B_0/μ_0 for $l = 1.0, \gamma l = 0.4$ and $D_0/\varepsilon_0 = 5\times10^8$ ($\tau_0(x) = p_3(x/l)^3$).

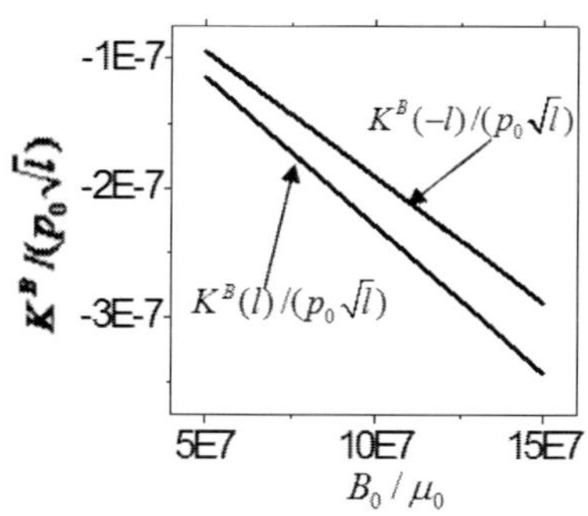

Fig. 3-29. Magnetic flux intensity factor versus B_0/μ_0 for $l = 1.0, \gamma l = 0.4$ and $D_0/\varepsilon_0 = 5\times10^8$ ($\tau_0(x) = p_0$).

For symmetric or anti-symmetric loading as shown in Figs. (3-29)-(3-32), the magnitudes of the magnetic flux intensity factors at both crack tips increase linearly with the increase in B_0/μ_0. This indicates that the magnetic flux intensity factors at the crack tips increase with the decrease in the magnetic permeability, μ_0, of the air inside the crack. We also conclude that the effects of the magnetic permeability, μ_0, of the air inside the crack on the electric displacement and magnetic flux fields near the crack tips cannot be ignored.

For symmetric or anti-symmetric loading as shown in Figs. (3-25)-(3-28), the magnitudes of electric displacement intensity factors at both crack tips increase linearly with the increase in B_0/μ_0. This indicates that the electric displacement intensity factors at the crack tips increase with the decrease in the magnetic permeability, μ_0, of the air inside the crack.

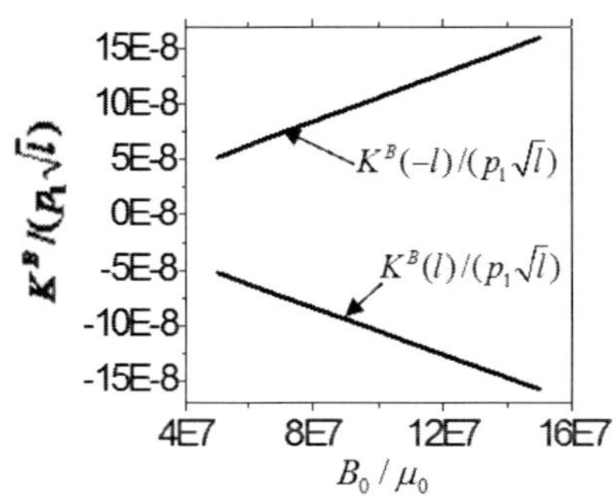

Fig. 3-30. Magnetic flux intensity factor versus B_0/μ_0 for $l = 1.0$, $\gamma l = 0.4$ and $D_0/\varepsilon_0 = 5\times10^8$ ($\tau_0(x) = p_1(x/l)$).

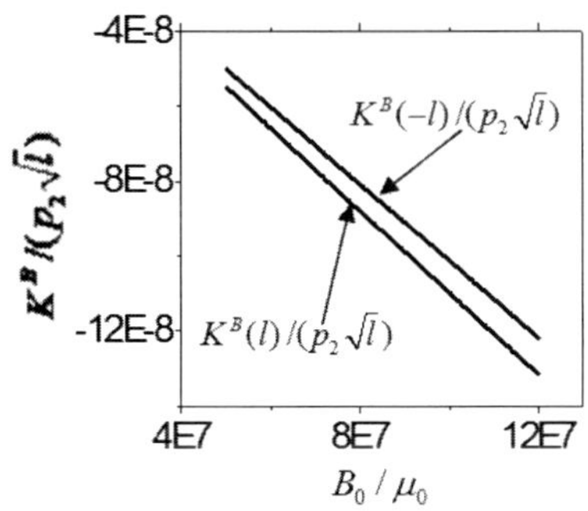

Fig. 3-31. Magnetic flux intensity factor versus B_0/μ_0 for $l = 1.0$, $\gamma l = 0.4$ and $D_0/\varepsilon_0 = 5\times10^8$ ($\tau_0(x) = p_2(x/l)^2$).

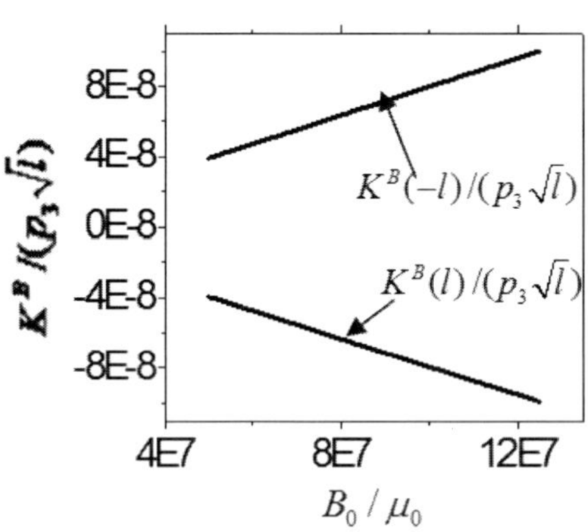

Fig. 3-32. Magnetic flux intensity factor versus B_0/μ_0 for $l = 1.0$, $\gamma l = 0.4$ and $D_0/\varepsilon_0 = 5\times10^8$ ($\tau_0(x) = p_3(x/l)^3$).

Appendix III: Coefficients

$$d_1 = c_{44}(-2e_{33}f_{33}g_{33} + c_{33}g_{33}^2 + e_{33}^2\mu_{33} - f_{33}^2\varepsilon_{33} + c_{33}\mu_{33}\varepsilon_{33}),$$

$$d_2 = -\{e_{31}^2 f_{33}^2 + 2c_{33}e_{31}f_{15}g_{33} + 2c_{33}e_{31}f_{31}g_{33} - 2c_{13}e_{31}f_{33}g_{33} - 2c_{44}e_{31}f_{33}g_{33} - 2c_{33}c_{44}g_{11}g_{33}$$
$$+c_{13}^2 g_{33}^2 - c_{11}c_{33}g_{33}^2 + 2c_{13}c_{44}g_{33}^2 - c_{33}e_{31}^2\mu_{33} + e_{33}^2(f_{15}^2 + 2f_{15}f_{31} + f_{31}^2 - c_{44}\mu_{11} - c_{11}\mu_{33})$$

$$+e_{15}^2(f_{33}^2-c_{33}\mu_{33})+2e_{15}\{[c_{33}(f_{15}+f_{31})-c_{13}f_{33}]g_{33}+e_{31}(f_{33}^2-c_{33}\mu_{33})\}$$
$$-2e_{33}\{-c_{44}f_{33}g_{11}+c_{13}f_{15}g_{33}+c_{13}f_{31}g_{33}+c_{44}f_{31}g_{33}-c_{11}f_{33}g_{33}$$
$$+e_{15}(f_{15}f_{33}+f_{31}f_{33}-c_{13}\mu_{33})+e_{31}[f_{15}f_{33}+f_{31}f_{33}-\mu_{33}(c_{13}+c_{44})]\}$$
$$+c_{44}f_{33}^2\varepsilon_{11}-c_{33}c_{44}\mu_{33}\varepsilon_{11}+c_{33}f_{15}^2\varepsilon_{33}+2c_{33}f_{15}f_{31}\varepsilon_{33}+c_{33}f_{31}^2\varepsilon_{33}-2c_{13}f_{15}f_{33}\varepsilon_{33}-2c_{13}f_{31}f_{33}\varepsilon_{33}$$
$$-2c_{44}f_{31}f_{33}\varepsilon_{33}+c_{11}f_{33}^2\varepsilon_{33}-c_{33}c_{44}\mu_{11}\varepsilon_{33}+c_{13}^2\mu_{33}\varepsilon_{33}-c_{11}c_{33}\mu_{33}\varepsilon_{33}+2c_{13}c_{44}\mu_{33}\varepsilon_{33}\},$$
$$d_3=-\gamma\{e_{31}^2f_{33}^2+2c_{33}e_{31}f_{15}g_{33}+2c_{33}e_{31}f_{31}g_{33}-2c_{13}e_{31}f_{33}g_{33}-2c_{44}e_{31}f_{33}g_{33}-c_{33}c_{44}g_{11}g_{33}$$
$$+c_{13}^2g_{33}^2-c_{11}c_{33}g_{33}^2+2c_{13}c_{44}g_{33}^2-c_{33}e_{31}^2\mu_{33}+e_{33}^2(f_{15}^2+2f_{15}f_{31}+f_{31}^2-c_{44}\mu_{11}-c_{11}\mu_{33})$$
$$+e_{15}^2(f_{33}^2-c_{33}\mu_{33})+2e_{15}g_{33}(c_{33}f_{15}+c_{33}f_{31}-c_{13}f_{33})+2e_{15}e_{31}(f_{33}^2-c_{33}\mu_{33})$$
$$+e_{33}[c_{44}f_{33}g_{11}-2c_{13}f_{15}g_{33}-2c_{13}f_{31}g_{33}-2c_{44}f_{31}g_{33}+2c_{11}f_{33}g_{33}$$
$$-2e_{15}(f_{15}f_{33}+f_{31}f_{33}-c_{13}\mu_{33})-2e_{31}(f_{15}f_{33}+f_{31}f_{33}-c_{13}\mu_{33}-c_{44}\mu_{33})]$$
$$+c_{44}f_{33}^2\varepsilon_{11}-c_{33}c_{44}\mu_{33}\varepsilon_{11}+c_{33}f_{15}^2\varepsilon_{33}+2c_{33}f_{15}f_{31}\varepsilon_{33}+c_{33}f_{31}^2\varepsilon_{33}-2c_{13}f_{15}f_{33}\varepsilon_{33}-2c_{13}f_{31}f_{33}\varepsilon_{33}$$
$$-2c_{44}f_{31}f_{33}\varepsilon_{33}+c_{11}f_{33}^2\varepsilon_{33}-c_{33}c_{44}\mu_{11}\varepsilon_{33}+c_{13}^2\mu_{33}\varepsilon_{33}-c_{11}c_{33}\mu_{33}\varepsilon_{33}+2c_{13}c_{44}\mu_{33}\varepsilon_{33}\},$$
$$d_4=\gamma^2[-e_{33}^2f_{15}f_{31}-c_{33}e_{31}f_{15}g_{33}+c_{44}e_{31}f_{33}g_{33}-c_{13}c_{44}g_{33}^2-c_{33}f_{15}f_{31}\varepsilon_{33}+c_{13}f_{15}f_{33}\varepsilon_{33}+c_{44}f_{31}f_{33}\varepsilon_{33}$$
$$-c_{13}c_{44}\mu_{33}\varepsilon_{33}+e_{33}(e_{31}f_{15}f_{33}+e_{15}f_{31}f_{33}+c_{13}f_{15}g_{33}+c_{44}f_{31}g_{33}-c_{13}e_{15}\mu_{33}-c_{44}e_{31}\mu_{33})$$
$$+e_{15}(-e_{31}f_{33}^2-c_{33}f_{31}g_{33}+c_{13}f_{33}g_{33}+c_{33}e_{31}\mu_{33})]$$
$$d_5=2c_{13}e_{33}f_{15}g_{11}+2c_{13}e_{33}f_{31}g_{11}+2c_{44}e_{33}f_{31}g_{11}-2c_{11}e_{33}f_{33}g_{11}+c_{33}c_{44}g_{11}^2-2c_{11}e_{33}f_{15}g_{33}$$
$$-2c_{13}^2g_{11}g_{33}+2c_{11}c_{33}g_{11}g_{33}-4c_{13}c_{44}g_{11}g_{33}+c_{11}c_{44}g_{33}^2+c_{11}e_{33}^2\mu_{11}$$
$$+e_{15}^2(2f_{31}f_{33}+c_{33}\mu_{11}-2c_{13}\mu_{33})+e_{31}^2(-2f_{15}f_{33}+c_{33}\mu_{11}+c_{44}\mu_{33})$$
$$-2e_{15}[c_{33}(f_{15}+f_{31})g_{11}-c_{13}f_{33}g_{11}-2c_{13}f_{15}g_{33}-c_{13}f_{31}g_{33}+c_{11}f_{33}g_{33}$$
$$+e_{33}(f_{15}f_{31}+f_{31}^2+c_{13}\mu_{11}-c_{11}\mu_{33})]$$
$$+2e_{31}\{-c_{33}f_{15}g_{11}-c_{33}f_{31}g_{11}+c_{13}f_{33}g_{11}+c_{44}f_{33}g_{11}+c_{13}f_{15}g_{33}-c_{44}f_{31}g_{33}$$
$$+e_{33}[f_{15}^2+f_{15}f_{31}-(c_{13}+c_{44})\mu_{11}]+e_{15}(-f_{15}f_{33}+f_{31}f_{33}+c_{33}\mu_{11}-c_{13}\mu_{33})\}$$
$$-c_{33}f_{15}^2\varepsilon_{11}-2c_{33}f_{15}f_{31}\varepsilon_{11}-c_{33}f_{31}^2\varepsilon_{11}+2c_{13}f_{15}f_{33}\varepsilon_{11}+2c_{13}f_{31}f_{33}\varepsilon_{11}+2c_{44}f_{31}f_{33}\varepsilon_{11}$$
$$-c_{11}f_{33}^2\varepsilon_{11}+c_{33}c_{44}\mu_{11}\varepsilon_{11}-c_{13}^2\mu_{33}\varepsilon_{11}+c_{11}c_{33}\mu_{33}\varepsilon_{11}-2c_{13}c_{44}\mu_{33}\varepsilon_{11}+2c_{13}f_{15}^2\varepsilon_{33}$$
$$+2c_{13}f_{15}f_{31}\varepsilon_{33}-c_{44}f_{31}^2\varepsilon_{33}-2c_{11}f_{15}f_{33}\varepsilon_{33}-c_{13}^2\mu_{11}\varepsilon_{33}+c_{11}c_{33}\mu_{11}\varepsilon_{33}-2c_{13}c_{44}\mu_{11}\varepsilon_{33}+c_{11}c_{44}\mu_{33}\varepsilon_{33}$$
$$d_6=2\gamma\{2c_{13}e_{33}f_{15}g_{11}+2c_{13}e_{33}f_{31}g_{11}+2c_{44}e_{33}f_{31}g_{11}-2c_{11}e_{33}f_{33}g_{11}+c_{33}c_{44}g_{11}^2-2c_{11}e_{33}f_{15}g_{33}$$
$$-2c_{13}^2g_{11}g_{33}+2c_{11}c_{33}g_{11}g_{33}-4c_{13}c_{44}g_{11}g_{33}+c_{11}c_{44}g_{33}^2+c_{11}e_{33}^2\mu_{11}$$
$$+e_{15}^2(2f_{31}f_{33}+c_{33}\mu_{11}-2c_{13}\mu_{33})+e_{31}^2(-2f_{15}f_{33}+c_{33}\mu_{11}+c_{44}\mu_{33})$$
$$-2e_{15}[c_{33}(f_{15}+f_{31})g_{11}-c_{13}f_{33}g_{11}-2c_{13}f_{15}g_{33}-c_{13}f_{31}g_{33}+c_{11}f_{33}g_{33}$$
$$+e_{33}(f_{15}f_{31}+f_{31}^2+c_{13}\mu_{11}-c_{11}\mu_{33})]$$
$$+2e_{31}[-c_{33}f_{15}g_{11}-c_{33}f_{31}g_{11}+c_{13}f_{33}g_{11}+c_{44}f_{33}g_{11}+c_{13}f_{15}g_{33}-c_{44}f_{31}g_{33}$$
$$+e_{33}(f_{15}^2+f_{15}f_{31}-c_{13}\mu_{11}-c_{44}\mu_{11})+e_{15}(-f_{15}f_{33}+f_{31}f_{33}+c_{33}\mu_{11}-c_{13}\mu_{33})]$$
$$-c_{33}f_{15}^2\varepsilon_{11}-2c_{33}f_{15}f_{31}\varepsilon_{11}-c_{33}f_{31}^2\varepsilon_{11}+2c_{13}f_{15}f_{33}\varepsilon_{11}+2c_{13}f_{31}f_{33}\varepsilon_{11}+2c_{44}f_{31}f_{33}\varepsilon_{11}-c_{11}f_{33}^2\varepsilon_{11}$$
$$+c_{33}c_{44}\mu_{11}\varepsilon_{11}-c_{13}^2\mu_{33}\varepsilon_{11}+c_{11}c_{33}\mu_{33}\varepsilon_{11}-2c_{13}c_{44}\mu_{33}\varepsilon_{11}+2c_{13}f_{15}^2\varepsilon_{33}+2c_{13}f_{15}f_{31}\varepsilon_{33}-c_{44}f_{31}^2\varepsilon_{33}$$

$$-2c_{11}f_{15}f_{33}\varepsilon_{33}-c_{13}^2\mu_{11}\varepsilon_{33}+c_{11}c_{33}\mu_{11}\varepsilon_{33}-2c_{13}c_{44}\mu_{11}\varepsilon_{33}+c_{11}c_{44}\mu_{33}\varepsilon_{33}\},$$

$$d_7=\gamma^2\{3c_{13}e_{33}f_{15}g_{11}+2c_{13}e_{33}f_{31}g_{11}+3c_{44}e_{33}f_{31}g_{11}-2c_{11}e_{33}f_{33}g_{11}+c_{33}c_{44}g_{11}^2-2c_{11}e_{33}f_{15}g_{33}$$
$$-2c_{13}^2g_{11}g_{33}+2c_{11}c_{33}g_{11}g_{33}-6c_{13}c_{44}g_{11}g_{33}+c_{11}c_{44}g_{33}^2+c_{11}e_{33}^2\mu_{11}+e_{15}^2(3f_{31}f_{33}+c_{33}\mu_{11}-3c_{13}\mu_{33})$$
$$+e_{31}^2(-2f_{15}f_{33}+c_{33}\mu_{11}+c_{44}\mu_{33})-e_{15}(2c_{33}f_{15}g_{11}+3c_{33}f_{31}g_{11}-3c_{13}f_{33}g_{11}-6c_{13}f_{15}g_{33}$$
$$-2c_{13}f_{31}g_{33}+2c_{11}f_{33}g_{33}+3e_{33}f_{15}f_{31}+2e_{33}f_{31}^2+3e_{33}c_{13}\mu_{11}-2e_{33}c_{11}\mu_{33})$$
$$+e_{31}[-3c_{33}f_{15}g_{11}-2c_{33}f_{31}g_{11}+2c_{13}f_{33}g_{11}+3c_{44}f_{33}g_{11}+2c_{13}f_{15}g_{33}-2c_{44}f_{31}g_{33}$$
$$+e_{33}(3f_{15}^2+2f_{15}f_{31}-2c_{13}\mu_{11}-3c_{44}\mu_{11})+e_{15}(-3f_{15}f_{33}+2f_{31}f_{33}+3c_{33}\mu_{11}-2c_{13}\mu_{33})]$$
$$-c_{33}f_{15}^2\varepsilon_{11}-3c_{33}f_{15}f_{31}\varepsilon_{11}-c_{33}f_{31}^2\varepsilon_{11}+3c_{13}f_{15}f_{33}\varepsilon_{11}+2c_{13}f_{31}f_{33}\varepsilon_{11}+3c_{44}f_{31}f_{33}\varepsilon_{11}-c_{11}f_{33}^2\varepsilon_{11}$$
$$+c_{33}c_{44}\mu_{11}\varepsilon_{11}-c_{13}^2\mu_{33}\varepsilon_{11}+c_{11}c_{33}\mu_{33}\varepsilon_{11}-3c_{13}c_{44}\mu_{33}\varepsilon_{11}+3c_{13}f_{15}^2\varepsilon_{33}+2c_{13}f_{15}f_{31}\varepsilon_{33}-c_{44}f_{31}^2\varepsilon_{33}$$
$$-2c_{11}f_{15}f_{33}\varepsilon_{33}-c_{13}^2\mu_{11}\varepsilon_{33}+c_{11}c_{33}\mu_{11}\varepsilon_{33}-3c_{13}c_{44}\mu_{11}\varepsilon_{33}+c_{11}c_{44}\mu_{33}\varepsilon_{33}\},$$

$$d_8=\gamma^3[c_{13}e_{33}f_{15}g_{11}+c_{44}e_{33}f_{31}g_{11}-2c_{13}c_{44}g_{11}g_{33}-e_{15}(e_{33}f_{15}f_{31}+c_{33}f_{31}g_{11}-c_{13}f_{33}g_{11}-2c_{13}f_{15}g_{33}$$
$$+c_{13}e_{33}\mu_{11})+e_{31}(-e_{15}f_{15}f_{33}-c_{33}f_{15}g_{11}+c_{44}f_{33}g_{11}+c_{33}e_{15}\mu_{11}+e_{33}f_{15}^2-e_{33}c_{44}\mu_{11})$$
$$+e_{15}^2f_{31}f_{33}-e_{15}^2c_{13}\mu_{33}-c_{33}f_{15}f_{31}\varepsilon_{11}+c_{13}f_{15}f_{33}\varepsilon_{11}+c_{44}f_{31}f_{33}\varepsilon_{11}-c_{13}c_{44}\mu_{33}\varepsilon_{11}+c_{13}f_{15}^2\varepsilon_{33}-c_{13}c_{44}\mu_{11}\varepsilon_{33}],$$

$$d_9=-2c_{11}e_{33}f_{15}g_{11}-c_{13}^2g_{11}^2+c_{11}c_{33}g_{11}^2-2c_{13}c_{44}g_{11}^2+2c_{11}c_{44}g_{11}g_{33}+e_{31}^2(-f_{15}^2+c_{44}\mu_{11})$$
$$+2e_{31}(e_{15}f_{15}f_{31}+c_{13}f_{15}g_{11}-c_{44}f_{31}g_{11}-c_{13}e_{15}\mu_{11})$$
$$+2e_{15}[c_{13}(2f_{15}+f_{31})g_{11}-c_{11}(f_{33}g_{11}+f_{15}g_{33}-e_{33}\mu_{11})]$$
$$-e_{15}^2(f_{31}^2+2c_{13}\mu_{11}-c_{11}\mu_{33})+2c_{13}f_{15}^2\varepsilon_{11}+2c_{13}f_{15}f_{31}\varepsilon_{11}-c_{44}f_{31}^2\varepsilon_{11}-2c_{11}f_{15}f_{33}\varepsilon_{11}$$
$$-c_{13}^2\mu_{11}\varepsilon_{11}+c_{11}c_{33}\mu_{11}\varepsilon_{11}-2c_{13}c_{44}\mu_{11}\varepsilon_{11}+c_{11}c_{44}\mu_{33}\varepsilon_{11}-c_{11}f_{15}^2\varepsilon_{33}+c_{11}c_{44}\mu_{11}\varepsilon_{33},$$

$$d_{10}=3\gamma\{-2c_{11}e_{33}f_{15}g_{11}-c_{13}^2g_{11}^2+c_{11}c_{33}g_{11}^2-2c_{13}c_{44}g_{11}^2+2c_{11}c_{44}g_{11}g_{33}+e_{31}^2(-f_{15}^2+c_{44}\mu_{11})$$
$$+2e_{31}(e_{15}f_{15}f_{31}+c_{13}f_{15}g_{11}-c_{44}f_{31}g_{11}-c_{13}e_{15}\mu_{11})$$
$$+2e_{15}[c_{13}(2f_{15}+f_{31})g_{11}-c_{11}(f_{33}g_{11}+f_{15}g_{33}-e_{33}\mu_{11})]$$
$$-e_{15}^2(f_{31}^2+2c_{13}\mu_{11}-c_{11}\mu_{33})+2c_{13}f_{15}^2\varepsilon_{11}+2c_{13}f_{15}f_{31}\varepsilon_{11}-c_{44}f_{31}^2\varepsilon_{11}-2c_{11}f_{15}f_{33}\varepsilon_{11}$$
$$-c_{13}^2\mu_{11}\varepsilon_{11}+c_{11}c_{33}\mu_{11}\varepsilon_{11}-2c_{13}c_{44}\mu_{11}\varepsilon_{11}+c_{11}c_{44}\mu_{33}\varepsilon_{11}-c_{11}f_{15}^2\varepsilon_{33}+c_{11}c_{44}\mu_{11}\varepsilon_{33}\},$$

$$d_{11}=\gamma^2\{-6c_{11}e_{33}f_{15}g_{11}-3c_{13}^2g_{11}^2+3c_{11}c_{33}g_{11}^2-7c_{13}c_{44}g_{11}^2+6c_{11}c_{44}g_{11}g_{33}-3e_{31}^2f_{15}^2+3e_{31}^2c_{44}\mu_{11}$$
$$+6e_{31}(e_{15}f_{15}f_{31}+c_{13}f_{15}g_{11}-c_{44}f_{31}g_{11}-c_{13}e_{15}\varepsilon_{11})+2e_{15}[c_{13}(7f_{15}+3f_{31})g_{11}$$
$$-3c_{11}(f_{33}g_{11}+f_{15}g_{33}-e_{33}\mu_{11})]+e_{15}^2(-3f_{31}^2-7c_{13}\mu_{11}+3c_{11}\mu_{33})$$
$$+7c_{13}f_{15}^2\varepsilon_{11}+6c_{13}f_{15}f_{31}\varepsilon_{11}-3c_{44}f_{31}^2\varepsilon_{11}-6c_{11}f_{15}f_{33}\varepsilon_{11}-3c_{13}^2\mu_{11}\varepsilon_{11}+3c_{11}c_{33}\mu_{11}\varepsilon_{11}-7c_{13}c_{44}\mu_{11}\varepsilon_{11}$$
$$+3c_{11}c_{44}\mu_{33}\varepsilon_{11}-3c_{11}f_{15}^2\varepsilon_{33}+3c_{11}c_{44}\mu_{11}\varepsilon_{11}\},$$

$$d_{12}=\gamma^3\{-2c_{11}e_{33}f_{15}g_{11}-c_{13}^2g_{11}^2+c_{11}c_{33}g_{11}^2-4c_{13}c_{44}g_{11}^2+2c_{11}c_{44}g_{11}g_{33}+e_{31}^2(-f_{15}^2+c_{44}\mu_{11})+2e_{13}(e_{15}f_{15}f_{31}$$
$$+c_{13}f_{15}g_{11}-c_{44}f_{31}g_{11}-c_{13}e_{15}\mu_{11})+2e_{15}[c_{13}(4f_{15}+f_{31})g_{11}-c_{11}(f_{33}g_{11}+f_{15}g_{33}-e_{33}\mu_{11})]$$
$$-e_{15}^2(f_{31}^2+4c_{13}\mu_{11}-c_{11}\mu_{33})+4c_{13}f_{15}^2\varepsilon_{11}+2c_{13}f_{15}f_{31}\varepsilon_{11}-c_{44}f_{31}^2\varepsilon_{11}-2c_{11}f_{15}f_{33}\varepsilon_{11}-2c_{13}^2\mu_{11}\varepsilon_{11}$$
$$+c_{11}c_{33}\mu_{11}\varepsilon_{11}-4c_{13}c_{44}\mu_{11}\varepsilon_{11}+c_{11}c_{44}\mu_{33}\varepsilon_{11}-c_{11}f_{15}^2\varepsilon_{33}+c_{11}c_{44}\mu_{11}\varepsilon_{33}\}$$

$$d_{13}=-c_{13}\gamma^4[-2e_{15}f_{15}g_{11}+e_{15}^2\mu_{11}-f_{15}^2\varepsilon_{11}+c_{44}(g_{11}^2+\mu_{11}\varepsilon_{11})],$$

$$d_{14}=-2c_{11}e_{15}f_{15}g_{11}+c_{11}c_{44}g_{11}^2+c_{11}e_{15}^2\mu_{11}-c_{11}f_{15}^2\varepsilon_{11}+c_{11}c_{44}\mu_{11}\varepsilon_{11},$$

$$d_{15}=\gamma(-8c_{11}e_{15}f_{15}g_{11}+4c_{11}c_{44}g_{11}^2+4c_{11}e_{15}^2\mu_{11}-4c_{11}f_{15}^2\varepsilon_{11}+4c_{11}c_{44}\mu_{11}\varepsilon_{11}),$$

$$d_{16} = c_{11}\gamma^2(-12e_{15}f_{15}g_{11} + 6c_{44}g_{11}^2 + 6e_{15}^2\mu_{11} - 6f_{15}^2\varepsilon_{11} + 6c_{44}\mu_{11}\varepsilon_{11}),$$

$$d_{17} = 4c_{11}\gamma^3(-2e_{15}f_{15}g_{11} + c_{44}g_{11}^2 + e_{15}^2\mu_{11} - f_{15}^2\varepsilon_{11} + c_{44}\mu_{11}\varepsilon_{11}),$$

$$d_{18} = c_{11}\gamma^4(-2e_{15}f_{15}g_{11} + c_{44}g_{11}^2 + e_{15}^2\mu_{11} - f_{15}^2\varepsilon_{11} + c_{44}\mu_{11}\varepsilon_{11}),$$

$$\alpha_1^{(1)} = -\gamma^3(-e_{31}f_{15}g_{11} - e_{15}f_{31}g_{11} + c_{13}g_{11}^2 + e_{15}e_{31}\mu_{11} - f_{15}f_{31}\varepsilon_{11} + c_{13}\mu_{11}\varepsilon_{11}),$$

$$\alpha_2^{(1)} = -\gamma^2[-3e_{31}f_{15}g_{11} + 3c_{13}g_{11}^2 + c_{44}g_{11}^2 + e_{15}^2\mu_{11}$$
$$+e_{15}(-2f_{15}g_{11} - 3f_{31}g_{11} + 3e_{31}\mu_{11}) - f_{15}^2\varepsilon_{11} - 3f_{15}f_{31}\varepsilon_{11} + 3c_{13}\mu_{11}\varepsilon_{11} + c_{44}\mu_{11}\varepsilon_{11}],$$

$$\alpha_3^{(1)} = -\gamma[-3e_{31}f_{15}g_{11} + 3c_{13}g_{11}^2 + 2c_{44}g_{11}^2 + 2e_{15}^2\mu_{11} + e_{15}(-4f_{15}g_{11} - 3f_{31}g_{11} + 3e_{31}\mu_{11})$$
$$-2f_{15}^2\varepsilon_{11} - f_{15}f_{31}\varepsilon_{11} + 3c_{13}\mu_{11}\varepsilon_{11} + 2c_{44}\mu_{11}\varepsilon_{11}],$$

$$\alpha_4^{(1)} = -[-e_{31}f_{15}g_{11} + c_{13}g_{11}^2 + c_{44}g_{11}^2 + e_{15}^2\mu_{11} + e_{15}(-2f_{15}g_{11} - f_{31}g_{11} + e_{31}\mu_{11})$$
$$-f_{15}^2\varepsilon_{11} - f_{15}f_{31}\varepsilon_{11} + c_{13}\mu_{11}\varepsilon_{11} + c_{44}\mu_{11}\varepsilon_{11}],$$

$$\alpha_5^{(1)} = \gamma^2(e_{33}f_{15}g_{11} + e_{31}f_{33}g_{11} + e_{31}f_{15}g_{33} + e_{15}f_{31}g_{33} - 2c_{13}g_{11}g_{33} - e_{31}e_{33}\mu_{11} - e_{15}e_{31}\mu_{33}$$
$$+f_{31}f_{33}\varepsilon_{11} - c_{13}\mu_{33}\varepsilon_{11} + f_{15}f_{31}\varepsilon_{33} - c_{13}\mu_{11}\varepsilon_{33}),$$

$$\alpha_6^{(1)} = -\gamma(-2e_{31}f_{33}g_{11} - 2e_{31}f_{15}g_{33} + 4c_{13}g_{11}g_{33} + 2c_{44}g_{11}g_{33} - e_{33}f_{15}g_{11} - 2e_{33}f_{31}g_{11} + e_{33}e_{15}\mu_{11} + 2e_{33}e_{31}\mu_{11}$$
$$+e_{15}^2\mu_{33} - e_{15}f_{33}g_{11} - 2e_{15}f_{15}g_{33} - 2e_{15}f_{31}g_{33} + 2e_{15}e_{31}\mu_{33} - f_{15}f_{33}\varepsilon_{11} - 2f_{31}f_{33}\varepsilon_{11}$$
$$+2c_{13}\mu_{33}\varepsilon_{11} + c_{44}\mu_{33}\varepsilon_{11} - f_{15}^2\varepsilon_{33} - 2f_{15}f_{31}\varepsilon_{33} + 2c_{13}\mu_{11}\varepsilon_{33} + c_{44}\mu_{11}\varepsilon_{33}),$$

$$\alpha_7^{(1)} = e_{31}f_{33}g_{11} + e_{31}f_{15}g_{33} - 2c_{13}g_{11}g_{33} - 2c_{44}g_{11}g_{33} + e_{33}[f_{15}g_{11} + f_{31}g_{11} - (e_{15} + e_{31})\mu_{11}]$$
$$-e_{15}^2\mu_{33} + e_{15}(f_{33}g_{11} + 2f_{15}g_{33} + f_{31}g_{33} - e_{31}\mu_{33}) + f_{15}f_{33}\varepsilon_{11} + f_{31}f_{33}\varepsilon_{11} - c_{13}\mu_{33}\varepsilon_{11} - c_{44}\mu_{33}\varepsilon_{11}$$
$$+f_{15}^2\varepsilon_{33} + f_{15}f_{31}\varepsilon_{33} - c_{13}\mu_{11}\varepsilon_{33} - c_{44}\mu_{11}\varepsilon_{33},$$

$$\alpha_8^{(1)} = -\gamma(-e_{33}f_{31}g_{33} - e_{31}f_{33}g_{33} + c_{13}g_{33}^2 + e_{31}e_{33}\mu_{33} - f_{31}f_{33}\varepsilon_{33} + c_{13}\mu_{33}\varepsilon_{33})$$

$$\alpha_9^{(1)} = -\{-e_{15}f_{33}g_{33} - e_{31}f_{33}g_{33} + c_{13}g_{33}^2 + c_{44}g_{33}^2 + e_{33}[-f_{15}g_{33} - f_{31}g_{33} + (e_{15} + e_{31})\mu_{33}]$$
$$-f_{15}f_{33}\varepsilon_{33} - f_{31}f_{33}\varepsilon_{33} + c_{13}\mu_{33}\varepsilon_{33} + c_{44}\mu_{33}\varepsilon_{33}\},$$

$$\alpha_1^{(2)} = c_{11}\gamma^3(g_{11}^2 + \mu_{11}\varepsilon_{11}\gamma),\ \alpha_2^{(2)} = 3c_{11}\gamma^2(g_{11}^2 + \mu_{11}\varepsilon_{11}),\ \alpha_3^{(2)} = 3c_{11}\gamma(g_{11}^2 + \mu_{11}\varepsilon_{11}\gamma),$$

$$\alpha_4^{(2)} = c_{11}(g_{11}^2 + \mu_{11}\varepsilon_{11}),\ \alpha_5^{(2)} = \gamma^3(-e_{31}f_{15}g_{11} - e_{15}f_{31}g_{11} + e_{15}e_{31}\mu_{11} - f_{15}f_{31}\varepsilon_{11}),$$

$$\alpha_6^{(2)} = \gamma^2(-2e_{15}f_{15}g_{11} - 3e_{31}f_{15}g_{11} - 3e_{15}f_{31}g_{11} - 2e_{31}f_{31}g_{11} + c_{44}g_{11}^2 + 2c_{11}g_{11}g_{33} + e_{15}^2\mu_{11} + 3e_{15}e_{31}\mu_{11}$$
$$+e_{31}^2\mu_{11} - f_{15}^2\varepsilon_{11} - 3f_{15}f_{31}\varepsilon_{11} - f_{31}^2\varepsilon_{11} + c_{44}\mu_{11}g_{11} + c_{11}\mu_{33}\varepsilon_{11} + c_{11}\mu_{11}\varepsilon_{33}),$$

$$\alpha_7^{(2)} = \gamma(-4e_{15}f_{15}g_{11} - 4e_{31}f_{15}g_{11} - 4e_{15}f_{31}g_{11} - 4e_{31}f_{31}g_{11} + 2c_{44}g_{11}^2 + 4c_{11}g_{11}g_{33} + 2e_{15}^2\mu_{11} + 4e_{15}e_{31}\mu_{11}$$
$$+2e_{31}^2\mu_{11} - 2f_{15}^2\varepsilon_{11} - 4f_{15}f_{31}\varepsilon_{11} - 2f_{31}^2\varepsilon_{11} + 2c_{44}\mu_{11}\varepsilon_{11} + 2c_{11}\mu_{33}\varepsilon_{11} + 2c_{11}\mu_{11}\varepsilon_{33})$$

$$\alpha_8^{(2)} = -2e_{31}(f_{15} + f_{31})g_{11} + c_{44}g_{11}^2 + 2c_{11}g_{11}g_{33} + e_{15}^2\mu_{11} + e_{31}^2\mu_{11} - 2e_{15}(f_{15}g_{11} + f_{31}g_{11} - e_{31}\mu_{11})$$
$$-f_{15}^2\varepsilon_{11} - 2f_{15}f_{31}\varepsilon_{11} - f_{31}^2\varepsilon_{11} + c_{44}\mu_{11}\varepsilon_{11} + c_{11}\mu_{33}\varepsilon_{11} + c_{11}\mu_{11}\varepsilon_{33},$$

$$\alpha_9^{(2)} = \gamma^2(-e_{31}f_{15}g_{33} - e_{15}f_{31}g_{33} + e_{15}e_{31}\mu_{33} - f_{15}f_{31}\varepsilon_{33})$$

$$\alpha_{10}^{(2)} = \gamma(-2e_{15}f_{15}g_{33} - 2e_{31}f_{15}g_{33} - 2e_{15}f_{31}g_{33} - 2e_{31}f_{31}g_{33} + 2c_{44}g_{11}g_{33} + c_{11}g_{33}^2 + e_{15}^2\mu_{33} + 2e_{15}e_{31}\mu_{33}$$
$$+e_{31}^2\mu_{33} + c_{44}\mu_{33}\varepsilon_{11} - f_{15}^2\varepsilon_{33} - 2f_{15}f_{31}\varepsilon_{33} - f_{31}^2\varepsilon_{33} + c_{44}\mu_{11}\varepsilon_{33} + c_{11}\mu_{33}\varepsilon_{33}),$$

$$\alpha_{11}^{(2)} = -2e_{31}(f_{15} + f_{31})g_{33} + 2c_{44}g_{11}g_{33} + c_{11}g_{33}^2 + e_{15}^2\mu_{33} + e_{31}^2\mu_{33} - 2e_{15}(f_{15}g_{33} + f_{31}g_{33} - e_{31}\mu_{33})$$
$$+c_{44}\mu_{33}\varepsilon_{11} - f_{15}^2\varepsilon_{33} - 2f_{15}f_{31}\varepsilon_{33} - f_{31}^2\varepsilon_{33} + c_{44}\mu_{11}\varepsilon_{33} + c_{11}\mu_{33}\varepsilon_{33},$$

$\alpha_{12}^{(2)} = c_{44}(g_{33}^2 + \mu_{33}\varepsilon_{33})$, $\alpha_1^{(3)} = -\gamma^3 c_{11}(f_{15}g_{11} - e_{15}\mu_{11})$,
$\alpha_2^{(3)} = -3\gamma^2 c_{11}(f_{15}g_{11} - e_{15}\mu_{11})$,
$\alpha_3^{(3)} = -3\gamma c_{11}(f_{15}g_{11} - e_{15}\mu_{11})$, $\alpha_4^{(3)} = -c_{11}(f_{15}g_{11} - e_{15}\mu_{11})$,
$\alpha_5^{(3)} = -\gamma^3 c_{13}(-f_{15}g_{11} + e_{15}\mu_{11})$,
$\alpha_6^{(3)} = -\gamma^2(-e_{31}f_{15}f_{31} + e_{15}f_{31}^2 - 3c_{13}f_{15}g_{11} - c_{13}f_{31}g_{11} + c_{11}f_{33}g_{11} + c_{11}f_{15}g_{33} + 3c_{13}e_{15}\mu_{11}$
$+ c_{13}e_{31}\mu_{11} - c_{11}e_{33}\mu_{11} - c_{11}e_{15}\mu_{33})$,
$\alpha_7^{(3)} = -\gamma(-e_{31}f_{15}^2 + e_{15}f_{15}f_{31} - 2e_{31}f_{15}f_{31} + 2e_{15}f_{31}^2 - 3c_{13}f_{15}g_{11} - 2c_{13}f_{31}g_{11} - c_{44}f_{31}g_{11} + 2c_{11}f_{33}g_{11}$
$+ 2c_{11}f_{15}g_{33} + 3c_{13}e_{15}\mu_{11} + 2c_{13}e_{31}\mu_{11} + c_{44}e_{31}\mu_{11} - 2c_{11}e_{33}\mu_{11} - 2c_{11}e_{15}\mu_{33})$,
$\alpha_8^{(3)} = -\{-c_{13}f_{15}g_{11} - c_{13}f_{31}g_{11} - c_{44}f_{31}g_{11} + c_{11}f_{33}g_{11} + c_{11}f_{15}g_{33} - c_{11}e_{33}\mu_{11}$
$+ e_{31}[-f_{15}^2 - f_{15}f_{31} + (c_{13} + c_{44})\mu_{11}] + e_{15}(f_{15}f_{31} + f_{31}^2 + c_{13}\mu_{11} - c_{11}\mu_{33})\}$,
$\alpha_9^{(3)} = -\gamma^2(e_{33}f_{15}f_{31} - e_{15}f_{31}f_{33} - c_{13}f_{15}g_{33} + c_{13}e_{15}\mu_{33})$,
$\alpha_{10}^{(3)} = -\gamma(e_{33}f_{15}^2 + 2e_{33}f_{15}f_{31} + e_{33}f_{31}^2 - e_{15}f_{15}f_{33} - 2e_{15}f_{31}f_{33} - e_{31}f_{31}f_{33} + c_{44}f_{33}g_{11} - 2c_{13}f_{15}g_{33}$
$- c_{13}f_{31}g_{33} + c_{11}f_{33}g_{33} - c_{44}e_{33}\mu_{11} + 2c_{13}e_{15}\mu_{33} + c_{13}e_{31}\mu_{33} - c_{11}e_{33}\mu_{33})$,
$\alpha_{11}^{(3)} = -[-e_{15}f_{15}f_{33} - e_{31}f_{15}f_{33} - e_{15}f_{31}f_{33} - e_{31}f_{31}f_{33} + c_{44}f_{33}g_{11} - c_{13}f_{15}g_{33}$
$- c_{13}f_{31}g_{33} - c_{44}f_{31}g_{33} + c_{11}f_{33}g_{33} + c_{13}e_{15}\mu_{33} + c_{13}e_{31}\mu_{33} + c_{44}e_{31}\mu_{33}$
$+ e_{33}(f_{15}^2 + 2f_{15}f_{31} + f_{31}^2 - c_{44}\mu_{11} - c_{11}\mu_{33})]$,
$\alpha_{12}^{(3)} = -c_{44}(f_{33}g_{33} - e_{33}\mu_{33})$, $\alpha_1^{(4)} = \gamma^3 c_{11}(e_{15}g_{11} + f_{15}\varepsilon_{11})$,
$\alpha_2^{(4)} = 3\gamma^2 c_{11}(e_{15}g_{11} + f_{15}\varepsilon_{11})$,
$\alpha_3^{(4)} = 3\gamma c_{11}(e_{15}g_{11} + f_{15}\varepsilon_{11})$, $\alpha_4^{(4)} = c_{11}(e_{15}g_{11} + f_{15}\varepsilon_{11})$,
$\alpha_5^{(4)} = \gamma^3(-c_{13}e_{15}g_{11} - c_{13}f_{15}\varepsilon_{11})$,
$\alpha_6^{(4)} = \gamma^2(e_{31}^2 f_{15} - e_{15}e_{31}f_{31} - 3c_{13}e_{15}g_{11} - c_{13}e_{31}g_{11} + c_{11}e_{33}g_{11} + c_{11}e_{15}g_{33} - 3c_{13}f_{15}\varepsilon_{11}$
$- c_{13}f_{31}\varepsilon_{11} + c_{11}f_{33}\varepsilon_{11} + c_{11}f_{15}\varepsilon_{33})$,
$\alpha_7^{(4)} = \gamma(e_{15}e_{31}f_{15} + 2e_{31}^2 f_{15} - e_{15}^2 f_{31} - 2e_{15}e_{31}f_{31} - 3c_{13}e_{15}g_{11} - 2c_{13}e_{31}g_{11} - c_{44}e_{31}g_{11} + 2c_{11}e_{33}g_{11} + 2c_{11}e_{15}g_{33}$
$- 3c_{13}f_{15}\varepsilon_{11} - 2c_{13}f_{31}\varepsilon_{11} - c_{44}f_{31}\varepsilon_{11} + 2c_{11}f_{33}\varepsilon_{11} + 2c_{11}f_{15}\varepsilon_{33})$,
$\alpha_8^{(4)} = -\{-e_{31}^2 f_{15} + e_{15}^2 f_{31} + (c_{13} + c_{44})e_{31}g_{11} - c_{11}e_{33}g_{11} + e_{15}[e_{31}(-f_{15} + f_{31}) + c_{13}g_{11} - c_{11}g_{33}]$,
$+ c_{13}f_{15}\varepsilon_{11} + c_{13}f_{31}\varepsilon_{11} + c_{44}f_{31}\varepsilon_{11} - c_{11}f_{33}\varepsilon_{11} - c_{11}f_{15}\varepsilon_{33}\}$,
$\alpha_9^{(4)} = \gamma^2(-e_{31}e_{33}f_{15} + e_{15}e_{31}f_{33} - c_{13}e_{15}g_{33} - c_{13}f_{15}\varepsilon_{33})$,
$\alpha_{10}^{(4)} = \gamma(-e_{15}e_{33}f_{15} - 2e_{31}e_{33}f_{15} - e_{31}e_{33}f_{31} + e_{15}^2 f_{33} + 2e_{15}e_{31}f_{33} + e_{31}^2 f_{33} + c_{44}e_{33}g_{11} - 2c_{13}e_{15}g_{33} - c_{13}e_{31}g_{33}$
$+ c_{11}e_{33}g_{33} + c_{44}f_{33}\varepsilon_{11} - 2c_{13}f_{15}\varepsilon_{33} - c_{13}f_{31}\varepsilon_{33} + c_{11}f_{33}\varepsilon_{33})$,
$\alpha_{11}^{(4)} = e_{15}^2 f_{33} + e_{31}^2 f_{33} + c_{44}e_{33}g_{11} + c_{11}e_{33}g_{33} - e_{15}[e_{33}(f_{15} + f_{31}) - 2e_{31}f_{33} + c_{13}g_{33}]$
$- e_{31}[e_{33}(f_{15} + f_{31}) + (c_{13} + c_{44})g_{33}] + c_{44}f_{33}\varepsilon_{11} - c_{13}f_{15}\varepsilon_{33} - c_{13}f_{31}\varepsilon_{33} - c_{44}f_{31}\varepsilon_{33} + c_{11}f_{33}\varepsilon_{33}$,
$\alpha_{12}^{(4)} = c_{44}(e_{33}g_{33} + f_{33}\varepsilon_{33})$.

The solving processes of the known functions $\beta_1(s)$, $\beta_2(s)$, $\beta_3(s)$, $\beta_4(s)$, $\beta_5(s)$ and $\beta_6(s)$.

For $\lambda_1^2 \neq \lambda_2^2 \neq \lambda_3^2 \neq \lambda_4^2 > 0$, we can write the matrices $[X_i](i=1,2)$ as follows:

$$[X_1]=\begin{bmatrix} \beta_1^{(1)} & \beta_2^{(1)} & \beta_3^{(1)} & \beta_4^{(1)} \\ \chi_1^{(1)} & \chi_2^{(1)} & \chi_3^{(1)} & \chi_4^{(1)} \\ \chi_1^{(3)} & \chi_2^{(3)} & \chi_3^{(3)} & \chi_4^{(3)} \\ \chi_1^{(4)} & \chi_2^{(4)} & \chi_3^{(4)} & \chi_4^{(4)} \end{bmatrix}, \ [X_2]=\begin{bmatrix} \beta_1^{(2)} & \beta_2^{(2)} & \beta_3^{(2)} & \beta_4^{(2)} \\ \beta_1^{(3)} & \beta_2^{(3)} & \beta_3^{(3)} & \beta_4^{(3)} \\ \beta_1^{(4)} & \beta_2^{(4)} & \beta_3^{(4)} & \beta_3^{(4)} \\ \chi_1^{(2)} & \chi_2^{(2)} & \chi_3^{(2)} & \chi_4^{(2)} \end{bmatrix} \quad \text{(A-1)}$$

From Eqs. (35) and (36), we can obtain

$$a=\frac{1}{2}[X_1]^{-1}\begin{bmatrix} \overline{f}_1 \\ 0 \\ 0 \\ 0 \end{bmatrix}+\frac{1}{2}[X_2]^{-1}\begin{bmatrix} \overline{f}_2 \\ \dfrac{D_0\overline{f}_2}{\varepsilon_0} \\ \dfrac{B_0\overline{f}_2}{\mu_0} \\ 0 \end{bmatrix}, \ b=\frac{1}{2}[X_1]^{-1}\begin{bmatrix} \overline{f}_1 \\ 0 \\ 0 \\ 0 \end{bmatrix}-\frac{1}{2}[X_2]^{-1}\begin{bmatrix} \overline{f}_2 \\ \dfrac{D_0\overline{f}_2}{\varepsilon_0} \\ \dfrac{B_0\overline{f}_2}{\mu_0} \\ 0 \end{bmatrix} \quad \text{(A-2)}$$

where $a=\begin{bmatrix} A_1 \\ A_2 \\ A_3 \\ A_4 \end{bmatrix}, \ b=\begin{bmatrix} B_1 \\ B_2 \\ B_3 \\ B_4 \end{bmatrix}$.

So the unknown functions A_i and B_i can be expressed as follows:

$$A_i=\frac{1}{2}[m_{i1}\overline{f}_1+(n_{i1}+\frac{D_0}{\varepsilon_0}n_{i2}+\frac{B_0}{\mu_0}n_{i3})\overline{f}_2],$$
$$B_i=\frac{1}{2}[m_{i1}\overline{f}_1-(n_{i1}+\frac{D_0}{\varepsilon_0}n_{i2}+\frac{B_0}{\mu_0}n_{i3})\overline{f}_2] \quad \text{(A-3)}$$

where $[m_{ij}]_{4\times4}=[X_1]^{-1}$, $[n_{ij}]_{4\times4}=[X_2]^{-1}$.

Substituting Eqs. (A-3) into Eqs. (3-34) and (3-35), we have

$$\frac{\overline{f}_1}{2}\begin{bmatrix} \beta_1^{(1)} & \beta_2^{(1)} & \beta_3^{(1)} & \beta_4^{(1)} \\ \chi_1^{(1)} & \chi_2^{(1)} & \chi_3^{(1)} & \chi_4^{(1)} \\ \chi_1^{(3)} & \chi_2^{(3)} & \chi_3^{(3)} & \chi_4^{(3)} \\ \chi_1^{(4)} & \chi_2^{(4)} & \chi_3^{(4)} & \chi_4^{(4)} \end{bmatrix}\begin{bmatrix} m_{11} \\ m_{21} \\ m_{31} \\ m_{41} \end{bmatrix}+\frac{\overline{f}_1}{2}\begin{bmatrix} \beta_1^{(1)} & \beta_2^{(1)} & \beta_3^{(1)} & \beta_4^{(1)} \\ \chi_1^{(1)} & \chi_2^{(1)} & \chi_3^{(1)} & \chi_4^{(1)} \\ \chi_1^{(3)} & \chi_2^{(3)} & \chi_3^{(3)} & \chi_4^{(3)} \\ \chi_1^{(4)} & \chi_2^{(4)} & \chi_3^{(4)} & \chi_4^{(4)} \end{bmatrix}\begin{bmatrix} m_{11} \\ m_{21} \\ m_{31} \\ m_{41} \end{bmatrix}=\begin{bmatrix} \overline{f}_1 \\ 0 \\ 0 \\ 0 \end{bmatrix} \quad \text{(A-4)}$$

$$\frac{\overline{f}_2}{2}\begin{bmatrix} \beta_1^{(2)} & \beta_2^{(2)} & \beta_3^{(2)} & \beta_4^{(2)} \\ \beta_1^{(3)} & \beta_2^{(3)} & \beta_3^{(3)} & \beta_4^{(3)} \\ \beta_1^{(4)} & \beta_2^{(4)} & \beta_3^{(4)} & \beta_3^{(4)} \\ \chi_1^{(2)} & \chi_2^{(2)} & \chi_3^{(2)} & \chi_4^{(2)} \end{bmatrix}\begin{bmatrix} n_{11}^* \\ n_{21}^* \\ n_{31}^* \\ n_{41}^* \end{bmatrix} + \frac{\overline{f}_2}{2}\begin{bmatrix} \beta_1^{(2)} & \beta_2^{(2)} & \beta_3^{(2)} & \beta_4^{(2)} \\ \beta_1^{(3)} & \beta_2^{(3)} & \beta_3^{(3)} & \beta_4^{(3)} \\ \beta_1^{(4)} & \beta_2^{(4)} & \beta_3^{(4)} & \beta_3^{(4)} \\ \chi_1^{(2)} & \chi_2^{(2)} & \chi_3^{(2)} & \chi_4^{(2)} \end{bmatrix}\begin{bmatrix} n_{11}^* \\ n_{21}^* \\ n_{31}^* \\ n_{41}^* \end{bmatrix} = \begin{bmatrix} \overline{f}_2 \\ \dfrac{D_0 \overline{f}_2}{\varepsilon_0} \\ \dfrac{B_0 \overline{f}_2}{\mu_0} \\ 0 \end{bmatrix} \tag{A-5}$$

where $n_{i1}^* = n_{i1} + \dfrac{D_0}{\varepsilon_0} n_{i2} + \dfrac{B_0}{\mu_0} n_{i3}$.

So we obtain

$$\begin{bmatrix} \beta_1^{(1)} & \beta_2^{(1)} & \beta_3^{(1)} & \beta_4^{(1)} \\ \chi_1^{(1)} & \chi_2^{(1)} & \chi_3^{(1)} & \chi_4^{(1)} \\ \chi_1^{(3)} & \chi_2^{(3)} & \chi_3^{(3)} & \chi_4^{(3)} \\ \chi_1^{(4)} & \chi_2^{(4)} & \chi_3^{(4)} & \chi_4^{(4)} \end{bmatrix}\begin{bmatrix} m_{11} \\ m_{21} \\ m_{31} \\ m_{41} \end{bmatrix} = \begin{bmatrix} 1 \\ 0 \\ 0 \\ 0 \end{bmatrix} \Rightarrow \begin{cases} \sum_{i=1}^{4} \chi_i^{(1)} m_{i1} = 0 \\ \sum_{i=1}^{4} \chi_i^{(3)} m_{i1} = 0 \\ \sum_{i=1}^{4} \chi_i^{(4)} m_{i1} = 0 \end{cases} \tag{A-6}$$

$$\begin{bmatrix} \beta_1^{(2)} & \beta_2^{(2)} & \beta_3^{(2)} & \beta_4^{(2)} \\ \beta_1^{(3)} & \beta_2^{(3)} & \beta_3^{(3)} & \beta_4^{(3)} \\ \beta_1^{(4)} & \beta_2^{(4)} & \beta_3^{(4)} & \beta_3^{(4)} \\ \chi_1^{(2)} & \chi_2^{(2)} & \chi_3^{(2)} & \chi_4^{(2)} \end{bmatrix}\begin{bmatrix} n_{11}^* \\ n_{21}^* \\ n_{31}^* \\ n_{41}^* \end{bmatrix} = \begin{bmatrix} 1 \\ \dfrac{D_0}{\varepsilon_0} \\ \dfrac{B_0}{\mu_0} \\ 0 \end{bmatrix} \Rightarrow \sum_{i=1}^{4} \chi_i^{(2)} n_{i1}^* = 0 \tag{A-7}$$

Substituting Eqs. (A-3) into Eqs. (3-31) and applying Eqs. (A-6)-(A-7) , we have:

$$\begin{aligned} \sigma_{zz}^{(1)}(x,0) &= \frac{e^{\gamma x}}{4\pi}\int_{-\infty}^{\infty}\sum_{i=1}^{4}\chi_i^{(1)}(s) m_{i1}\overline{f}_1(s) e^{isx} ds + \frac{e^{\gamma x}}{4\pi}\int_{-\infty}^{\infty}\sum_{i=1}^{4}\chi_i^{(1)}(s) n_{i1}^* \overline{f}_2(s) e^{isx} ds \\ &= \frac{e^{\gamma x}}{4\pi}\int_{-\infty}^{\infty}\sum_{i=1}^{4}\chi_i^{(1)}(s) n_{i1}^* \overline{f}_2(s) e^{isx} ds = \frac{e^{\gamma x}}{2\pi}\int_{-\infty}^{\infty}\beta_1(s)\overline{f}_2(s) e^{isx} ds \end{aligned} \tag{A-8}$$

$$\begin{aligned} \sigma_{xz}^{(1)}(x,0) &= \frac{e^{\gamma x}}{4\pi}\sum_{i=1}^{4}\int_{-\infty}^{\infty}\chi_i^{(2)}(s) m_{i1}\overline{f}_1(s) e^{isx} ds + \frac{e^{\gamma x}}{4\pi}\sum_{i=1}^{4}\int_{-\infty}^{\infty}\chi_i^{(2)}(s) n_{i1}^* \overline{f}_2(s) e^{isx} ds \\ &= \frac{e^{\gamma x}}{4\pi}\sum_{i=1}^{4}\int_{-\infty}^{\infty}\chi_i^{(2)}(s) m_{i1}\overline{f}_1(s) e^{isx} ds = \frac{e^{\gamma x}}{2\pi}\sum_{i=1}^{4}\int_{-\infty}^{\infty}\beta_2(s)\overline{f}_1(s) e^{isx} ds \end{aligned} \tag{A-9}$$

$$\begin{aligned} D_z^{(1)}(x,0) &= \frac{e^{\gamma x}}{4\pi}\int_{-\infty}^{\infty}\sum_{i=1}^{4}\chi_i^{(3)}(s)m_{i1}\overline{f}_1(s)e^{isx}ds + \frac{e^{\gamma x}}{4\pi}\int_{-\infty}^{\infty}\sum_{i=1}^{4}\chi_i^{(3)}(s)n_{i1}^{*}\overline{f}_2(s)e^{isx}ds \\ &= \frac{e^{\gamma x}}{4\pi}\int_{-\infty}^{\infty}\sum_{i=1}^{4}\chi_i^{(3)}(s)n_{i1}^{*}\overline{f}_2(s)e^{isx}ds = \frac{e^{\gamma x}}{2\pi}\int_{-\infty}^{\infty}\beta_3(s)\overline{f}_2(s)e^{isx}ds \end{aligned} \tag{A-10}$$

$$\begin{aligned} D_x^{(1)}(x,0) &= \frac{e^{\gamma x}}{4\pi}\int_{-\infty}^{\infty}\sum_{i=1}^{4}\chi_i^{(5)}(s)m_{i1}\overline{f}_1(s)e^{isx}ds + \frac{e^{\gamma x}}{4\pi}\int_{-\infty}^{\infty}\sum_{i=1}^{4}\chi_i^{(5)}(s)n_{i1}^{*}\overline{f}_2(s)e^{isx}ds, \quad (\overline{f}_1(s)=0) \\ &= \frac{e^{\gamma x}}{4\pi}\int_{-\infty}^{\infty}\sum_{i=1}^{4}\chi_i^{(5)}(s)n_{i1}^{*}\overline{f}_2(s)e^{isx}ds = \frac{e^{\gamma x}}{2\pi}\int_{-\infty}^{\infty}\beta_4(s)\overline{f}_2(s)e^{isx}ds \end{aligned} \tag{A-11}$$

$$\begin{aligned} B_z^{(1)}(x,0) &= \frac{e^{\gamma x}}{4\pi}\int_{-\infty}^{\infty}\sum_{i=1}^{4}\chi_i^{(4)}(s)m_{i1}\overline{f}_1(s)e^{isx}ds + \frac{e^{\gamma x}}{4\pi}\int_{-\infty}^{\infty}\sum_{i=1}^{4}\chi_i^{(4)}(s)n_{i1}^{*}\overline{f}_2(s)e^{isx}ds \\ &= \frac{e^{\gamma x}}{4\pi}\int_{-\infty}^{\infty}\sum_{i=1}^{4}\chi_i^{(4)}(s)n_{i1}^{*}\overline{f}_2(s)e^{isx}ds = \frac{e^{\gamma x}}{2\pi}\int_{-\infty}^{\infty}\beta_5(s)\overline{f}_2(s)e^{isx}ds \end{aligned} \tag{A-12}$$

$$\begin{aligned} B_x^{(1)}(x,0) &= \frac{e^{\gamma x}}{4\pi}\int_{-\infty}^{\infty}\sum_{i=1}^{4}\chi_i^{(6)}(s)m_{i1}\overline{f}_1(s)e^{isx}ds + \frac{e^{\gamma x}}{4\pi}\int_{-\infty}^{\infty}\sum_{i=1}^{4}\chi_i^{(6)}(s)n_{i1}^{*}\overline{f}_2(s)e^{isx}ds \\ &= \frac{e^{\gamma x}}{4\pi}\int_{-\infty}^{\infty}\sum_{i=1}^{4}\chi_i^{(6)}(s)n_{i1}^{*}\overline{f}_2(s)e^{isx}ds = \frac{e^{\gamma x}}{2\pi}\int_{-\infty}^{\infty}\beta_6(s)\overline{f}_2(s)e^{isx}ds \end{aligned} \tag{A-13}$$

REFERENCES

Abramowitz, M., Stegun, I. A., Handbook of Mathematical Functions, Dover, New York, 1968, p.778.

Avellaneda, M., Harshe, G., Magnetoelectric effect in piezoelectric/magnetostrictive multiplayer (2-2) composites. *Journal of Intelligent material Systems and Structures* 1994. Vol 5, 501-513.

Benveniste, Y., Magnetoelectric effect in fibrous composites with piezoelectric and magnetostrictive phases. *Physical Review B* 1995. *Vol* 51, 16424-16427.

Chen, J., Liu Z. X. and Zou, Z. Z., Electriomechanical impact of a crack in a functionally graded piezoelectric medium, *Theoretical and Applied Fracture Mechanics* 2003. *Vol* 39, 47-60.

Chen, W. Q., Lee, K.Y., Ding, H. J., General solution for transversely isotropic magneto-electro-thermo-elasticity and the potential theory method. International Journal of Engineering Science 2004. Vol 42, 1361-1379.

Chung, M.Y., Ting, T.C.T., The Green function for a piezoelectric piezomagnetic anisotropic elastic medium with an elliptic hole or rigid inclusion, Philosophical Magazine Letters 1995. Vol 72, 405-410.

Dascalu, D., Homentcovschi, D., An intermediate crack model for flaws in piezoelectric solids. ACTA Mechanica, 2002, Vol 154(1-4) 85-100.

Deeg, W.E.F., The analysis of dislocation, crack and inclusion problems in piezoelectric solids, Ph.D. thesis, 1980, Stanford University.

Delale and Erdogan, F., On the mechanical modeling of the interfacial region in bonded half-planes. *ASME Journal of Applied Mechanics* 1988. Vol 55, 317-324.

Ding, H.J.; Chen, B., Liang, J., General solutions for coupled equations for piezoelectric media, *International Journal of Solids and Structures* 1996. Vol 33(16) 2283-2296.

Erdelyi, A., 1954. Tables of Integral Transforms, Vol.1, McGraw-Hill, New York.

Fildis, H., Yahsi, O.S., The axisymmetric crack problem in a non-homogeneous interfacial region between homogeneous half-spaces. *International Journal of Fracture* 1996. Vol 78, 139-163.

Gao, C.F., Kessler, H., Balke, H., Fracture analysis of electromagnetic thermoelastic solids. European Journal of Mechanics. A/Solid 2003. Vol 22(3) 433-442.

Gao, C.F., Kessler, H., Balke, H., Crack problems in magnetoelectroelastic solids. Part I: exact solution of a crack. *International Journal of Engineering* Science 2003. Vol 41(9) 969-981.

Gao, C.F., Kessler, H., Balke, H., Crack problems in magnetoelectroelastic solids. Part II: general solution of collinear cracks. International Journal of Engineering Science 2003. Vol 41(9) 983-994.

Gao, C.F., Tong, P., Zhang, T.Y., Interfacial crack problems in magneto-electroelastic solids. *International Journal of Engineering Science* 2003. Vol 41(18) 2105-2121.

Gao, H.J., Zhang, T.Y., Tong, P., Local and global energy release rates for an electrically yielded crack in a piezoelectric ceramics. *Journal of Mechanics and Physics of Solids* 1997, Vol 45(4) 491-510.

Gradshteyn, I.S., Ryzhik IM. Table of integral, series and products. Academic Press, 1980. New York.

Hao, T.H., Shen, Z.Y., A new electric boundary condition of electric fracture mechanics and its applications, *Engineering Fracture Mechanics* 1994. Vol 47(6) 793-802.

Hao, T. H., Multiple collinear cracks in a piezoelectric material, International Journal of Solids and Structures 2001. Vol 38(50-51) 9201-9208.

Harshe, G., Dougherty, J.P., Newnham, R. E, Theoretical modeling of 3-0/0-3 magnetoelectric composites. *International Journal of Applied Electromagnetics in Materials* 1993. Vol 4, 161-171.

Huang, J.H., Kuo, W.S., The analysis of piezoelectric/piezomagnetic composite materials containing ellipsoidal inclusions. *Journal of Applied Physics* 1997. Vol 81(3) 1378-1386.

Itou, S., Three dimensional waves propagation in a cracked elastic solid, Journal of Applied Mechanics 1978. Vol 45(6) 807-811.

Jin, B. and Zhong, Z., A moving mode-III crack in functionally graded piezoelectric material: permeable problem, *Mechanics Research Communications* 2002. Vol 29, 217-224.

Jin, D. R., Functionally graded PZT/ZnO piezoelectric composites, *Journal of Materials Science Letters* 2003. Vol 22, 971-974.

Li, C. Y. and Weng, G. J., Antiplane crack problem in functionally graded piezoelectric materials, *Journal of Applied Mechanics* 2002. Vol 69(4) 481-488.

Li, J.Y., Magnetoelectroelastic multi-inclusion and inhomogeneity problems and their applications in composite materials. *International Journal of Engineering Science* 2000, Vol 38, 1993-2011.

Liu, J.X., Liu, X.L., Zhao, Y.B., Green's functions for anisotropic magnetoelectroelastic solids with an elliptical cavity or a crack. *International Journal of Engineering Science 2001*. Vol 39(12) 1405-1418.

McMeeking, R.M., On mechanical stress at cracks in dielectrics with application to dielectric breakdown. *Journal of Applied Physics* 1989. Vol 62, 3316-3122.

Mikhailov, G. K., Parton, V. S., 1990. Electromagnetoelasticity. Hemisphere, New York.

Morse, P.M., Feshbach, H., Methods of Theoretical Physics, Vol.1, McGraw-Hill, 1958. New York.

Nan, C.W., Magnetoelectric effect in composites of piezoelectric and piezomagnetic phases. Physical Review B 1994. Vol 50, 6082-6088.

Ozturk, M. and Erdogan, F., Mode I crack problem in an inhomogeneous orthotropic medium, *International Journal of Engineering Science* 1997. Vol 35, 869-883.

Pak, Y.E., Crack extension force in a piezoelectric material. *Journal of Applied mechanics* 1990. Vol 57, 647-653.

Pan, E., Three-dimensional Green's functions in anisotropic magneto-electro-elastic bimaterails. Zeitschrift fur Angewandte Mathematik und Physik 2002, Vol 53, 815-838.

Parton, V.S., Fracture mechanics of piezoelectric materials, ACTA Astronautra 1976. Vol 3, 671-683.

Shbeeb, N.I. and Binienda, W.K., Analysis of an interface crack for a functionally graded strip sandwiched between two homogeneous layers of finite thickness. *Engineering Fracture Mechanics* 1999, Vol 64, 693-720.

Sih, G.C., Song, Z.F., Magnetic and electric poling effects associated with crack growth in BaTiO3-CoFe2O4 composite. *Theoretical and Applied Fracture Mechanics* 2003. Vol 39, 209-227.

Soh, A.K., Fang, D.N. and Lee, K.L., Analysis of a bi-piezoelectric ceramic layer with an interfacial crack subjected to anti-plane shear and in-plane electric loading, European Journal of Mechanics. A/Solid 2000. Vol 19, 961-977.

Song, Z.F., Sih, G.C., Crack initiation behavior in magnetoelectroelastic composite under in-plane deformation. *Theoretical and Applied Fracture Mechanics* 2003. Vol 39, 189-207.

Soon Man Kwon, 2003. Electrical nonlinear anti-plane shear crack in a functionally graded piezoelectric strip, *International Journal of Solids and Structures* Vol 40, 5649-5667.

Sosa, H., On the fracture mechanics of piezoelectric solids. *International Journal of Solids and Structures* 1992. Vol 29, 2613-2622.

Spyropoulos, C, P., Sih, G. C., Song, Z. F., Magnetoelectroelastic composite with poling parallel to plane of line crack under out-of-plane deformation. *Theoretical and Applied Fracture Mechanics* 2003. Vol 39(3) 281-289.

Srivastava K.N., Palaiya K.N. and Karaulia, D.S., Interaction of shear waves with two coplanar Griffith cracks situated in an infinitely long elastic strip, *International Journal of Fracture*, 1983, Vol 23, 3-14.

Suo, Z., Kuo, C.M., Barnett, D.M., Willis, J.R., Fracture mechanics for piezoelectric ceramics. *Journal of the Mechanics and Physics of Solids* 1992. Vol 40(4) 739-765.

Takagi, K., Li J. F., Yokoyama, S. and Watanabe, R., Fabrication and evaluation of PZT/Pt piezoelectric composites and functionally graded actuators, *Journal of the European Ceramic Society* 2003. Vol 10, 1577-1583.

Tian, W.Y., Gabbert, U., Multiple crack interaction problem in magnetoelectroelastic solids, European Journal of Mechanics. A/Solid 2004. Vol 23, 599-614.

Ueda, S., Transient response of a center crack in a functionally graded piezoelectric strip under electromechanical impact, *Engineering Fracture Mechanics* 2006. Vol 73, 1455-1471.

Van, Suchtelen, J., Product properties: a new application of composite materials. Phillips Research Reports 1972. Vol 27, 28-37.

Wang, B.L., Mai, Y.W., Crack tip field in piezoelectric/piezomagnetic media. *European Journal of Mechanics.* A/Solid 2003. Vol 22(4) 591-602.

Wang, B.L., Mai, Y.W., Fracture of piezoelectromagnetic materials. Mechanics Research Communications 2004. Vol 31(1) 65-73.

Wang, X., Shen, Y.P., The general solution of three-dimensional problems in magnetoelectroelastic media. *International Journal of Engineering Science* 2002, Vol 40, 1069-1080.

Wu, T.L., Huang, J.H., Closed-form solutions for the magnetoelectric coupling coefficients in fibrous composites with piezoelectric and piezomagnetic phases. *International Journal of Solids and Structures 2000.* Vol 37, 2981-3009.

Yang, F.Q., Fracture mechanics for a Mode I crack in piezoelectric materials. *International Journal of Solids and Structures* 2001. Vol 38, 3813-3830.

Zhang, T.Y., Qian, C.F., Tong, P., Linear electro-elastic analysis of a cavity or a crack in a piezoelectric material. *International Journal of Solids and Structures* 1998. Vol 35, 2121-2149.

Zhang, T. Y., Tong, P., Fracture mechanics for a mode III crack in a piezoelectric material. *International Journal of Solids and Structures* 1996. Vol 33, 343-359.

Zhong, Z., Meguid, S.A., Analysis of a circular arc-crack in piezoelectric materials, *International Journal of Fracture* 1997. Vol 84, 143-158.

Zhou, Z. G., Chen, Z. T., Basic solution of a Mode-I limited-permeable crack in functionally graded piezoelectric/piezomagnetic materials, *International Journal of Solids and Structures,* 2008, Vol.45, No.7-8, 2265-2296.

Zhou, Z. G.; Wang, B., Two parallel symmetry permeable cracks in functionally graded piezoelectric/piezomagnetic materials under anti-plane shear loading, *International Journal of Solids and Structures*, 2004. Vol 41, 4407-4422.

Zhou, Z. G.; Wang, B., An interface crack between two dissimilar functionally graded piezoelectric/piezomagnetic material half infinite planes subjected to the harmonic anti-plane shear stress waves, *International Journal of Applied Electromagnetics and Mechanics*, 2008, Vol 27, n 1-2, 117-132.

Zhou, Z.G., Wang, B., Sun, Y.G., Two collinear interface cracks in magneto-electro-elastic composites, International Journal of Engineering Science 2004. Vol 42, 1157-1167.

Zhou, Z.G., Wu, L.Z., Wang, B., The dynamic behavior of two collinear interface cracks in magneto-electro-elastic composites. *European Journal of Mechanics A/ Solids 2005.* Vol 24(2) 253-262.

Zhou, Z.G., Wu, L.Z., Wang, B., The behavior of a crack in functionally graded piezoelectric/piezomagnetic materials under anti-plane shear loading, Archive of Applied Mechanics 2005. Vol 74(8), 526-535.

Zhou, Z.G.; Zhang, P.W. and Wu, L.Z., The closed form solution of a Mode-I crack in the piezoelectric/piezomagnetic materials, *International Journal of Solids and Structures,* 2007. Vol 44, 419-435.

In: Mechanics of Functionally Graded Materials and Structures ISBN 978-1-61668-920-9
Editors: Z. Zhong, L. Wu and W. Chen

Chapter 9

ELASTODYNAMIC CRACK PROBLEM OF FUNCTIONALLY GRADED MATERIALS AND STRUCTURE

Li Ma[1] and Linzhi Wu
Center for Composite Materials, Harbin Institute of Technology
Harbin, China

ABSTRACT

Fracture mechanics evaluation provides a deeper insight into the physics of crack initiation and propagation in FGMs, which can aid in the design, optimization and application of these novel materials. Stationary crack problem in FGMs have received considerable attention from a number of researchers around the world. However, it should be mentioned that most FGMs will be used in critical situations, such as dynamic loading. Because one additional dimension, time, enters when going from a static to a corresponding dynamic case, analysis of the latter is more difficult. Therefore, the dynamic fracture responses of FGMs have received little attention from the scientific community. This chapter summarizes the recent developments of elastodynamic crack problem of functionally graded materials and structure. It covers crack moving at constant velocity in FGMs; Scattering waves about stationary cracks in FGMs and Numerical analysis of dynamic crack problems in FGMs.

1. INTRODUCTION

Advanced materials play a vital role in emerging technologies, particularly in the aerospace and biological arenas, and have drawn the attention of the industrial and scientific communities from multiple disciplines. Work on advanced materials has focused on processing a variety of alloys, ceramics, composites and coatings. Conventional composites in general are of a discrete, piecewise nature with sharp interfaces. This results in abrupt

[1] e-mail: mali@hit.edu.cn.

transitions in their thermal, mechanical and physical properties. Often, these interfaces contain flaws because of imperfect bonding, residual stresses, etc. Even otherwise, the mismatch in properties across the interface leads to stress concentrations and consequently the interface becomes a favoured site for cracking, debonding and spallation during normal operating cycles. To improve the interfacial bonding strength and to take fuller advantage of composite technology, a possible solution is functionally graded materials (FGMs) (Shukla et al., 2007). FGMs permit spatial tailoring of both the microstructure and/or composition so that the maximum possible benefits can be derived from their inhomogeneity. Controlled tailoring of the composition can significantly reduce the magnitude of the thermal stresses, suppress plastic flow and cracking, and improve interfacial bonding, provided that the graded microstructure is thermally stable during the operating cycle (Suresh and Mortensen, 1998). The distinct advantages of being able to withstand severe temperature gradients while simultaneously maintaining structural integrity are the most important arguments for choosing FGMs in critical structural applications, such as for spacecraft heat shields and in nuclear reactors and power plants.

Fracture mechanics evaluation provides a deeper insight into the physics of crack initiation and propagation in FGMs, which can aid in the design, optimization and application of these novel materials. Analytical studies of crack problems in FGMs date back to the early 1980s. Delale and Erdogan (1983) and Eischen (1987) showed that, for a continuous and piecewise-differentiable material property variation, the inverse square root stress singularity at the crack tip is the same in graded materials as in homogeneous materials, because of the similarly structured integral equations that define the crack problems for each case. These results are of significance for the fracture mechanics community as this implies that the stress intensity factor (SIF) concept remains valid as a governing fracture parameter for FGMs. Confirmation of the validity of SIF as a governing fracture parameter in graded materials and the subsequent prediction of SIF values for various loading and gradient situations have constituted the major area for theoretical work regarding FGMs to date. A variety of approaches have been used, generally involving specialized mathematical techniques, including integral equation methods, asymptotic analysis and finite element modeling. A more detailed review of the most significant previous research efforts in the field of fracture of FGMs can be found in an excellent review article written by Erdogan (1995).

Stationary crack problem in FGMs have received considerable attention from a number of researchers around the world. However, it should be mentioned that most FGMs will be used in critical situations, such as dynamic loading. Because one additional dimension, time, enters when going from a static to a corresponding dynamic case, analysis of the latter is more difficult. Therefore, the dynamic fracture responses of FGMs have received little attention from the scientific community. An excellent review paper on dynamic fracture studies in FGMs has been written by Shukla et al. (2007).

Dynamic processes in a cracked FGM body fall into two basic categories: dynamic crack propagation and dynamic loading of bodies with stationary cracks. Often a mixture of these two kinds of processes appears, for instance when dynamic loading on a body containing a stationary crack causes dynamic crack propagation, or when dynamic crack propagation generates waves, which, after reflection at an outer boundary of the body, impinge on the crack, perhaps even after the crack, for some reason, has been arrested.

Although Atkinson's (1975) study of steady-state crack growth in non-homogeneous solids in the late 1970s preceded the more numerous investigations of quasi-static fracture behavior of the 1990s, it was not until the quasi-static problems began to be understood better that researchers turned their attention to considerations of dynamic crack growth. The primary reason for this deferred attention was probably the mathematical complexity associated with developing the necessary solutions for the dynamic crack equations. Examples of the dynamic analysis include the study of the steady crack propagation in an interphase with spatially varying elastic properties under anti-plane loading conditions reported by Wang and Meguid (1994/1995). Parmeswaran and Shukla (1999) developed the expressions for the first stress invariant and studied the effect of non-homogeneity on the same, and the steady state dynamic fracture of FGMs under the in-plane loading with the material properties that vary along the crack propagation direction was reported. Meguid et al. (2002) and Jiang and Wang (2002) studied the dynamic crack propagation in FGMs under plane elastic deformation using Fourier transform technique. Bi et al. (2003) investigated the Yoffe crack problem in an infinite strip of FGMs subjected to the anti-plane shear. Ma et al. (2004a, 2005a) studied crack propagating problem in functionally graded (orthotropic) strip under plane loading. Itou (2005) analysized the stress intensity factor for a moving cylindrical crack in a FGMs cylindrical layer in composite materials using Fourier transform technique and Schmidt method.

Babaei and Lukasiewicz (1998) obtained the dynamic stress intensity factor (DSIF) for the crack in a non-homogeneous layer between isotropic half-spaces under the anti-plane shear impact load. Wang et al. (2000) solved the dynamic crack problem for the non-homogeneous composite materials but they considered the FGM layer as multi-layered homogeneous media. Ma et al. analyzed the scattering problem of elastic wave in FGMs (2002, 2004b, 2005b, 2007). The torsional impact problems have been investigated by Li and his coworkers (2001 and 2002). In their papers, Li et al. (2001) considered the problem of a cylindrical crack located in the FGM interlayer between two coaxial elastic dissimilar homogeneous cylinders under the torsional impact loading. The dynamic stress intensity factor is found to increase rapidly to the peak value and then decrease and tend to the static value with almost no oscillations. Guo et al. (2004, 2005a, 2005b) studied the transient fracture behavior of a functionally graded coating-substrate system and a layered structure with a functionally graded interfacial layer containing a crack vertical to the boundary under in-plane impact loading, respectively. In their papers, it was found that the DSIF for the internal crack usually increases rapidly to a peak value and then tend to the static value with minimal oscillations. Considering the antiplane shear impact condition, Choi (2004, 2006) analyzed the elastodynamic crack problem for a layered structure with a functionally graded interfacial zone. In their investigations, a crack that was vertical or inclined to the interface was located in the homogeneous medium related to the graded interfacial zone.The transient internal crack problem for a functionally graded strip and electromechanical impact of a cracked functionally graded piezoelectric medium was investigated by Chen et al. (2002, 2003a, 2003b, 2005). Wang and Mai (2006) considered a periodic array of cracks in an infinite functionally graded material under transient mechanical loading. Huang and Wang (2004) developed a multilayered model to conduct the fracture analysis of a functionally graded interfacial zone under harmonic antiplane loading. Recently, Chalivendra and Shukla (2005) and Chalivendra (2007) developed the transient crack-tip field expansions for straight and curved cracks propagating in an FGM. Most recently, Jain and Shukla (2007) investigated the transient behavior of mixed-mode cracks in FGMs, both analytically and experimentally.

Nakagaki et al. (1995) addresses a numerical treatment of fracture occurring in an FGM under dynamic load. By using the boundary integral equation method, Zhang et al. (2003) analyzed the effects of material gradients on dynamic stress intensity factor. Jin et al. (2002, 2003) investigated elastic-plastic mode I crack growth in TiB/Ti FGMs by using three-dimensional interface cohesive elements. The dynamic fracture of FGMs has been modeled using an explicit cohesive volumetric finite element scheme that incorporates spatially varying constitutive and failure properties by Kandula et al. (2005). Zhang and Paulino (2005) used cohesive zone models to simulate two-dimensional mixed-mode dynamic crack propagation in FGMs. Moon et al. (2002) investigated crack growth resistance (R-curve) behavior of multilayer graded alumina-zirconia FGMs considering a crack parallel to the material gradation. Fujimoto and Noda (2000) investigated propagation of a single crack in a partially stabilized zirconia (PSZ) and T1-6A1-4V FGMs under transient thermal loads. Gu and Asaro (1997) studied the mixed-mode fracture of FGMs with consideration of a crack perpendicular to the material gradient. Considering local homogeneity and maximum energy release rate criterion, they found that the crack kinked towards the compliant region due to the greater release of elastic potential energy. Anlas et al. (2002) and Kim and Paulino (2003) have incorporated the material property variation within the finite element formulation. Recently, Wang and Nakamura (2004) reviewed some of the existing models for crack growth simulations and chose a cohesive methodology to simulate dynamic crack propagation in elastic–plastic FGMs. Tilbrook et al. (2005, 2006) predicted the energy release rate for a curved crack using finite element simulations. Recently, there have been some studies of the fracture mechanics of anisotropic, functionally graded materials. Sladek et al. (2005) proposed a meshless method based on the local Petrov-Galerkin approach for crack analysis in anisotropic FGMs. Kim and Paulino (2004a) presented an interaction integral formulation for evaluating the elastic T-stress for a mixed-mode crack in orthotropic non-homogeneous materials. Kim and Paulino (2004b) simulated the crack propagation path in FGMs under mixed non-proportional loading by using finite element method. Dolbow and Gosz (2002) considered the plane problem of an arbitrarily oriented crack using extended finite method (XFEM). Elastic wave propagation in cracked, functionally graded materials with elastic parameters that are exponential functions of a single spatial co-ordinate is studied by using boundary integral equation method (Dineva et al., 2007).

Although several well-established experimental techniques have been used to characterise the crack behavior in FGMs, including photoelasticity (Parameswaran and Shukla,1998), digital image correlation (DIC) (Li et al., 2000) and coherent gradient sensing (CGS) (Rousseau and Tippur, 2002), comparing with the theoretical and numerical efforts, activities in the experimental characterisation of crack behavior in FGMs have been limited mainly by the complexity of the experiments and the cost of manufacturing FGMs. One other reason for the limited experimental studies of dynamic crack propagation behavior in FGMs stems from the inability to analytically describe the stress and deformation field at the tip of a dynamically growing crack. These field expressions are needed because the usual interpretation procedure involves fitting the experimental data to the crack-tip stress-field expressions to estimate fracture parameters like the SIF, T-stress, etc. Parameswaran and Shukla (1998) conducted experimental studies of the dynamic fracture of FGMs with discrete property variation using the photoelasticity technique. Lin et al. (1994) investigated mode I fracture of aluminium alloy 2124/SiC FGMs considering a crack parallel to the material gradation. Carpenter et al. (1999) performed fracture testing and analysis of a layered functionally graded Ti/TiB beam subjected to three-point bending. Parameswaran and Shukla (1998) used a gravity casting technique to manufacture a particulate epoxy composite with a spatially varying reinforcement concentration. Rousseau and Tippur (2000) used coherent gradient sensing (CGS) technique to investigate the fracture behavior of the glass particle/epoxy FGMs with a crack perpendicular to the elastic gradient and proved experimentally that the crack kinked towards the compliant region. And they pointed out that both maximum hoop stress and vanishing K_{II} criteria can be applied to predict the kinking angle for FGMs. Similar problem was also studied by Tilbrook et al. (2006) for the alumina/epoxy FGMs using crack growth experiment, phase-shifted Moiré interferometry technique and finite element method. They found that the elastic gradient had a dominant influence on the mode mixity of the crack-tip and the crack propagation direction. Abanto-Bueno and Lambros (2006) experimentally studied the fracture response of the FGMs (ultraviolet-irradiation-hardened polymer) under different mixed-mode loading conditions which were generated by applying an inclined external in-plane loading, inclining the crack at an angle to the material gradient, and a combination of both cases, respectively. In addition, they extracted the fracture parameters including the stress intensity factors K_I, K_{II} and T-stress using digital image correlation (DIC) technique, and discussed the effect of T-stress on the crack kink. For the dynamic fracture of FGMs, Kirugulige and Tippur (2006) studied the influence of material gradient on the crack propagation during a mixed-mode dynamic fracture test. They applied CGS technique coupled with a high-speed photography to map the transient crack-tip deformation before and after the crack initiation, and found that the crack situated on the compliant side of the specimen kinked less than the one situated on the stiff side. Lambros et al. (1999) used an UV-sensitive material whose stiffness and fracture toughness depend strongly on the amount of UV exposure. Jin et al. investigated mixed-mode fracture response of ZrO_2/NiCr functionally graded materials by using DIC technique and extended finite element method. (Jin et al., 2009a, 2009b).

This chapter summarizes the recent developments of elastodynamic crack problem of functionally graded materials and structure. It covers crack moving at constant velocity in FGMs; Scattering waves about stationary cracks in FGMs and Numerical analysis of dynamic crack problems in FGMs.

2. Crack Moving at Constant Velocity in FGMs

The first solution of a dynamic steady state crack problem was given by Yoffe (1951). She studied a crack of constant length, moving with constant velocity in an infinite solid. Thus, the crack has one leading edge, where the process is *tearing*, and one trailing edge, where the process is *healing*. The solution shows most of the features pertinent to dynamically moving cracks in general. There are a few researchers studied the Yoffe crack problem in FGMs. Wang and Meguid (1994) proposed a theoretical and numerical treatment of a finite crack propagating in an interfacial layer with spatially varying elastic properties under the anti-plane loading condition. Bi et al. (2003) investigated the Yoffe crack problem in an infinite strip of FGMs subjected to the anti-plane shear. Meguid et al. (2002) studied the dynamic crack propagation in FGMs under the plane elastic deformation using Fourier transform technique. Jiang and Wang (2002) analyzed the dynamic plane behavior of a Yoffe type crack propagating in a functionally graded interlayer bonded to dissimilar half planes. Ma et al. (2004, 2005a) studied crack propagating problem in functionally graded strip by use of Schmidt method. It will be demonstrated here for Yoffe crack problem in orthotropic FGMs strip under the plane elastic deformation which has been studied by Ma et al. (2005a).

2.1. Statement of the Yoffe Crack Problem in Functionally Graded Strip

Following the work of Ma et al. (2005a), here we consider a functionally graded orthotropic strip of width $2h$ with a crack (or a slipping region) of length $2a$ traveling at a constant speed V along a straight line bisecting the strip, as show in Figure 2-1 (Ma et al., 2005a). Plane strain or plane stress is assumed. The crack is subjected to prescribed normal and shear stresses along its surface. For the present problem, it is convenient to divide the functionally graded orthotropic strip into two regions, namely material 1 in the upper half strip and material 2 located in the lower half strip. c_{11}, c_{12}, c_{22}, c_{66} and ρ stand for the elastic stiffness constants and mass density, respectively. To make the analysis tractable, the material properties of the functionally graded orthotropic medium are supposed to be one-dimensionally dependent as:

$$c_{11} = c_{110}e^{\beta y},\ c_{12} = c_{120}e^{\beta y},\ c_{22} = c_{220}e^{\beta y},\ c_{66} = c_{660}e^{\beta y},\ \rho = \rho_0 e^{\beta y} \tag{2-1}$$

where β is the gradient of material properties and is a positive or negative constant. In this case, the longitudinal and shear wave speeds of the medium are constants. This model includes the effects of varying elastic stiffness constants and mass density, but ignores the effect of the spatial variation of wave speed. It should be mentioned that this effect is second order in nature if the spatial variation of wave speed is not significant in comparison with that of the elastic stiffness constants and mass density (Meguid et al. 2002, Jiang and Wang, 2002).

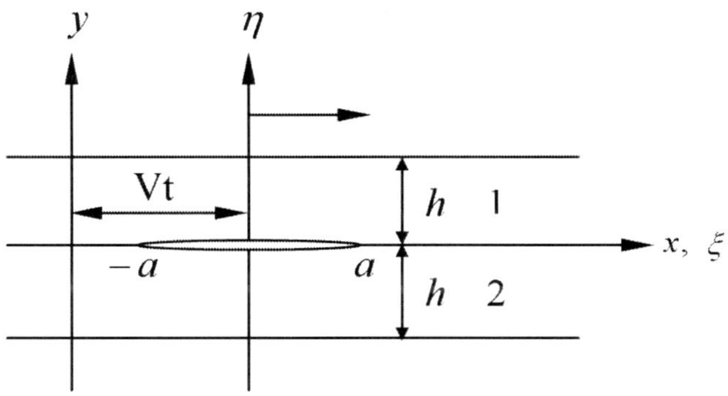

Figure 2.1. Moving crack in the functionally graded orthotropic strip (Ma et al., 2005a).

The basic equations which govern the plane dynamic behavior of the medium can be expressed in a fixed Cartesian coordinate system (x, y) as,

$$\sigma^{(j)}_{xx,x} + \sigma^{(j)}_{xy,y} = \rho u^{(j)}_{,tt},\quad j = 1,2 \tag{2-2a}$$

$$\sigma^{(j)}_{xy,x} + \sigma^{(j)}_{yy,y} = \rho v^{(j)}_{,tt},\quad j = 1,2 \tag{2-2b}$$

and

$$\sigma^{(j)}_{xx} = c_{11}u^{(j)}_{,x} + c_{12}v^{(j)}_{,y},\quad j = 1,2 \tag{2-3a}$$

$$\sigma^{(j)}_{yy} = c_{12}u^{(j)}_{,x} + c_{22}v^{(j)}_{,y},\quad j = 1,2 \tag{2-3b}$$

$$\sigma^{(j)}_{xy} = c_{66}(u^{(j)}_{,y} + v^{(j)}_{,x}),\quad j = 1,2 \tag{2-3c}$$

where u, v are the x and y components of the displacement vector, σ_{ij} ($i, j = x, y$) are the stress components, and a comma indicates partial derivatives.

For the present problem, it is convenient to introduce the following *Galilean* transformation

$$\xi = x - Vt\ ,\ \eta = y \tag{2-4}$$

with ξ and η being a translating coordinate system, which is attached to the propagating crack. It is, however, assumed that the propagation of the crack has prevailed for such a long time that the stress distribution around its tip is time invariant in the translating reference frame. Accordingly, the equilibrium equation (2-2) can be written in term of the moving coordinate system (ξ, η) as

$$\sigma^{(j)}_{\xi\xi,\xi} + \sigma^{(j)}_{\xi\eta,\eta} = \rho V^2 u^{(j)}_{,\xi\xi}\ ,\ j = 1,2 \tag{2-5a}$$

$$\sigma^{(j)}_{\xi\eta,\xi} + \sigma^{(j)}_{\eta\eta,\eta} = \rho V^2 v^{(j)}_{,\xi\xi}\ ,\ j = 1,2 \tag{2-5b}$$

which govern the steady state solution of the problem.

From equation (2-1), it can be known that, in the translating coordinate system (ξ, η), the materials properties can be written as

$$c_{11}(\eta) = c_{110}e^{\beta\eta}\ ,\ c_{12}(\eta) = c_{120}e^{\beta\eta}\ ,\ c_{22}(\eta) = c_{220}e^{\beta\eta}\ ,\ c_{66}(\eta) = c_{660}e^{\beta\eta}\ ,\ \rho(\eta) = \rho_0 e^{\beta\eta} \tag{2-6}$$

For a general nonhomogeneous medium with material constants, which are continuous functions of η, the governing equation of motion can be expressed by displacement components in the moving coordinate system as

$$c_{11}u^{(j)}_{,\xi\xi} + c_{66}u^{(j)}_{,\eta\eta} + (c_{12} + c_{66})v^{(j)}_{,\xi\eta} + c_{66,\eta}\left[u^{(j)}_{,\eta} + v^{(j)}_{,\xi}\right] = \rho V^2 u^{(j)}_{,\xi\xi}\ ,\ j = 1,2 \tag{2-7a}$$

$$c_{22}v^{(j)}_{,\eta\eta} + c_{66}v^{(j)}_{,\xi\xi} + (c_{12} + c_{66})u^{(j)}_{,\xi\eta} + \left[c_{12,\eta}u^{(j)}_{,\xi} + c_{22,\eta}v^{(j)}_{,\eta}\right] = \rho V^2 v^{(j)}_{,\xi\xi}\ ,\ j = 1,2 \tag{2-7b}$$

Substituting equations (2-6) into equations (2-7) results in the following governing equations of the problem,

$$\alpha_1 u^{(j)}_{,\xi\xi} + u^{(j)}_{,\eta\eta} + \alpha_2 v^{(j)}_{,\xi\eta} + \beta\left[u^{(j)}_{,\eta} + v^{(j)}_{,\xi}\right] = 0\ ,\ j = 1,2 \tag{2-8a}$$

$$\alpha_3 v^{(j)}_{,\eta\eta} + v^{(j)}_{,\xi\xi} + \alpha_4 u^{(j)}_{,\xi\eta} + \alpha_3\beta\left[v^{(j)}_{,\eta} + \alpha_5 u^{(j)}_{,\xi}\right] = 0\ ,\ j = 1,2 \tag{2-8b}$$

with $\alpha_1 = \dfrac{c_{110}}{c_{660}} - c^2$, $\alpha_2 = \dfrac{c_{120}}{c_{660}} + 1$, $\alpha_3 = \dfrac{c_{220}}{c_{660}(1-c^2)}$ and $\alpha_4 = \dfrac{c_{120} + c_{660}}{c_{660}(1-c^2)}$, $\alpha_5 = \dfrac{c_{120}}{c_{220}}$,

where $c = V/c_{sh}$ and c_{sh} is the shear wave speed of the nonhomogeneous medium given by

$$c_{sh} = \sqrt{c_{660}/\rho_0} \tag{2-9}$$

The system of above governing equations is solved by using the Fourier transform technique. The general expression for displacement components are written as

$$\begin{cases} u^{(1)}(\xi,\eta) = \dfrac{2}{\pi}\displaystyle\int_0^\infty \sum_{j=1}^{4} A_j(s)e^{\lambda_j\eta}\sin(s\xi)ds \\ v^{(1)}(\xi,\eta) = \dfrac{2}{\pi}\displaystyle\int_0^\infty \sum_{j-1}^{4} m_j(s)A_j(s)e^{\lambda_j\eta}\cos(s\xi)ds \end{cases} \tag{2-10a}$$

$$\begin{cases} u^{(2)}(\xi,\eta) = \dfrac{2}{\pi}\displaystyle\int_0^\infty \sum_{j=5}^{8} A_j(s)e^{\lambda_{j-4}\eta}\sin(s\xi)ds \\ v^{(2)}(\xi,\eta) = \dfrac{2}{\pi}\displaystyle\int_0^\infty \sum_{j=5}^{8} m_{j-4}(s)A_j(s)e^{\lambda_{j-4}\eta}\cos(s\xi)ds \end{cases} \tag{2-10b}$$

From equations (2-3), stress components are obtained as

$$\begin{cases} \sigma_{\eta\eta}^{(1)}(\xi,\eta) = \dfrac{2e^{\beta\eta}}{\pi}\displaystyle\int_0^\infty \sum_{j=1}^{4} \left[c_{220}m_j(s)\lambda_j + c_{120}s\right]A_j(s)e^{\lambda_j\eta}\cos(s\xi)ds \\ \tau_{\xi\eta}^{(1)}(\xi,\eta) = \dfrac{2e^{\beta\eta}}{\pi}\displaystyle\int_0^\infty \sum_{j=1}^{4} c_{660}\left[\lambda_j - sm_j(s)\right]A_j(s)e^{\lambda_j\eta}\sin(s\xi)ds \end{cases} \tag{2-11a}$$

$$\begin{cases} \sigma_{\eta\eta}^{(2)}(\xi,\eta) = \dfrac{2e^{\beta\eta}}{\pi}\displaystyle\int_0^\infty \sum_{j=5}^{8} \left[c_{220}m_{j-4}(s)\lambda_{j-4} + c_{120}s\right]A_j(s)e^{\lambda_{j-4}\eta}\cos(s\xi)ds \\ \tau_{\xi\eta}^{(2)}(\xi,\eta) = \dfrac{2e^{\beta\eta}}{\pi}\displaystyle\int_0^\infty \sum_{j=5}^{8} c_{660}\left[\lambda_{j-4} - sm_{j-4}(s)\right]A_j(s)e^{\lambda_{j-4}\eta}\sin(s\xi)ds \end{cases} \tag{2-11b}$$

where s is the transform variable. $A_1(s)$, $A_2(s)$, $A_3(s)$ … and $A_8(s)$ are arbitrary unknowns, $\lambda_1(s)$, $\lambda_2(s)$, $\lambda_3(s)$ and $\lambda_4(s)$ are the roots of the characteristic equation

$$\lambda^4 + 2\beta\lambda^3 + \frac{1}{\alpha_3}\left[s^2(\alpha_2\alpha_4 - \alpha_1\alpha_3 - 1) + \beta^2\alpha_3\right]\lambda^2 + \frac{\beta s^2}{\alpha_3}(\alpha_4 + \alpha_2\alpha_3\alpha_5 - \alpha_1\alpha_3 - 1)\lambda + \alpha_5 s^2\beta^2 + \frac{\alpha_1}{\alpha_3}s^4 = 0 \tag{2-12}$$

Here, λ_1 and λ_2 have positive real parts and λ_3 and λ_4 have negative real parts. The coefficients $m_j(s)$ $(j = 1,2,3,4)$ are expressed for each root $\lambda_j(s)$ as

$$m_j(s) = s\frac{\alpha_4\lambda_j + \alpha_3\alpha_5\beta}{s^2 - \alpha_3\lambda_j^2 - \alpha_3\beta\lambda_j} \tag{2-13}$$

For the present problem, since the normal and shear stress and displacement are continuous across the interface and traction free along both the upper boundary ($\eta = h$) and the lower boundary ($\eta = -h$), the following conditions must be satisfied.

$$\sigma_{\eta\eta}^{(1)}(\xi,0^+) = \sigma_{\eta\eta}^{(2)}(\xi,0^-) = -\sigma_0(\xi), \quad \tau_{\xi\eta}^{(1)}(\xi,0^+) = \tau_{\xi\eta}^{(2)}(\xi,0^-) = -\tau_0(\xi), \quad |\xi| \le a \tag{2-14a}$$

$$\sigma_{\eta\eta}^{(1)}(\xi,0^+) = \sigma_{\eta\eta}^{(2)}(\xi,0^-), \quad \tau_{\xi\eta}^{(1)}(\xi,0^+) = \tau_{\xi\eta}^{(2)}(\xi,0^-), \quad |\xi| > a \tag{2-14b}$$

$$u^{(1)}(\xi,0^+) = u^{(2)}(\xi,0^-), \quad v^{(1)}(\xi,0^+) = v^{(2)}(\xi,0^-), \quad |\xi| > a \tag{2-14c}$$

$$\sigma_{\eta\eta}^{(1)}(\xi,h) = 0, \quad \tau_{\xi\eta}^{(1)}(\xi,h) = 0, \quad |\xi| \le \infty \tag{2-14d}$$

$$\sigma_{\eta\eta}^{(2)}(\xi,-h) = 0, \quad \tau_{\xi\eta}^{(2)}(\xi,-h) = 0, \quad |\xi| \le \infty \tag{2-14e}$$

From equations (2-14), it can be seen that there are four unknown constants A_1, A_2, A_3, A_4 which can be obtained from the boundary conditions. To solve the present problem, the jumps of displacements across the crack surfaces can be defined as follows:

$$f_1(\xi) = u^{(1)}(\xi,0^+) - u^{(2)}(\xi,0^-), \quad |\xi| \le a \tag{2-15a}$$

$$f_2(\xi) = v^{(1)}(\xi,0^+) - v^{(2)}(\xi,0^-), \quad |\xi| \le a \tag{2-15b}$$

Applying the Fourier transform and boundary conditions (2-14), we have

$$[X_1]\begin{bmatrix} A_1(s) \\ A_2(s) \end{bmatrix} + [X_2]\begin{bmatrix} A_3(s) \\ A_4(s) \end{bmatrix} = 0 \tag{2-16a}$$

$$[X_3]\begin{bmatrix} A_5(s) \\ A_6(s) \end{bmatrix} + [X_4]\begin{bmatrix} A_7(s) \\ A_8(s) \end{bmatrix} = 0 \tag{2-16b}$$

$$[X_5]\begin{bmatrix} A_1(s) \\ A_2(s) \end{bmatrix} + [X_6]\begin{bmatrix} A_3(s) \\ A_4(s) \end{bmatrix} = [X_7]\begin{bmatrix} A_5(s) \\ A_6(s) \end{bmatrix} + [X_8]\begin{bmatrix} A_7(s) \\ A_8(s) \end{bmatrix} \tag{2-16c}$$

$$[X_9]\begin{bmatrix} A_1(s) \\ A_2(s) \end{bmatrix} + [X_{10}]\begin{bmatrix} A_3(s) \\ A_4(s) \end{bmatrix} - [X_{11}]\begin{bmatrix} A_5(s) \\ A_6(s) \end{bmatrix} - [X_{12}]\begin{bmatrix} A_7(s) \\ A_8(s) \end{bmatrix} = \begin{bmatrix} \bar{f}_1(s) \\ \bar{f}_2(s) \end{bmatrix} \tag{2-16d}$$

Below, we denote the Fourier transform of functions by using a superposed bar. If $f(x)$ is an even function, the Fourier transform is defined as follows:

$$\bar{f}(s) = \int_0^\infty f(x)\cos(sx)dx, \quad f(x) = \frac{2}{\pi}\int_0^\infty \bar{f}(s)\cos(sx)ds \tag{2-17}$$

When $f(x)$ is an add function, the Fourier transform is defined as follows:

$$\bar{f}(s) = \int_0^\infty f(x)\sin(sx)dx \quad , \quad f(x) = \frac{2}{\pi}\int_0^\infty \bar{f}(s)\sin(sx)ds \tag{2-18}$$

By solving equations (2-16) with four unknown functions, substituting the solutions into equations (2-10) and (2-11) and applying the boundary conditions, it can be obtained that

$$\sigma_{\eta\eta}^{(1)}(\xi,0^+) = \frac{2}{\pi}\int_0^\infty [d_1(s)\bar{f}_1(s) + d_2(s)\bar{f}_2(s)]\cos(s\xi)ds = -\sigma_0(\xi) \quad , \quad 0 \le |\xi| \le a \tag{2-19a}$$

$$\int_0^\infty \bar{f}_1(s)\cos(s\xi)ds = 0 \quad , \quad |\xi| > a \tag{2-19b}$$

$$\tau_{\xi\eta}^{(1)}(\xi,0^+) = \frac{2}{\pi}\int_0^\infty [d_3(s)\bar{f}_1(s) + d_4(s)\bar{f}_2(s)]\sin(s\xi)ds = -\tau_0(\xi) \quad , \quad 0 \le |x| \le a \tag{2-20a}$$

$$\int_0^\infty \bar{f}_2(s)\sin(s\xi)ds = 0 \quad , \quad |\xi| > a \tag{2-20b}$$

where $d_1(s)$, $d_2(s)$, $d_3(s)$ and $d_4(s)$ are known functions. To determine the unknown functions $\bar{f}_1(s)$ and $\bar{f}_2(s)$, the above two pairs of dual integral equations (2-19) and (2-20) can be solved numerically. Then the crack-tip stress intensity factor can be obtained. The detailed process of solution can be found in the paper of Ma et al. (2005a).

2.2. Numerical Results and Discussions

Numerical calculations are carried out for three kinds of materials (i.e. isotropic, orthotropic I and II), whose mechanical properties are listed in Table 2-1 (Chen et al. 2002). The only difference between orthotropic I and II is that their reinforced directions are perpendicular to each other.

Table 2-1. Mechanical properties used for analysis (Chen et al. 2002)

	$c_{110}(N/m^2)$	$c_{120}(N/m^2)$	$c_{220}(N/m^2)$	$c_{660}(N/m^2)$	$\rho_0(N/m^2)$
Isotropic	2.198×10^{11}	6.593×10^{10}	2.198×10^{11}	7.692×10^{10}	7.840×10^{3}
Orthotropic I	1.578×10^{10}	3.248×10^{9}	1.048×10^{10}	7.070×10^{9}	1.580×10^{3}
Orthotropic II	1.048×10^{10}	3.248×10^{9}	1.578×10^{10}	7.070×10^{9}	1.580×10^{3}

First, we restrict our attention to the static crack problem in the isotropic homogeneous strip under the normal loading $\sigma_0(\xi,0) = \sigma_0$ along the crack surface. The normalized static

SIFs ($k_I = K_I / \sigma_0 \sqrt{\pi a}$, $k_{II} = K_{II} / \sigma_0 \sqrt{\pi a}$) given by Sih (1977) are compared with the results of Ma et al. (2005a) in Figure 2-2. A very good agreement can be observed in the figure.

In the present paper, we consider the case where the crack is subjected to a constant normal stress $\sigma_0(\xi) = \sigma_0$, $\tau_0(\xi) = 0$ along its surfaces.

The results of the normalized static SIF k_I for the cracked functionally graded isotropic strip are compared with those for the functionally graded orthotropic strip under $a/h = 1.0$ in Figure 2-3 (Ma et al., 2005a). In general, k_I increases as the relative gradient parameter βa increases. The curve corresponding to the isotropic material is slightly higher than that of the orthotropic material I and lower than that of the orthotropic material II.

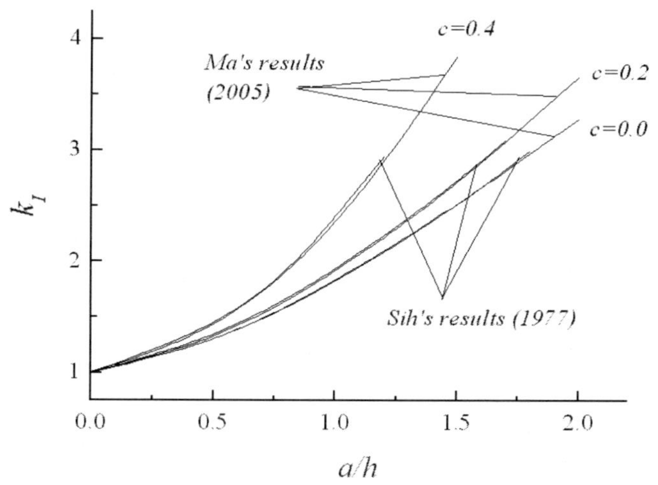

Figure 2.2. Comparison of normalized static stress intensity factors of the isotropic homogeneous strip for Sih's results (1977) and Ma's results (2005a).

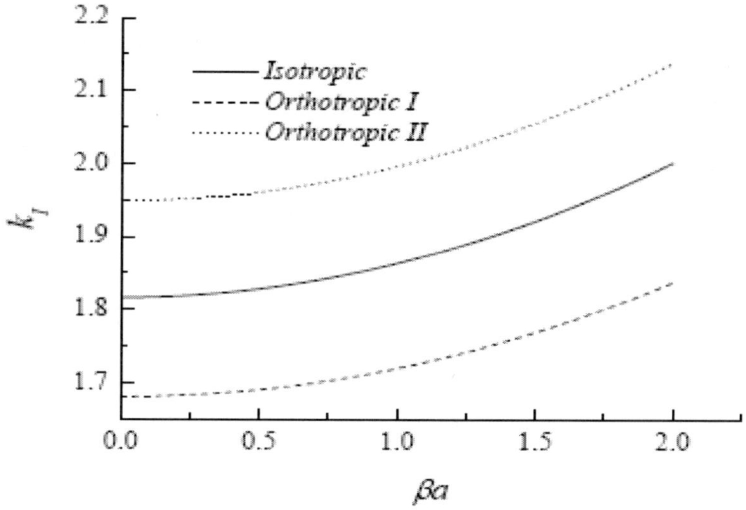

Figure 2.3. Comparison of normalized static stress intensity factors for the functionally graded isotropic material with those for the functionally graded orthotropic materials (Ma et al. 2005a).

Figure 2-4 shows the effects of gradient parameter βa upon normalized dynamic stress intensity factor (DSIF) k_I for a crack with different crack propagation speeds c ($c = V / c_{sh}$, c_{sh} is the shear wave speed of the medium) in the functionally graded isotropic material (Ma et al. 2005a). It is observed that k_I increases with the increasing of the material gradient parameter βa.

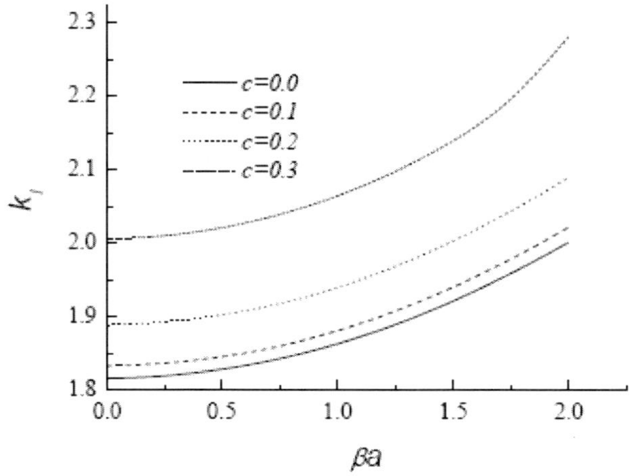

Figure 2.4. Variation of normalized SIF k_I with the gradient parameter βa under the mode I loading for functionally graded isotropic material (Ma et al. 2005a).

Figure 2-5 shows the effects of gradient parameter βa upon normalized DSIF k_I with different crack propagation speeds $c = V / c_{sh}$ in the functionally graded orthotropic material I and II, respectively (Ma et al., 2005a). The general feature of these curves is that the DSIF k_I increases with the increasing of the material gradient parameter βa. The dynamic stress intensity factor increases significantly with the increasing of c can be observed. From these two figures, it can be found that, the difference of the DSIFs between the orthotropic I and II becomes more obvious as the value of βa increases.

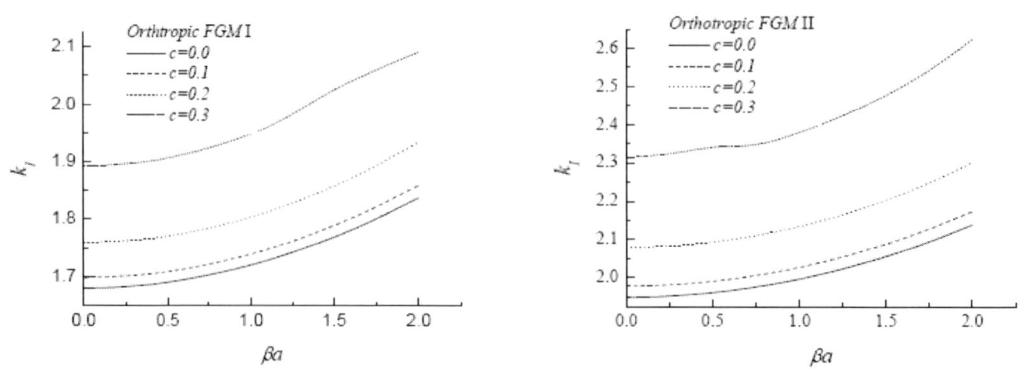

Figure 2.5. Variation of normalized SIF k_I with the gradient parameter βa under the mode I loading for functionally graded orthotropic material I and II (Ma et al. 2005a).

Figure 2-6 shows the effect of the thickness of the strip upon the normalized DSIF k_I under normal loading for different crack speeds for the case where $\beta a = 1.0$ in the functionally graded isotropic strip and the corresponding values for functionally graded orthotropic materials I and II, respectively (Ma et al., 2005a). Here, the dynamic stress intensity factor increases with the increasing of a/h. The curves in these figures are seen to be steeper higher crack velocities. And the effect of narrowing the strip width is to increase the load transfer to the crack tip region.

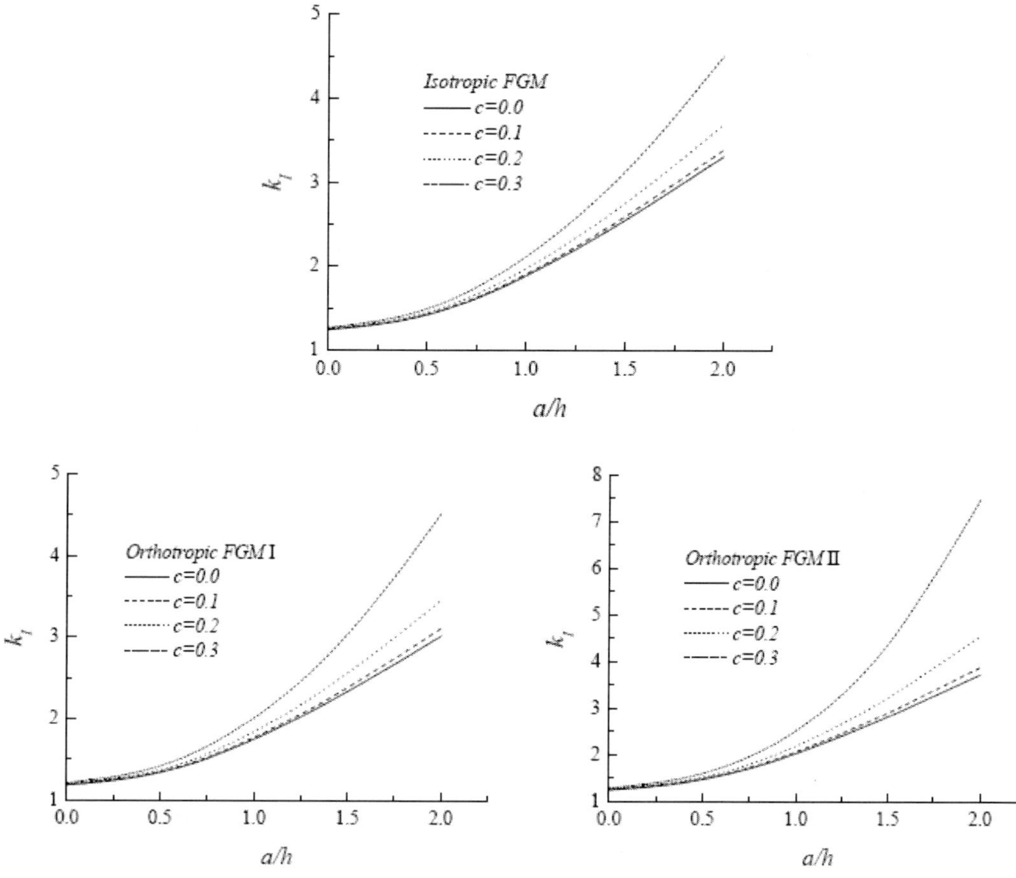

Figure 2.6. Variation of normalized SIF k_I with the gradient parameter a/h under the mode I loading for functionally graded isotropic, orthotropic material I and II (Ma et al. 2005a).

3. Scattering Waves about Stationary Cracks in FGMs

Stress wave are capable of creating stress intensity factors which may be larger than those produced by static stresses of the same magnitudes. However, a general feature is that the stress intensity factor does not even experience a step increase for an incoming step pulse, but it is gradually increasing from zero. Typically, the stress intensity factor produced by the impact of a step pulse of constant magnitude, overshoots the corresponding static stress intensity factor after a while, whereupon oscillations occur around this static value.

The present section shows the dynamic behavior of a finite crack in the functionally graded materials subjected to the normally incident elastic time harmonic waves which was investigated by Ma et al. (2005b). The crack arbitrarily oriented with respect to the direction of property gradient is considered. The problem is solved under plane strain or generalized plane stress conditions. By using the Fourier transform and defining the jumps of the displacements across the crack surfaces as unknown functions, two pairs of dual integral equations are derived. Numerical examples are provided to show the effect of the gradient parameter and the crack configuration on the dynamic stress intensity factors of the functionally graded materials with a crack.

3.1. Formulation of the Problem

As show in Figure 3-1 given by Ma et al. (2005b), consider the plane scattering problem of a normally incident longitudinal wave to a Griffith crack in the functionally graded medium. The crack is assumed to occupy the region $|x| \le a$, $y = 0$, $|z| < \infty$. The shear modulus μ and mass density ρ are approximated by

$$\mu(x_1) = \mu_0 \exp(\delta x_1), \quad \mu(x, y) = \mu_0 \exp(\beta x + \gamma y) \tag{3-1a}$$

$$\rho(x_1) = \rho_0 \exp(\delta x_1), \quad \rho(x, y) = \rho_0 \exp(\beta x + \gamma y) \tag{3-2b}$$

where

$$\beta = \frac{\delta}{\sqrt{1+\tan^2\theta}}, \quad \gamma = \frac{\delta \tan\theta}{\sqrt{1+\tan^2\theta}} \tag{3-2}$$

where μ_0, ρ_0 and δ are material constants. In this case, the longitudinal and shear wave speeds of the medium are constant. This model includes the effects of varying shear modulus and mass density, but ignores the effect of the spatial variation of wave speed. Similar to section 2, it is convenient to divide the functionally graded medium into two regions, namely material 1 in the upper half plane and material 2 in the lower half plane as show in Figure 3-1.

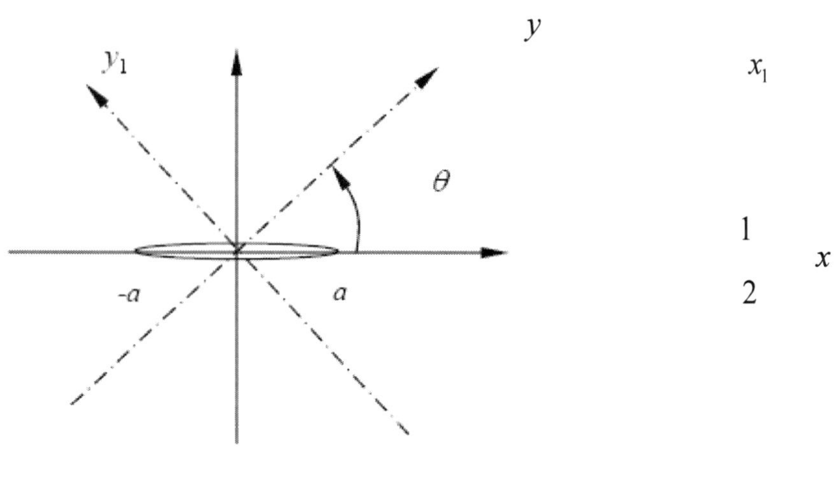

Figure 3.1. Geometry of a finite crack in functionally graded materials (Ma et al., 2005b).

In the present work, the time harmonic stress wave is normally incident. Let ω be the circular frequency of the incident wave, σ_0 is a magnitude of the incident wave. Since the time-factor $\exp(i\omega t)$ is a common factor in all equations, it is dropped out hereafter. By denoting $u^{(j)}(x,y)$ and $v^{(j)}(x,y)$ (The superscript $j=1,2$ corresponds to the upper half plane and the lower half plane throughout this section) as the amplitude of displacements components in the x- and y-directions, $\sigma_{xx}^{(j)}(x,y)$, $\sigma_{yy}^{(j)}(x,y)$ and $\tau_{xy}^{(j)}(x,y)$ as the amplitude of stress components, respectively. If only the stress intensity factors are considered, the mixed boundary value problem shown in Figure 3-1 must be solved under the following conditions

$$\sigma_{yy}^{(1)}(x,0)=\sigma_{yy}^{(2)}(x,0)=-\sigma_0,\quad \tau_{xy}^{(1)}(x,0)=\tau_{xy}^{(2)}(x,0)=0,\quad |x|\le a \tag{3-3a}$$

$$\sigma_{yy}^{(1)}(x,0)=\sigma_{yy}^{(2)}(x,0),\quad \tau_{xy}^{(1)}(x,0)=\tau_{xy}^{(2)}(x,0),\quad |x|>a \tag{3-3b}$$

$$u^{(1)}(x,0)=u^{(2)}(x,0),\quad v^{(1)}(x,0)=v^{(2)}(x,0),\quad |x|>a \tag{3-3c}$$

The constitutive relations for the FGMs are written as

$$\sigma_{xx}^{(j)}(x,y)=\frac{\mu_0 e^{\beta x+\gamma y}}{\kappa-1}\left[(1+\kappa)\frac{\partial u^{(j)}}{\partial x}+(3-\kappa)\frac{\partial v^{(j)}}{\partial y}\right] \tag{3-4a}$$

$$\sigma_{yy}^{(j)}(x,y)=\frac{\mu_0 e^{\beta x+\gamma y}}{\kappa-1}\left[(1+\kappa)\frac{\partial v^{(j)}}{\partial y}+(3-\kappa)\frac{\partial u^{(j)}}{\partial x}\right] \tag{3-4b}$$

$$\tau_{xy}^{(j)}(x,y)=\mu_0 e^{\beta x+\gamma y}\left[\frac{\partial u^{(j)}}{\partial y}+\frac{\partial v^{(j)}}{\partial x}\right],\quad (j=1,2) \tag{3-4c}$$

where $\kappa=3-4\upsilon$ for the state of plane strain, $\kappa=(3-\upsilon)/(1+\upsilon)$ for the state of generalized plane stress and υ is the Poisson's ratio. Since the variation of Poisson's ratio υ within a practical range has the rather insignificant influence on the stress intensity factor (Delae and Erdogan, 1983; Konda and Erdogan, 1994; Delae and Erdogan, 1988; Chen, 1990), it can be taken as a constant. $\delta\neq 0$ corresponds to the functionally graded materials, while $\delta=0$ does to the homogenous materials. In the absence of body forces, the elastic behavior of the medium with the variable shear modulus and density in equations (3-1) are governed by following equations

$$(1+\kappa)\frac{\partial^2 u^{(j)}}{\partial x^2}+(\kappa-1)\frac{\partial^2 u^{(j)}}{\partial y^2}+2\frac{\partial^2 v^{(j)}}{\partial x\partial y}+\beta\left[(1+\kappa)\frac{\partial u^{(j)}}{\partial x}+(3-\kappa)\frac{\partial v^{(j)}}{\partial y}\right]$$
$$+(\kappa-1)\gamma\left[\frac{\partial u^{(j)}}{\partial y}+\frac{\partial v^{(j)}}{\partial x}\right]=-\frac{(\kappa-1)\rho_0\omega^2}{\mu_0}u^{(j)} \tag{3-5a}$$

$$(1+\kappa)\frac{\partial^2 v^{(j)}}{\partial y^2}+(\kappa-1)\frac{\partial^2 v^{(j)}}{\partial x^2}+2\frac{\partial^2 u^{(j)}}{\partial x\partial y}+(\kappa-1)\beta\left[\frac{\partial u^{(j)}}{\partial y}+\frac{\partial v^{(j)}}{\partial x}\right]$$

$$+\gamma\left[(1+\kappa)\frac{\partial v^{(j)}}{\partial y}+(3-\kappa)\frac{\partial u^{(j)}}{\partial x}\right]=-\frac{(\kappa-1)\rho_0\omega^2}{\mu_0}v^{(j)}, \quad (j=1,2) \tag{3-5b}$$

The system of above governing equations (3-5) is solved by using the Fourier transform technique. The general expressions for displacements components can be written as

$$\begin{cases} u^{(1)}(x,y)=\dfrac{1}{2\pi}\displaystyle\int_{-\infty}^{\infty}\sum_{j=1}^{2}A_j(s)e^{-\lambda_j y}e^{-isx}ds \\ v^{(1)}(x,y)=\dfrac{1}{2\pi}\displaystyle\int_{-\infty}^{\infty}\sum_{j=1}^{2}m_j(s)A_j(s)e^{-\lambda_j y}e^{-isx}ds \end{cases} \tag{3-6a}$$

$$\begin{cases} u^{(2)}(x,y)=\dfrac{1}{2\pi}\displaystyle\int_{-\infty}^{\infty}\sum_{j=3}^{4}A_j(s)e^{-\lambda_j y}e^{-isx}ds \\ v^{(2)}(x,y)=\dfrac{1}{2\pi}\displaystyle\int_{-\infty}^{\infty}\sum_{j=3}^{4}m_j(s)A_j(s)e^{-\lambda_j y}e^{-isx}ds \end{cases} \tag{3-6b}$$

According to equations (3-4), the stress components are written as

$$\begin{cases} \sigma_{yy}^{(1)}(x,y)=\dfrac{\mu_0 e^{\beta x+\gamma y}}{2\pi(\kappa-1)}\displaystyle\int_{-\infty}^{\infty}\sum_{j=1}^{2}\left[-(\kappa+1)m_j(s)\lambda_j-is(3-\kappa)\right]A_j(s)e^{-\lambda_j y}e^{-isx}ds \\ \tau_{xy}^{(1)}(x,y)=\dfrac{\mu_0 e^{\beta x+\gamma y}}{2\pi}\displaystyle\int_{-\infty}^{\infty}\sum_{j-1}^{2}\left[-\lambda_j-im_j(s)s\right]A_j(s)e^{-\lambda_j y}e^{-isx}ds \end{cases} \tag{3-7a}$$

$$\begin{cases} \sigma_{yy}^{(2)}(x,y)=\dfrac{\mu_0 e^{\beta x+\gamma y}}{2\pi(\kappa-1)}\displaystyle\int_{-\infty}^{\infty}\sum_{j=3}^{4}\left[-(\kappa+1)m_j(s)\lambda_j-is(3-\kappa)\right]A_j(s)e^{-\lambda_j y}e^{-isx}ds \\ \tau_{xy}^{(2)}(x,y)=\dfrac{\mu_0 e^{\beta x+\gamma y}}{2\pi}\displaystyle\int_{-\infty}^{\infty}\sum_{j=3}^{4}\left[-\lambda_j-im_j(s)s\right]A_j(s)e^{-\lambda_j y}e^{-isx}ds \end{cases} \tag{3-7b}$$

where s is the transform variable, $A_j(s)$, ($j=1,2,3,4$) are unknowns, and $\lambda_j(s)$ (j=1, 2, 3, 4) are the roots of the characteristic equation

$$\lambda^4+\Delta_1\lambda^3+\Delta_2\lambda^2+\Delta_3\lambda+\Delta_4=0 \tag{3-8}$$

where

$$\Delta_1=2\gamma \tag{3-9a}$$

$$\Delta_2 = -2s^2 - 2i\beta s + \frac{2c^2\kappa + (\kappa-3)\beta^2 + (1+\kappa)\gamma^2}{1+\kappa} \tag{3-9b}$$

$$\Delta_3 = -2\gamma s^2 - \frac{8i\beta\gamma s}{1+\kappa} + \frac{2c^2\kappa\gamma}{1+\kappa} \tag{3-9c}$$

$$\Delta_4 = s^4 + 2i\beta s^3 - \frac{2c^2\kappa + (\kappa+1)\beta^2 + (\kappa-3)\gamma^2}{\kappa+1}s^2 - \frac{2ic^2\kappa\beta}{1+\kappa}s + \frac{\kappa-1}{1+\kappa}c^4 \tag{3-9d}$$

For each root $\lambda_j(s)$, $m_j(s)$ (j=1, 2, 3, 4) are expressed as

$$m_j(s) = \frac{-(\kappa+1)s^2 + (\kappa-1)\lambda_j^2 - is\beta(\kappa+1) - \gamma(\kappa-1)\lambda_j + (\kappa-1)c^2}{-2is\lambda_j + \beta(3-\kappa)\lambda_j + is\gamma(\kappa-1)} \tag{3-10}$$

where $c = \omega / c_{sh}$ and c_{sh} is the shear wave speed of the non-homogeneous medium given by $c_{sh} = \sqrt{\mu/\rho}$. From equations (3-6) to (3-7), it can be seen that there are four unknown constants $A_1(s)$, $A_2(s)$, $A_3(s)$ and $A_4(s)$ which can be obtained from the boundary conditions.

To solve the present problem, the jumps of the displacement across the crack surfaces can be defined as follows

$$f_1(x) = u^{(1)}(x,0) - u^{(2)}(x,0) \tag{3-11a}$$

$$f_2(x) = v^{(1)}(x,0) - v^{(2)}(x,0) \tag{3-11b}$$

Applying the Fourier transform and according to the boundary conditions (3-4)-(3-6), we have

$$[X_1]\begin{bmatrix} A_1(s) \\ A_2(s) \end{bmatrix} = [X_2]\begin{bmatrix} A_3(s) \\ A_4(s) \end{bmatrix} \tag{3-12a}$$

$$[X_3]\begin{bmatrix} A_1(s) \\ A_2(s) \end{bmatrix} - [X_4]\begin{bmatrix} A_3(s) \\ A_4(s) \end{bmatrix} = \begin{bmatrix} \bar{f}_1(s) \\ \bar{f}_2(s) \end{bmatrix} \tag{3-12b}$$

where

$$[X_1] = \begin{bmatrix} \dfrac{-(\kappa+1)m_1(s)\lambda_1 - is(3-\kappa)}{\kappa-1} & \dfrac{-(\kappa+1)m_2(s)\lambda_2 - is(3-\kappa)}{\kappa-1} \\ -\lambda_1 - im_1(s)s & -\lambda_2 - im_2(s)s \end{bmatrix} \tag{3-13a}$$

$$[X_2]=\begin{bmatrix} \dfrac{-(\kappa+1)m_3(s)\lambda_3-is(3-\kappa)}{\kappa-1} & \dfrac{-(\kappa+1)m_4(s)\lambda_4-is(3-\kappa)}{\kappa-1} \\ -\lambda_3-im_3(s)s & -\lambda_4-im_4(s)s \end{bmatrix} \tag{3-13b}$$

$$[X_3]=\begin{bmatrix} 1 & 1 \\ m_1(s) & m_2(s) \end{bmatrix} \tag{3-13c}$$

$$[X_4]=\begin{bmatrix} 1 & 1 \\ m_3(s) & m_4(s) \end{bmatrix} \tag{3-13d}$$

A superposed bar indicates the Fourier transform in this paper. It is defined as follows

$$\bar{f}(s)=\int_{-\infty}^{\infty} f(x)e^{isx}dx \quad , \quad f(x)=\frac{1}{2\pi}\int_{-\infty}^{\infty}\bar{f}(s)e^{-isx}ds \tag{3-14}$$

By solving four equations (3-12) with four unknown functions, substituting the solutions into equations (3-7a) and utilizing the boundary conditions, we can obtain

$$\sigma_{yy}^{(1)}(x,0)=\frac{\mu_0 e^{\beta x}}{2\pi}\int_{-\infty}^{\infty}\left[d_1(s)\bar{f}_1(s)+d_2(s)\bar{f}_2(s)\right]e^{-isx}ds=-\sigma_0(x) \quad , \quad 0\le|x|\le a \tag{3-15a}$$

$$\int_{-\infty}^{\infty}\bar{f}_1(s)e^{-isx}ds=0\,,\ |x|>a \tag{3-15b}$$

$$\tau_{xy}^{(1)}(x,0)=\frac{\mu_0 e^{\beta x}}{2\pi}\int_{-\infty}^{\infty}\left[d_3(s)\bar{f}_1(s)+d_4(s)\bar{f}_2(s)\right]e^{-isx}ds=0 \quad , \quad 0\le|x|\le a \tag{3-16a}$$

$$\int_{-\infty}^{\infty}\bar{f}_2(s)e^{-isx}ds=0\,,\ |x|>a \tag{3-16b}$$

where d_1 (s), d_2 (s), d_3 (s) and d_4 (s) are known functions, and can be expressed as follows

$$[X_5]=[X_3]-[X_4][X_2]^{-1}[X_1] \tag{3-17a}$$

$$\begin{bmatrix} d_1(s) & d_2(s) \\ d_3(s) & d_4(s) \end{bmatrix}=[X_1][X_5]^{-1} \tag{3-17b}$$

To determine the unknown functions $\bar{f}_1(s)$ and $\bar{f}_1(s)$, the above two pairs of dual integral equations (3-15) and (3-16) must be solved. The process of solution can be found in the paper of Ma et al. (2005b) in detail. Then the crack-tip stress intensity factor can be obtained.

3.2. Numerical Results and Discussions

This section presents some numerical results given by Ma et al. (2005b) which disussed the dynamic behavior of the functionally graded materials with a crack for the plane strain state and $\upsilon = 0.3$ was assumed. The normalized stress intensity factors are defined by

$$k_j(\pm a) = K_j(\pm a)/\sigma_0\sqrt{\pi a}\ ,\ (\ j = I, II\ ) \tag{3-18}$$

The variations of the normalized dynamic stress intensity factor (DSIF) with the angle θ for $\delta a = 1.0$, $\omega = 0$ are shown in Figure 3-2 (Ma et al., 2005b). These results are very similar to Konda's results (Konda and Erdogan, 1994). The interesting and surprising result from these figures is that the maximum values of K_I and K_{II} do not generally correspond to the limiting crack orientations $\theta = 0$ and $\theta = \pi/2$. Duo to the complexity of the problem, two special cases, namely, $\theta = 0$ and $\theta = \pi / 2$ are discussed by Ma et al. (2005).

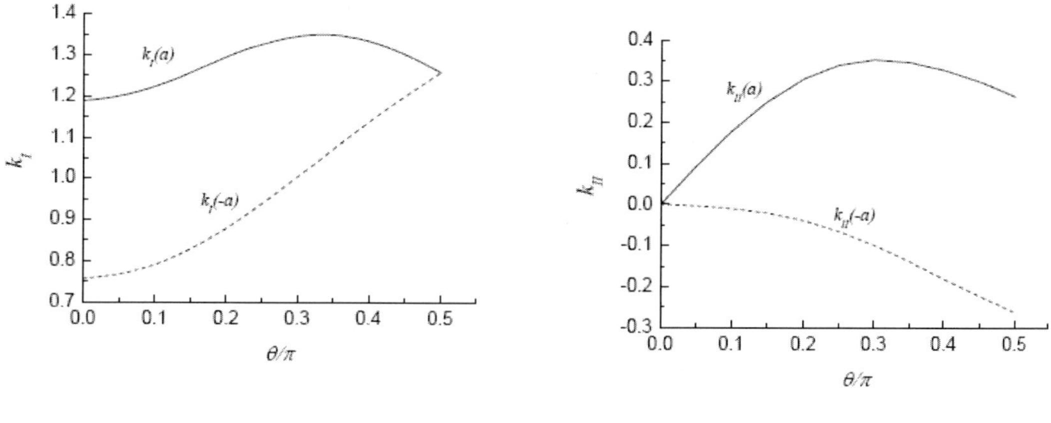

Figure 3.2. Variation of normalized stress intensity factor with the crack orientation in at $\delta a = 1.0$ and $\omega = 0$ (Ma et al. 2005b).

Figures 3-3 show the variations of the normalized stress intensity factor k_I and k_{II} with normal wave number $\omega a/c_{sh}$ for $\theta = \pi/2$ (Ma et al., 2005b). In this case, $x = 0$ is a plane of symmetry, the crack problem is one of mixed mode and the mode II component becomes quit significant for larger values of δa. Fluctuation in the dynamic stress intensity factors of FGMs for the harmonic stress wave exhibits a similar behavior to the homogeneous materials, but the peak values of DSIFs appear earlier than those corresponding to the homogeneous materials. And with the increase of gradient parameter δa, the peak values of DSIF will increase.

Figures 3-4 show the variations of the normalized stress intensity factor k_I with normal wave number $\omega a/c_{sh}$ for $\theta = 0$ at $\delta a = 0.1$ and $\delta a = 0.25$ (Ma et al., 2005b). In this case, the mode II stress intensity factor k_{II} is so smaller than k_I and can be omitted. These conclusions can also be found in Konda's paper (Konda and Erdogan, 1994). It is also apparent from these figures that the predominant term of the DSIFs of the crack tip in the stiffer side $k_I(a)$ is greater than that in the softer side $k_I(-a)$. And the peak values of DSIFs are much larger than the corresponding static values (i.e. $\omega = 0$).

From the figures 3-2 to 3-4, it can also be seen that, the normalization stress intensity factors k_I and k_{II} in increase until reaching a peak value and then come to decrease with the increase in the circular frequency of the incident wave.

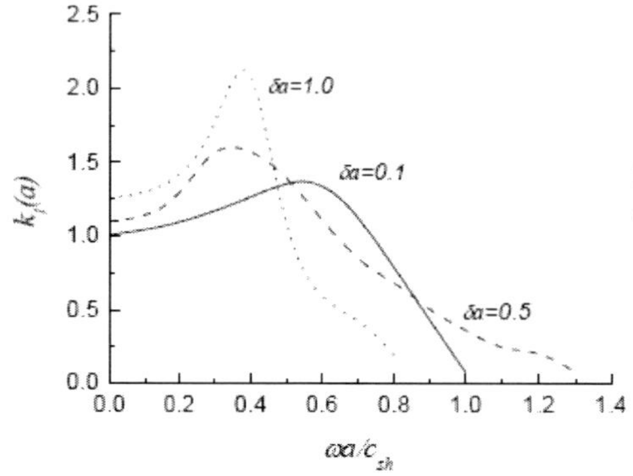

Figure 3.3. Variation of normalized stress intensity factor with normal wave number for $\theta = \pi / 2$ (Ma et al. 2005b).

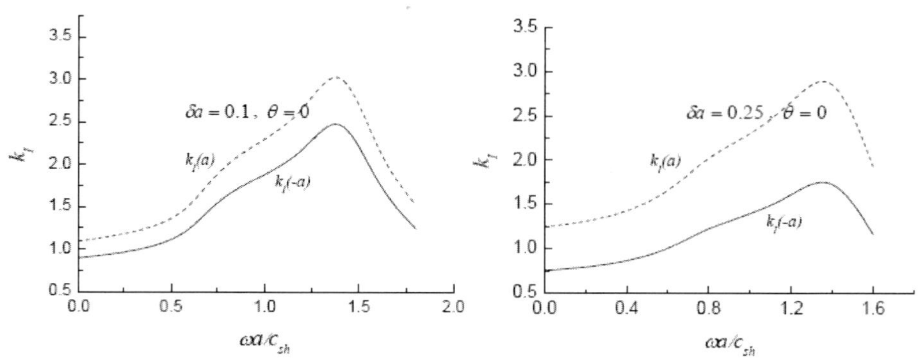

Figure 3.4. Variation of normalized stress intensity factor k_I with normal wave number $\omega a / c_{sh}$ (Ma et al. 2005b).

4. Numerical Analysis of Dynamic Crack Problems in FGMs

The eXtended Finite Element Method (XFEM) is a methodology for modeling cracks of arbitrary geometry in a finite element method without remeshing. It originated in Belytschko and Black (1999) and Moës et al. (1999). The state-of-the-art of X-FEM for computational fracture mechanics can be found in excellent review papers of Belytschko et al. (2009) and Abdelaziz et al. (2009).

Dynamic crack propagation is an application domain for which XFEM is particularly suitable because the most prevalent method for treating crack growth, remeshing, is very awkward for these problems. In most dynamic crack propagation problems, the crack advances over a large part of the mesh, so that remeshing would need to be performed many times. Remeshing is especially daunting for fragmentation simulations, since such simulations would entail very extensive remeshing. Furthermore, the continuity of the solution is quite important in any numerical model, and even with excellent projection schemes, significant discontinuities can be introduced in the velocities, stresses and displacements by remeshing.

The present section addresses some computational issues concerning the failure analysis of FGMs in a two-dimensional setting (Ma et al., 2010). Focusing on quasi-brittle FGMs, endowed with spatially varying elastic and fracture properties, the XFEM is used to simulate the crack propagation. The XFEM is capable of modeling crack growth without remeshing and possesses advantages over conventional method. This feature is especially important for FGMs, since the gradation of the mechanical properties may lead to complex propagation paths also in simple symmetric tests

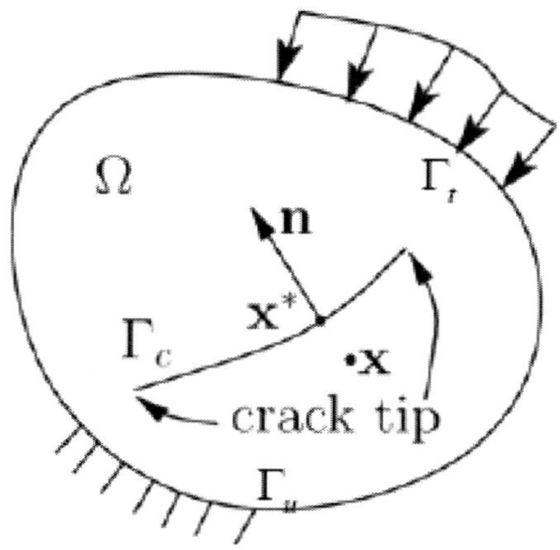

Figure 4.1. Domain Ω of a solid body containing a crack Γ_c .

4.1. Extended Finite Element Method

The extended finite element method (XFEM) is a numerical technique for modeling discontinuities by local enrichment functions in the area of interest. For the sake of subsequent discussion, let us consider a solid body occupying the domain Ω and containing a crack Γ_c (Figure 4-1). In the XFEM formulation by Moës et al. (1999), two types of enrichment, that is, the Heaviside and crack-tip enrichment are included to represent the displacement jump across the crack and the singular near-tip displacement field around the crack-tip.

The enriched interpolation function for the displacement is written as

$$u^h(\mathrm{x}) = \sum_{I\in N} N_I(\mathrm{x})\mathrm{u}_I + \sum_{J\in N^\Gamma} \tilde{N}_J(\mathrm{x})H_J(\mathrm{x})\mathrm{a}_J + \sum_{K\in N^\Lambda} \tilde{N}_K(\mathrm{x})\sum_{\alpha=1}^{4} B_{\alpha K}(\mathrm{x})\mathrm{b}_{\alpha K} \tag{4-1}$$

where N is the set of conventional (not enriched) nodes, N^Γ is the set of nodes enriched with discontinuous enrichment and N^Λ is the set of nodes enriched with asymptotic enrichment. The determination of N^Γ and N^Λ is done according to the following rules: N^Γ is the set of nodes whose support is entirely split by the crack; N^Λ is the set of nodes which contain the crack tip in the support of their shape functions. $N_I(\mathrm{x})$ is the standard interpolation function used in conventional finite elements, $H(\mathrm{x})$ is the Heaviside step function and the functions B span the space of the asymptotic crack tip fields. The u_I' are the unknown standard displacement degree of freedom associated with node I, the a_J's are the unknown enriched coefficients associated with the discontinuous enrichment function H_J active on node J and defined by

$$H_J(\mathrm{x}) = H(\mathrm{x}) - H(\mathrm{x}_J) \tag{4-2}$$

where, $H(\mathrm{x})$ is the Heaviside step function which takes on the vale $+1$ above the crack and -1 below the crack:

$$H(x) = \begin{cases} +1 \\ -1 \; if \; (\mathrm{x} - \mathrm{x}^*) \cdot \mathrm{n} \geq 0 \\ \quad otherwise \end{cases} \tag{4-3}$$

where x is a sample point, x^* (lies on the crack) is the closest point projection of x, and n is the unit outward normal to the crack at x^*.

Finally, $\mathrm{b}_{\alpha K}$ are additional, enrichment degrees of freedom associated with the enrichment function $B_{\alpha K}$ active at node K and defined by

$$B_{\alpha K}(\mathrm{x}) = B_\alpha(\mathrm{x}) - B_\alpha(\mathrm{x}_K) \tag{4-4}$$

The crack tip enrichment functions B_α are obtained from the asymptotic displacement fields. Based on the assumption of local homogenization of asymptotic crack tip fields in FGMs (Kim and Paulino, 2005, Eischen, 1987), B_α can be written as

$$B \equiv [B_1, B_2, B_3, B_4] = \left[\sqrt{r}\sin\frac{\theta}{2}, \sqrt{r}\cos\frac{\theta}{2}, \sqrt{r}\sin\frac{\theta}{2}\cos\theta, \sqrt{r}\cos\frac{\theta}{2}\cos\theta \right] \tag{4-5}$$

Here, r and θ are polar coordinates in the local crack tip coordinate system.

Four-node quadrilateral elements are used exclusively in Ma's (2010) implementation for two-dimensional situations, for which more elaborate considerations are necessary since an element can be cut by a crack either completely or partially. For an element that is completely cut by a crack, all the nodes of the element are enriched. An element that is partially cut by a crack contains the crack tip in its interior. For such an element, the nodes on the edge toward which the crack is heading are not enriched an only the remaining node is enriched.

For elements cut by a crack, the standard Gauss quadrature is not adequate due to discontinuities that exist inside the elements (Sukumar, Prevost, 2003). To accurately evaluate the integrals resulting from the virtual work statement, we divide the element domain into subdomains of triangular shape and then apply a quadrature formula to each subdomain. Depiction in detail can be obtained in the master thesis of Phu (2005).

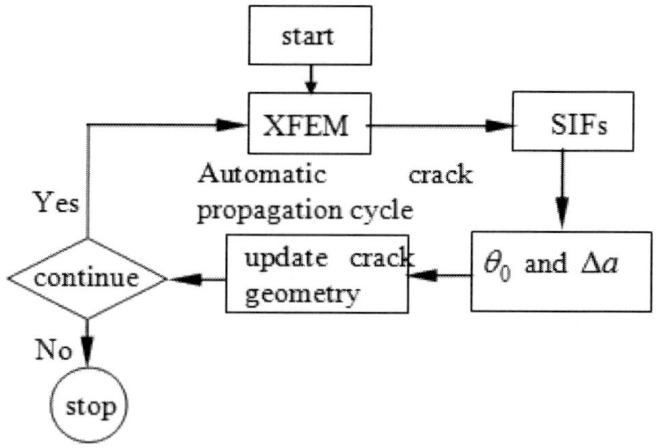

Figure 4.2. Automatic crack propagation procedure used in the XFEM code (Ma et al., 2010).

XFEM based crack growth simulation involves a series of steps. Figure 4-2 illustrates crack propagation procedure at each step (Ma et al., 2010). At each crack propagation cycle, crack initiation condition must be assessed for determining whether the crack will growth or not. If the crack driving force is big enough for the crack tip fields to reach the fracture envelope, then the crack will grow.

4.2. Crack Propagation Criterion

The singularity and angular functions of asymptotic crack tip field for FGMs are the same as for homogeneous materials Eischen (1987). Thus local homogenization arguments may allow the use of fracture criteria originally developed for homogeneous materials. Here we adopt the maximum hoop stress criterion (Erdogan and Sih, 1963, Kim and Paulino, 2007) to gives the crack growth direction. This criterion states that the crack will propagate from its tip in the direction θ_0 where the circumferential (hoop) stress $\sigma_{\theta\theta}$ is maximum. Therefore the crack initiation angle θ_0 is obtained from

$$\partial\sigma_{\theta\theta}/\partial\theta = 0 \, , \; \partial^2\sigma_{\theta\theta}/\partial\theta^2 < 0 \Rightarrow \theta = \theta_0 \tag{4-6}$$

Then the critical angle θ_0 is given by

$$\theta_0 = 2\tan^{-1}\left[\frac{1}{4}\left(\frac{K_I}{K_{II}} \pm \sqrt{\left(\frac{K_I}{K_{II}}\right)^2 + 8}\right)\right] \tag{4-7}$$

If $K_{II} = 0$ then $\theta_0 = 0$ (pure mode I) and by noting that if $K_{II} > 0$, the crack growth angle $\theta_0 < 0$, and if $K_{II} < 0$ then $\theta_0 > 0$.

Once the crack initiation angle is determined, the crack initiation condition is obtained as

$$\sqrt{2\pi r}(\sigma_{\theta\theta}) = \cos\frac{\theta_0}{2}\left[K_I\cos^2\frac{\theta_0}{2} - \frac{3}{2}K_{II}\sin\theta_0\right] \tag{4-8}$$

For pure mode I, when K_{II} and θ_0 are equal to zero, K_I can be replaced by mode I fracture toughness function of FGMs $K_{Ic}(\mathrm{x})$, therefore

$$\sqrt{2\pi r}(\sigma_{\theta\theta}) = K_{Ic}(\mathrm{x}) \tag{4-9}$$

Substitution of Eq.(4-8) into Eq.(4-9) yield the crack initiation condition (Erdogan and Sih, 1963, Kim and Paulino, 2007)

$$\cos\frac{\theta_0}{2}\left[K_I\cos^2\frac{\theta_0}{2} - \frac{3}{2}K_{II}\sin\theta_0\right] = K_{Ic}(\mathrm{x}) \tag{4-10}$$

4.3. The Interaction Integral Methods

The interaction integral method is an accurate scheme to evaluate stress intensity factors (SIFs) in FGMs (Dolbow and Gosz, 2002). Here we adopt a non-equilibrium formulation, which uses displacement and strain fields developed for homogeneous materials, and employ the non-equilibrium stress fields $\sigma_{ij}^{aux} = C_{ijkl}(x)\varepsilon_{kl}^{aux}$, where $C_{ijkl}(x)$ is the FGM stiffness tensor, σ_{ij}^{aux} is the auxiliary stress, and ε_{kl}^{aux} is the auxiliary strain. The interaction integral is derived from the path-independent J-integral for two admissible states (actual and auxiliary) of a cracked elastic FGM body. The interaction integral, based on the non-equilibrium formulation, is obtained as (Dolbow and Gosz, 2002)

$$I = \int_\Omega (\sigma_{ij}u_{i,1}^{aux} + \sigma_{ij}^{aux}u_{i,1} - \sigma_{ik}\varepsilon_{ik}^{aux}\delta_{1j})q_{,j}d\Omega + \int_\Omega \left[\sigma_{ij}(u_{i,1j}^{aux} - \varepsilon_{ij,1}^{aux})\right]q d\Omega - \int_\Omega \left[C_{ijkl,1}\varepsilon_{kl}\varepsilon_{ij}^{aux}\right]q d\Omega \tag{4-11}$$

where u_i, ε_{ij} and σ_{ij} are the actual displacement, strain and stress fields, respectively; q is a weight function which varies from 1 on the inner contour to 0 on the outer contour. The relationship between interaction integral and SIFs (K_I and K_{II}) is given by

$$I = 2(K_I K_I^{aux} + K_{II}K_{II}^{aux})/E_{tip}^* \tag{4-12}$$

where $E_{tip}^* = E_{tip}$ for plane stress and $E_{tip}^* = E_{tip}/(1-\upsilon^2)$ for plane strain. The stress intensity factor K_I and K_{II} associated with the actual fields can therefore be obtained through a judicious choice of the auxiliary fields. For example, if we choose the auxiliary fields such that $K_I^{aux} = 1.0$ and $K_{II}^{aux} = 0.0$, we have

$$K_I = I^{(1)}E_{tip}^*/2 \tag{4-13}$$

The mode II stress intensity factor is determined in a similar fashion. The interaction integral therefore provide a straightforward means for decoupling K_I and K_{II} in a general mixed-mode problem.

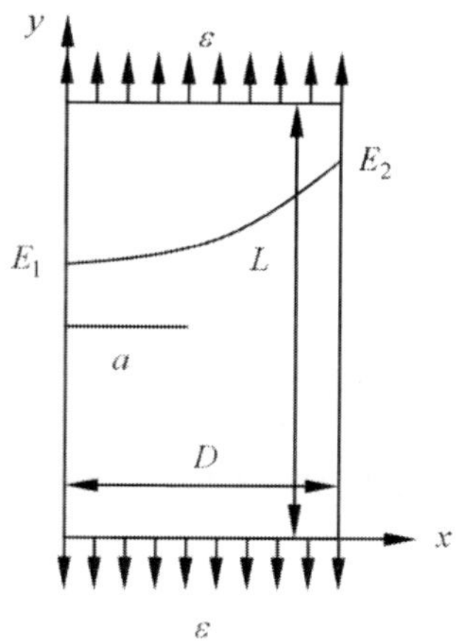

Figure 4.3. Edge-cracked FGM plate for the benchmark problem (Ma et al., 2010).

The benchmark problem considered by Erdogan and Wu (1997) was used in the work of Ma et al. (2010) to verify the XFEM implementation. In the problem depicted in Figure 4-3 (Ma et al., 2010), a rectangular FGM plate of height L and width D is subjected to a far-field uniaxial tensile strain ε. The plate contains, at its edge, a crack of length a, which is perpendicular to y-axis. Poisson's ratio was assumed as constant and the spatial variation of Young's modulus is given by

$$E(x) = C_1 e^{C_2 x} \tag{4-14}$$

where $C_1 = E_1$, $C_2 = \ln(E_2 / E_1) / D$.

Table 4.1. Normalized K_I values for various crack lengths and material properties (Ma et al., 2010)

E_2/E_1	$a/D = 0.4$		
	$K_I / \sigma\sqrt{\pi a}$	Erdogan and Wu (1997)	Error (%)
0.2	1.8379	1.8751	1.98
0.5	1.9615	1.9872	1.29
0.8	2.0452	2.0673	1.07
2.0	2.2587	2.2764	0.78
3.0	2.3791	2.3953	0.68
5.0	2.5583	2.5730	0.57
10.0	2.8608	2.8736	0.45

The problem models a FGM whose properties transition from material 1 to material 2. The quantities E_1 and E_2 in Eq.(4-14) are the Young's moduli for the two different materials. In the following, all results are reported for an uniform mesh of 50×90 quadrilateral elements. Stress intensity factors were calculated for crack lengths of $a/D = 0.4$ and various ratio E_2/E_1. The results are summarized in Table 4-1 (Ma et al., 2010). the table indicates a good comparison between the numerical and analytical results. The error ranges from 0.45% to 1.98% over all trials.

4.5. Crack Growth in Four Point Bending Fgms Beam

The efficiency and capability of the proposed procedure will be shown by a complex example. Rousseau and Tippur (2000) investigated crack growth behaviour of a crack normal to the material gradient in a graded glass/epoxy subjected to four point bending. Figure 4-4 shows specimen geometry and boundary conditions of the FGMs beam with a crack.

All the parameters (include the geometrical parameters, the material properties and the spatially varying fracture toughness) adopted in the work of Ma et al. (2010) were as same as reported by Rousseau and Tippur (2000) to simulated the crack propagating in four point bending FGMs beam. Figure 4-5 shows the experimental results given by Rousseau and Tippur (2000) and the crack trajectories obtained by the extended finite element method (Ma et al., 2010) at the initial and final steps, respectively, of crack propagation considering $\Delta a = 1.5mm$.

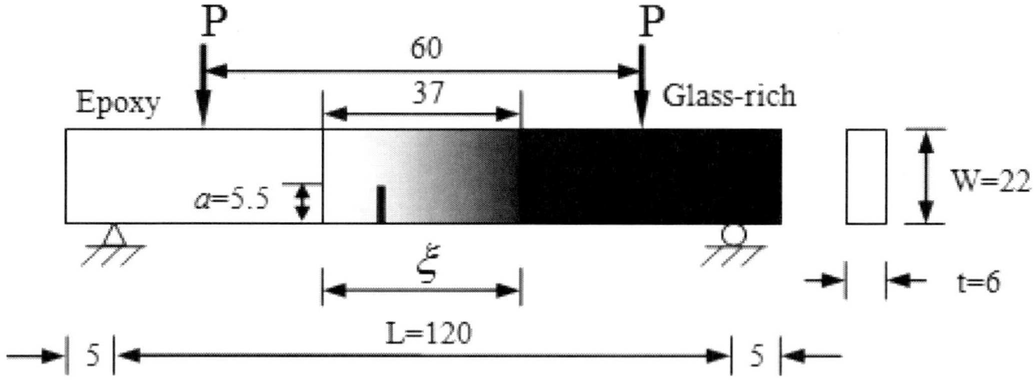

Figure 4.4. Geometry and boundary conditions of the epoxy/glass FGMs beam.

There is good agreement between two results. Moreover, experimental and numerical results for crack initiation angle at the initial step are in good agreement, i.e. $\theta_{\exp} = 7^\circ$ and $\theta_{num} = 6.86^\circ$, respectively

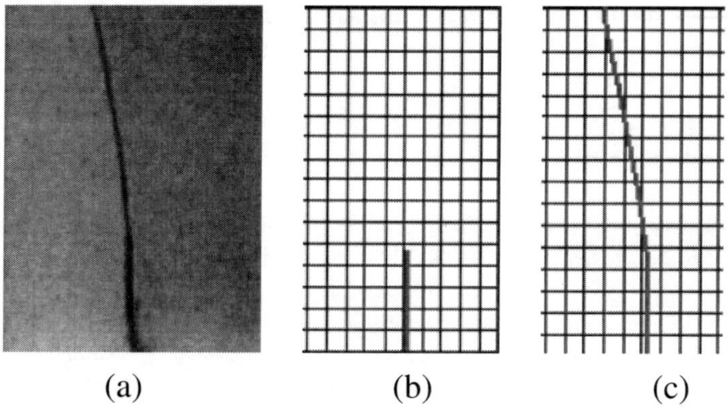

Figure 4-5. Crack trajectories obtained by experimental and numerical method: (a) experimental results given by Rousseau and Tippur (2000); (b) initial step of crack propagation given by numerical simulation; (c) final step of crack propagation given by numerical simulation (Ma et al., 2010).

ACKNOWLEDGMENTS

The present work is supported by National Science Foundation of China under grant Nos. 10432030 and 10502017, the Program of Excellent Team in Harbin Institute of Technology. LM acknowledges the Program for New Century Excellent Talents in University under grant No. NCET-08-0152.

REFERENCES

Abanto-Bueno, J., and Lambros, J. Investigation of crack growth in functionally graded materials using digital image correlation. Engineering Fracture Mechanics 2002 Vol 69, 1695-1711.

Abanto-Bueno J, Lambros J. An experimental study of mixed mode crack initiation and growth in functionally graded materials. Experimental Mechanics 2006 Vol 46,179-196.

Abdelaziz Y., Nabbou A. and Hamouine A. A state-of-the-art review of the X-FEM for computational fracture mechanics, Applied Mathematical Modelling. 2009 (in press), doi:10.1016/j.apm.2009.02.010

Anlas, G., Lambros, J. and Santare M. H. Dominance of asymptotic crack tip fields in elastic functionally graded materials. International Journal of Fracture 2002 Vol 115, 193-204.

Atkinson, C. Some results on crack propagation in media with spatially varying elastic moduli. International Journal of Fracture 1975 Vol 11, 619-628.

Babaei, R., Lukasiewicz, S.A. Dynamic response of a crack in a functionally graded material between two dissimilar half places under anti-plane impact load. Engineering Fracture Mechanics 1998 Vol 60, 479-487.

Bao, G., Wang, L. Multiple crack in functionally graded ceramic/metal coatings. International Journal of Solids and Structures 1995 Vol 32:2853-2871

Belytschko, T. and Black, T. Elastic crack growth in finite elements with minimal remeshing. International Journal for Numerical Methods in Engineering 1999 Vol 45, 601–620.

Belytschko, T., Gracie, R. and Ventura, G. A review of extended/generalized finite element methods for material modeling Modelling and Simulation in Materials Science and Engineering 2009. Vol 17(4):043001

Bi, X.S., Cheng, J., Chen, X.L. Moving crack for functionally graded material in an infinite length strip under antiplane shear. Theoretical and Applied Fracture Mechanics 2003 Vol 39, 89-97.

Carpenter, R. D., Liang,W.W., Paulino, G. H., Gibeling, J. C., and Munir, Z. A. Fracture testing and analysis of a layered functionally graded Ti/TiB beam in 3-point bending. Materials Science Forum 1999 Vol 308-311, 837-842.

Chalivendra, V. B. and Shukla, A. Transient elastodynamic crack growth in functionally graded materials. ASME Journal of Applied Mechanics 2005 Vol 72, 237-248.

Chalivendra, V. B. Asymptotic analysis of transient curved crack in functionally graded materials. Inernational Journal of Solids and Structures 2007 Vol 16, 465-479.

Chen, J., Liu, Z., Zou, Z. Transient internal crack problem for a nonhomogeneous orthotropic strip (mode I). International Journal of Engineering Science 2002 Vol 40, 1761-1774.

Chen, J., Liu, Z., Zou, Z. Dynamic response of a crack in a functionally graded interface of two dissimilar piezoelectric half-planes. Archive of Applied Mechanics 2003 Vol 72, 686-696.

Chen, J., Liu, Z., Zou, Z. Electromechanical impact of a crack in a functionally graded piezoelectric medium. Theoretical and Applied Fracture Mechanics 2003 Vol 39, 47-60.

Chen, J. and Liu Z.-X. Transient response of a mode III crack in an orthotropic functionally graded strip. European Journal of Mechanics A/Solids. 2005. Vol 24, 325-336.

Choi, H.J. Impact response of a surface crack in a coating/substrate system with a functionally graded interlayer: antiplane deformation. International Journal of Solids and Structures 2004 Vol 41, 5631-5645.

Choi, H.J. 2006 Elastodynamical analysis of a crack at an arbitrary angle to the graded interfacial zone in bonded half-planes under antiplane shear impact. Mechanics Research Communications Vol 33, 636-650.

Delae, F., Erdogan, F. The crack problem for a nonhomogeneous plane. ASME Journal of Applied Mechanics 1983 Vol 50, 609-614.

Dineva, P.S., Rangelov, T.V. and Manolis, G.D. Elastic Wave Propagation in a Class of Cracked, Functionally Graded Materials by BIEM. Computational Mechanics 2007 Vol 39:293-308.

Dolbow, J. and Gosz, M. On the computation of mixed-mode stress intensity factors in functionally graded materials. International Journal of Solids and Structures 2002 Vol 39:2557-2574.

Eischen, J.W. Fracture of nonhomogeneous materials. International Journal of Fracture 1987 Vol 34, 3-22.

Erdelyi, A. (ed.) Table of the Integral Transforms Vol. 1, McGraw-Hill, 1954 New York.

Erdogan, F. and Sih G.C. On the crack extension in plates under plane loading and transverse shear. Journal of Basic Engineering 1963 Vol 85, 519-527.

Erdogan, F. Fracture mechanics of functionally graded materials, Journal of Composites Engineering 1995 Vol 5, 753-770.

Erdogan, F. and Wu, B.H. The surface crack problem for a plate with functionally graded properties. ASME Journalof Applied Mechanics 1997 Vol 64, 449-456.

Forth, S. C., Favrow, L. H., Keat, W. D., and Newman, J. A. Three dimensional mixed-mode fatigue crack growth in a functionally graded titanium alloy. Engineering Fracture Mechanics 2003 Vol 70, 2175-2185.

Fujimoto, T., and Noda, N. Crack propagation in a functionally graded plate under thermal shock. Archive of Applied Mechanics 2000 Vol 70, 377-386.

Gradshteyn, I.S. and Ryzhik, I.M. Table of Integral, Series and Products. Academic Press, 1980 New York.

Gu, P, Asaro, R.J. Crack deflection in functionally graded materials. International Journal of Solids and Structures 1997 Vol 34, 3085-3098.

Guo, L.-C., Wu, L.-Z., Zeng, T., Ma, L. Fracture analysis of a functionally graded coating-substrate structure with a crack perpendicular to the interface-part II: transient problem. International Journal of Fracture 2004 Vol 127, 39-59.

Guo, L.-C., Wu, L.-Z., Sun, Y.-G., Zeng, T., and Wang, L.-C. The transient response of a crack in a functionally graded strip under an in-plane impact load. ZAMM Zeitschrift für Angewandte Mathematik und Mechanik 2005 Vol 85, 711-720

Guo, L.-C., Wu, L.-Z., Sun, Y.-G. The transient fracture behavior for a functionally graded layered structure subjected to an in-plane impact load. Acta Mechanica Sinica 2005 Vol 21, 257-266.

Huang, G.Y., Wang, Y.S. New model for fracture analysis of a functionally graded interfacial zone under harmonic anti-plane loading. Engineering Fracture Mechanics 2004 Vol 71, 1841-1851.

Itou, S. Three dimensional waves propagation in a cracked elastic solid. ASME Journal of Applied Mechanics 1978 Vol 45, 807-811.

Itou, S. Stress intensity factor for a moving cylindrical crack in a nonhomogeneous cylindrical layer in composite materials. Archive of Applied Mechanics 2005 Vol 75, 18-30.

Jain, N. and Shukla, A. Asymptotic analysis and reflection photoelasticity to study transient crack propagation in graded materials. Journal of Mechanics of Materials and Structures 2007 Vol 2, 595-612.

Jiang, L.Y., Wang, X.D. On the dynamic propagation in an interphase with spatially varying elastic properties under inplane loading. International Journal of Fracture 2002 Vol 114, 225-244.

Jin, X., Wu, L.-Z., Guo, L.-C. and Sun, Y.-G. Prediction of the variation of elastic modulus in ZrO2/NiCr functionally graded materials. Composite Science and Technology 2009 Vol 69, 1587-1591.

Jin, X., Wu, L.-Z., Guo, L.-C., Yu, H.-J. and Sun, Y.-G. Experimental investigation of the mixed-mode crack propagation in ZrO2/NiCr functionally graded materials. Engineering Fracture Mechanics 2009 Vol 76, 1800-1810

Jin, Z.-H. and Batra R.C. Some basic fracture mechanics concepts in functionally graded materials. Journal of the Mechanics and Physics of Solids Vol 44:1221–34.

Jin, Z.-H., Paulino, G. H., and Dodds Jr., R. H. 2002 Finite element investigation of quasi-static crack growth in functionally graded materials using a novel cohesive zone fracture model. ASME Journal of Applied Mechanics 1996 Vol 69, 370-379.

Jin, Z.-H., Paulino, G. H. and Dodds Jr., R. H. Cohesive fracture modeling of elastic-plastic crack growth in functionally graded materials. Engineering Fracture Mechanics 2003 Vol 70, 1885-1912.

Kandula, S.S.V, Abanto-Bueno, J., Geubelle, P.H. and Lambros, J. Cohesive modeling of dynamic fracture in functionally graded materials. International Journal of Fracture 2005 Vol 132:275-296.

Kim J. -H. and Paulino G. H. T-stress, mixed-mode stress intensity factors, and crack initiation angles in functionally graded materials: a unified approach using the interaction integral method. Computer Methods in Applied Mechanics and Engineering 2003 Vol 192, 1463-1494.

Kim, J. -H. and Paulino, G. H. T-stress in orthotropic functionally graded materials: Lekhnitskii and Stroh formalisms. International Journal of Fracture 2004 Vol 126, 345-384.

Kim, J.-H., and Paulino, G. H. Simulation of crack propagation in functionally graded materials under mixed-mode and non-proportional loading. International Journal of Mechanics and Materials in Design 2004 Vol 1, 63-94.

Kim, J.-H. and Paulino, G.H. Mixed-mode crack propagation in functionally graded materials. Materials Science Forum 2005 Vol 492-493:409-414.

Kim, J.-H., Paulino, G.H. On Fracture Criteria for Mixed-Mode crack propagation in Functionally Graded Materials. Mechanics of Advanced Materials and Structures 2007 Vol 14, 227-244.

Kirugulige M.S. and Tippur H.V. Mixed-mode dynamic crack growth in functionally graded glass-filled epoxy. Experimental Mechanics 2006 Vol 6, 269–281.

Konda, N. and Erdogan, F. The mixed mode crack problem in a nonhomogeneous elastic medium. Engineering Fracture Mechanics 1994 Vol 47, 533-545.

Lambros J., Santare M.H., Li H., Sapna G.H. A novel technique for the fabrication of laboratory scale model functionally graded materials. Experimental Mechanics 1999 Vol 39, 184-190.

Lambros, J., Cheeseman, B. A. and Santare, M. H. Experimental investigation of the quasi-static fracture of functionally graded materials. International Journal of Solids and Structures 2000 Vol 37, 3715-3732.

Li, C.Y., Weng, G.J., Duan, Z.P. Dynamic behavior of a cylindrical crack in a functionally graded interlayer under torsional loading. International Journal of Solids and Structures 2001 Vol 38, 7473-7485.

Li, C.Y., Duan, Z.P., Zou, Z.Z. Torsional impact response of a penny-shaped interface crack in bonded materials with a graded materials interlayer. ASME Journal of Applied Mechanics 2002 Vol 69, 303-308.

Li, H., Lambros, J., Cheeseman, B. and Santare M. H. Experimental investigation of quasi-static fracture of functionally graded materials. International Journal of Solids and Structures 2000 Vol 37, 3715-3732.

Lin, C. Y., McShane, B., Lin H., and Rawlings, R. D. Structure and properties of functionally gradient aluminium alloy 2124/sic composites. Materials Science and Technology 1994 Vol 10, 659-664.

Ma, L., Wu, L.-Z., Guo, L.-C. Dynamic behavior of two collinear anti-plane shear cracks in a functionally graded layer bonded to dissimilar half planes. Mechanics Research Communications 2002 Vol 29, 207-215.

Ma, L., Wu, L.-Z., Zhou, Z.-G., Zeng, T. Crack propagating in a functionally graded strip under the plane loading. International Journal of Fracture 2004 Vol 126, 39-55.

Ma, L., Wu, L.-Z. and Zhou, Z.-G. Dynamic stress intensity factors around two parallel cracks in a functionally graded layer bonded to dissimilar half planes subjected to anti-plane incident time-harmonic stress wave. International Journal of Engineering Science 2004 Vol 42, 187-202.

Ma, L., Wu, L.-Z and Guo, L.-C. On the moving Griffith crack in a nonhomogeneous orthotropic strip. International Journal of Fracture 2005 Vol 136, 187-205.

Ma, L.,Wu, L.-Z., Guo, L.-C., Zhou, Z.-G. Dynamic behavior of a finite crack in the functionally graded materials. Mechanics of Materials 2005 Vol 37, 1153-1165.

Ma, L., Li, J., Abdelmoula, R. and Wu, L.-Z. Dynamic stress intensity factor for cracked functionally graded orthotropic medium under time-harmonic loading. European Journal of Mechanics A/Solids 2007 Vol 26, 325-336.

Ma, L., Wang Z.-Y. and Wu L.-Z. Wu. Numerical simulation of mixed-mode crack propagation in Functionally Graded Materials. Materials Science Forum 2010 Vol 631-632, 121-126.

Meguid, S.A., Wang, X.D., Jiang, L.Y. On the dynamic propagation of a finite crack in functionally graded materials. Engineering Fracture Mechanics 2002 Vol 69, 1753-1768.

Moës N, Dolbow J, Belytschko T A finite element method for crack growth without remeshing. International Journal for Numerical Methods in Engineering 1999 Vol 46(1):131-156

Moon, R. J., Hoffman, M., Hilden, J., Bowman, K. J., Trumble, K. P., and Rodel, J. R-curve behavior in alumina-zirconia composites with repeating graded layers. Engineering Fracture Mechanics 2002 Vol 69, 1647-1665.

Morse, P.M. and Feshbach, H. Methods of Theoretical Physics, Vol. 1. McGraw-Hill, 1958 New York.

Nakagaki, M., Sasaki, H., Hagihara, S. A study of crack in functionally graded material under dynamic loading. ASME Pressure Vessels and Piping Division (Publication) PVP, Dynamic Fracture, Failure and Deformation 1995 Vol 300, 1-6.

Parameswaran, V. and Shukla, A. Dynamic fracture of a functionally gradient material having discrete property variation. Journal of Materials Science 1998 Vol 33, 3303-3311.

Parameswaran, V. and Shukla, A. Processing and characterization of a model functionally graded material. Journal of Material Science 1998 Vol 35, 21-29.

Parameswaran, V. and Shukla, A. Crack-tip stress fields for dynamic fracture in functionally gradient materials. Mechanics of Materials 1999 Vol 31, 579-596.

Phu, N. V. An object oriented approach to the extended finite element method with applications to fracture mechanics. Master Thesis. Hochiminh City University of Technology 2005.

Rousseau, C. -E. and Tippur, H. V. Influence of elastic variations on crack initiation in functionally graded glass-filled epoxy. Engineering Fracture Mechanics 2002 Vol 69, 1679-1693.

Rousseau, C.-E. and Tippur, H.V. Compositionally graded materials with cracks normal to the elastic gradient. ACTA Materialia 2000 Vol 48, 4021-4033.

Shukla A., Jain N. and Chona R. A Review of Dynamic Fracture Studies in Functionally Graded Materials. Strain 2007 Vol 43, 76-95.

Sih, G.C. Mechanics of Fracture 4: Elastodynamic Crack Problems. Noordhoff International Publishing, 1977 Leyden.

Sladek J., Sladek V. and Zhang C. The MLPG method for crack analysis in anisotropic functionally graded mmaterials. Structure Integrity Durability 2005 Vol 1, 131-144.

Sukumar N, Chopp DL, Moës N, Belytschko T Modeling holes and inclusions by level sets in the extended finite-element method. Computer Methods in Applied Mechanics and Engineering 2001 Vol 190:6183–6200

Sukumar, N. and Prevost, J. H. Modeling quasi-static crack growth with the extended finite element method Part I: Computer implementation. International Journal of Solids and Structures 2003 Vol 40(26), 7513-7537.

Sukumar N, Huang ZY, Prevost J-H, Suo Z Partition of unity enrichment for bimaterial interface cracks. International Journal for Numerical Methods in Engineering 2004 Vol 59:1075–1102

Suresh, S. and Mortensen, A. Functionally Graded Materials. The Institute of Materials, IOM Communications Ltd, 1998 London.

Tilbrook, M. T., Moon, R. J. and Hoffman, M. Curved crack propagation in homogeneous and graded materials. Fatigue and Fracture of Engineering Materials and Structures 2005. Vol 28, 939-950.

Tilbrook, M,T,, Rozendurg, K., Steffler, E.D., Rutgers, L. and Hoffman, M. Crack propagation path in layered, graded composites. Composites Part B 2006 Vol 37:490-498

Uzun, H., Lindley, T. C., McShane, H. B., and Rawlings, R. D. Fatigue crack growth behavior of 2124/SiC/1Op functionally graded materials. Metallurgical and Materials Transactions A 2001 Vol 32A, 1831-1839.

Wang, Z. and Nakamura, T. Simulations of crack propagation in elastic-plastic graded materials. Mechanics of Materials 2004 Vol 36, 601-622.

Wang, B.L., Han, J.C., Du, S.Y. Crack problem for non-homogeneous composite materials subjected to dynamic loading. Inernational Journal of Solids and Structures 2000 Vol 37, 1251-1274.

Wang, B.L., Mai, Y.W. A periodic array of cracks in functionally graded materials subjected to transient loading. International Journal of Engineering Science 2006 Vol 44, 351-364.

Wang, X.D., Meguid, S.A. On the dynamic crack propagation in an interface with spatially varying elastic properties. International Journal of Fracture 1994/1995 Vol 69, 87-99.

Yau, W.F. Axisymmetric slipless indentation of an infinite elastic cylinder. SIAA Journal of Applied Mathematics 1967 Vol 15, 219-227.

Yoffe, E.H. The moving Griffith crack. Philosophical Magazine 1951 Vol 7, 739-750.

Zhang, C. Sladek, J. and Sladek, V. Effects of material gradients on transient dynamic mode-III stress intensity factors in a FGM. Inernational Journal of Solids and Structures 2003 Vol 40, 5251-5270.

Zhang, Z., and Paulino, G. H. Cohesive zone modeling of dynamic failure in homogeneous and functionally graded materials. International Journal of Plasticity 2005 Vol 21, 1195-1254.

ABOUT THE AUTHORS

Dr. Li MA received his Ph.D. degree in Engineering Mechanics from Harbin Institute of Technology, P.R. China in 2004. He is an Associate Professor of Center for Composite Materials, Harbin Institute of Technology, China and has been on the faculty there since 2003. Dr. MA's current research interests are focused on mechanical behavior of advanced composite materials and structures, fracture mechanics in functionally graded materials and piezoelectric materials. He has authored over 20 peer-reviewed scientific articles on mechanics of materials.

Linzhi Wu received the Ph.D. degree from Harbin Institute of Technology, China, 1992. He is a Full Professor of the Center for Composite Materials at Harbin Institute of Technology, Harbin, P.R.China. His research interests include meso-mechanics, fracture mechanics of functionally graded materials, design and analysis of composite structures, braided composites, ultra-light composite truss structures. He is the editors of the following journals: Acta Mechanica Sinica, Acta Mechanica Solida Sinica (in Chinese), Advances in Mechanics (in Chinese).

In: Mechanics of Functionally Graded Materials and Structures ISBN 978-1-61668-920-9
Editors: Z. Zhong, L. Wu and W. Chen

Chapter 10

MODELING METHOD FOR THE STATIC CRACK PROBLEMS OF FUNCTIONALLY GRADED MATERIALS WITH GENERAL PROPERTIES

Licheng Guo[1], Linzhi Wu and Hongjun Yu

Department of Astronautic Science and Mechanics and Center for Composite Materials, Harbin Institute of Technology, Harbin, China

ABSTRACT

In this chapter, the modeling methods for the crack problems of functionally graded materials (FGMs) with general (or arbitrary) properties are reviewed and discussed. The focus of this chapter is on the analytical and numerical modeling methods for the crack problem of FGMs with general properties. Firstly, the basic problem (a "particular" crack problem of a functionally graded layered structure with exponential properties and a crack crossing the interface) for establishing the piecewise-exponential model (PE model) is presented. Then the PE model for FGMs with general mechanical properties is introduced. In the PE model, the FGM is divided into some nonhomogeneous layers along the gradient direction of the properties, with each layer's properties varying exponentially. By this method, the real material properties can be approached by a series of exponential functions. Moreover, considering the significance of the investigation of the thermal fracture behavior for FGMs, the PE model is extended to the crack problems of FGMs with general thermomechanical properties. It should be mentioned that: in past decades, there are very limited literatures on the analytical study of crack problems in the FGMs with non-exponential properties. However, for the FGMs with exponential properties, there have been lots of investigations regarding many types of crack problems. These achievements will provide a wide support for the PE model to be applied in solving many types of crack problems in FGMs with general properties. On the other hand, the analytical modeling methods for the crack problems of FGMs with general properties are very difficult to consider the influences of the complex internal structures (such as the component particles and interfaces) on the fracture behavior of FGMs. Therefore, numerical modeling methods for the crack problems of FGMs with general properties should be developed together with the analytical methods. The key problem in

[1] e-mail: guolc@hit.edu.cn.

the numerical modeling methods lies in how to obtain the crack tip quantities with high efficiency. An interaction integral and its associated domain form without any derivatives of the material properties are introduced. It can be used to analyze the crack problems of FGMs with general properties, considering the effects of the component particles and internal interfaces on the fracture behavior.

1. INTRODUCTION

In the past decades, FGMs with continuous variation in composition have attracted a significant amount of attention. Compared with the traditional homogeneous materials, FGMs can be used to reduce the mismatch of material properties and improve the bonding strength and toughness in some important structures such as the coating-substrate systems. At the same time, lots of challenging problems have been met in the fracture behaviors of functionally graded materials. Since the actual material properties of FGMs may be arbitrarily distributed in spatial position, the fracture study of the modeling methods for FGMs with arbitrary properties is especially significant.

Many crack problems in FGMs have been studied in the past years. The square root character of the crack-tip stress field in FGMs with continuously varying properties has been revealed by Delale and Erdogan (1983) and Jin and Noda (1994). By assuming the material moduli to be exponential functions, Erdogan and his coworkers (seeing Delale and Erdogan, 1983; Chen and Erdogan, 1996; Erdogan and Wu, 1997) studied a series of crack problems in FGMs. Noda and Jin (1993) and Jin and Noda (1994) investigated the thermal fracture behaviors of FGMs. In functionally graded structures, mainly the crack parallel to the surface or interface (including the interface crack) and that vertical to the interface or surface have been investigated. For FGMs with crack(s) parallel to the surface or interface, many literatures can be found, e.g. Chen and Erdogan (1996), Li et al. (2001), Guo et al. (2004a), Chen (2005). For FGMs with crack(s) vertical to the interface or free surface of FGM structures, some important studies have been conducted: Erdogan and Wu (1996, 1997) investigated the isotropic functionally graded strip under thermal and mechanical loading, respectively; Chen et al. (2002) and Guo et al.(2004b) studied the dynamic and static crack problem for a functionally graded orthotropic strip. The functionally graded layered structures with crack(s) vertical to the interface have been studied by Choi et al. (1996 and 1998), Ueda (2001) and Guo et al.(2005). Besides these, Konda and Erdogan (1994) and Choi et al. (2001) studied the arbitrarily oriented crack in FGMs. There are also some other types of researches. Ueda and Shindo (2000) investigated the crack kinking problem in FGMs subjected a constant initial strain resulting from stress relaxation. The cylindrical FGM layer with a circumferential crack normal to the surfaces was investigated by Kadioglu et al. (1998) by introducing an elastic foundation. By means of the Schmidt method, the elastic wave problems in FGMs have been investigated by Zhou et al.(2004) and Ma et al. (2005). A periodic array of cracks in an infinite functionally graded material under transient mechanical loading was considered by Wang and Mai (2006).

It should be mentioned that to solve the crack problems analytically, most of the above theoretical investigations assumed the material properties to be exponential functions. From the viewpoint of fracture mechanics, the governing partially differential equations for FGMs are usually very difficult to solve analytically when the material properties vary spatially. Actually, very few functions can be used as material properties so that the analytical solutions can be obtained. When the properties are assumed to be exponential functions, the governing partially differential equations can be reduced to the general partially differential equations with constant coefficients and the analytical solutions are easy to obtain. That is why most of the theoretical analysis of crack problems in FGMs has been conducted by assuming the properties to be exponential functions. Furthermore, a wider range of crack problem can be dealt with by using exponential functions than using other functions to depict the material properties.

Since it is nearly impossible to obtain the analytical solutions for the crack problems in FGMs with properties arbitrarily distributed in spatial position, some approximated methods must be developed. For the crack problems of FGMs with general properties, there are mainly several analytically approximated methods. 1) Homogenous multi-layered model (HM model): it was adopted by Wang et al. (2000, 2002) and Itou (2001). In their work, FGMs are divided into a number of homogeneous layers with constant properties along the thickness direction. 2) Linear multi-layered model (LM model): it was developed by Huang et al. (2004) and Wang and Gross (2000). In their model, the nonhomogeneous materials are divided into some sub-layers along the thickness direction with the material properties varying linearly in each layer. 3) Piecewise exponential model (PE mode): it was proposed by (Guo and Noda, 2007 and Guo et al., 2008a). In the PE model, the functionally graded material is divided into some nonhomogeneous layers along the gradient direction of the properties with each layer having properties varying exponentially. By this method, the real material properties can be approached by a series of exponential functions. 4) A multi-layered model was employed by Zhong and Cheng (2008) to model arbitrary variations of material properties based on two linear-distributed material softness parameters. These methods are very significant for solving the crack problems of FGMs with arbitrary properties. For HM model, the advantage is simple and convenient to deal with many types of crack problems of FGMs. However, in the HM model the material properties are still discontinuous between each sub-layer. It is well known that there is usually large stress gradient in the stress fields of crack-tip zone and therefore the stress gradient may be strongly influenced by the variation of local nonhomogeneity of the properties, especially for the FGMs with large gradient of material properties. Since the crack tips in the HM model are located in homogenous materials, for the case with large gradient of the properties, the influences of local material nonhomogeneity on the crack-tip fields may not be adequately allowed for and at the same time the calculating efficiency may decrease. For the other three models, the influences of the local nonhomogeneity on the crack-tip fields can be fully considered.

Except analytical investigations, Butcher et al. (1999), Rousseau and Tippur (2001) and Yao et al. (2007) conducted experimental studies of the fracture behaviors of FGMs. On the other hand, the numerical technique is one of the most convenient and reliable methods to determine the fracture parameters. Some important numerical simulations regarding the fracture behaviors of FGMs have been conducted by many researchers, including elastic two dimensional crack problems (Kim and Paulino, 2002; Chi and Chung, 2003), semi-elliptical surface crack problems in three-dimensional FGMs (Walters et al., 2004), numerical simulations of crack propagation problems in elastic-plastic FGMs (Wang and Nakamura, 2004). Among the available numerical methods, J contour integral (Rice, 1968) has generated great interest for its convenience in solving the parameters characterizing crack-tip fields. The contour integral has been proved to be a conservation integral for homogeneous materials. Eischen (1987) proved that for nonhomogeneous materials with continuous and generally differentiable properties, the stress and strain singularity near a crack tip is identical as the well-known inverse square root stress singularity in homogeneous materials. A modified integral J_1^* (Eischen, 1987) and a path-independent J_e-integral (Honein and Herrmann, 1997) were introduced for nonhomogeneous materials. Smelser and Gurtin (1977) proved J-integral to be still path-independent for a straight interface crack. Rice (1988) and Hutchinson and Suo (1992) conducted a lot of investigations on the interface crack. Jin and Sun (2007) provided a mathematically rigorous proof that for nonhomogeneous materials, both J_1^* and J_e are the (potential) energy release rates.

In order to obtain mode I and mode II stress intensity factors (SIFs) separately, an interaction (energy) contour integral method (Stern et al., 1976) was derived from the J-integral by considering a composition of two admissible states (the actual fields and known auxiliary fields). Nakamura (1991) used an equivalent domain integral of the interaction integral to evaluate mixed-mode SIFs along straight three-dimensional interface cracks. The same method was employed to deal with curved three-dimensional interface crack problems (Nahta and Moran, 1993; Gosz et al., 1998). Gosz and Moran (2002) developed the interaction integral method for non-planar three-dimensional crack problems. Dolbow and Gosz (2002) introduced the interaction integral method to compute mixed-mode SIFs for two-dimensional crack problems in FGMs with their properties varying continuously. In comparison with modified J-integral for nonhomogeneous materials, they found that the interaction integral is more convenient since it does not require the evaluation of strain energy densities along the traction-free crack faces. For isotropic FGMs, Kim and Paulino (2003) computed the mixed-mode SIFs and T-stress by means of the interaction integral method in combination with the finite element method (FEM). Rao and Rahman (2003) employed the method to analyze mixed-mode crack problems in orthotropic FGMs. Kim and Paulino (2005) gave a summary on three definitions of the auxiliary fields and discussed how to extract mixed-mode SIFs and T-stress for isotropic and orthotropic FGMs, respectively. Krysl and Belytschko (1999) utilized the interaction integral method in conjunction with the element free Galerkin method to investigate three-dimensional stationary and propagating crack problems. Johnson and Qu (2007) extended the interaction integral to calculate the SIFs of three dimensional curvilinear cracks in a homogeneous body or on a bimaterial interface under non-uniform temperatures. Yu et al. (2009) proposed a new domain expression of the interaction integral which does not contain any derivatives of material properties parameters and hence it is still valid even when the integral domain contains material interfaces.

In the following parts, firstly, the basic problem (a "particular" crack problem of a functionly graded layered structure with exponential properties and a crack crossing the interface) for establishing the PE model is presented. Then the PE model for FGMs with general mechanical properties is introduced. Furtherly, the PE model is extended to the crack problems of FGMs with general thermomechanical properties. On the other hand, in the analytical modeling methods for the crack problems of FGMs with general properties, it is very difficult to consider the influences of the complex internal structures (such as the component particles and interfaces) on the fracture behavior of FGMs. Therefore, numerical modeling methods for the crack problems of FGMs with general properties should be developed together with the analytical methods. An interaction integral and its associated domain form without any derivatives of the material properties is paid attention to. Finally, some representative examples are analyzed.

2. A Particular Model: The Crack Problem for a Functionally Graded Layered Structure with a Crack Crossing the Interface

2.1. Analytical Procedure for a Functionally Grade Layered Structure with a Crack Crossing the Interface

In the past decades, many investigations have been conducted on the crack problems in functionally graded structures. However, almost all of them have been focused on the crack(s) along the interface or in either a functionally graded layer or a homogeneous layer related to the functionally graded medium. Very few analytical studies have been conducted on the crack crossing the interface in functionally graded structures. Several numerical analyses have been conducted on this problem (Bleeck et al., 1998 and Chi and Chung, 2003). For actual cases, the fracture related failures in FGMs may take place in various ways including the cracking across the interface.

In this part, the aims to study a functionally grade layered structure with a crack crossing the interface include two points:

1) It is a significant problem related to the fracture behavior of functionally graded structure. A crack propagating across the interface between different layers in PSZ/superalloy FGMs has been observed by Kawasaki and Watanabe (1996). This type of crack also was met in thermal shock performance experiments of W/Cu FGM, see Ling et al.(2003). Actually, cracks vertical to the interface or free surface have been frequently met as one of the main factors resulting in fracture (Kawasaki and Watanabe, 2002). When the crack tips get to the interface, it is possible for the crack to cross the interface and expand into the next layer.

2) It is a basic problem for establishing a general model for the crack problems of FGMs with general properties. In order to establish a general fracture mechanics model for FGMs with general properties, namely the piecewise exponential model (PE model), the crack problem for a functionally graded layered structure with a crack crossing the interface must be solved firstly.

This problem has been studied by Guo and Noda (2008). As stated in part 1, it is more convenient to solve the crack problem analytically when the material properties are descibed by exponential functions than by other types of functions. Therefore, the moduli are still assumed to be exponential functions in this basic problem as shown in Figure 2.1 (Guo and Noda, 2008). Due to this assumption, the model for functionally graded layered structure is "particular". In Figure 2.1, both of two layers may be nonhomogeneous and the mechanical properties are continuous across the interface. The layers are infinite in the y direction.

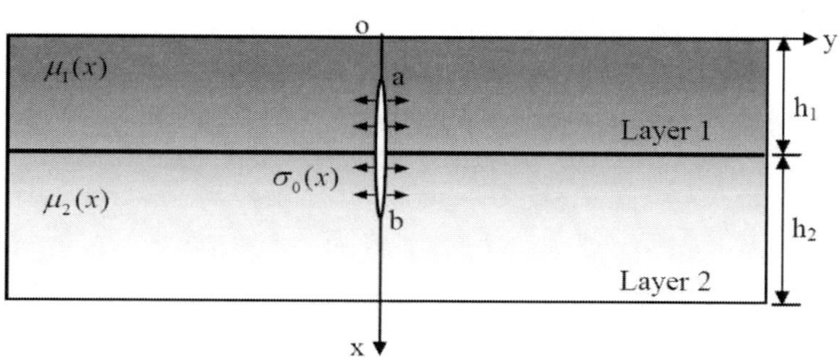

Figure 2.1. Schematic of a functionally graded layered structure with a crack crossing the interface (Guo and Noda, 2008).

The constitutive equations and governing equations for both layrs can be written as

$$\begin{cases} \sigma_{nxx}(x,y)=\dfrac{\mu_n(x)}{\kappa_n(x)-1}\left\{[1+\kappa_n(x)]\dfrac{\partial u_n(x,y)}{\partial x}+[3-\kappa_n(x)]\dfrac{\partial v_n(x,y)}{\partial y}\right\} \\ \sigma_{nyy}(x,y)=\dfrac{\mu_n(x)}{\kappa_n(x)-1}\left\{[1+\kappa_n(x)]\dfrac{\partial v_n(x,y)}{\partial y}+[3-\kappa_n(x)]\dfrac{\partial u_n(x,y)}{\partial x}\right\} \\ \sigma_{nxy}(x,y)=\mu_n(x)\left[\dfrac{\partial u_n(x,y)}{\partial y}+\dfrac{\partial v_n(x,y)}{\partial x}\right] \end{cases} \quad n=1,2 \tag{2-1}$$

$$\begin{cases} \dfrac{\partial\sigma_{nxx}(x,y)}{\partial x}+\dfrac{\partial\sigma_{nxy}(x,y)}{\partial y}=0 \\ \dfrac{\partial\sigma_{nyy}(x,y)}{\partial y}+\dfrac{\partial\sigma_{nxy}(x,y)}{\partial x}=0 \end{cases} \quad n=1,2 \tag{2-2}$$

where, $\mu_n(x)$ ($n=1,2$) are shear moduli of two FGM layers, $\kappa_n(x)=3-4\upsilon_n(x)$ for plane strain and $\kappa_n(x)=[3-\upsilon_n(x)]/[1+\upsilon_n(x)]$ for plane stress, $\upsilon_n(x)$ are the Poisson's ratio. According to the previous work (Delale and Erdogan, 1983; Erdogan and Wu, 1997), the Poisson's ratio is assumed to be a constant.

The crack-face loading condition is given as

$$\sigma_{nyy}(x,0)=-\sigma_0(x) \quad a<x<b \tag{2-3}$$

Here, $\sigma_0(x)$ is the crack face traction described by a known function. It should be noted that in condition (2-3), $n=1$ for the part of the crack lying in Layer 1 and $n=2$ for the part of the crack lying in Layer 2.

The shear moduli of both FGM layers can be written as, respectively,

$$\mu_1(x)=\mu_{01}e^{\delta_1 x} \qquad 0\le x\le h_1 \tag{2-4}$$

$$\mu_2(x)=\mu_{02}e^{\delta_2 x} \qquad h_1\le x\le h_1+h_2 \tag{2-5}$$

Here, μ_{01}, μ_{02}, δ_1 and δ_2 are material constants. Due to the continuity of the material properties at the interface, we have

$$\mu_{02}=\mu_{01}e^{(\delta_1-\delta_2)h_1} \tag{2-6}$$

Then, the displacement and stress components can be obtained as

$$\begin{cases} u_n(x,y)=\dfrac{1}{2\pi}\displaystyle\int_{-\infty}^{\infty}\sum_{j=1}^{2}E_{n1j}(s)A_{n1j}(s)e^{\lambda_{n1j}(s)y-isx}ds+\dfrac{2}{\pi}\int_0^{\infty}\sum_{j=1}^{4}E_{n2j}(\alpha)A_{n2j}(\alpha)e^{\lambda_{n2j}(\alpha)y}\cos(\alpha y)d\alpha \\ v_n(x,y)=\dfrac{1}{2\pi}\displaystyle\int_{-\infty}^{\infty}\sum_{j=1}^{2}F_{n1j}(s)A_{n1j}(s)e^{\lambda_{n1j}(s)y-isx}ds+\dfrac{2}{\pi}\int_0^{\infty}\sum_{j=1}^{4}F_{n2j}(\alpha)A_{n2j}(\alpha)e^{\lambda_{n2j}(\alpha)y}\sin(\alpha y)d\alpha \end{cases} \tag{2-7}$$

$$\begin{cases} \sigma_{nxx}(x,y) = e^{\delta_n x}\left[\dfrac{1}{2\pi}\int_{-\infty}^{\infty}\sum_{j=1}^{2} B_{n1j}(s)A_{n1j}(s)e^{\lambda_{n1j}(s)y-isx}ds + \dfrac{2}{\pi}\int_0^{\infty}\sum_{j=1}^{4} B_{n2j}(\alpha)A_{n2j}(\alpha)e^{\lambda_{n2j}(\alpha)x}\cos(\alpha y)d\alpha\right] \\ \sigma_{nyy}(x,y) = e^{\delta_n x}\left[\dfrac{1}{2\pi}\int_{-\infty}^{\infty}\sum_{j=1}^{2} C_{n1j}(s)A_{n1j}(s)e^{\lambda_{n1j}(s)y-isx}ds + \dfrac{2}{\pi}\int_0^{\infty}\sum_{j=1}^{4} C_{n2j}(\alpha)A_{n2j}(\alpha)e^{\lambda_{n2j}(\alpha)x}\cos(\alpha y)d\alpha\right] \\ \sigma_{nxy}(x,y,p) = e^{\delta_n x}\left[\dfrac{1}{2\pi}\int_{-\infty}^{\infty}\sum_{j=1}^{2} D_{n1j}(s)A_{n1j}(s)e^{\lambda_{n1j}(s)y-isx}ds + \dfrac{2}{\pi}\int_0^{\infty}\sum_{j=1}^{4} D_{n2j}(\alpha)A_{n2j}(\alpha)e^{\lambda_{n2j}(\alpha)x}\sin(\alpha y)d\alpha\right] \end{cases} \tag{2-8}$$

Here, s and α are the Fourier variables, and $A_{n1j}(s)$ ($n=1,2$; $j=1,2$) and $A_{n2j}(\alpha)$ ($n=1,2$; $j=1,\cdots,4$) are unknowns to be solved.

For further derivation, the following auxiliary function is introduced.

$$g(x) = \begin{cases} \dfrac{\partial v_1(x,0)}{\partial x} & 0 < x < h_1 \\ \dfrac{\partial v_2(x,0)}{\partial x} & h_1 \le x < h_1 + h_2 \end{cases} \tag{2-9}$$

By applying the boundary and continuity conditions, the unknowns A_{n1j} and $\{A_{n2}(\alpha)\}$ $(n=1,2)$ can be solved and expressed by the auxiliary function $g(u)$ as

$$A_{n1j}(s) = q_{nj}(s)\int_a^b g(u)e^{isu}du,\ n=1,2 \text{ and } j=1,2 \tag{2-10}$$

$$\begin{Bmatrix} \{A_{12}(\alpha)\} \\ \{A_{22}(\alpha)\} \end{Bmatrix}_{8\times1} = [\Phi(\alpha)]_{8\times8}^{-1}\int_a^b \begin{Bmatrix} \{\Gamma_0(u,\alpha)\} \\ \{\Gamma_2(u,\alpha)\} \\ \{\Gamma_1(u,\alpha)\} \end{Bmatrix}_{8\times1} g(u)du \tag{2-11}$$

Here $q_{nj}(s)$ ($n=1,2$) and $[\Phi(\alpha)]_{8\times8}$ are known matrix, and

$$\{A_{12}(\alpha)\} = \{A_{121}(\alpha) \quad A_{122}(\alpha) \quad A_{123}(\alpha) \quad A_{124}(\alpha)\}^T \tag{2-12}$$

$$\{A_{22}(\alpha)\} = \{A_{221}(\alpha) \quad A_{222}(\alpha) \quad A_{223}(\alpha) \quad A_{224}(\alpha)\}^T \tag{2-13}$$

$$\{\Gamma_0(u,\alpha)\} = \{R_{01}(u,\alpha) \quad R_{02}(u,\alpha)\}^T \tag{2-14}$$

$$\{\Gamma_1(u,\alpha)\} = \{R_{03}(u,\alpha) \quad R_{04}(u,\alpha)\}^T \tag{2-15}$$

$$\{\Gamma_2(u,\alpha)\} = \begin{Bmatrix} R_{13}(u,\alpha) - R_{23}(u,\alpha) \\ R_{14}(u,\alpha) - R_{24}(u,\alpha) \\ R_{11}(u,\alpha) - e^{(\delta_2-\delta_1)h_1}R_{21}(u,\alpha) \\ R_{12}(u,\alpha) - e^{(\delta_2-\delta_1)h_1}R_{22}(u,\alpha) \end{Bmatrix} \tag{2-16}$$

In the above expressions, it is necessary to simplify $R_{01}(u,\alpha)$, …, $R_{04}(u,\alpha)$ and $R_{n1}(u,\alpha)$, …, $R_{n4}(u,\alpha)$ so that further derivation can be continued.

Substituting $A_{n1j}(s)$ and $A_{n2j}(\alpha)$ into the crack face loading condition, a singular integral equation regarding the unknown auxiliary function $g(u)$ can be obtained as

$$\frac{1}{\pi}\int_a^b\left[\frac{w_{n11}}{x-u}+k_{n1}(u,x)+\pi h_{n2}(u,x)\right]g(u)du=-e^{\delta_n x}\sigma_0(x) \tag{2-17}$$

$g(u)$ can be solved numerically from the above singular integral equation. Then the crack-tip stress intensity factor can be obtained. More details can be found in Guo and Noda (2008).

2.2. Results and Analyses

The Young's moduli at the upper surface of Layer 1, the interface and the lower surface of Layer 2 are defined as E_{10}, E_{20}, and E_{30}, respectively. The Young's moduli of both layers can be written as

$$E_1(x)=E_{10}e^{\delta_1 x} \quad 0\le x\le h_1 \tag{2-18}$$

$$E_2(x)=E_{20}e^{\delta_2(x-h_1)} \quad h_1\le x\le h_1+h_2 \tag{2-19}$$

Thus, $\delta_1 h_1=\ln(E_{20}/E_{10})$ and $\delta_2 h_2=\ln(E_{30}/E_{20})$. A crack subjected to a uniform load σ_0 on the crack surface and plane strain state are considered. The SIFs are normalized by $k_0=\sigma_0\sqrt{a_0 h_1}$ with $a_0=(b-a)/(2h_1)$. The normalized x-coordinate of the crack center is defined as $c_0=(b+a)/(2h_1)$. $\nu=0.3$.

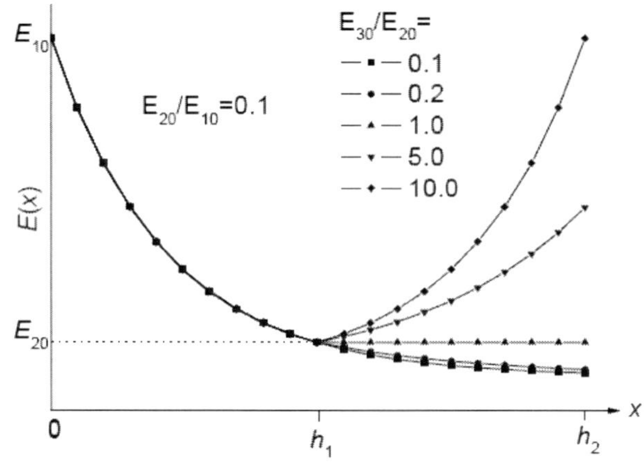

Figure 2.2(a). Schematic of a group of properties when E_{20}/E_{10}=0.1 and E_{30}/E_{20} increases from 0.1 to 10.0.

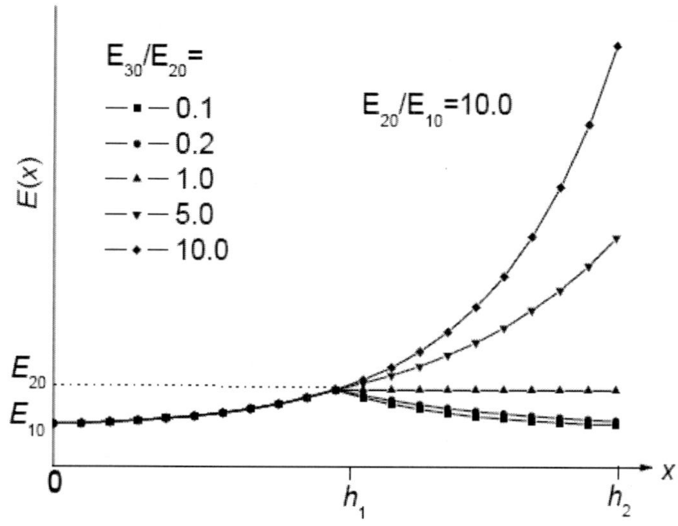

Figure 2.2(b). Schematic of a group of properties when E_{20} / E_{10} =10.0 and E_{30} / E_{20} increases from 0.1 to 10.0.

In Guo and Noda (2008), following three groups of nonhomogeneity parameters representing different distributing characteristics of the properties are used, see Figure 2.2. In Figure 2.2(a), $E_{20} / E_{10} = 0.1$ and E_{30} / E_{20} varies from 0.1 to 10.0; in Figure 2.2(b), E_{20} / E_{10} = 10.0 and E_{30} / E_{20} varies from 0.1 to 10.0; in Figure 2.2(c), E_{30} / E_{20} =1.0 and E_{20} / E_{10} varies from 0.1 to 10.0.

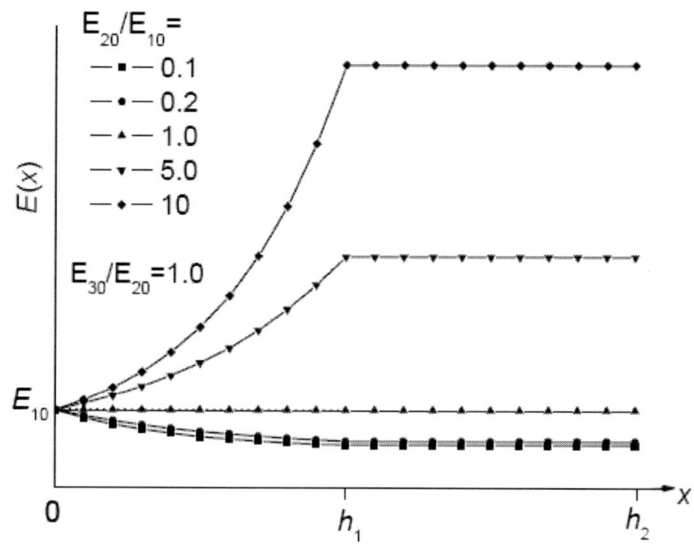

Figure 2.2(c). Schematic of a group of properties when E_{20} / E_{10} increases from 0.1 to 10.0 and E_{30} / E_{20} =1.0.

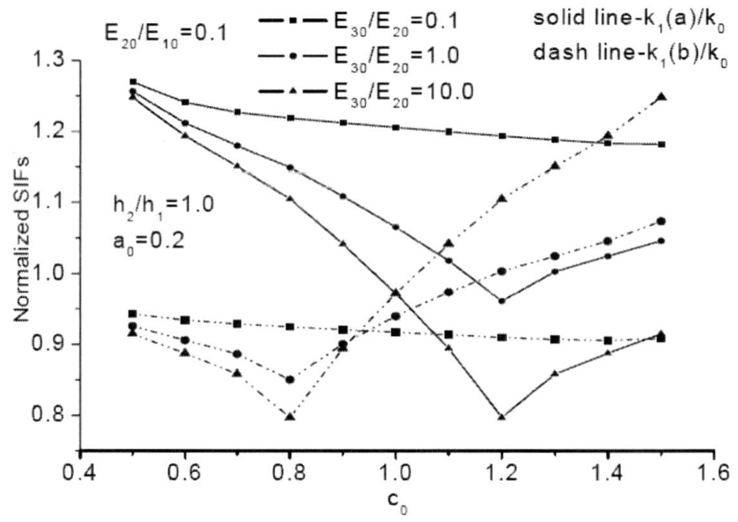

Figure 2.3. Variations of the SIFs with the crack center position c_0 when the crack moves from Layer 1 to Layer 2. (Guo and Noda, 2008)

In the following part, a significant problem, namely the crack-tip characteristics when the crack goes through the interface from the Layer 1 into Layer 2, is considered. The corresponding results are shown in Figure 2.3 and Figure 2.4 (Guo and Noda, 2008). To make the results understood more easily, some schematics of the structure for different crack center location are provided in Table 2.1 (Guo and Noda, 2008) to depict some important crack locations as the crack crosses the interface.

Figure 2.3 shows the variation of the SIFs with the crack center coordinate c_0 when E_{20}/E_{10}=0.1 and E_{30}/E_{20} increases from 0.1 to 10.0. The range of c_0 is chosen as 0.5 to 1.5. It means the crack moves from the center of Layer 1 ($c_0 = 0.5$) to the center of Layer 2 ($c_0 = 1.5$). Since $h_2/h_1 = 1$ and $a_0 = 0.2$, the crack-tip (b, 0) gets to the interface when $c_0 = 0.8$ and the crack-tip (a, 0) gets to the interface when $c_0 = 1.2$, see the schematics in Table 2.1. From Figure 2.3, when $E_{20}/E_{10} = E_{30}/E_{20}$=0.1, since Layer 1 and 2 have the same nonhomogeneity constant or material distribution, the SIFs have no discontinuities in slope when the crack tips go through the interface. When $E_{20}/E_{10} \neq E_{30}/E_{20}$, it can be observed that: $k_1(a)/k_0$ and $k_1(b)/k_0$ get to their extremum and kink when the crack tips get to the interface.

Table 2.1. Schematics of some particular crack locations in the structure for different crack center position when $a_0 = 0.2$ (Guo and Noda, 2008)

$c_0 = 0.5$	$c_0 = 0.8$	$c_0 = 1.0$	$c_0 = 1.2$	$c_0 = 1.5$
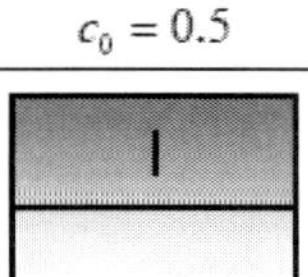	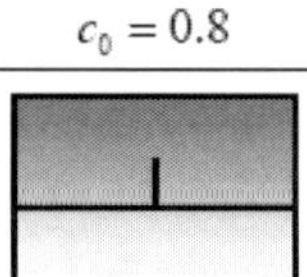	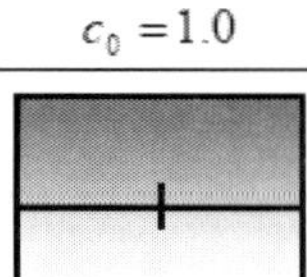	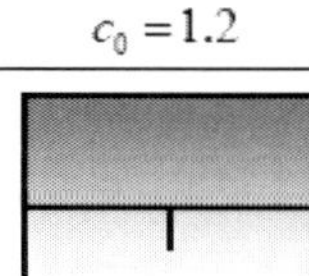	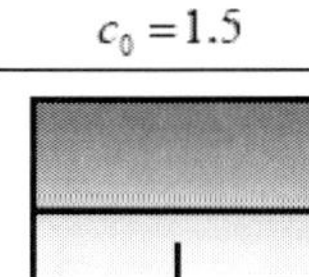

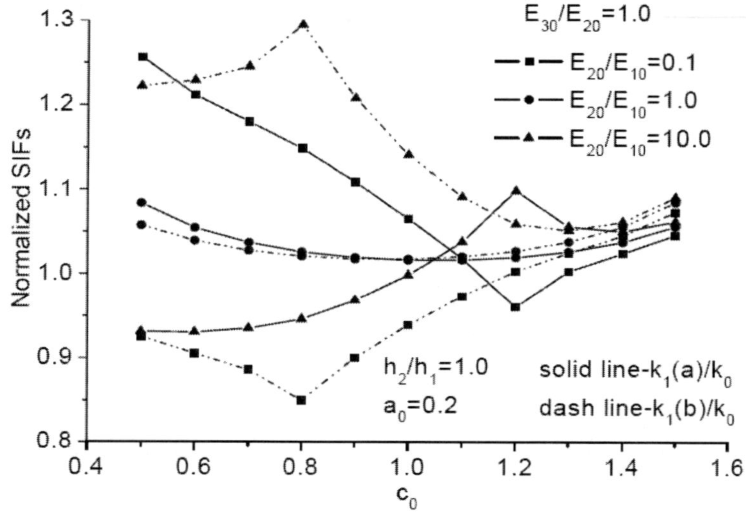

Figure 2.4. Variations of the SIFs with the crack center position c_0 as the crack moves from Layer 1 to Layer 2. (Guo and Noda, 2008)

Figure 2.4 depicts the variation of the SIFs with c_0 for a functionally graded coating-substrate structure. E_{30}/E_{20}=1.0 and E_{20}/E_{10} increases from 0.1 to 10.0. The range of c_0 is also assumed to be between 0.5 and 1.5. Firstly, when $E_{20}/E_{10}=E_{30}/E_{20}$=1.0, both Layer 1 and Layer 2 are the same homogeneous material. Under this circumstance, the SIFs have no discontinuities in slope when the crack tips go through the interface. Moreover, the SIF is larger for the crack tip closer to the free surface. Secondly, when $E_{20}/E_{10} \neq E_{30}/E_{20}$, it is observed that: The SIFs get to the extremum and kink when the crack tips get to the interface;

As a summary, an interesting phenomenon is found that if $E_{20}/E_{10} \neq E_{30}/E_{20}$, the SIFs usually get to the extremum and kink when the crack tips get to the interface. It is related to the distribution characteristics of the mechanical properties at the interface. We can believe that the crossing-interface crack problem is not only a basic problem for establishing the following piecewise-exponential model but also a particular interesting problem worth studying furtherly.

3. A General Model for FGMs with General Mechanical Properties: Piecewise-Exponential Model (PE Model)

3.1. Formulation of the Piecewise-Exponential Model (PE Model)

It is known that most of the analytical solutions for the crack problems of FGMs have been obtained by assuming the mechanical properties to be exponential functions. On the contrary, due to the difficulty in solving the differential equations analytically, very limited analytical solutions can be found for the crack problems of FGMs with non-exponential properties. Then, whether or not is it possible to solve the crack problem for FGMs with general properties with the help of the "particular" solutions for FGMs with exponential properties? Actually, it is possible to realize this if the general properties of FGMs can be approached by some piecewise exponential functions. In Part 2, the basic problem has been solved analytically for a crack crossing the interface of a functionally graded layered structure with exponential properties. Guo and Noda (2007) has applied this solutiong to establish the PE model.

The PE mode is originally proposed by Guo and Noda (2007). The schematic of a functionally graded strip with arbitrarily distributed properties along the thickness direction is shown in Figure 3.1. An internal crack is located perpendicularly to the free surfaces. The thicknesses of the strip is h. The crack length is c.

The constitutive equations can be written as

$$\begin{cases} \sigma_{xx}(x,y) = \dfrac{\mu(x)}{\kappa(x)-1}\left\{[1+\kappa(x)]\dfrac{\partial u(x,y)}{\partial x} + [3-\kappa(x)]\dfrac{\partial v(x,y)}{\partial y}\right\} \\ \sigma_{yy}(x,y) = \dfrac{\mu(x)}{\kappa(x)-1}\left\{[1+\kappa(x)]\dfrac{\partial v(x,y)}{\partial y} + [3-\kappa(x)]\dfrac{\partial u(x,y)}{\partial x}\right\} \\ \sigma_{xy}(x,y) = \mu(x)[\dfrac{\partial u(x,y)}{\partial y} + \dfrac{\partial v(x,y)}{\partial x}] \end{cases} \tag{3-1}$$

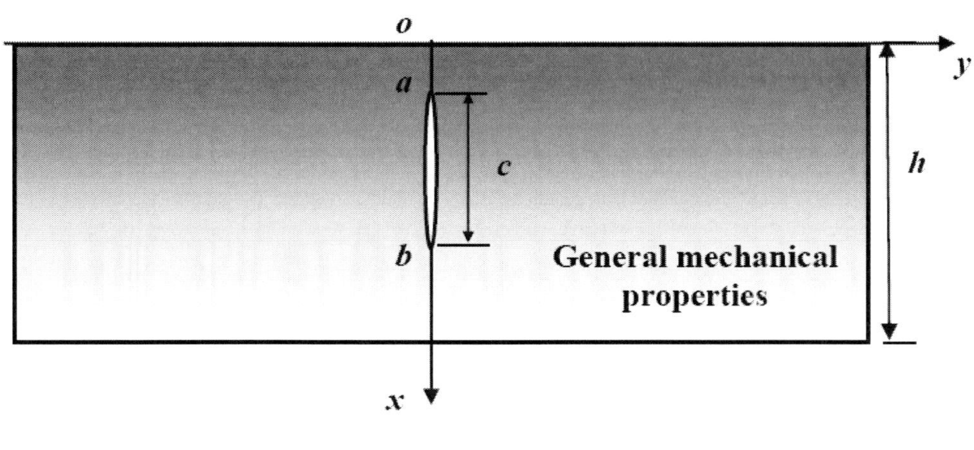

Figure 3.1. Schematic of a functionally graded strip with general mechanical properties.

The equabrium equations are

$$\sigma_{ij,j} = 0,\ i,j = x,y \tag{3-2}$$

When the actual material properties of the FGMs are arbitrarily distributed in the thickness direction, Eqs. (3-1) and (3-2) are nearly impossible to solve analytically. The PE model is proposed as shown in Figure 3.2. The graded medium is divided into M nonhomogeneous layers along the thickness direction with each layer's properties varying exponentially. By this way, the real material properties can be approached by a series of exponential functions. It should be noted that the real material properties are used on both surfaces of each nonhomogeneous layer in the PE model. Therefore, the nature of continuously varying properties of FGMs can be approached accurately. Figure 3.2 gives the schematic view of the PE model (Guo and Noda, 2007). In Figure 3.2(a), the functionally graded region is divided into M layers. The layers containing the crack tips are denoted by n_1 and n_2, respectively.

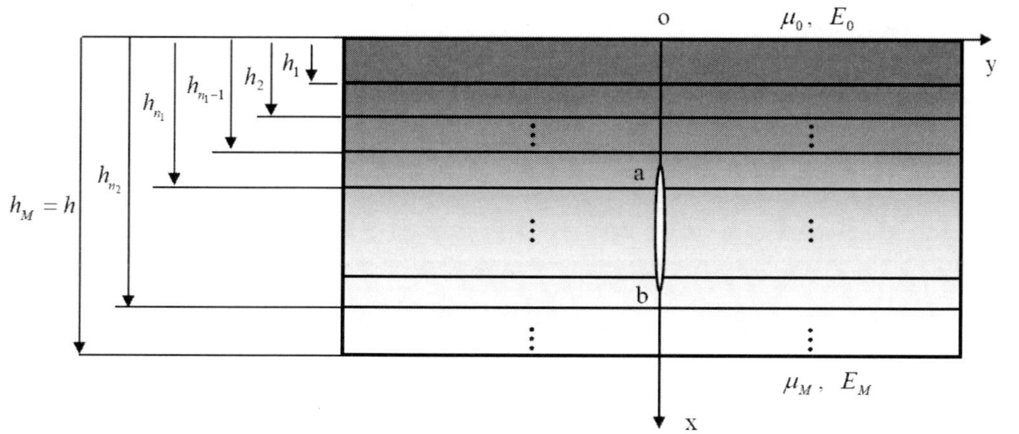

Figure 3.2(a). Schematic of the layers in the PE Model.

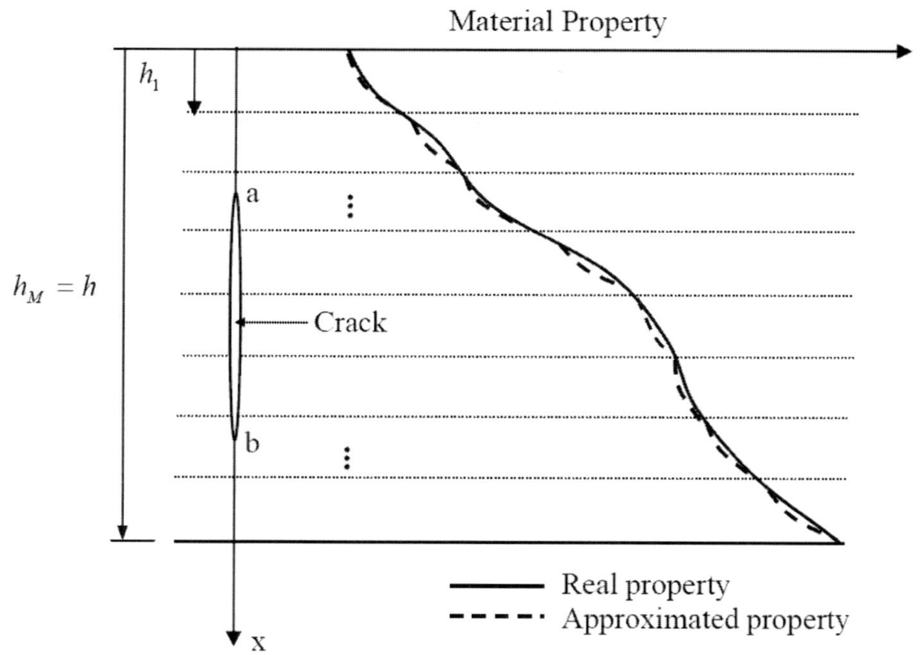

Figure 3.2(b). Schematic of the approximated mechanical property in the PE Model.

Figure 3.2 Piecewise-exponential model (PE Model) for a functionally graded strip with arbitrarily distributed mechanical properties. (Guo and Noda, 2007)

According to Eqs. (3-1) and (3-2), the constitutive relations and governing equations for each layer is same as those in part 2. The Poisson's ratio of each layer can be assumed to be a constant. Namely, $\kappa_n = \kappa = 3-4\upsilon$ ($n=1,2,\cdots,M$) (plane strain) and $\kappa_n = \kappa = (3-\upsilon)/(1+\upsilon)$ ($n=1,2,\cdots,M$) (plane stress).

The real shear modulus of the functionally graded strip with general mechanical properties is described as

$$\mu(x) = f(x) \tag{3-3}$$

where $f(x)$ is a known continuous real function. Assuming the shear modulus of each layer in Figure 3.2 to be

$$\mu_n(x) = \mu_{n0} e^{\delta_n x}, \quad n=1,2,\cdots,M \tag{3-4}$$

and applying the real material properties on both surfaces of each layer, μ_{n0} and δ_n can be solved as

$$\delta_n = \frac{1}{h_n - h_{n-1}} \ln\left[\frac{f(h_n)}{f(h_{n-1})}\right] \tag{3-5}$$

$$\mu_{n0} = f(h_n) e^{-\delta_n h_n} \tag{3-6}$$

In the PE model, the boundary and continuity conditions can be written as:

$$\sigma_{1xx}(0,y) = 0 \quad -\infty < y < \infty \tag{3-7}$$

$$\sigma_{1xy}(0,y) = 0 \quad -\infty < y < \infty \tag{3-8}$$

$$\sigma_{Mxx}(h_M = h, y) = 0 \quad -\infty < y < \infty \tag{3-9}$$

$$\sigma_{Mxy}(h_M = h, y) = 0 \quad -\infty < y < \infty \tag{3-10}$$

$$\sigma_{nxy}(x,0) = 0, \quad n=1,\cdots,M \tag{3-11}$$

$$\sigma_{nyy}(x,0) = -\sigma_0(x), \quad a<x<b, \quad n=n_1,\cdots,n_2 \tag{3-12}$$

$$v_n(x,0) = 0, \quad 0<x<a \text{ or } b<x<h \tag{3-13}$$

$$\begin{cases} u_n(h_n,y) = u_{n+1}(h_n,y) \\ v_n(h_n,y) = v_{n+1}(h_n,y) \end{cases}, \quad n=1,2,\cdots,M-1 \tag{3-14}$$

$$\begin{cases} \sigma_{nxx}(h_n,y) = \sigma_{(n+1)xx}(h_n,y) \\ \sigma_{nxy}(h_n,y) = \sigma_{(n+1)xy}(h_n,y) \end{cases}, \quad n=1,2,\cdots,M-1 \tag{3-15}$$

Here $\sigma_0(x)$ is a known function depicting the loading condition on the crack face.

In the PE model, the auxiliary function can be introduced through the following form:

$$g(x) = \frac{\partial v_n(x,0)}{\partial x},\ a < x < b,\ n = n_1, \cdots, n_2 \tag{3-16}$$

with the following conditions

$$g(x) = 0\ \ x \notin (a,b),\quad \int_a^b g(x)dx = 0 \tag{3-17}$$

Now, each layer in the PE model can be analyzed by the method in Part 2 and hence the displacement and stress expressions have the same forms as those given in Part 2. By using the boundary and continuity conditions, the unkowns in the displacement and stress expressions can be solved and expressed by the auxiliary conditions. They are

$$A_{n1j} = q_{nj}\int_a^b g(u)e^{isu}du,\ n = 1,2,\cdots,M \text{ and } j = 1,2 \tag{3-18}$$

$$\begin{Bmatrix} \{A_{12}\} \\ \{A_{22}\} \\ \vdots \\ \{A_{M2}\} \end{Bmatrix}_{4M\times 1} = [\Phi]^{-1}_{4M\times 4M}\int_a^b \begin{Bmatrix} \{\Gamma_0\} \\ \{\Gamma_1\} \\ \vdots \\ \{\Gamma_M\} \end{Bmatrix}_{4M\times 1} g(u)du \tag{3-19}$$

Here

$$[\Phi]_{4M\times 4M} = \begin{bmatrix} [X_1(0)] & 0 & 0 & \cdots & 0 & 0 & 0 \\ [\Theta_1(h_1)] & -[\Theta_2(h_1)] & 0 & \cdots & 0 & 0 & 0 \\ 0 & [\Theta_2(h_2)] & -[\Theta_3(h_2)] & \cdots & 0 & 0 & 0 \\ \vdots & \vdots & \vdots & \vdots & \vdots & \vdots & \vdots \\ 0 & 0 & 0 & \cdots & [\Theta_{M-2}(h_{M-2})] & -[\Theta_{M-1}(h_{M-2})] & 0 \\ 0 & 0 & 0 & \cdots & 0 & [\Theta_{M-1}(h_{M-1})] & -[\Theta_M(h_{M-1})] \\ 0 & 0 & 0 & \cdots & 0 & 0 & [X_M(h)] \end{bmatrix} \tag{3-20}$$

The other expressions in $\{A_{n2}\}$ $(n = 1,\cdots,M)$ need to be simplified by using the theory of residues. Through complex mathematic manipulations, the following singular integral equation can be obtained:

$$\frac{1}{\pi}\int_a^b \left[\frac{4\mu_{n0}}{1+\kappa}\frac{1}{x-u} + k_{n1}(u,x) + \pi H_{n1}(u,x)\right]g(u)du = -e^{\delta_n x}\sigma_0(x),\ n = n_1,\cdots,n_2 \tag{3-21}$$

It should be noted that the number n in the singular integral equation (3-21) is corresponding to the layer in which x lies.

The above singular integral equations can be solved numerically by the method of Erdogan and Gupta (1972). The intervals can be normalized by setting

$$u = \frac{b-a}{2} r + \frac{b+a}{2}, \quad x = \frac{b-a}{2} s + \frac{b+a}{2} \tag{3-22}$$

The solution of equation (3-21) for the present internal crack problem may be written as

$$g(r) = \frac{f(r)}{\sqrt{(1-r^2)}} \tag{3-23}$$

Thus, the singular integral equation (3-21) can be reduced to the following standard form,

$$\sum_{k=1}^{m} \frac{1}{m} f(r_k) \left[\frac{4\mu_{n0}}{1+k} \frac{1}{s_j - r_k} + k_{n1}(r_k, s_j) + \pi h_{n2}(r_k, s_j) \right] = -e^{\delta_n s_j} \sigma_0(s_j), \quad j = 1, \ldots, m-1 \tag{3-24}$$

Here

$$r_k = \cos\left[\frac{(2k-1)\pi}{2m} \right] \quad k = 1, \ldots, m \tag{3-25}$$

$$s_j = \cos\left[\frac{r\pi}{m} \right] \quad j = 1, \ldots, m-1 \tag{3-26}$$

When the unknowns $f(r_k)$ are solved, the stress intensity factor for an internal crack can be obtained from the following relations:

$$k_1(a) = \lim_{x \to a} \sqrt{2(a-x)} \sigma_{yy}(x,0) = -\frac{4\mu_{n_1 0}}{1+\kappa} e^{\delta_{n_1} a} \sqrt{\frac{b-a}{2}} f(-1) \tag{3-27}$$

$$k_1(b) = \lim_{x \to b} \sqrt{2(x-b)} \sigma_{yy}(x,0) = \frac{4\mu_{n_2 0}}{1+\kappa} e^{\delta_{n_2} b} \sqrt{\frac{b-a}{2}} f(1) \tag{3-28}$$

For more details, please see the reference (Guo and Noda, 2007)

3.2. Analyses of the Crack Problem with Some Representative Mechanical Properties

In Guo and Noda (2007), several types of mechanical properties defined by some representative functions including the exponential functions, linear functions, general power-law functions) will be considered. Tthe real shear modulus $f(x)$ defined by an exponential function can be written as

$$f(x) = f_0 e^{\delta(x/h)} \tag{3-29}$$

The real shear modulus $f(x)$ defined by a power-law function can be expresses as

$$f(x) = f_0 + (f_M - f_0)(x/h)^g \tag{3-30}$$

In the following analysis, the representative powers $g = 1$, 1/2 and 2 are considered. When the power $g = 1$, the modulus is reduced to a linear function.

In order to check whether the above reperesentative functions can be approached accurately by a series of exponential functions, some coomparison is provided in Figures 3.3. When the real moduli are defined by exponential and linear functions, the approximated properties are obtained by using 6 layers. When the moduli are defined by power-law functions with g=1/2 and 2, the approximated properties are obtained by using 8 layers. From Figure 3.3, it can be found that the real properties can be approached well.

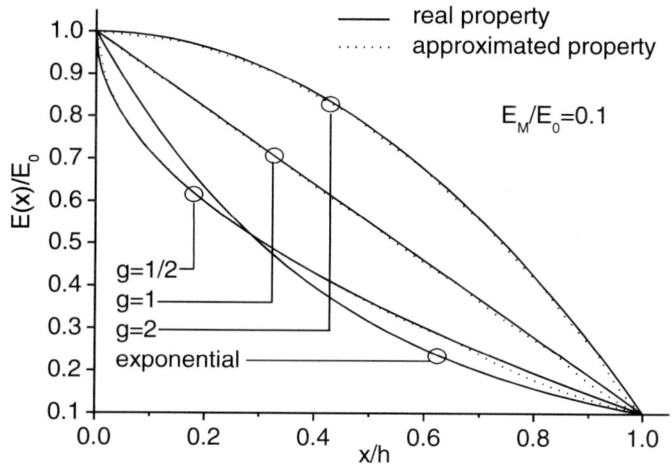

Figure 3.3. Comparisons between the real properties and the approximated properties for different types of properties.

To verify the PE model, some comparisons between the theoretical results and those calculated by using finite element method (FEM) were provided by Guo and Noda (2007). A functionally graded strip with a central crack subjected to membrane loading is chosen. It was observed that the results from two methods agree well. Moreover, it is usually enough to use M=6~8 for converged results.

In order to reveal the characteristics of SIFs for different types of properties, a central crack subjected to a constant crack surface loading $-\sigma_0$ is considered. The stress intensity factors are normalized by $k_0 = \sigma_0\sqrt{c/2}$.

1). When the real modulus is defined by an exponential function and a linear function, the results are shown in Figure 3.4 and 3.5, respectively. For these two cases, it can be found that $k_1(a)/k_0$ decreases with the increasing of $\ln(E_M/E_0)$ and $k_1(b)/k_0$ increases with the increasing of $\ln(E_M/E_0)$. Moreover, $k_1(a)/k_0 > k_1(b)/k_0$ when $\ln(E_M/E_0) < 0$; $k_1(a)/k_0 < k_1(b)/k_0$ when $\ln(E_M/E_0) > 0$; $k_1(a)/k_0 = k_1(b)/k_0$ when $\ln(E_M/E_0) = 0$. Note that $\ln(E_M / E_0) = 0$ corresponds to the homogeneous materials and $\ln(E_M / E_0) > 0$ ($\ln(E_M / E_0) < 0$) means that the medium gets stiffer (softer) from $x = 0$ to $x = h$. From Figures 3.4 and 3.5, another common characteristic can be found that $k_1(a)/k_0$ corresponding to $\ln(E_M / E_0)$ is equal to $k_1(b)/k_0$ corresponding to - $\ln(E_M / E_0)$.

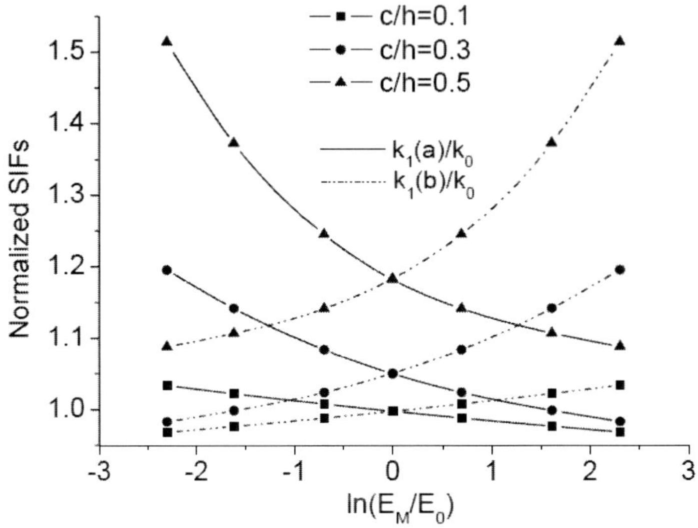

Figure 3.4. Variations of SIFs with $\ln(E_M / E_0)$ when the modulus is defined by an exponential function. (Guo and Noda, 2007)

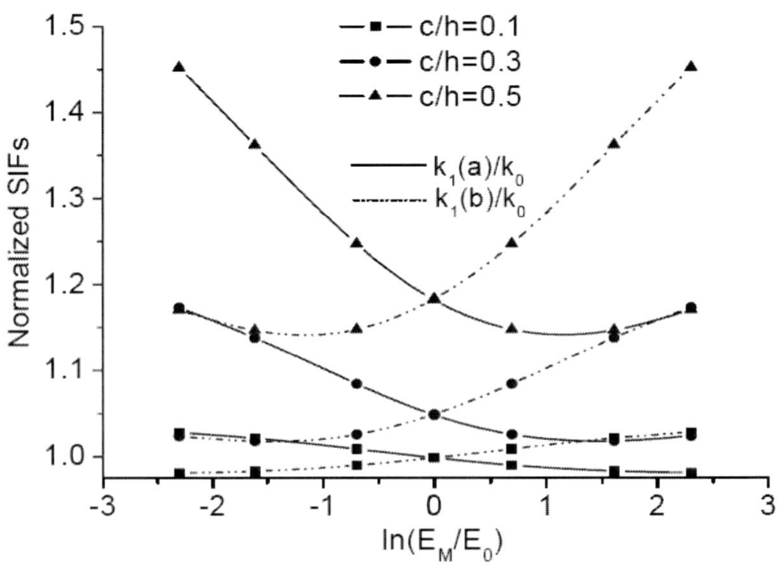

Figure 3.5. Variations of SIFs with $\ln(E_M / E_0)$ when the modulus is defined by a linear function. (Guo and Noda, 2007)

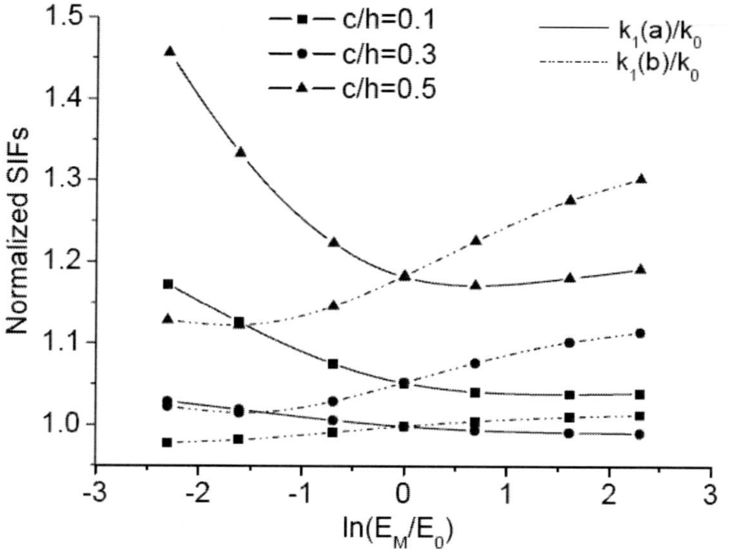

Figure 3.6. Variations of SIFs with $\ln(E_M / E_0)$ when the modulus is defined by a power-law function with g=1/2. (Guo and Noda, 2007)

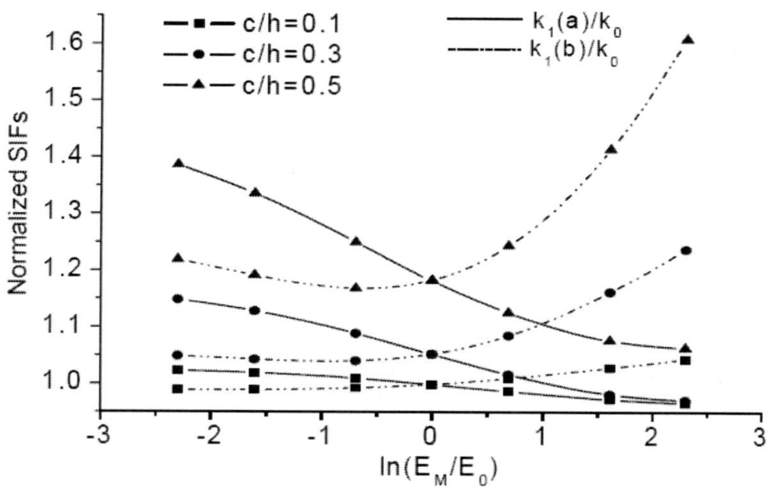

Figure 3.7. Variations of SIFs with $\ln(E_M / E_0)$ when the modulus is defined by a power-law function with g=2. (Guo and Noda, 2007)

2). When the real property is defined by power-law functions with *g*=1/2 and 2, seeing Figures 3.6 and 3.7. When *g*=1/2 (Figure 3.6), it can be observed that $k_1(a)/k_0$ decreases obviously with the increasing of $\ln(E_M / E_0)$ when $\ln(E_M / E_0) < 0$ and then varies slightly with the variation of $\ln(E_M / E_0)$ when $\ln(E_M / E_0) > 0$. On the other hand, $k_1(b)/k_0$ increases with the increasing of $\ln(E_M / E_0)$. But compared with $k_1(a)/k_0$, the varying range of $k_1(b)/k_0$ is smaller. Differently from Figure 3.6, when *g*=2 (Figure 3.7): $k_1(b)/k_0$

varies slightly with the increasing of $\ln(E_M / E_0)$ when $\ln(E_M / E_0) < 0$ and then increasing obviously with the increasing of $\ln(E_M / E_0)$ when $\ln(E_M / E_0) > 0$; different from $k_1(b)/k_0$, $k_1(a)/k_0$ decreases with the increasing of $\ln(E_M / E_0)$. But compared with $k_1(b)/k_0$, the varying range of $k_1(a)/k_0$ is smaller. Morever, for the power-law functions with g=1/2 and 2, there is no symmetry characteristic of SIFs as that shown in Figures 3.4 and 3.5.

3.3. Summary on the PE Model

In the PE model, the functionally graded material is divided into some nonhomogeneous layers along the gradient direction of the properties, with each layer's properties varying exponentially. Thus, the real material properties can be approached by a series of exponential functions. Since the real material properties are used on both surfaces of each nonhomogeneous layer, the nature of continuously varying properties of FGMs can be approached accurately. A summary on PE model is provided through the following points:

The influences of the local material nonhomogeneity on the crack-tip fields can be fully considered.

Since the crack problems in a nonhomogeneous layer with exponential function properties are easier to solve than those with properties expressed by other types of functions, there have been very limited literatures on the theoretical analysis of crack problems in FGMs with non-exponential properties in past decades. For FGMs with exponential properties, there have been lots of investigations regarding many types of crack problems. These work will provide a wide support for the present PE model to be applied in solving many types of crack problems in the FGMs or nonhomogeneous matertials with arbitrary properties.

In recent years, some published analytical methods assuming the property to be one predefined exponential function have been thought unrealistic. However, the PE model indicates that these methods are significant as they form the base of the PE model so that the PE model can be applied to solve the fracture behavior of FGMs with general or practical properties.

Some further work needs to be done so that the PE model can be applied to more crack problems (such as dynamic crack problems and thermal shock problems) for FGMs with general mechanical properties, thermal properties, electric properties, even stochastic properties and so on.

4. Solution of the Crack Problems for FGMs with General Thermomechanical Properties

In this part, the PE model is extended to investigate a functionally graded plate (FGP) with arbitrary thermomechanical properties and a crack normal to the surfaces (Guo et al., 2008a). The thermal stress intensity factors (TSIFs) for some representative samples with different types of thermomechanical properties are presented.

4.1. Temperature Fields and Thermal Stresses

An infinite FGP with arbitrarily distributed thermomechanical properties along the gradient direction is shown in Figure 4.1. The thickness is h. A crack normal to the surfaces will be analyzed. Define the heat conduction, thermal expansion coefficient, shear modulus, Young's modulus, and Poisson's ratio as $\lambda(x)$, $\alpha(x)$, $\mu(x)$, $E(x)$ and υ, respectively.

Firstly, the temperature field and thermal stresses for the FGP without the crack will be obtained. Since the heat conduction is one-dimensional for the present problem, by solving the following steady state equation

$$\partial[\lambda(x)\partial T/\partial x]/\partial x = 0 \tag{4-1}$$

the temperature field can be obtained as

$$T(x) = T_1 + (T_2 - T_1)[\int_0^x 1/\lambda(x)dx]/[\int_0^h 1/\lambda(x)dx] \tag{4-2}$$

where T_1 and T_2 are the temperatures at the upper surface and lower surface of the FGP, respectively.

According to Erdogan and Wu (1996), when the FGP is only subjected to the temperature distribution $T(x)$, the thermal stress in the absence of the crack can be obtained as

$$\sigma_{yy}^T(x) = E(x)\{Ax + B - \alpha(x)(1+\upsilon)[T(x) - T_0]\}/(1-\upsilon^2) \text{ (for plane strain state)} \tag{4-3}$$

$$\sigma_{yy}^T(x) = E(x)\{Ax + B - \alpha(x)[T(x) - T_0]\} \text{ (for plane stress state)} \tag{4-4}$$

Here, T_0 is the initial temperature, A and B are coefficients determined from the boundary conditions of the structure. If the FGP is only subjected to the temperature loading without constraining along its boundaries, A and B can be solved from the following relations

$$\int_0^h \sigma_{yy}^T(x)dx = 0, \quad \int_0^h \sigma_{yy}^T(x)xdx = 0 \tag{4-5}$$

The crack problem can be solved by superposition method. The present crack problem can be reduced to a perturbation problem and the crack surface loading resulting from the temperature distribution will be equal and opposite to the thermal stress in Eq.(4-3) or (4-4). Therefore, when applying the PE model in Part 3 to the steady thermal problem in Figure 4.1, only the crack-face loading condition needs to be replaced by the following one:

$$\sigma_{nyy}(x,0) = -\sigma_{yy}^T(x), \quad a < x < b, \quad n = n_1, \cdots, n_2 \tag{4-6}$$

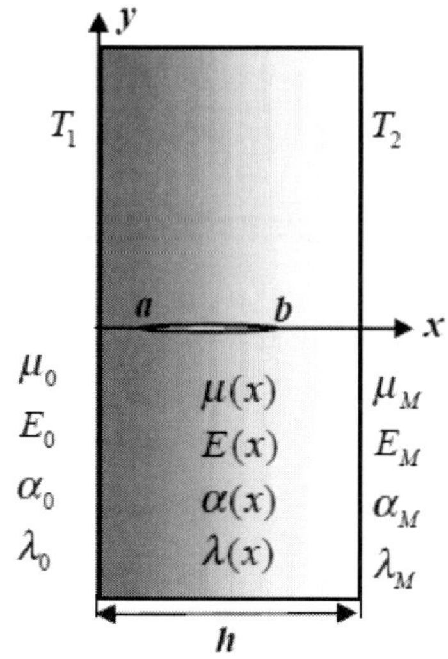

Figure 4.1(a) Embedded crack

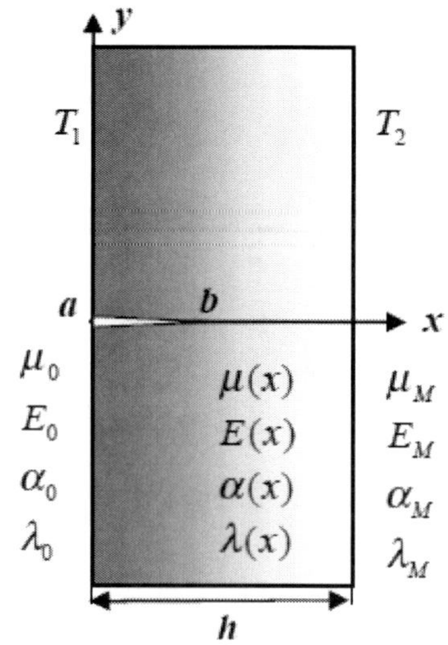

Figure 4.1(b) Edge crack

Figure 4.1. A FGP with a crack and arbitrary thermomechanical properties along the gradient direction.

4.2. Analyses of the Crack Problem with Some Representative Thermomechanical Properties

To state the validity of the PE model for the thermal crack problem, different properties defined by some representative functions will be considered. When the real thermomechanical properties are exponential functions, they can be described as

$$\mu(x) = \mu_0 e^{\delta(x/h)},\ \alpha(x) = \alpha_0 e^{\omega(x/h)},\ \lambda(x) = \lambda_0 e^{\eta(x/h)} \tag{4-7}$$

When the real thermo-mechanical properties are power-law functions, they can be written as

$$\begin{aligned} \mu(x) &= \mu_0 + (\mu_M - \mu_0)(x/h)^{g_1} \\ \lambda(x) &= \lambda_0 + (\lambda_M - \lambda_0)(x/h)^{g_2} \\ \alpha(x) &= \alpha_0 + (\alpha_M - \alpha_0)(x/h)^{g_3} \end{aligned} \tag{4-8}$$

When the power $g_i = 1$, the corresponding property is reduced to a linear function. The ratios of the shear modulus, Young's modulus, heat conduction and thermal expansion coefficient at both surfaces of the FGP in Figure 4.1 are μ_M / μ_0, E_M / E_0, λ_M / λ_0 and α_M / α_0, respectively. For convenience, plane strain state is considered in the following analyses. $a_0 = (b-a)/2$ and $\mu_M / \mu_0 = E_M / E_0$.

Following are some representative samples provided by Guo et al. (2008a). The TSIF is normalized by $k_e = \sigma_c \sqrt{a_0}$ (for the embedded crack) and $k_e = \sigma_c \sqrt{b}$ (for the edge crack), respectively. Here,

$$\sigma_c = E_0 \alpha_0 |T - T_0| / (1 - \nu) \text{ when } T(x) = T_1 = T_2 \tag{4-9}$$

$$\sigma_c = E_0 \alpha_0 |T_2 - T_1| / (1 - \nu) \text{ when } T_1 \neq T_2 \tag{4-10}$$

For convenience, the word "exponential" in Figure 4.3 through Figure 4.5 means all the thermomechanical properties are exponential functions.

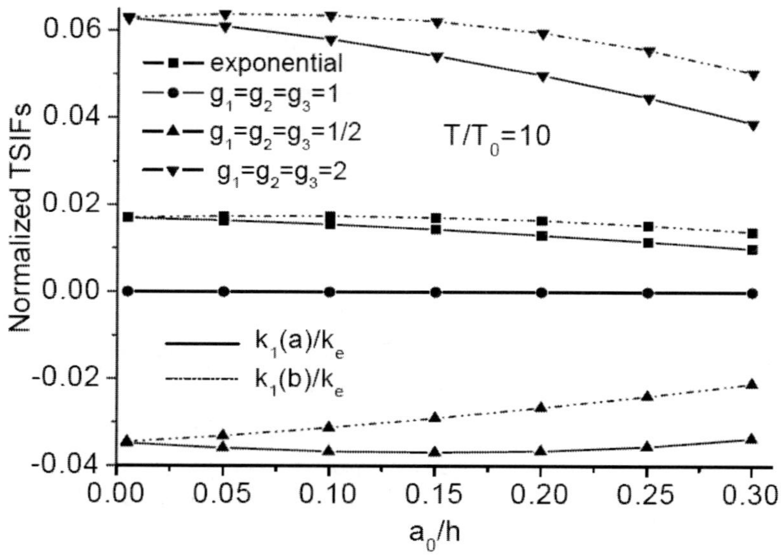

Figure 4.2. Comparison between the TSIFs for an embedded crack in a FGP with different types of properties. (Guo et al., 2008a)

Figure 4.2 displays the TSIFs for an embedded crack in a (Rene-41)/Zirconia FGP with different types of properties and the common ratios $E_M / E_0 = 1.455$, $\lambda_M / \lambda_0 = 12.2$ and $\alpha_M / \alpha_0 = 1.67$. The uniform temperature distribution $T(x) = T_1 = T_2 = 10T_0$ is considered. It is observed that:

The difference is obvious for the TSIFs corresponding to different types of thermomechanical properties.

When the thermomechanical properties vary linearly ($g_1 = g_2 = g_3 = 1$), the TSIFs are zero.

It should be noted that the distribution of the thermal stresses may result in the compression state in some zones. In the compression zones, the calculated TSIFs may become negative (such as $g_1 = g_2 = g_3 = 1/2$), which implies the crack-tip part has been closed. The negative TSIFs could shed some light on the degree of closure occurring near the crack tip region.

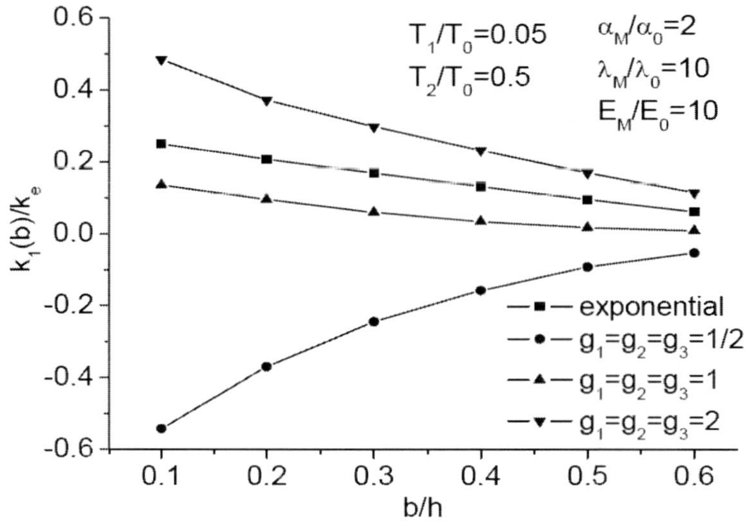

Figure 4.3. Comparison between the TSIFs for an edge crack in a FGP with different types of properties when $T_1 \neq T_2$. (Guo et al., 2008a)

Figure 4.3 displays the TSIFs for an edge crack in a cooled FGP with different types of properties when $E_M / E_0 = 10$, $\lambda_M / \lambda_0 = 10$ and $\alpha_M / \alpha_0 = 2$. Figure 4.3 shows the results corresponding to $T_1 = 0.05T_0$ and $T_2 = 0.5T_0$. It can be found that:

The difference is obvious for the TSIFs corresponding to different types of thermomechanical properties. It indicates that the analytical methods by assuming the properties only as exponential functions to analyze the crack problems of FGMs may be not practical;

The results for the exponential properties are usually between those corresponding to the power-law properties, which implies the exponential properties may be used to study the crack problem of FGMs qualitatively;

The calculated TSIFs become negative when $g_1 = g_2 = g_3 = 1/2$, which implies the crack-tip part has been closed.

The PE model can also take into account variations in the Poisson's ratio by specifying a different υ for each layer according to its centroidal coordinate. It should be mentioned that the present model is not limited to the properties used above. The properties can be arbitrarily distributed along the gradient direction.

5. A Numerical Modeling Method for the Static Crack Problems of FGMs with General Properties

FGMs are usually two or multi-phase particulate composites in which material composition and microstructure vary spatially (Rahman and Chakraborty, 2007) or the volume fraction of particles varies in one or several directions (Birman and Byrd, 2007). In the analytical modeling method for the crack problems of FGMs with general properties, it is very difficult to consider the influences of the complex internal structures and interfaces on the fracture behavior of FGMs. Therefore, numerical modeling method for the crack problems of FGMs with general properties should be developed together with the analytical methods.

In the following part, the derivation of an interaction integral and its associated domain form without any derivatives of material properties is presented. It can be used to analysize the crack problems of FGMs with general properties, considering the effects of the component particles and internal interfaces on the fracture behavior (Yu et al., 2009).

5.1. An Interaction Integral without any Derivatives of Material Properties

Since the interaction integral is derived from the J-integral (Rice, 1968) for two admissible states (actual and auxiliary fields), the auxiliary fields used in the interaction integral is introduced firstly. The auxiliary fields in the local polar coordinate system shown in Figure 5.1 can be defined as (Kim and Paulino, 2005)

$$u_i^{aux} = K_I^{aux} f_i^I (r^{1/2}, \theta, \mu_{tip}, \kappa_{tip}) + K_{II}^{aux} f_i^{II} (r^{1/2}, \theta, \mu_{tip}, \kappa_{tip}), \tag{5-1}$$

$$\sigma_{ij}^{aux} = K_I^{aux} g_{ij}^I (r^{-1/2}, \theta) + K_{II}^{aux} g_{ij}^{II} (r^{-1/2}, \theta), \tag{5-2}$$

$$\varepsilon_{ij}^{aux} = S_{ijkl}(x)\sigma_{kl}^{aux}, \quad (i, j, k, l = 1, 2), \tag{5-3}$$

where, K_I^{aux} and K_{II}^{aux} are the auxiliary mode I and II SIFs, respectively, and the representative functions f_i^I, f_i^{II}, g_{ij}^I and g_{ij}^{II} are given by Williams (1957) as

$$\begin{aligned}
f_1^I &= \frac{1}{2\mu_{tip}}\sqrt{\frac{r}{2\pi}}\cos\frac{\theta}{2}\left(\kappa_{tip} - 1 + 2\sin^2\frac{\theta}{2}\right) \\
f_1^{II} &= \frac{1}{2\mu_{tip}}\sqrt{\frac{r}{2\pi}}\sin\frac{\theta}{2}\left(\kappa_{tip} + 1 + 2\cos^2\frac{\theta}{2}\right) \\
f_2^I &= \frac{1}{2\mu_{tip}}\sqrt{\frac{r}{2\pi}}\sin\frac{\theta}{2}\left(\kappa_{tip} + 1 - 2\cos^2\frac{\theta}{2}\right) \\
f_2^{II} &= -\frac{1}{2\mu_{tip}}\sqrt{\frac{r}{2\pi}}\cos\frac{\theta}{2}\left(\kappa_{tip} - 1 - 2\sin^2\frac{\theta}{2}\right)
\end{aligned} \tag{5-4a,b,c,d}$$

$$\begin{aligned}
g_{11}^I &= \frac{1}{\sqrt{2\pi r}}\cos\frac{\theta}{2}\left(1 - \sin\frac{\theta}{2}\sin\frac{3\theta}{2}\right) \\
g_{11}^{II} &= -\frac{1}{\sqrt{2\pi r}}\sin\frac{\theta}{2}\left(2 + \cos\frac{\theta}{2}\cos\frac{3\theta}{2}\right) \\
g_{22}^I &= \frac{1}{\sqrt{2\pi r}}\cos\frac{\theta}{2}\left(1 + \sin\frac{\theta}{2}\sin\frac{3\theta}{2}\right) \\
g_{22}^{II} &= \frac{1}{\sqrt{2\pi r}}\sin\frac{\theta}{2}\cos\frac{\theta}{2}\cos\frac{3\theta}{2} \\
g_{12}^I &= g_{21}^I = g_{22}^{II}
\end{aligned}$$

$$g_{12}^{II} = g_{21}^{II} = g_{11}^{I} \tag{5-5a,b,c,d,e,f}$$

Here, $\kappa_{tip} = 3 - 4\nu_{tip}$ for plane strain and $\kappa_{tip} = (3-\nu_{tip})/(1+\nu_{tip})$ for plane stress, μ_{tip} and ν_{tip} are the shear modulus and the Poisson's ratio at crack tip location, respectively. It should be noted that except at the crack tip (Kim and Paulino, 2003)

$$\varepsilon_{ij}^{aux} \neq (u_{i,j}^{aux} + u_{j,i}^{aux})/2 \tag{5-6}$$

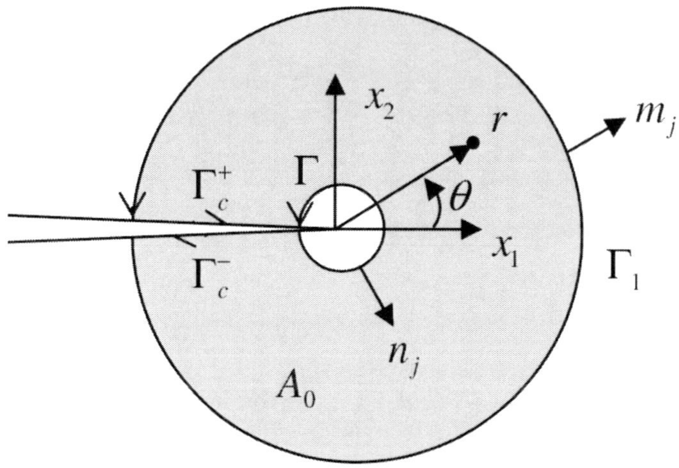

Figure 5.1. Schematic of the integral path and domain for the contour integrals and the related equivalent domain integrals with the domain A_0 enclosed by Γ_0 ($\Gamma_0 = \Gamma_1 + \Gamma^- + \Gamma_c^+ + \Gamma_c^-$). (Yu et al., 2009)

According to Kim and Paulino (2005), the interaction integral I, the interactional part of the superimposed load of the actual field and the auxiliary field, can be obtained as

$$I = \lim_{\Gamma\to 0} \oint_{\Gamma_0} (\sigma_{ij}^{aux} u_{j,1} + \sigma_{ij} u_{j,1}^{aux} - \frac{1}{2}(\sigma_{jk}^{aux} \varepsilon_{jk} + \sigma_{jk} \varepsilon_{jk}^{aux})\delta_{1i}) m_i q d\Gamma \tag{5-7}$$

According to Eqs. (5-1)-(5-3), it can be found that $\sigma_{ij}\varepsilon_{ij}^{aux} = \sigma_{kl}^{aux}\varepsilon_{kl}$ and thus, Eq.(5-7) can be simplified as

$$I = \lim_{\Gamma\to 0} \oint_{\Gamma_0} (\sigma_{ij}^{aux} u_{j,1} + \sigma_{ij} u_{j,1}^{aux} - \sigma_{jk}^{aux} \varepsilon_{jk} \delta_{1i}) m_i q d\Gamma \tag{5-8}$$

Generally, the contour integral is converted into an equivalent domain integral to avoid the potential source of inaccuracy in numerical calculations (Moran and Shih, 1987a,b). The material properties are assumed to be continuous in the domain enclosed by the contour Γ_0. Let A_0 and A denote the domains enclosed by the contours Γ_0 and Γ_1, respectively. Taking the limit $\Gamma \to 0$ leads to $A_0 \to A$. By applying divergence theorem, $\sigma_{ij,i} u_{j,1}^{aux} = 0$,

$\sigma_{ij,i}^{aux} u_{j,1} = 0$, and the symmetry of the auxiliary stress tensor, , the interaction integral I can be simplified as

$$I = \int_A (\sigma_{ij}^{aux} u_{j,1} + \sigma_{ij} u_{j,1}^{aux} - \sigma_{jk}^{aux} \varepsilon_{jk} \delta_{1i}) q_{,i} dA + \int_A (\sigma_{ij} u_{j,i1}^{aux} - \sigma_{ij,1}^{aux} \varepsilon_{ij}) q dA \tag{5-9}$$

It should be noted that Kim and Paulino (2005) derived the same expression as Eq. (5-9) by assuming the material properties to be exponential functions. In Yu et al.(2009), it is proved that Eq. (5-9) is valid for the material with arbitrarily continuous properties. Considering the symmetry of σ_{ij}, it can be found that

$$\sigma_{ij} u_{j,i1}^{aux} = \sigma_{ij} \frac{1}{2}(u_{j,i1}^{aux} + u_{i,j1}^{aux}) = \sigma_{ij} (S_{ijkl}^{tip} \sigma_{kl}^{aux})_{,1} = \sigma_{ij} S_{ijkl}^{tip} \sigma_{kl,1}^{aux} \tag{5-10}$$

Substituting $\varepsilon_{ij} = S_{ijkl}(x)\sigma_{kl}$ and Eq. (5-10) into Eq. (5-9), we have

$$I = \int_A (\sigma_{ij}^{aux} u_{j,1} + \sigma_{ij} u_{j,1}^{aux} - \sigma_{jk}^{aux} \varepsilon_{jk} \delta_{1i}) q_{,i} dA + \int_A \sigma_{ij} [S_{ijkl}^{tip} - S_{ijkl}(x)] \sigma_{kl,1}^{aux} q dA \tag{5-11}$$

It can be found that the derivatives of material properties ($C_{ijkl,1}$) exist unavoidably in the traditional J-integral (Dolbow and Gosz, 2002). Differently, the above interaction integral (Yu et al., 2009) does not involve any derivatives of material properties. Therefore, the interaction integral does not need the material properties to be differentiable. Since it may be difficult to obtain the derivatives of material properties or there are no derivatives in many actual cases, the applicable range of the present interaction integral is wider than that of the traditional J-integral for nonhomogeneous materials.

Next, whether the continuity condition of material properties in the interaction integral is necessary will be discussed.

As shown in Figure 5.2, there is a bimaterial interface $\Gamma_{\text{interface}}$ in the domain enclosed by the integral contour Γ_0 and the interface is assumed to be perfectly bonded. Thus, the domain A (the domain in Γ_0 with $\Gamma \to 0$) is divided by $\Gamma_{\text{interface}}$ into A_1 and A_2 enclosed by the contour Γ_{01} and Γ_{02}, respectively. As a result, $A = A_1 + A_2$, $\Gamma_{01} = \Gamma_{11} + \Gamma_{\text{interface}} + \Gamma_{13} + \Gamma_c^+ + \Gamma^- + \Gamma_c^-$ and $\Gamma_{02} = \Gamma_{12} + \Gamma_{\text{interface}}^-$. Thus, the interaction integral can be written as

$$I = \lim_{\Gamma \to 0} \oint_{\Gamma_{01}} (\sigma_{ij}^{aux} u_{j,1} + \sigma_{ij} u_{j,1}^{aux} - \sigma_{jk}^{aux} \varepsilon_{jk} \delta_{1i}) m_i q d\Gamma$$
$$+ \oint_{\Gamma_{02}} (\sigma_{ij}^{aux} u_{j,1} + \sigma_{ij} u_{j,1}^{aux} - \sigma_{jk}^{aux} \varepsilon_{jk} \delta_{1i}) m_i q d\Gamma + I_{\text{interface}}^* \tag{5-12}$$

Here

$$I^{*}_{\text{interface}} = \int_{\Gamma_{\text{interface}}} (\sigma^{aux}_{ij} u_{j,1} + \sigma_{ij} u^{aux}_{j,1} - \sigma^{aux}_{jk} \varepsilon_{jk} \delta_{1i})^{②} m_i q d\Gamma$$
$$+ \int_{\Gamma^{-}_{\text{interface}}} (\sigma^{aux}_{ij} u_{j,1} + \sigma_{ij} u^{aux}_{j,1} - \sigma^{aux}_{jk} \varepsilon_{jk} \delta_{1i})^{①} m_i q d\Gamma \quad (5\text{-}13)$$

Here, the variables or expressions on the interface marked by the superscripts ①and ② means that they belong to the domains A_1 and A_2, respectively.

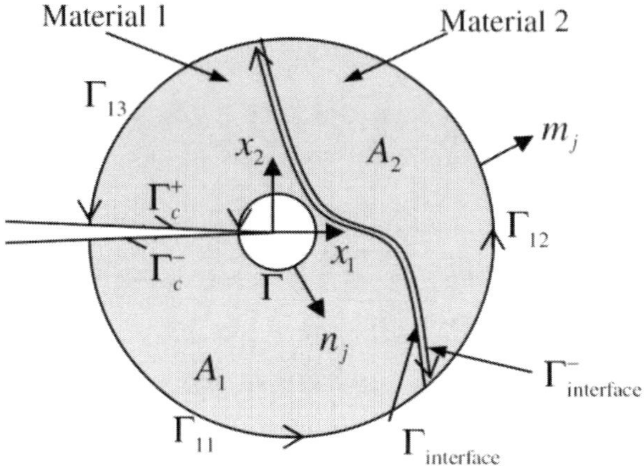

Figure 5.2. The integral domain A cut by a bimaterial interface $\Gamma_{\text{interface}}$ with domains A, A_1 and A_2 enclosed by Γ_0, Γ_{01} and Γ_{02} when $\Gamma \to 0$. (Yu et al., 2009)

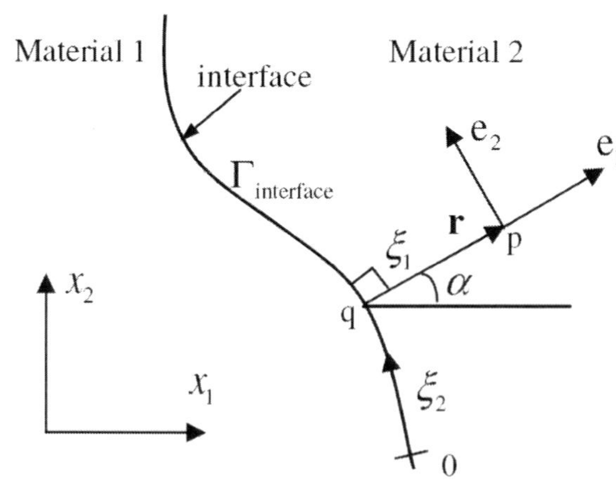

Figure 5.3. The curvilinear coordinate system on the interface. (Yu et al., 2009)

Figure 5.3 shows a curved bimaterial interface $\Gamma_{\text{interface}}$. The auxiliary stresses and displacements and their derivatives are continuous on the interface. Therefore, $(\boldsymbol{\sigma}^{aux})^{①} = (\boldsymbol{\sigma}^{aux})^{②} = \boldsymbol{\sigma}^{aux}$ and $(\frac{\partial \mathbf{u}^{aux}}{\partial x_1})^{①} = (\frac{\partial \mathbf{u}^{aux}}{\partial x_1})^{②} = \frac{\partial \mathbf{u}^{aux}}{\partial x_1}$. The tensor form of the integral $I^{*}_{\text{interface}}$ can be written as

$$I^*_{\text{interface}} = \int_{\Gamma_{\text{interface}}} \{\mathbf{n}\cdot\boldsymbol{\sigma}^{aux}\cdot[(\frac{\partial \mathbf{u}}{\partial x_1})^{②} - (\frac{\partial \mathbf{u}}{\partial x_1})^{①}] + \mathbf{n}\cdot(\boldsymbol{\sigma}^{②} - \boldsymbol{\sigma}^{①})\cdot\frac{\partial \mathbf{u}^{aux}}{\partial x_1} - \boldsymbol{\sigma}^{aux} : (\boldsymbol{\varepsilon}^{②} - \boldsymbol{\varepsilon}^{①})n_1\} q d\Gamma \quad (5\text{-}14)$$

where n_1 is the component of $\mathbf{n}$ in x_1 direction as shown in Figure 5.3. According to the equilibrium condition on the bimaterial interface, the tractions on both sides of the interface should be equal. Therefore

$$\boldsymbol{\sigma}^{②}\cdot\mathbf{n} = \boldsymbol{\sigma}^{①}\cdot\mathbf{n} \quad (5\text{-}15)$$

Since the interface is perfectly bonded, the derivatives of actual displacements with respect to the curvilinear coordinate ξ_2 are equal on both sides of the interface. We have

$$(\frac{\partial \mathbf{u}}{\partial \xi_2})^{①} = (\frac{\partial \mathbf{u}}{\partial \xi_2})^{②} \quad (5\text{-}16)$$

Thus, it can be proved that (Yu et al., 2009)

$$I^*_{\text{interface}} = 0 \quad (5\text{-}17)$$

Similarly, the same result can be obtained for the crack faces penetrated by the interface. Substituting $I^*_{\text{interface}}$ into (5-12) and applying the divergence theorem, the same expression in Eq. (5-11) can be obtained. It indicates that the material continuity condition at interface is not necessary for the interaction integral in Yu et al.(2009).

The interaction integral can be obtained as (Kim and Paulino, 2003)

$$I = \frac{2}{E'_{tip}}(K_I K_I^{aux} + K_{II} K_{II}^{aux}) \quad (5\text{-}18)$$

where $E'_{tip} = E_{tip}$ for generalized plane stress and $E'_{tip} = E_{tip}/(1-\nu_{tip}^2)$ for plane strain. Here, E_{tip} and ν_{tip} are the Young's modulus and the Poisson's ratio at the crack tip. By setting $K_I^{aux} = 1$, $K_{II}^{aux} = 0$ and $K_I^{aux} = 0, K_{II}^{aux} = 1$, mode I and mode II SIFs can be decoupled using the above equation.

5.2. Application of the Interaction Integral in the Extended Finite Element Method

In order to calculate the above interaction integral, the extended finite element method (XFEM) developed by Belytschko and Black (1999) and Moës et al. (1999) can be used. More information about the XFEM can be found in some published papers (e.g., Belytschko et al., 2001). In order to employ the interaction integral method in the XFEM, Eq. (5-11) should be discretized first (Yu et al., 2009). For nonhomogeneous materials, actual material properties at integration points are adopted when the element stiffness matrix is formed (Yu et al., 2007 and Guo et al.2008b).

Figure 5.4(a) shows a two dimensional plate of length L and width W with a horizontal crack of length $2a$ in the neighbourhood of a circular inclusion of radius R. The problem of an infinite plate with such a configuration was investigated by Wang and Chau (2001). The inclusion is offset by a distance of b horizontally and c vertically from the origin which coincides with the center of the crack. E_1 and E_2 are the Young's moduli of the matrix and the inclusion, and ν_1 and ν_2 are the corresponding Poisson's ratios, respectively. The following data are used for numerical analysis: $L = W = 100$; $a/W = 0.01$; $b/a = (2.8, 3, 4, 6, 8, 10)$; $R = 2a$; $E_1 = 1$; $E_2/E_1 = 10^4$; $\nu_1 = \nu_2 = 0.35$; $\sigma(x_1) = \sigma_0 = 1$; $K_0 = \sigma_0\sqrt{\pi a}$; generalized plane strain. The mesh configurations are shown in Figure 5.4(b).

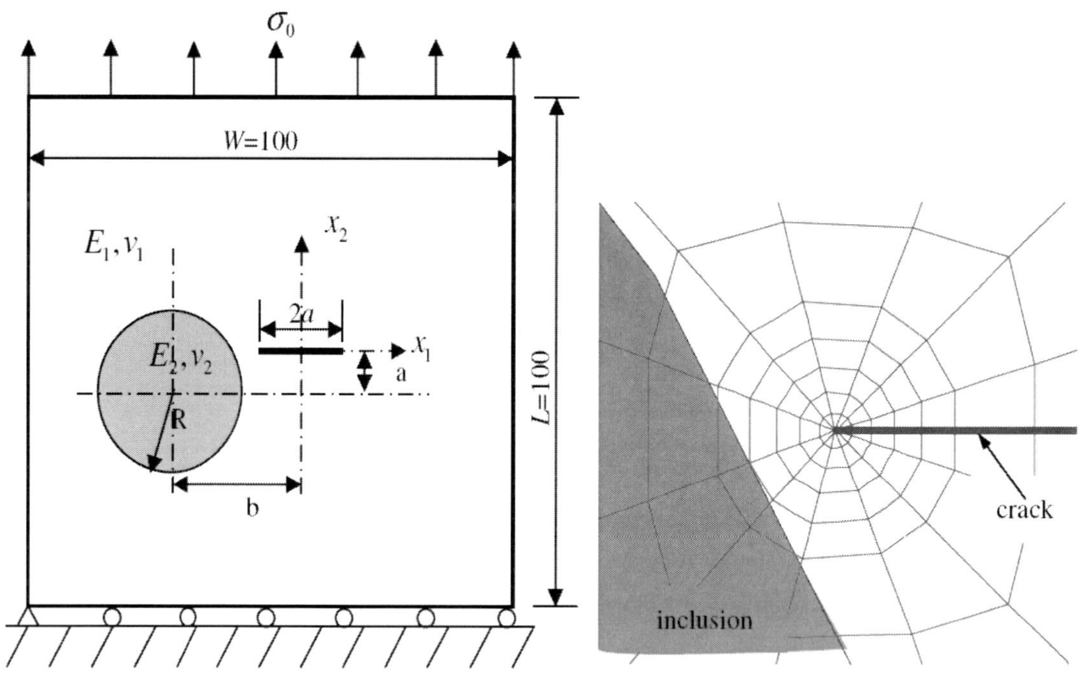

(a) Geometry and boundary conditions (b) Mesh around the left crack tip.

Figure 5.4. Schematic of a plate with a horizontal crack in the neighborhood of a circular rigid inclusion subjected to far field tension. (Yu et al., 2009)

Table 5.1. Normalized stress intensity factors for a plate with a crack and an inclusion under far field tension σ_0

$\frac{b}{a}$	Results (Yu et al., 2009) ($E_2/E_1 = 10000$, $\nu = 0.35$)				Wang and Chau (2001) ($E_2/E_1 = \infty$, $\nu = 0.35$)			
	$\frac{K_I(-a)}{K_0}$	$\frac{K_{II}(-a)}{K_0}$	$\frac{K_I(a)}{K_0}$	$\frac{K_{II}(a)}{K_0}$	$\frac{K_I(-a)}{K_0}$	$\frac{K_{II}(-a)}{K_0}$	$\frac{K_I(a)}{K_0}$	$\frac{K_{II}(a)}{K_0}$

3.0	0.5908	0.0652	0.8011	-0.0711	0.5810	0.0636	0.7995	-0.0733
6.0	0.9505	-0.0373	0.9687	-0.0255	0.9506	-0.0368	0.9684	-0.0252
8.0	0.9785	-0.0179	0.9844	-0.0129	0.9787	-0.0173	0.9842	-0.0125
10	0.9877	-0.0101	0.9903	-0.0076	0.9878	-0.0091	0.9901	-0.0069

The comparison between the normalized SIFs by the above method and those reported by Wang and Chau (2001) is given in Table 5.1 given by Yu et al.(2009). It can be observed that the relative errors of the mode I SIFs are all within 2% compared with those of Wang and Chau (2001). Compared with the mode I SIFs, the mode II ones are very small and especially, the mode II ones can be omitted when $b/a=10$. That implys the interaction integral method is valid for the materials with discontinuous properties.

The influences of the material continuity on the SIFs are investigated by selecting four types of mechanical properties. The model shown in Figure 5.5 is adopted. Four types of material properties are chosen:

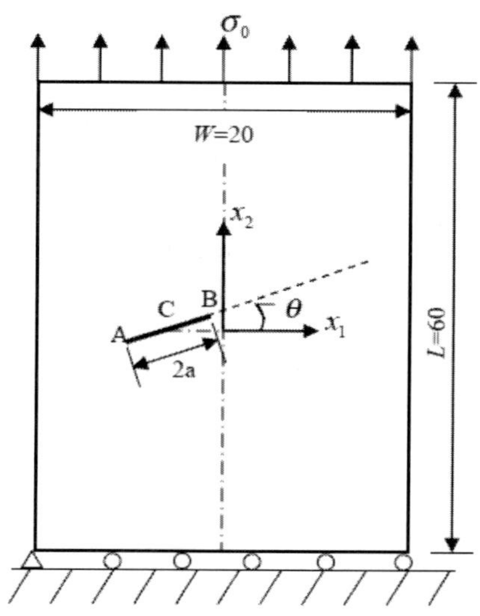

Figure 5.5. Schematic of a nonhomogeneous material plate with a rectilinear crack. (Yu et al., 2009)

Case 1: The Young's modulus is discontinuous at $x_1=0$.

$$E(x_1)=\begin{cases} E_0\,\mathrm{e}^{\beta x_1} & (x_1\le 0) \\ 2E_0 & (x_1>0) \end{cases} \tag{5-19}$$

Case 2: The Young's modulus is continuous but its derivatives are discontinuous at $x_1=0$:

$$E(x_1)=\begin{cases} E_0\,\mathrm{e}^{\beta x_1} & (x_1\le 0) \\ E_0 & (x_1>0) \end{cases} \tag{5-20}$$

Case 3: The Young's modulus and its first-order derivatives are continuous at $x_1=0$, but its high-order derivatives are discontinuous.

$$E(x_1) = \begin{cases} E_0 \, e^{\beta x_1} & (x_1 \le 0) \\ E_0(1+\beta x_1) & (x_1 > 0) \end{cases} \tag{5-21}$$

Case 4: The mechanical properties and their derivatives are continuous at $x_1 = 0$.

$$E(x_1) = E_0 \, e^{\beta x_1} \tag{5-22}$$

For all the cases, the Poisson's ratio is constant. In Figure 5.5, $L = 60$; $W = 20$; $\theta/\pi = 0$; $E_0 = 1$; $\beta = \ln(10)/W$; $\nu = 0.3$; $\sigma_0 = 1$; $K_0 = \sigma_0\sqrt{\pi a}$; $2a/W = 0.1$; generalized plane strain. According to Yu et al.(2009), the distance from the crack tip to the interface is restricted to be more than 2.5 percent of the crack length ($R_\delta \ge 0.025$) in numerical calculations. Figures 5.6(a) and (b) show the variation of the normalized mode I SIFs with crack center coordinate $2c/W$. The range of $2c/W$ is from -0.5 to 0.5, which means the crack moves from the center of left half plate to the center of right half plate. It can be found that:

1) For Case 1, it is found that when the crack tip passes the interface, the SIFs usually have great variations;

2) For Case 2, it is found that that the SIFs have a kinking behavior at the interface. The same phenomenon has been observed in Part 1 (Guo and Noda, 2008);

3) For Case 3 and Case 4, the SIFs vary smoothly and the SIFs has no kinking behavior. Therefore, the continuity of high-order derivatives of mechanical properties has slight effects on the SIFs.

From the above four cases, it can be found that the improvement of the continuity of material properties can reduce the varying range of the SIFs and smooth their varying trends.

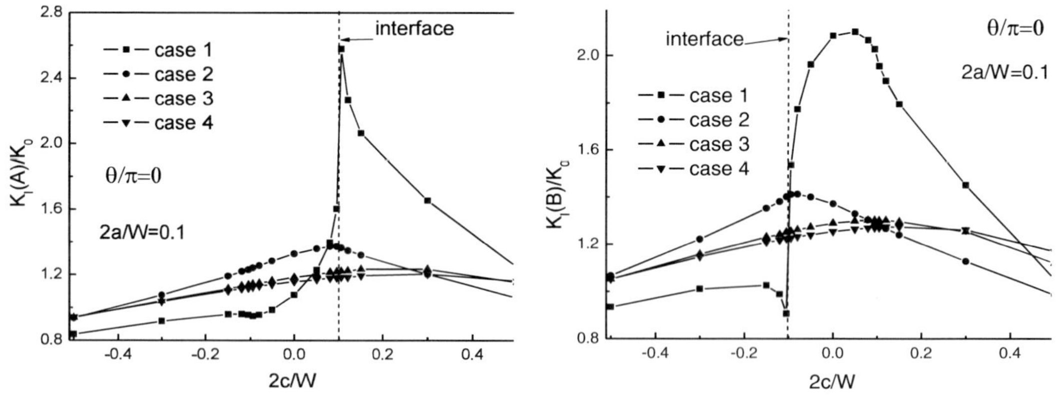

(a) SIFs at crack tip A (b) SIFs at crack tip B

Figure 5.6. Variation of the normalized SIFs with crack center $2c/W$ for 4 types of material properties. (Yu et al., 2009)

More results and analyses can be found in Yu et al. (2009).

5.3. Conclusion on the Interaction Integral Containing no Derivatives of Material Properties

The domain expression of the interaction integral derived by Yu et al.(2009) does not contain any derivatives of material properties, and it is still valid even when the integral

domain contains material interfaces. Therefore, it is no necessary to limit the material properties to be continuous. Due to this advantage, the application range of the interaction integral method can be greatly enlarged so that the crack problems for FGMs with general properties can be analyzed even with the consideration of the effects of material particles and internal interfaces on the fracture behavior. The interaction integral method is combined with the XFEM. Numerical results show that the interaction integral method exhibits good convergence and high efficiency for nonhomogeneous and discontinuous material properties.

ACKNOWLEDGMENTS

This work is sponsored by NSFC (No.10432030, No.10872056) and LC2009C01.

REFERENCES

Belytschko, T., and Black, T. Elastic crack growth in finite elements with minimal remeshing. *Int. J. Numer. Meth. Eng.* 1999, *Vol* 45, 601-620.

Belytschko, T., Moës, N., Usui, S., and Parimi, C. Arbitrary discontinuities in finite elements. *Int. J. Numer. Meth. Eng.* 2001, *Vol* 50, 993-1013.

Birman, V., and Byrd, L.W. Modeling and analysis of functionally graded materials and structures. *Appl. Mech. Rev.* 2007, *Vol* 60, 195-216.

Bleeck, O., Munz, D., Schaller, W., and Yang, Y.Y. Effect of a graded interlayer on the stress intensity factor of cracks in a joint under thermal loading. *Eng. Fract. Mech.* 1998, *Vol* 60, 615-623.

Butcher, R.J., Rousseau, C.E., and Tippur, H.V. A functionally graded particulate composite: preparation, Measurements and Failure Analysis. *Acta Mater.* 1999, *Vol* 47, 259-268.

Chen, J. Determination of thermal stress intensity factors for an interface crack in a graded orthotropic coating-substrate structure. *Int. J. Fract.* 2005, *Vol* 133, 303-328.

Chen, J., Liu, Z.X., and Zou, Z.Z. Transient internal crack problem for a nonhomogeneous orthotropic strip Mode I. *Int. J. Eng. Sci.* 2002. *Vol* 40, 1761-1774.

Chen, Y.F., and Erdogan, F. The interface problem for a nonhomogeneous coating bonded to a homogeneous substrate. *J. Mech. Phys. Solids.* 1996, *Vol* 44, 771-787.

Chi, S.H., and Chung, Y.L. Cracking in coating-substrate composites with multi-layered and FGM coatings. *Eng. Fract. Mech.* 2003, *Vol* 70, 1227-1243.

Choi, H.J. Bonded dissimilar strips with a crack perpendicular to the functionally graded interface. *Int. J. Solids Struct.* 1996. *Vol* 33, 4101-4117.

Choi, H.J. The problem for bonded half-planes containing a crack at an arbitrary angle to th e graded interfacial zone. *Int. J. Solids Struct.* 2001. *Vol* 38, 6559-6588.

Choi, H.J., Jin, T.E., and Lee, K.Y. Collinear cracks in a layered half-plane with a graded nonhomogeneous interfacial zone-Part I: Mechanical response. *Int. J. Fract.* 1998, *Vol* 94, 103-122.

Delale, F., and Erdogan, F. The crack problem for a nonhomogeneous plane. *J. Appl. Mech.* 1983, *Vol* 50, 609-614.

Dolbow, J.E., and Gosz, M. On the computation of mixed-mode stress intensity factors in functionally graded materials. *Int. J. Solids Struct.* 2002, *Vol* 39, 2557-2574.

Eischen, J.W. Fracture of nonhomogeneous materials. *Int. J. Fract.* 1987, *Vol* 34, 3-22.

Erdogan, F., and Gupta, G.D. On the numerical solution of singular integral equations. *Quarterly of Applied Mathematics* 1972, *Vol* 30, 525-534.

Erdogan, F., Gupta, G.D., and Ratwani, M. Interaction between a circular inclusion and an arbitrarily oriented crack. *J. Appl. Mech.* 1974, *Vol* 47, 1007-1013.

Erdogan, F., and Wu, B.H. Crack problem in functionally graded material layers under thermal stresses. *J. Therm. Stresses* 1996, *Vol* 19, 237-265.

Erdogan, F., and Wu, B.H. The surface crack problem for a plate with functionally graded properties. *J. Appl. Mech.* 1997, *Vol* 64, 449-456.

Gosz, M., Dolbow, J., and Moran, B. Domain integral formulation for stress intensity factor computation along curved three-dimensional interface cracks. *Int. J. Solids Struct.* 1998, *Vol* 3515, 1763-1783.

Gosz, M., and Moran, B. An interaction energy integral method for computation of mixed-mode stress intensity factors along non-planar crack fronts in three dimensions. *Eng. Fract. Mech.* 2002, *Vol* 69, 299-319.

Guo, L.C., and Noda, N. Modeling method for a crack problem of functionally graded materials with arbitrary properties- piecewise-exponential model. *Int. J. Solids Struct.* 2007, *Vol* 44, 6768-6790.

Guo, L.C., and Noda N. Fracture mechanics analysis of functionally graded layered structures with a crack crossing the interface. *Mech. Mater.* 2008, *Vol* 40, 81-99.

Guo, L.C., Noda, N., and Wu, L.Z. Thermal fracture model for a functionally graded plate with a crack normal to the surfaces and arbitrary thermomechanical properties. *Comp Sci Tech.* 2008a, *Vol* 68, 1034-1041.

Guo, L.C., Wu, L.Z., and Ma, L. The interface crack problem under a concentrated load for a functionally graded coating-substrate composite system. *Comp. Struct.* 2004a, Vol 63, 397-406.

Guo, L.C., Wu, L.Z., Zeng, T., and Ma, L. Mode I crack problem for a functionally graded orthotropic strip. *Eur. J. Mech.A-Solids.* 2004b, *Vol* 23, 219-234.

Guo, L.C., Wu, L.Z., and Yu H.J. Numerical Analysis of Stress Intensity Factors in Three-Dimensional Functionally Graded Materials. *AIP Conf. Proc.* 2008b, *Vol* 973: 115-120.

Honein, T., and Herrmann, G. Conservation laws in nonhomogeneous plane elastostatics. J Mech. Phy. Solids 1997, Vol 45(5), 789-805.

Huang, G.Y., Wang, Y.S., and Yu, S.W. Fracture analysis of a functionally graded interfacial zone under plane deformation. *Int. J. Solids Struct.* 2004, *Vol* 41, 731-743.

Hutchinson, J.W., and Suo, Z., Mixed mode cracking in layered materials. *Advances in Applied Mechanics* 1992, *Vol* 29, 63-191.

Itou, S. Stress intensity factors around a crack in a nonhomogeneous interfacial layer between two dissimilar elastic half-planes. *Int. J. Fract.* 2001, *Vol* 110, 123-135.

Jin, Z.H., and Noda, N. Crack-tip singular fields in nonhomogeneous materials. *J. Appl. Mech.* 1994, *Vol* 61, 738-740.

Jin, Z.H., and Sun, C.T. Integral Representation of Energy Release Rate in Graded Materials. *J. Appl. Mech.* 2007, *Vol* 74, 1046-1048.

Johnson, J., and Qu, J.M. An interaction integral method for computing mixed mode stress intensity factors for curved bimaterial interface cracks in non-uniform temperature fields. *Eng. Fract. Mech.* 2007, *Vol* 74, 2282-2291.

Kadioglu, S., Dag, S., and Yahsi, S. Crack problem for a functionally graded layer on an elastic foundation. *Int. J. Fract.* 1998, *Vol* 94, 63-77.

Kawasaki, A., and Watanabe, R. Effect of gradient microstructure o thermal shock crack extension in metal/ceramic functionally graded materials. Proceedings of the 4th international symposium on functionally graded materials, Japan, 1996, 143-148.

Kawasaki, A., and Watanabe, R. Thermal fracture behavior of metal/ceramic functionally graded materials. *Eng. Fract. Mech.* 2002, *Vol* 69, 1713-1728.

Kim, J.H., and Paulino, G. H. Finite element evaluation of mixed mode stress intensity factors in functionally graded materials. *Int. J. Numer. Meth. Eng.* 2002, *Vol* 53, 1903-1935.

Kim, J.H., and Paulino, G.H. T-stress, mixed-mode stress intensity factors, and crack initiation angles in functionally graded materials: a unified approach using the interaction integral method. *Comput. Meth. Appl. Mech. Eng.* 2003, *Vol* 192, 1463-1494.

Kim, J.H., and Paulino, G.H. Consistent formulations of the interaction integral method for fracture of functionally graded materials. *J. Appl. Mech.* 2005, *Vol* 72, 351-364.

Konda, N., and Erdogan, F. The mixed mode crack problem in a homogeneous elastic plane. *Eng. Frac.Mech.* 1994, *Vol* 47, 533-545

Krysl, P., and Belytschko, T. The element free Galerkin method for dynamic propagation of arbitrary 3-D cracks. *Int. J. Numer. Meth. Eng.* 1999, *Vol* 44, 767-800.

Li, C.Y., Weng, G.J., Duan, and Z.P. Dynamic behavior of a cylindrical crack in a functionally graded interlayer under torsional loading. *Int. J. Solids Struct.* 2001, *Vol* 38, 7473-7485.

Ling, Y.H., Bai, X.D, and Ge, C.C. Design and thermal shock performance of W/Cu functionally graded material used as plasma facing component. *Materials Science Forum.* 2003, *Vol* 423-425, 49-54

Ma, L., Wu L.Z., Guo L.C., and Zhou, Z.G. Dynamic behavior of a finite crack in the functionally graded materials. *Mech. Mater.* 2005, *Vol* 37, 1153-1165.

Moran, B., Shih, C.F. A general treatment of crack tip contour integrals. *Int. J. Fract.* 1987a, *Vol* 35, 295-310.

Moran, B., and Shih, C.F. Crack tip and associated domain integrals from momentum and energy balance. *Eng. Fract. Mech.* 1987b, *Vol* 27(6), 615-642.

Moës, N., Dolbow, J., Belytschko, T. A finite element method for crack growth without remeshing. *Int. J. Numer. Meth. Eng.* 1999, *Vol* 46, 131-150.

Nahta, R., and Moran, B. Domain integrals for axisymmetric interface crack problems. *Int. J. Solids Struct.* 1993, *Vol* 30(15), 2027-2040.

Nakamura, T. Three-dimensional stress fields of elastic interface cracks. *J. Appl. Mech.* 1991, *Vol* 58, 939-946.

Noda, N., and Jin, Z.-H. Thermal Stress Intensity Factors for a Crack in a Strip of Functionally Gradient Material. *Int. J. Solids Struct.* 1993, *Vol* 30, 1039-1056.

Rahman, S., and Chakraborty, A., A stochastic micromechanical model for elastic properties of functionally graded materials. *Mech. Mater.* 2007, *Vol* 39, 548-563.

Rao, B.N., and Rahman, S. An interaction integral method for analysis of cracks in orthotropic functionally graded materials. *Computational Mechanics* 2003, *Vol* 32, 40-51.

Rice, J.R. A path independent integral and the approximate analysis of strain concentration by notches and cracks. *J. Appl. Mech.* 1968, *Vol* 35, 379-386.

Rice, J.R. Elastic fracture mechanics concepts for interfacial cracks. *J. Appl. Mech.* 1988, *Vol* 55, 98-103.

Rousseau, C.-E., and Tippur, H.V. Dynamic fracture of compositionally graded materials with cracks along the elastic gradient: experiments and analysis. *Mech. Mater.* 2001, *Vol* 33, 403-421.

Smelser, R.E., and Gurtin. M.E. On the J-integral for bi-material bodies. *Int. J. Fract.* 1977, *Vol* 13, 382-384.

Stern, M., Becker, E.B., and Dunham, R.S. A contour integral computation of mixed-mode stress intensity factors. *Int. J. Fract.* 1976, *Vol* 12(3), 359-368.

Ueda, S. The surface crack problem for a layered plate with a functionally graded nonhomogeneous interface. *Int. J. Fract.* 2001, *Vol* 110, 189-204.

Ueda, S., and Shindo, Y. Crack kingking in functionally graded materials due to an initial strain resulting from stress relation. *J. Therm. Stresses* 2000, *Vol* 23, 285-290.

Walters, M.C., Paulino, G.H., and Dodds, R.H. Stress-intensity factors for surface cracks in functionally graded materials under mode-I thermomechanical loading. *Int. J. Solids Struct.* 2004, *Vol* 41, 1081-1118.

Wang, B.L., Han, J.C., and Du, S.Y. Crack problems for non-homogeneous composites subjected to dynamic loading. *Int. J. Solids Struct.* 2000, *Vol* 37, 1251-1274.

Wang, B.L., and Mai, Y.W. A periodic array of cracks in functionally graded materials subjected to transient loading. *Int. J. Eng. Sci.* 2006, *Vol* 44, 351-364.

Wang, B.L., Mai, Y.W., and Noda, N. Fracture mechanics analysis model for functionally graded materials with arbitrarily distributed properties. *Int. J. Fract.* 2002, *Vol* 116, 161-177.

Wang, Y.B., and Chau, K.T. A new boundary element method for mixed boundary value problems involving cracks and holes: Interactions between rigid inclusions and cracks. *Int. J. Fract.* 2001, *Vol* 110, 387-406.

Wang, Y.S., and Gross, D. Analysis of a crack in a functionally gradient interface layer under static and dynamic loading. *Key Eng. Mater.* 2000, *Vol* 183-187, 331-336.

Wang Z.Q., and Nakamura, T. Simulations of crack propagation in elastic-plastic graded materials. *Mech. Mater.* 2004, *Vol* 36, 601-622.

Williams, M.L. On the stress distribution at the base of a stationary crack. *J. Appl. Mech.* 1957, *Vol* 24(1), 109-114.

Yao, X.F., Xu, W., and Yeh, H.Y. Investigation of crack tip evolution in functionally graded materials using optical caustics. *Polymer Testing* 2007, *Vol* 26, 122-131.

Yu, H.J., Guo, L.C., and Wu, L.Z., Numerical investigation on the fracture behaviors of three-dimensional functionally graded materials. *Key Engineering Materials* 2007, *Vol* 353-358, 1098-1101.

Yu, H.J., Wu, L.Z., Guo, L.C., Du, S.Y., and He, Q.L. Investigation of mixed-mode stress *Int. J. Solids Struct.* 2009, *Vol* 46(20), 3710-3724

Zhong Z., Cheng Z.Q. Fracture analysis of a functionally graded strip with arbitrary distributed material properties. *Int. J. Solids Struct.* 2008, *Vol* 45(13), 3711-3725.

Zhou, Z.G., Wang, B., and Sun, Y.G. Investigation of the dynamic behavior of a finite crack in the functionally graded materials by use of the Schmidt method. *Wave Motion* 2004, *Vol* 39, 213-225.

ABOUT THE AUTHORS

Guo Licheng received the Ph.D. degree from Harbin Institute of Technology, China, 2004. He is a Professor of the Department of Astronautic Science and Mechanics at the Harbin Institute of Technology, Harbin, P.R. China. He was Associate Professor of Center for Composite Materials at Harbin Institute of Technology from 2005 to 2008. During June, 2006 ~ May, 2008, he was Postdoctoral Fellow of Japan Society for the Promotion of Science, Shizuoka University, Japan. He has published about fifty papers. His research interests include the fracture mechanics of nonhomogeneous materials (including functionally graded materials), composite mechanics, and computational mechanics of composite structures.

Linzhi Wu received the Ph.D. degree from Harbin Institute of Technology, China, 1992. He is a Full Professor of the Center for Composite Materials at Harbin Institute of Technology, Harbin, P.R.China. His research interests include meso-mechanics, fracture mechanics of functionally graded materials, design and analysis of composite structures, braided composites, ultra-light composite truss structures. He is the editor of the following journals: Acta Mechanica Sinica, Acta Mechanica Solida Sinica (in Chinese), Advances in Mechanics (in Chinese).

INDEX

#

A

B

C

D

E

F

G

H

I

J

L

M

N

O

P

Q

R

S

T

U

V

W

Y

Z

L

M

N

O

P

Q

R

S

T

U

V

W

Y

Z